C$_4$ Grasses and Cereals

C$_4$ Grasses and Cereals

GROWTH, DEVELOPMENT, AND STRESS RESPONSE

C. Allan Jones

A Wiley-Interscience Publication

JOHN WILEY & SONS

New York • Chichester • Brisbane • Toronto • Singapore

Copyright © 1985 by John Wiley & Sons, Inc.

Library of Congress Cataloging in Publication Data:
Jones, C. Allan.
    C⁴ grasses and cereals.

    "A Wiley-Interscience publication."
    Includes index.
    1. Grasses. 2. Grain. 3. Grasses—Physiology.
4. Grain—Physiology. I. Title.

QK495.G74J66 1985   584′.9   85-6311
ISBN 0-471-82409-7

Printed in the United States of America

10 9 8 7 6 5 4 3 2 1

To
Jef, Erin, and Anna

PREFACE

More than a century has elapsed since Haberlandt (1882) recognized that the grasses can be divided into two major groups differing in internal leaf anatomy. Most tropical and warm-season grasses and cereals exhibit "Kranz" (halo or wreath, in German) leaf anatomy consisting of a prominent bundle sheath of closely packed cells with bright green chloroplasts. Others, principally cool-adapted grasses, have no such sheath. The significance of Kranz anatomy remained unclear until the 1960s, when it was recognized that in grasses with Kranz anatomy the first products of photosynthetic CO_2 fixation are four-carbon organic acids, rather than the three-carbon compounds produced by most other plants. The term C_4 was coined to distinguish these species, with their distinctive leaf anatomy and CO_2 fixation pathway, from other (C_3) species. It was soon evident that some of the world's most important field crops, forage grasses, and weeds are C_4 grasses, including maize (*Zea mays*), sugarcane (*Saccharum* spp. hybrid), grain sorghum (*Sorghum bicolor*), and the millets (*Pennisetum americanum, Eleusine corcana*, and others). Research during the last 20 years has revealed many differences between C_3 and C_4 grasses as well as many similarities among C_4 species. It is now clear that C_4 grasses are a quite homogeneous group with many characteristics in common.

The current demand for sophisticated, highly specialized research often requires that agricultural scientists concentrate on only one or a few species or, alternatively, on one or a few aspects of crop growth and development. It is difficult, if not impossible, to remain abreast of research on several species or many processes affecting growth. However, during the past eight years I have been fortunate to work with a number of C_4 grasses: forage grasses in Colombia, sugarcane in Hawaii, and maize in Texas. The work has involved many factors affecting plant growth and development, including genotypic variation in shoot and root growth, response to drought stress and flooding, aluminum toxicity, soil compaction, nitrogen and phosphorus fertility, and defoliation by clipping, grazing, and fire. During this time I realized that C_4 grasses as different as sugarcane, maize, and forage grasses grow, develop, and respond to stresses in a very similar manner. In contrast, seemingly minor differences in morphology, development, and response to stresses can dictate which ecotype or cultivar will prove successful under specific conditions.

The objectives of this book are to describe the common patterns of growth, development, and response to stresses of C_4 grasses; how these patterns differ among genotypes; and how the differences are responsible for variation in adaptation to particular environments. Thus, the book serves as a general reference for both the similarities and differences among C_4 grasses. As a general reference, it cannot provide an exhaustive literature review on all topics discussed; however, the approximately 2000 citations provide a balanced, up-to-date sample of the research on each topic.

The book can be divided into four parts. Chapters 1, 2, and 3 give a brief overview of the importance of C_4 grasses as well as their distinctive anatomy, physiology, taxonomy, and distribution. They introduce the reader to the group, show how it differs from C_3 grasses, and describe its biochemical and anatomical subtypes. Chapters 4, 5, and 6 describe the normal growth and development of C_4 grasses, including shoot, grain, and root growth. The effects of non-nutritional factors such as temperature, light intensity, CO_2 concentration, defoliation, drought, and soil strength and aeration are described in Chapters 7 through 11. Finally, Chapters 12 through 16 deal with nutritional deficiencies and toxicities.

C_4 Grasses and Cereals is a book that cuts across scientific disciplines while restricting itself to an ecologically and physiologically coherent group. It is designed to serve as a supplementary text or reference for courses in agrostology, crop physiology, range science, and plant nutrition. It will also be used as a reference by postgraduate students, agronomists, range scientists, crop physiologists, plant breeders, soil scientists, and engineers working on C_4 grasses. It provides a broader perspective of C_4 grasses and the factors affecting their growth than most scientists are likely to encounter in their areas of specialization.

The book is relevant to researchers in temperate as well as tropical areas. For example, maize is routinely grown as far north as southern Canada, Austria, West Germany, and Czechoslovakia. C_4 grasses are important components of the grasslands as far north as Montana, southeastern Alberta, and southwestern Saskatchewan. *Spartina* species are important marsh grasses in Canada and northern Europe.

I hope that this book will stimulate specialists to look outside their specific areas of research and use research with other species and processes better to understand the growth and development of this important group of plants—the C_4 grasses.

I gratefully acknowledge the help of Jefflyn Jones, Brenda Ganger, and Gail Robinson for typing the manuscript; Norman Erskine and Frances Essig for preparing figures and line drawings; and Mike Moncur for providing the excellent photographs of inflorescence development in C_4 grasses. The encouragement of colleagues in the Agricultural Research Service of the United States Department of Agriculture, Texas Agricultural Experiment Station, Hawaiian Sugar Planters' Association, University of Hawaii, and the Centro Internacional de Agricutura Tropical is gratefully acknowledged. Finally, the book could not have been written without the strong encouragement of my wife, Jef.

I am currently a Plant Physiologist, United States Department of Agriculture; Agricultural Research Service; Grassland, Soil and Water Research Laboratory; P.O. Box 748; Temple, Texas 76503. The views expressed here are mine and may not reflect the policy of the U.S. Department of Agriculture.

C. ALLAN JONES

Temple, Texas
July 1985

CONTENTS

1

C$_4$ GRASSES AND HUMANS

The Gramineae or grass family is one of the largest and by far the most important family of flowering plants. According to Gould (1968) it contains about 600 genera and 7500 species, placing it fifth in number of species behind the Compositeae, Orchidaceae, Leguminoseae, and Rubiaceae. However, in terms of importance to humans, it is far more important than any other family. The Gramineae contains cereals such as wheat, rice, maize (corn), barley, grain sorghum, and pearl millet. It contains the world's most important source of sucrose, sugarcane, and most of the forage and concentrated feeds consumed by domestic animals are produced by grasses.

WHAT ARE C_4 GRASSES?

More than a century ago, Haberlandt (1882) recognized that grasses can be divided into two major groups differing in internal leaf anatomy and ecological adaptation. Certain grasses, primarily of tropical origins or affinities, have a type of leaf anatomy that is fundamentally different from that of other grasses, most notably the winter cereals and other grasses of temperate origins. Most tropical and many warm-season temperate grasses have leaf mesophyll cells radially arranged around a prominent bundle sheath made up of closely packed cells with bright green chloroplasts. In contrast, the winter cereals and most temperate grasses have leaf mesophyll cells that are not radially arranged, and the chloroplasts of the bundle sheath cells are similar to those of the mesophyll cells. Certain notable exceptions to this rule were soon found. For example, rice and the bamboos, which are tropical in origin, have leaf anatomy more like the temperate than the tropical group.

The importance of these early anatomical studies was unclear until the 1960s, when biochemists discovered that the photosynthetic pathways of the two groups are fundamentally different. In the temperate group, the first stable product of photosynthetic CO_2 fixation is a three-carbon compound, 3-

phosphoglyceric acid. In contrast, the first stable products of the tropical group are four-carbon organic acids. Thus, the grasses of temperate affinities, as well as most other green plants, are termed C_3 plants. It is now clear that the C_4 photosynthetic pathway requires the distinctive leaf anatomy discovered by Haberlandt, and the C_4 photosynthetic pathway confers important advantages under certain ecological conditions.

WHY STUDY C_4 GRASSES?

Some of the world's most important crop species are C_4 grasses. These include maize (*Zea mays*), grain sorghum (*Sorghum bicolor*), pearl millet (*Pennisetum americanum*), and sugarcane (*Saccarum* spp. hybrid). In addition, in the developing countries of the tropics, the lesser millets play an important nutritional role in restricted areas. These species include finger millet (*Eleusine coracana*), Japanese barnyard millet (*Echinochloa frumentaceae*), tef (*Eragrostis tef*), common millet (*Panicum miliaceum*), little millet (*Panicum sumatrense*), kodo millet (*Paspalum scrobiculatum*), and foxtail millet (*Setaria italica*).

C_4 grasses are also among the world's most important forage and range grasses. They are especially important in the tropics and subtropics at elevations up to about 2500 m. They are also important warm-season forage and range species in temperate zones. In addition, eight C_4 grasses are among the world's 18 most important annual and perennial weeds (Holm et al., 1977).

The C_4 photosynthetic pathway generally enables C_4 grasses to grow rapidly and make efficient use of water in warm environments (see Chapter 3). Their similar temperature responses and their economic importance make them an important and coherent group to study. Over and over, scientists have found that principles discovered in one member of the group apply to other members as well. The most convenient way to illustrate the importance as well as the diversity of the C_4 grasses is to describe briefly several important and representative species. These and many other species are described in more detail in the following books and review articles: Bogdan (1977), Falvey (1981), Harlan (1976), Holm et al. (1977), Jones (1979), McCosker and Teitzel (1975), Quinlan et al. (1976), Simmonds (1976), Whiteman (1980), and Whyte et al. (1959).

CEREALS

C_4 grasses include some of the world's most important cereals. Five of the most important are described below.

Zea mays

Maize, also known as Indian corn or corn (in the United States), is the world's third most important cereal after wheat and rice. About 440 million tons were produced worldwide in 1982/83. It is a robust annual member of the subfamily

Panicoideae, tribe Paniceae (sometimes placed in a separate tribe, Maydeae). Maize is grown primarily for grain and secondarily for fodder and raw material for industrial processes. The grain is used for both human and animal consumption. The vegetative parts of the plant are cut green and either dried or made into silage for animal feed. The stem is solid, 1–6 (usually 2–3) m tall, and 3–4 cm in diameter. A single stem is usually produced, but some cultivars routinely produce one or more tillers from the base of the primary stem. The first few internodes are short but subsequent internodes elongate. Leaves are alternate, and 12–24 are normally produced. The plant is monoecious with male and female spikelets borne on separate inflorescences. The male inflorescence (tassel) occurs at the apex of the stem and consists of a single rachis and several lateral branches. The female inflorescence (ear) is a modified spike produced from a short lateral branch in the axil of one of the largest leaves about half way up the stem. One or two ears are normally produced. Each consists of a short, modified axillary branch (cob) bearing spikelets in longitudinal rows covered and protected by modified leaves (husks). The spikelet produces two florets, the upper pistillate and the lower sterile and reduced. The upper floret produces a long, threadlike style (silk) that grows up to 45 cm in length and emerges between the husks at the apex of the ear, usually one or more days after the tassel sheds its pollen. Maize is primarily cross-pollinated, though some self-pollination usually occurs. Fruit (kernel, grain) size, shape, color, and number vary greatly among cultivars. However, they normally mature 40–60 days after fertilization (80–140 days after emergence). They vary from white to yellow, red, and purple. Their size ranges from about 100 to 500 mg each (usually 200 to 400). Cultivars can be divided into groups based on grain structure. "Pod corn" is the most primitive form, and the grains are enclosed in floral bracts or glumes. It is not grown commercially. "Popcorn" has small grains with a high proportion of hard endosperm and little soft starch in the center. This type of grain is also considered primitive. "Flint" maize grains are larger than popcorn, but they also have hard endosperm and little starch. It has many color forms, resists insect attack, and is widely grown in Europe, Asia, Central and South America, and tropical Africa. "Dent" maize grains have hard endosperm on the sides, but have more white starch than flint grains. The starch extends to the apex of the grain. This starch shrinks when the grain dries, producing a characteristic apical dent. Dent maize is usually yellow or white, but colored forms occur. "Soft or flour" maize grains have endosperm consisting of soft starch, and the grain usually has no dent and has many color forms. It is easily ground or chewed and is grown in the southwestern United States, western South America, and South Africa. "Sweet" corn has a recessive gene that inhibits the conversion of sugar to starch. It is harvested in the milk stage and is popular as a vegetable. It is also used in the Mexican and South American beer, chicha. "Waxy" maize results from the presence of a gene controlling the type of starch in the grain. Waxy cultivars have only amylopectin, while normal cultivars have a mixture of amylose and amylopectin.

Maize, as mentioned earlier, is mainly cross-pollinated. It exhibits great heterozygosity, and the advantages of hybrid maize were recognized in the early twentieth century. Commercial production began in the 1930s. In 1935 only about 1% of U.S. maize growing area was planted to hybrids, and the average yield was about 1500 kg/ha. By 1944, about 60% of the area was planted to hybrids, and the average yield was slightly over 2000 kg/ha. By 1948–1955, over 90% of U.S. maize was produced by hybrids, and average yields were about 2500 kg/ha. Improved hybrids, higher fertilizer rates, and better insect and disease control led to a gradual increase in average yields, which reached 4500 kg/ha in 1965 and 6000 kg/ha in 1972. Experimental plot yields in excess of 18,000 kg/ha and farmers' yields in excess of 12,000 kg/ha confirm the high yield potential of hybrid maize. In developing countries where the cost of hybrid seed is a barrier, improved synthetic and open-pollinated composite cultivars are also being developed.

The early history of maize evolution and domestication is still quite controversial, but it is clear that maize (*Zea mays* ssp. *mays*) is closely related to teosinte (*Zea mays* ssp. *mexicana*, also known as *Zea mexicana* and *Euchlaena mexicana*), which produces fertile hybrids with maize and has terminal male and axillary female inflorescences. Goodman (1976) concludes that maize evolved from teosinte, possibly under the direct influence of man. At any rate, the domestication and selection of maize probably began in central or southwestern Mexico about 5000 years ago. Distinguishable groups of cultivars arose in Mexico and Central America, in the northeastern United States, on the northern coast of South America, in the Andes, and in central Brazil. The Spanish and Portugese quickly distributed maize throughout the world in the sixteenth century.

Maize has many assets. These include its high yield per unit labor and per unit land area. It is a compact, easily transportable source of nutrition. Its husks give protection from birds and rain. It can be harvested over a long period, stores well, and can even be left dried in the field until harvesting is convenient. It provides numerous useful food products, and it is frequently preferred to sorghum and the millets.

Major diseases affecting maize include leaf blight (*Helminthosporium turcicum*), which causes large brown lesions on the blades, and Southern leaf blight (*Helminthosporium maydis*) causing reddish brown elongated lesions between the veins. Leafspot (*Helminthosporium carbonum*) causes irregular brown spots on the leaves and black mouldy growth on the grain. Rust (*Puccinia sorgi* and *P. polysora*) causes small brown pustules on the leaves. Downy mildew (*Sclerospora* spp.) causes pale yellow streaks and necrosis of the leaves. Stalk and ear blight (*Diplodia macrospora*) cause stalks to break and ears to rot. Maize smut (*Ustilago maydis*) and head smut (*Sphacelotheca reiliana*) produce masses of black spores on the shoot and inflorescence, respectively.

Several virus diseases such as corn stunt, sugarcane mosaic, and streak are fairly widespread and are transmitted by insects. Numerous insects, including stem borers, earworms, cutworms, and others also attack maize.

Sorghum bicolor

Cultivated sorghum (formerly *S. vulgare*, *Holcus sorghum*, and *Andropogon sorghum*) is also known as grain sorghum; great millet; guinea corn (in West Africa); kafir corn (in South Africa); milo, sorgo, and sometimes maize (in the United States); kaoliang (in China); durra (in Sudan); mtama (in East Africa); and jola, jawa, and cholam (in India). Grain sorghum is the fourth most important world cereal, following wheat, rice, and maize. It is an erect, robust, tillering, usually annual member of the subfamily Panicoideae, tribe Andropogoneae. It is grown primarily for grain and secondarily for brooms, syrup, building materials, fodder, and forage. The grain is consumed by both humans and animals. In the United States, Europe, and Japan, it is used extensively for animal feed. Some cultivars (usually called sorgo) have large, juicy, sweet stems that can be used for syrup. Others (broomcorns) have inflorescences with long, straight side branches that can be used as brooms.

Cultivars range from 0.5 to 6 m in height. Dwarf cultivars have been developed for combine harvest, while taller cultivars are hand harvested for grain and other products. Short-season cultivars can mature in 100 days or less, while long-season types can take more than 180 days. The grain is produced on compact panicles that are usually erect but that can be recurved (pendant). The seeds vary from pale yellow to deep purple or brown, are 4–8 mm in diameter, and weigh 15–40 mg each. Seeds normally mature 25–55 days after anthesis. Dwarf cultivars developed for combine harvesting have rather synchronous ripening, while tillers of some hand-harvested types mature over a period of time making multiple harvests necessary.

Cultivated sorghum is predominately self-pollinated, but cytoplasmic male sterility has been used to produce hybrids with much greater yield potential than earlier cultivars. The development of adapted hybrids was largely responsible for doubling sorghum grain yields in the United States between 1954–1956 and 1962–1964.

Cultivated sorghum arose in northeastern Africa (probably Ethiopia) at least 5000 years ago. From there it was taken to western and central Africa. It probably reached India at about the same time as pearl millet and finger millet, 3000–4000 years ago. From there it reached China, perhaps 1500–2000 years ago.

Harlan and de Wet (1972) devised a simple classification system for cultivated sorghum based primarily on spikelet morphology. They describe five races of cultivated *S. bicolor*. The spikelet characteristics correlate quite well with inflorescence type and geographical distribution in Africa. The "bicolor" race has an elongated grain with clasping glumes and an open panicle. The race is heterogenous and inclues sorgo (sweet sorghum) and broomcorn. The bicolor race is widespread in Africa, but it is low yielding and not cultivated in large acreages.

The "guinea" race has grain that is flattened on both sides and at maturity is twisted nearly 90° between long gaping glumes. Like the bicolor race, it has an open panicle. The guinea race is concentrated in West Africa, with a secondary center in East Africa. It is well adapted to high-rainfall areas.

The "caudatum" race has asymmetrical grains that are flattened on one side and rounded on the other. The glumes are about half the length of the grain. Several inflorescence types are found in the caudatum race. It is very important in Sudan, Chad, Nigeria, and Uganda. It has high-yield potential and is an important source of germplasm for plant breeding programs.

The "kafir" race has symmetrical, nearly cylindrical grains with clasping glumes of variable length. It is a major race in East Africa south of the equator, and it is an important source of breeding materials.

The "durra" race has distinctive obovate grains that are narrow at the base and broad near the tip. The glumes are very wide. It is a Near Eastern race and is also very important in India. In addition to long-season cultivars, very short-season, drought-evading durra cultivars are grown in low-rainfall areas.

Intermediate races share the characteristics of two of the basic races. Guinea-candatum is a major race in Nigeria, Chad, and Sudan. Guinea-kafir is a major race in India. Almost all modern U.S. hybrids are kafir-caudatums.

Cultivated sorghum's greatest asset is its drought resistance and high yield potential. It is grown in areas too hot and dry for maize. It tolerates poorly drained soils and heavy rainfall better than pearl millet. Its grain yield potential is more than 6000 kg/ha.

Cultivated sorghum is quite susceptible to bird damage during grain filling, but types with recurved panicles and astringent seeds are more resistant. Insects such as sorghum shoot fly, chinch bug, and sorghum midge, stalk borers, grasshoppers, and other insects can damage cultivated sorghum. It is also quite susceptible to storage pests.

Major diseases include anthracnose (*Colletotrichum*), which damages leaves and stems. Leaf blight (*Helminthosporium*) is widespread under humid conditions. Charcoal rot (*Macrophomina*) causes damage in hot dry areas. Milo disease (*Periconia*), rust (*Puccinia*), downy mildew (*Peronosclerospora*), honeydew disease (*Sphacelia*), smut (*Sphacelotheca*), sooty stripe (*Ramulispora*), and other diseases are also quite widespread.

Witchweed (*Striga*) species, semiparasitic weeds that attach to the roots of grasses, are important pests in Africa and India.

Pennisetum americanum

Pearl millet (formerly *P. typhoides* or *P. glaucum*) is also known as bulrush millet, cattail millet, spiked millet, and (in India) bajra. It is an erect, tillering, annual member of the subfamily Panicoideae, tribe Paniceae. It is grown primarily for grain and secondarily for fodder in the drier parts of tropical Africa and India. The vegetative plant parts are used for animal feed, bedding, thatch, fencing, and fuel.

Cultivars range from 0.5 to over 3 m tall and mature in 60–180 days. The grain is produced on a stiff, compact, cylindrical inflorescence 2–3 cm in diameter and 15–45 cm long. The seeds are white, yellow, grey or light blue, weigh 3–15 mg each, with 900–3000 in each inflorescence. Flowering and maturity are not necessarily synchronous on different tillers, and two or more harvests may be necessary.

Pearl millet is protogynous (styles dry before pollen is shed); therefore, it is largely cross-pollinated. Because of this cross-pollination and the existence of many cultivars, most commercial fields have a variety of plant types. However, male-sterile lines are now available, and hybrids will doubtless play an important role in the future.

Pearl millet arose in tropical West Africa and was taken to East Africa and India at least 3000 years ago. It is the most important crop in the Sahel zone of Africa just south of the Sahara. It is as important as grain sorghum in the Sudan zone and is the fourth or fifth most important cereal in India. Pearl millet's greatest asset is that it can be grown economically in low-rainfall areas on sandy infertile soils too poor for other cereals. It also stores well.

Short-maturity genotypes can be grown in very low rainfall zones (250 mm). Longer-maturity types are grown in higher-rainfall zones. Grain yields of rain-fed pearl millet average 250–750 kg/ha in Africa and 770–1100 kg/ha in India. Yields of up to 5000 kg/ha can be obtained with adequate water and plant nutrition.

The major pest of pearl millet is birds, which find the seed very palatable and the head a convenient perch. Other pests include caterpillars, grasshoppers, stem borers, and earworms. Major diseases include green ear, a downy mildew (*Sclerospora*) which causes the glumes and reproductive organs to grow into a mass of leaf-like organs. Smut (*Tolyposporium*) can be severe in wet weather. Infected grains turn bright green then black. Honeydew or sugary disease (*Sphacelia*) cause young grain to produce pink secretions that later turn black. Rust (*Puccinia*) and leaf spot (*Curvularia, Helminthosporium, Piricularia*) are also relatively common.

Eleusine coracana

Finger millet is also known as African millet, koracan, ragi (in India), wimbi (Swahili), bulo (in Uganda), and telebun (in Sudan). It is a robust, freely tillering, tufted annual member of the subfamily Eragrostoideae, tribe Chlorideae. It is grown primarily for grain and secondarily for fodder in tropical Africa and India. The fodder is used for livestock feed and the fields are often grazed after harvest.

The plant is normally 0.4–1.0 m tall and matures in 105–180 days, depending on cultivar. The grain is produced in a terminal digitate inflorescence of 3–9 dense sessile spikes 5–15 cm long. There are 60–80 spikelets per spike and 4–7 seeds per spikelet. The seeds are usually reddish, orange, brown, or nearly black. They are quite small, weighing 4–5 mg each.

Cultivated finger millet is mainly self-pollinated. There are many cultivars, but two main groups are recognized. African highland types have long spikelets, long glumes, long lemmas, and grains enclosed within the florets. Afro-Asian types have short spikelets, glumes, and lemmas; mature grains protrude from the floret. Some improved cultivars have been selected in India.

Finger millet probably arose in the East African highlands and was taken to India over 3000 years ago. One of its greatest assets is that it can be stored for many years without deterioration or insect damage. Thus, it provides insurance against famine.

Finger millet is less drought tolerant than pearl millet. It is normally grown in areas receiving 300–2000 mm annual rainfall. It is very sensitive to weed competition at the seedling stage and will not tolerate waterlogging. In Africa it is sometimes grown immediately after cutting and burning woodland. The fire provides nutrients from the ash and stimulates nitrogen mineralization. The heat-sterilized soil also provides weed control. Average yields are quite variable, but over 5000 kg/ha have been obtained experimentally.

Finger millet has few diseases and pests. Birds are not a problem since they have difficulty perching on the small inflorescences. Blast (*Piricularia*) is a serious leaf and inflorescence disease. Blight (*Helminthosporium*) also causes damage in some countries. The stored seed is almost immune to insect damage.

Eragrostis tef

Tef is a small (20–90 cm), fine-stemmed, tufted, annual member of the subfamily Eragrostoideae, tribe Eragrosteae. It is grown as a cereal throughout Ethiopia, where it was domesticated before recorded history. It is also grown for hay or fodder in South Africa and India. The inflorescence is an open panicle and seeds are small (0.25–0.3 mg). Some cultivars have white seeds; others have red or brown seeds. In Ethiopia it is cultivated at altitudes between 1700 and 2800 m at up to 2500 mm annual rainfall. The crop matures in about 4 months and yields normally vary from 300 to 3000 kg/ha.

SUGARCANE

Sugarcane is the world's most important sugar crop. It is an erect, very robust, tillering, perennial member of the subfamily Panicoideae, tribe Andropogoneae. It is grown primarily for sugar (sucrose); however, secondary products such as molasses, ethyl alcohol (for rum or fuel), animal feed, and fiber (bagasse) are also economically important. In 1981/82 and 1982/83 annual world production was about 680 million tons of sugarcane, which yielded about 65 million tons of sugar. In contrast, annual production of sugarbeets was about 280 million metric tons, yielding approximately 35 million tons of sugar.

Sugarcane is vegetatively propagated by planting stem sections (setts) from which axillary buds grow to produce erect stems with internodes 5–25 cm long and 1.5–6 cm in diameter. Secondary and tertiary tillers are produced at the base of the stem. The stems can grow to lengths of 6 m or more, growing upward and eventually lodging if they are not harvested. Leaves are borne alternately with laminae 70–120 cm long and up to 10 cm wide. About 10 mature green leaves are usually present at any time. The sheath and lamina are usually shed when they senesce. Adventitious roots are produced at the base of the tillers, and after a short time they supply the major portion of a new tiller's water and nutrients.

The inflorescence, also known as the arrow or tassel, is produced under certain environmental conditions. Genotype, photoperiod, temperature, nutrition, and water stress all affect inflorescence initiation and growth. Since the tiller ceases to grow and eventually deteriorates after flowering, genotypes that flower readily under field conditions are selected against by breeders.

At harvest, the dry matter of a sugarcane tiller consists of 50–60% millable cane, 30–40% unusable tops (leaves and immature stem), and about 10% roots and stubble. Older crops may contain an higher percentage of millable cane. Of the millable cane's total weight, about 75% is water, and approximately 50% of the dry weight is sugar. Efficient factories can recover about 85% of the sucrose in the cane. Thus, depending on the sugar content of the cane and its recovery, 9–13 tons of raw sugar are recovered for each 100 tons of cane harvested.

Commercially grown sugarcane cultivars are complex hybrids of several *Saccharum* species. *Saccharum spontaneum* is a very variable wild species occurring in eastern and northern Africa, India, China, Southeast Asia, Taiwan, Malaysia, and the Pacific islands. It is an aggressive, rhizomatous, freely tillering perennial. In modern sugarcane breeding programs, it is used as a source of vigor, hardiness, and disease resistance. *Saccharum robustum* is found in Southeast Asia from New Guinea to Taiwan, in India, and in Africa. Since before recorded history, ecotypes have been selected and propagated for chewing on the basis of sweet juice and low fiber content. The "noble" canes, *S. officinarum*, probably were selected from *S. robustum* by stone-age cultures in New Guinea. They were spread throughout the Pacific and Southeast Asia prior to the arrival of European man. Natural hybrids of *S. officinarum* and *S. spontaneum* resulted in *S. sinense*, the vigorous "thin" canes of northeastern India and southern China, where they were used for syrup production. The first sugarcane to reach Europe was a cultivar of *S. sinense*. It was later taken by the Spanish and Portugese to the New World, where it was known as "Creole" or "cana criolla" and formed the basis of sugarcane culture in the sixteenth century. In the late eighteenth century Creole was replaced by improved cultivars of *S. officinarum* such as Bourbon (Otaheite), Cheribon, Tanna (Caledonia), and Badila. Modern sugarcane breeding began in Java and Barbados at the end of the nineteenth century due to the discovery of true viable seeds of sugarcane and the need to obtain increased disease resistance. Initially, only *S. officinarum* was used in breeding programs, but by the 1920s interspecific hybrids between *S. officinarum* and *S. spontaneum* were providing hardier, more disease-resistant cultivars that rapidly replaced earlier cultivars.

Sugarcane's greatest asset is its ability to produce sugar efficiently and in great quantity per unit land area. It is grown on a wide range of soils and tolerates moderate drought, flooding, and salinity stresses. Because of the difficulty in harvesting, transporting, and processing large amounts of stalks and sugar, most sugarcane is grown by large plantations or cooperatives with well-coordinated planting, harvesting, and processing.

Due to long and persistent efforts by plant breeders, most commercial sugarcane cultivars are quite resistant to diseases and insect pests. Rigorous phytosanitary measures also prevent the transfer of diseases from one area to another. Nevertheless, the following diseases can cause significant damage in susceptible cultivars. The bacterial "gumming disease" (*Xanthomonas*) produces yellowish stripes near the leaf tips, reddish vascular bundles of the stem, and exudation of yellowish gum when the stem is cut. Ratoon stunt disease is caused by an actinomycetelike organism that causes reddish discoloration of the vascular elements of the stalk. "Pokka boeng" disease, sett rot, and stalk rot are all caused by *Fusarium* fungal attack. Pokka boeng causes dark red discoloration of the stem parenchyma and vascular tissue. Downy mildew (*Peronosclerospora*) causes rapid elongation of thin, brittle stalks with twisted, shredded leaves. It can be serious in warm, humid weather.

Rust (*Puccinia*) is often an important foliar disease. Smut (*Ustilago*) causes production of many small tillers that fail to produce useful stalks. In addition, infected tillers develop a characteristic whiplike, spore-filled structure from the shoot apex. Chlorotic streak and sugarcane mosaic are viral diseases transmitted by insects.

Many sugarcane diseases also infect other grasses, which can provide a source of infection. Disease control in sugarcane is based on (1) strict phytosanitary precautions to prevent spread of diseases among sugarcane-producing areas, (2) development of resistant cultivars, (3) treatment of setts with fungicide and hot water, (4) sterilization of equipment used to cut setts, and (5) destruction of infected plants in the field.

FORAGE GRASSES

C_4 grasses are the basis of beef and milk production in most tropical and subtropical regions. They are also important species in warm temperate regions, especially during summer. Twenty of the most important C_4 forage grasses are described below.

Andropogon gayanus

Gambagrass, also known as sadabahar (India) and carimagua (South America), is a large, cross-pollinated, tufted, perennial member of the subfamily Panicoideae, tribe Andropogoneae. Fertile stems are 1–3 m tall. Leaves are up to 50 cm long and 15 mm wide. The inflorescence is a large spathate panicle with 2–18 primary branches, each terminating in a pair of racemes.

Gambagrass is a native of Africa but has shown promise as a forage grass throughout the tropics. It is adapted to a wide range of soils and is known for its drought tolerance, retention of green foliage well into the dry season, rapid production of high-quality forage at the beginning of the rainy season, tolerance of burning, growth on infertile acid soils, and ability to associate with legumes. It also tolerates light frost, and one variety grows on seasonally flooded soil.

Andropogon gerardii var. gerardii

Big bluestem is a robust, tufted perennial member of the subfamily Panicoideae, tribe Andropogoneae. The culms are usually 0.8–2 m tall. Leaf blades are about 30 cm long and up to 12 mm wide. Big bluestem is not rhizometous or has short, stout rhizomes, but other varieties of the same species have well-developed rhizomes. The inflorescence has 2–7 spikate branches 4–11 cm long.

Big bluestem is one of the four most important species in the North America "true" or "tall-grass" prairie, which occurs in higher-rainfall areas east of the short-grass prairie from North Dakota southward to Texas. It begins growth late in the spring and flowers from August through November. It is best adapted to moist, well-drained soils of relatively high fertility. It is very palatable to livestock, but due to its erect, nonrhizometous habit, it cannot tolerate overgrazing. Big bluestem is easily propagated by seed, and it is sometimes used to revegetate areas where severe grazing pressure is not anticipated.

Bouteloua gracilis

Blue grama is a low fine-leaved, rhizomatous, perennial member of the subfamily Eragrostoideae, tribe Chlorideae. The culms are usually 25–60 cm tall. The leaf blades are 8–15 cm long and 1–2.5 mm wide. The inflorescence has one to three pectinate (comblike) branches with 40–90 closely spaced spikelets per branch.

Blue grama is one of the most important species of the North American western "mixed" or "short-grass" prairie, which extends from southern Saskatchewan and Alberta southward to Texas and northern Mexico. It is a warm-season species that begins growth well after the associated C$_3$ grasses. It is adapted to a range of soils, but it is especially common on clayey upland soils. It is very palatable to livestock, but due to its low-growing rhizomatous habit, it is more tolerant of grazing than most associated species. Blue grama is easily propagated by seed and is used in range revegetation in its area of adaptation. Since ecotypes have adapted to their areas of origin, seed from other areas should not be used for range improvement.

Brachiaria decumbens

Surinamgrass is an apomictic, stolonoferous perennial member of the subfamily Panicoideae, tribe Paniceae. Flowering culms are usually 30–60 cm tall. Leaves are up to 14 cm long and up to 12 mm wide. The inflorescence has two to five racemes arising from a central axis. Pastures are established both from stem cuttings and seed.

Surinamgrass is a native of East Africa, but it has been introduced into many parts of the world. It, like pangolagrass, is best suited to warm, moist areas with annual rainfall exceeding 1000 mm. However, surinamgrass is more tolerant of very acid soils and drought stress than pangolagrass.

Surinamgrass produces good-quality, palatable forage that yields good animal gains. It responds well to fertilization, but it is aggressive and is difficult to combine with legumes. In many areas spittlebug is a serious pest that has reduced the area planted to surinamgrass.

Brachiaria mutica

Paragrass or angolagrass is a coarse, vigorous, stoloniferous, perennial member of the subfamily Panicoideae, tribe Paniceae. The flowering culms grow upward producing a canopy 1–2 m deep from long, freely rooting stolons up to 5 m long. The leaf blades are 10–30 cm long and 5–15 cm wide. The panicle produces 10–20 simple or sometimes branched branches 2.5–15 cm long with spikelets in two rows.

Paragrass probably arose in humid West Africa and has since been spread to many parts of the world as a valuable forage grass adapted to humid, fertile soils. It is especially well adapted to seasonally flooded soils, but it cannot be used on dry, well-drained soils in arid and semiarid regions. It can be grown in soils flooded with brackish water, and it responds well to fertilizer. In addition, it tolerates shade and grows well under tree crops.

Paragrass is readily established from stem cuttings, and since it sets few viable seed, this is the most common method of establishment. Once established, it is quite vigorous and productive. It combines well with legumes when it is grown on well-drained soils, but few forage legumes can tolerate the seasonally flooded conditions under which it is often grown. Paragrass runners are long and quite tender, making it somewhat susceptible to trampling and overgrazing.

Cenchrus ciliaris

Buffelgrass, or African foxtail, is a tufted or rhizomatous perennial member of the subfamily Eragrostoideae, tribe Chlorideae. Flowering culms are up to 140 cm tall with leaves up to 30 cm long and 2–5 mm wide. The inflorescence is a cylindrical spike bearing clusters of spikelets, each surrounded by an involucre of bristles.

Buffelgrass is a native of Africa, Arabia, India, and Pakistan that has been widely introduced to dry areas throughout the tropics and subtropics. It is very drought resistant and can grow in areas with as little as 300 mm annual rainfall in two rainy seasons or 400 mm in a single rainy season. It also tolerates high temperatures and some flooding, but it is not adapted to acid soils and is susceptible to frost.

It is normally apomictic, but a few sexual plants have been found, and these are very useful in breeding programs. Compared with other drought-tolerant species, buffelgrass is a high-quality forage that gives good animal gains.

Chloris gayana

Rhodesgrass is an erect, leafy, stoloniferous, cross-pollinated perennial member of the subfamily Eragrostoideae, tribe Chlorideae. Flowering culms are 60–140 cm tall. Leaves are up to 50 cm long and 20 mm wide. It has a typical digitate inflorescence with 3–20 dense spikate branches 4–15 cm long.

Rhodesgrass originated in southern and eastern Africa and is now grown throughout the tropics and subtropics. It grows on a wide range of soils, including alkaline and saline soils. It is best adapted to tropical and subtropical areas with annual rainfall ranging from 600 to 1150 mm. It is quite drought tolerant and will survive long dry periods, but it is not well adapted to high rainfall, very acid soils, or poorly drained clay soils. Some cultivars are quite frost resistant. Rhodesgrass is easily established from seed, can be used for hay or grazing, is tolerant of trampling, will crowd out most weeds, but will associate with some legumes. Fertilization is usually needed to extend the productive life of the sward more than 3 years. Because of its easy establishment from seed, rhodesgrass is often found along roadsides and abandoned fields. However, it is not a serious weed because it is easily destroyed by tillage.

Palatability and digestibility of the young grass are good, but they decline rapidly and, in general, are lower than other improved grasses.

Cynodon dactylon

Bermudagrass is a vigorous, stoloniferous, and rhizomatous perennial member of the subfamily Eragrostoideae, tribe Chlorideae. The stolons and rhizomes readily root at the nodes to form a dense turf. The culms are erect or ascending. The canopy is usually from 5 to 50 cm deep, depending on cultivar and management. The inflorescence normally is digitate with three to seven spikate branches 3–10 cm long. Though bermudagrass is an important forage grass, it is also one of the world's most serious grass weeds.

Bermudagrass originated in Africa and has since been spread to many tropical, subtropical, and warm temperate areas. In Africa, it is found from sea level to 2200 m and at mean annual temperatures between 6 and 29°C. Its shoots are killed by frost, but stolons and rhizomes overwinter in much of the United States and Europe. It tolerates a wide range of soils, from sandy to clayey and from acid to alkaline.

A wide range of ecotypes and hybrid cultivars are known. Some ecotypes produce seed, and the species is an important weed in many cultivated crops. However, sterile hybrids reproduced by cuttings have little weed potential and have become valuable pasture and lawn grasses. "Coastal" bermudagrass is a hybrid that is the most important forage grass in the southeastern United States. "Coastcross-1" is another important hybrid cultivar. It is more digestible than Coastal and produces better animal gains. It is also easier to eradicate since it produces no rhizomes; however, the lack of rhizomes makes it less winter hardy.

Digitaria decumbens

Pangolagrass is a fine-stemmed perennial member of the subfamily Panicoideae, tribe Paniceae. It produces long, creeping stolons that root from the nodes. The canopy can reach 120 cm but is normally much shorter. The leaves are 10–25 cm long and 2–7 mm wide. The inflorescence is a digitate panicle with 5–10 racemes in one or two whorls. It is highly sterile and is established with stem cuttings.

Pangolagrass is a native of South Africa but is now grown as a forage grass throughout the tropics. It is best suited to the warm tropics and subtropics where annual rainfall exceeds 1000 mm, but it can be grown in areas with lower rainfall. It does not tolerate frost, but it grows well on a wide range of soils.

Pangolagrass is a high-quality forage that produces good forage and animal yields. It responds well to fertilization but is too aggressive to combine well with legumes. A stunting virus transmitted by aphids has recently become a serious problem, and in many areas pangolagrass is being replaced by *Cynodon* and *Brachiaria* species.

Eragrostis curvula

Weeping lovegrass is a densely tufted perennial member of the subfamily Eragrostoideae, tribe Eragrosteae. It is apomictic, grows 30–120 cm tall, and has fine leaves up to 30 cm long and 1–5 mm wide. The inflorescence is a panicle ranging from loose and spreading to dense and narrow. It occurs naturally in the southern part of Africa and has been widely introduced into tropical, subtropical, and warm temperate countries.

Weeping lovegrass tolerates drought, salinity, frost, close grazing, and high temperatures. However, it is not adapted to acid or frequently waterlogged soils. Palatability varies widely among cultivars and with age, and burning may be required to remove old unpalatable forage.

Hemarthria altissima

Limpograss, also known as capim gamalote (Brazil), pasto clavel (Argentina), and baksha and panisharu (India), is a decumbent, rhizomatous perennial member of the subfamily Panicoideae, tribe Andropogoneae. It is native to tropical and subtropical areas of the Old World and has been introduced to the Americas. Leaves are up to 20 cm long and 6 mm wide. The inflorescences are racemes that appear singly or in groups of two to four from leaf axils.

Limpograss is well adapted to wet soils, but it can tolerate some seasonal drought stress. Its most important characteristics are its high palatability and digestibility.

Heteropogon contortus

Speargrass is an apomictic, tufted, perennial member of the subfamily Panicoideae, tribe Andropogoneae. It is widely distributed throughout the tropics and subtropics of the world, but is is especially common in India, West Africa, and Australia. It grows to a height of 40-100 cm, has leaf blades 7-20 cm long and 3-7 mm wide. The inflorescence is a raceme, and the lemma of the fertile spikelet produces a long (5-12 cm) twisted awn that aids in distribution and burial of seeds.

Speargrass is seldom cultivated, but it is often an important species in woodlands and savannas. It tolerates fire and light grazing, but animal gains on pure speargrass are generally low due to its poor quality. In Australia it combines well with improved legumes, and mixtures produce good animal gains.

Hyparrhenia rufa

Jaragua, also known as yaragua and puntero, is a tufted, perennial member of the subfamily Panicoideae, tribe Andropogoneae. It grows 0.3-2.5 m tall with leaves 30-60 cm long and 2-8 mm wide. The inflorescence has two racemic branches that emerge from a large spathelike leaf sheath. Jaragua is a native of Africa, but it is widely distributed in the New World, where it was introduced during the slave trade. In general, its quality as a forage grass is low due to its rapid production of flowering stems. However, it is quite tolerant of burning, low soil fertility, and acid soils; therefore, it is an important forage grass in much of tropical America

Panicum coloratum

Kleingrass, colored guinea, or makarikarigrass is a cross-pollinated, sexual, tufted, perennial member of the subfamily Panicoideae, tribe Paniceae. It produces erect or spreading stems 40-150 cm high. Leaves are up to 40 cm long and 14 mm wide. The inflorescence is a loose, erect panicle.

Kleingrass is a native of Africa, but it has been introduced to a number of tropical and subtropical countries. It is adapted to a variety of soils where annual rainfall is between 600 and 1200 mm. Variety *makarikariense* is especially well adapted to heavy clay soils that are seasonally waterlogged. Variety *coloratum* needs better drainage. Both varieties are quite tolerant of frost and form good associations with tropical legumes. In addition, forage quality and palatability are good.

Panicum maximum

Guineagrass (*Panicum maximum* var. *typica*) and the smaller green panic or slender guinea (*P. maximum* var. *trichoglume*) are apomictic, robust, tufted, perennial members of the subfamily Panicoideae, tribe Paniceae. They are natives of Africa but are now important pasture grasses (and occasionally weeds) between 10°N and 20°S worldwide. They are adapted to a wide range of soils and are common at elevations to 2000 m where annual rainfall exceeds about 1300 mm. They are not frost tolerant.

Plants are 0.5–4.5 m tall with leaves 15–100 cm long and up to 35 mm wide. The inflorescence is a loose panicle. Guineagrasses are quite variable morphologically, and many cultivars are available. Cultivars Hamil and Coloniao are tall (2.5–3.5 m) plants with thick stems. Cultivars Makueni and Sabi are smaller (1.0–1.5 m) with finer stems; cv. Embu has a creeping habit. Guineagrasses produce palatable, high-quality forage and combine well with many tropical legumes.

Paspalum dilatatum

Dallisgrass or Pasto miel is a spreading, tufted, perennial member of the subfamily Panicoideae, tribe Paniceae. It reaches 50–150 cm, has leaf blades up to 45 cm long and 3–13 cm wide. The panicle consists of several spikelike racemes widely spaced along a slender axis. It is a native of southeastern Brazil, Uruguay, and northern Argentina, but it has been introduced into many other tropical and subtropical countries.

Dallisgrass is best adapted to moist environments with over 1000 mm annual rainfall, but it also survives relatively long dry periods. It is more resistant to frost than rhodesgrass (*Chloris gayana*) and setariagrass (*Setaria anceps*). It is often grown in association with rapidly established grasses such as rhodesgrass and molassesgrass (*Melinis minutiflora*), but it is much more persistent and eventually replaces them. It is often an important component of bermudagrass (*C. dactylon*) pastures and is also sown with both tropical legumes and clovers. It is quite palatable to livestock and tolerates heavy grazing.

Paspalum notatum

Bahiagrass is a creeping perennial member of the subfamily Panicoideae, tribe Paniceae. It produces both rhizomes and stolons. The stolons are thick with short internodes, and they root freely from the nodes. The fertile stems are erect and up to 100 cm tall. Leaf blades are usually 5–20 cm long and 2–10 mm wide. The inflorescence is made up of two to five racemate branches.

Bahiagrass is native from the southern United States to Argentina and has been introduced to a number of other countries. It forms a dense sod and is used for forage, as a lawn grass, and as erosion protection. It grows on a variety of soils, including saline, acid, and flooded soils. It is more tolerant of frost than is dallisgrass, and it can survive quite severe drought in a dormant state. Bahia-grass is quite aggressive and low growing, and it tolerates heavy grazing pressure. However, it does not easily associate with legumes.

Pennisetum clandestinum

Kikuyugrass derives its common name from the Kikuyu tribe in Kenya, within whose lands it was first collected. It is a member of the subfamily Panicoideae, tribe Paniceae, and is a vigorous, creeping perennial that spreads by rhizomes and stolons, both of which root at the nodes. It also produces small, inconspicuous inflorescences borne in leaf axils on short, erect lateral culms. The inflorescence is almost completely enclosed by the leaf sheath, and the only parts visible are the stigma and, when present, the anthers.

Kikuyugrass is adapted to a wide range of environments, from sea level to almost 3000 m, and from 500 mm to more than 1000 mm rainfall. However, it is best adapted to areas with moderate-to-good soil fertility, cool temperatures, and annual rainfall greater than 1000 mm. It withstands light frosts, but it is killed by sustained freezing temperatures. It is very resistant to frequent close defoliation, and it is a valuable pasture and lawn grass where it is adapted. Frequent grazing or mowing stimulate seed production, and seed may be harvested mechanically or spread through the feces of grazing animals. However, vegetative propogation is the most common form of establishment.

Pennisetum purpureum

Napiergrass or elephantgrass is a very large, robust, tufted, perennial member of the subfamily Panicoideae, tribe Paniceae. It produces culms 2–7 m tall with leaf blades up to 90 cm long. The culms become woody and unpalatable when they mature. The inflorescence is a compact, cylindrical spike up to 30 cm long. It is a native of tropical Africa and has become widespread throughout the tropics and subtropics. It is best adapted to humid lowland regions but it grows at elevations up to 1500 m. The aboveground portion is killed by frost, but the rhizomes can survive freezing weather in the southern United States. It is often found along roadsides, waste areas, and water course. In addition, it is often grown in pure stand and can be cut and fed green to cattle or grazed. However, because of its rapid growth, it requires heavy fertilization and careful management to maintain high-quality forage.

Fertile hybrids are formed between napiergrass and pearl millet (*P. americanum*). This allows napierlike forage to be produced from seed. It also has somewhat higher forage quality than napiergrass and produces good grazing in the late summer in subtropical climates.

Napiergrass can become a weed of perennial crops such as citrus, coffee, oil palm, pineapple, rubber, sugarcane, and tea; but it is seldom a problem in annual crops.

Setaria anceps

Setariagrass, known as napierzinho in Brazil because of resemblance to napier-grass, is a tufted, perennial member of the subfamily Panicoideae, tribe Pani-ceae. It reaches 1–2 m in height, has leaves up to 40 cm long and 8–20 mm wide. Like other species of the genus, its inflorescence is a dense spike.

Setariagrass occurs naturally in tropical Africa between 600 and 2600 m where annual rainfall exceeds 750 mm, but it has been introduced into a number of tropical countries. It tolerates fire, occasional flooding, and light frosts, and it associates well with legumes. Important cultivars include Kazungula, Nandi, and Narok.

WEEDS

C_4 grasses are serious weeds in many crops, primarily due to their rapid growth and high water use efficiency at high temperatures. As pointed out above, many C_4 forage grasses can be serious weeds in waste areas and in cultivated crops. In addition, other C_4 grasses with little forage value are serious weeds. Four of the latter are discussed below.

Echinochloa cruzgalli

Barnyardgrass is a robust, tufted, annual member of the subfamily Panicoideae, tribe Paniceae. It and jungle rice (*Echinochloa colonum*) are the two most im-portant grass weeds of rice, worldwide. Barnyardgrass culms are erect or decum-bent and up to 1.5 m long with leaves up to 50 cm long and 15 mm wide. The panicle has numerous alternate racemate branches 2–4 cm long. It is morpho-logically variable.

Barnyardgrass is a native of Europe and India and is now a common weed of most agricultural areas of the world except in Africa. In temperate environments the seed germinates in the spring and early summer. It is a short-day plant that flowers in response to decreasing daylength. The critical photoperiod varies among genotypes, ensuring their adaptation to a wide range of latitudes.

Barnyardgrass is adapted to a wide range of soils, but it grows best in wet, or even submerged, soils. In rice, it competes strongly for nutrients and can reduce yields dramatically; however, it is also an important weed in many other crops, including cotton, maize, potato, and sugar beet.

Eleusine indica

Goosegrass or crowsfootgrass is an annual, tufted member of the subfamily Era-grostoideae, tribe Chlorideae. It is among the eight most important weeds of the world. Goosegrass culms are up to 60 cm tall with leaves up to 15 cm long. The panicle is digitate with two to seven spikate branches. It probably arose in Africa and is now found in most crop-growing areas of the world.

Goosegrass is adapted to a wide range of soils and is a serious weed in many cultivated crops as well as in footpaths, open ground, lawns, and pastures. Its phenological development is quite variable and is sensitive to photoperiod, temperature, and soil moisture. It flowers most readily at short photoperiods in warm moist environments, and it is a prolific seed producer. Long photoperiods stimulate vegetative growth, though some flowering may occur at photoperiods up to 16 hr. High light intensity increases the number and lateral spread of tillers as well as dry matter production. Low light intensity produces more erect plants with fewer tillers and less dry matter.

Goosegrass is quite palatable to livestock when it is young; however, it decreases in palatability as it matures. Under some circumstances it can also produce toxic amounts of prussic (hydrocyanic) acid.

Imperata cylindrica

Cogongrass or bladygrass is a widespread perennial member of the subfamily Panicoideae, tribe Andropogoneae. It is among the world's 18 most serious weeds, though it has some value as a forage in Southeast Asia. Cogongrass has many names throughout the tropics (lalang in Malasia, alang-alang in Indonesia, illuk in Sri Lanka, bladygrass in Australia, cogon in the Philippines, and hakha in Laos and Thailand).

Cogongrass grows in tufts with slender, erect culms 15–120 cm tall with blades up to 150 cm long. It spreads by seed and by long, tough rhizomes. The inflorescence is a compact, cylindrical panicle up to about 20 cm long. It has a silvery, fluffy appearance due to hairs that surround the spikelets. Cogongrass is very resistant to fire and is tolerant of infertile soils. It is found at altitudes up to 2700 m and in areas receiving between 500 and 5000 mm annual precipitation. It does not tolerate frost and is also intolerant of heavy shading or frequent defoliation. It is commonly recognized as a serious weed in plantation agriculture of the humid tropics, and it forms almost pure stands in abandoned, frequently burned agricultural land. It is palatable to animals only at early stages following burning, and improved grasses produce greater animal gains. However, in many areas cogongrass is an important source of thatch.

Sorghum halepense

Johnsongrass is a member of the subfamily Panicoideae, tribe Andropogoneae. It is an aggressive, perennial grass that is among the world's 18 most important weeds; however, it also produces valuable forage in many areas. Johnsongrass forms culms 0.5–3 m tall arising from robust, scaly rhizomes. Blades are up to 60 cm long and 5 cm wide. The large, open panicle is 15–50 cm long, and propagation is by seeds and rhizomes.

Johnsongrass arose in the Mediterranean region and has spread throughout the tropics and subtropics. It is best adapted to warm, humid, summer-rainfall areas of the subtropics. It is a common weed in cultivated land, waste places, roadsides, and field borders. The culms are sensitive to frost, but the rhizome overwinters in subtropical and warm temperate regions. Pure johnsongrass stands lose vigor in a few years, but they are readily rejuvenated by plowing, which causes a loss in apical dominance and rapid growth of culms from the broken rhizomes. Though infrequent plowing can increase weed infestation, frequent tillage can greatly reduce it in cultivated fields.

Johnsongrass can be a good forage grass when properly managed, but in dry weather and following light frosts it often produces prussic (hydrocyanic) acid, which can be lethal to cattle.

2

THE C$_4$–KRANZ SYNDROME

In the nineteenth century French and German botanists began to study the internal anatomy of grasses, particularly the leaf blades. They discovered the important structural and functional roles of the leaf veins (vascular bundles containing xylem, phloem, sclerenchyma, and chlorenchyma). These parallel longitudinal bundles are responsible for long-distance movement of water, nutrients, and assimilates; their sclerenchyma strengthens the leaf; and they are often important photosynthetic organs. However, these early botanists soon recognized that there are two contrasting types of internal leaf anatomy among grasses (Duval-Jouve, 1875). Certain grasses, primarily tropical in origin, have a type of leaf anatomy that is fundamentally different than that of most temperate grasses. These tropical grasses have leaf mesophyll cells arranged radially around a prominent bundle sheath made up of closely appressed cells with prominent bright green chloroplasts. In contrast, winter cereals and cool-season grasses usually have leaf mesophyll cells that are not radially arranged, and the chloroplasts of the bundle sheath cells are not distinctive (Haberlandt, 1882; Heinricher, 1884).

These early anatomists were puzzled by the distinctive leaf anatomy of tropical grasses. Haberlandt (1914, quoted in Crookston, 1980) noted that "It is uncertain whether this green inner (bundle) sheath merely represents an unimportant addition to the chlorophyll apparatus of the plant, or whether there exists some as yet undiscovered division of labor between chloroplasts in the sheath and those in the girdle (mesophyll) cells." More complete understanding of the "division of labor" suggested by Haberlandt had to await the discovery of the C_3 photosynthetic pathway in the late 1940s and the subsequent discovery and study of the C_4 photosynthetic pathway during the 1960s and 1970s. This work has revealed that C_4 plants, which include many tropical and warm-season temperate grasses, possess a group of anatomical, ultrastructural, physiological, and biochemical characteristics that distinguish them from C_3 grasses. Brown (1977) has proposed that the term *Kranz*, meaning wreath or halo in German, be applied to plants with the C_4 photosynthetic pathway and leaf mesophyll radially arranged around a prominent bundle sheath. In this chapter the important characteristics of the Kranz syndrome are described and the "division of labor" suggested by Haberlandt is discussed. This is one of the keys to understanding the outstanding success and wide adaptation of C_4 grasses.

DISCOVERY OF C_3 AND C_4 PHOTOSYNTHESIS

When radioactive ^{14}C became available in the late 1940s, several research groups began experiments to determine the biochemical pathway responsible for photosynthetic CO_2 fixation. The most rapid progress was made by Calvin and his colleagues, first using green algae and subsequently studying higher plants (Bassham 1964; Calvin and Benson, 1948). In all the algae and higher plants studied by this group, the first products of photosynthetic carbon fixation were the three-carbon compounds 3-phosphoglyceric acid, glyceraldehyde-3-phosphate, and dihydroxyacetone-3-phosphate.

From the late 1940s when the Calvin–Benson cycle (also known as the Calvin cycle and the reductive pentose phosphate pathway) was discovered until the late 1950s, most plant physiologists and biochemists assumed that the C_3 pathway was the only pathway of photosynthetic carbon fixation in plants. However, from 1954 through 1959 a series of studies by Kortschak, Hartt, and Burr at the Hawaiian Sugar Planters' Association indicated that four-carbon dicarboxylic acids are the first products of photosynthesis in sugarcane (Burr et al., 1957; Kortschak et al., 1954–1959). Similar results were reported in 1963 for maize (Tarchevskii and Karpilov, 1963). Kortschak et al. (1965) suggested that in sugarcane the first stable products of photosynthesis are malic and aspartic acids, and these products are subsequently converted to glucose via 3-phosphoglyceric acid (PGA). Subsequent work, especially that of Hatch and Slack (1966, 1968, 1970), provided extensive evidence that in many higher plants four-carbon (C_4) organic acids are, indeed, the first stable products of photosynthesis. In these C_4 plants photosynthesis is fundamentally different than in C_3 plants.

It soon became evident that a number of other plant characteristics such as leaf anatomy, the ratio of ^{12}C to ^{13}C in the tissue, the response of net photosynthetic rate to varying concentrations of O_2, and the level to which plants photosynthetically deplete the CO_2 concentration in the air of a closed container are highly correlated with the photosynthetic mechanism (C_3 or C_4) of the plants. By the early 1970s, it was clear that C_4 photosynthesis occurs in a large number of plant families, including the Gramineae, Cyperaceae, Amaranthaceae, Chenopodiaceae, Portulacaceae, Euphorbiaceae, Nyctaginaceae, Aizoaceae, Compositeae, and Zygophyllaceae. In addition, at least 19 genera in 11 families contain both C_3 and C_4 species (Downton, 1971; Hattersley and Roksandic, 1983; Raghavendra, 1980). Among the grasses the C_4 pathway was found in most members of the subfamilies Panicoideae and Erogrostoideae, but it was not found in the subfamilies Festucoideae, Bambusoideae, Arundinoideae, and Oryzoideae. Finally, by the mid-1970s it was evident that certain aspects of the C_4 dicarboxylic acid pathway differ among C_4 plants (Gutierrez et al., 1974a,b).

DIFFERENCES BETWEEN C_3 AND C_4 SPECIES

Following the early work demonstrating that four-carbon dicarboxylic acids are the first stable products of photosynthesis in C_4 plants, other consistent differences between C_3 and C_4 plants were discovered. The major biochemical, physiological, and anatomical differences between the two groups have been frequently reviewed (Edwards and Walker, 1983; Hatch and Slack, 1970; Zelitch, 1973), and are summarized in Table 2.1. These differences are discussed in greater detail in the following sections. For more detail concerning C_3 and C_4 photosynthesis, Edwards and Walker (1983) should be consulted.

Leaf Anatomy

Leaf anatomy is a good indicator of photosynthetic pathway in grasses. In cross section, the veins of the grass leaf are separated by mesophyll cells that usually contain chloroplasts. Two concentric cell layers, called the bundle sheath, normally surround the xylem and phloem bundles of the veins. Mesophyll cells are found between bundle sheaths of adjacent veins.

The two layers of cells making up the bundle sheath are often quite different. The inner (mestome) layer is derived from the procambium and is typically made up of small thick-walled cells that lack chloroplasts. The outer (parenchyma) sheath is derived from the ground parenchyma and may have thick- or thin-walled cells of average to very large size. These cells contain chloroplasts.

The thick cell walls of the mestome sheath have Casparian strips and resist degradation by concentrated H$_2$SO$_4$. The veins of all C$_3$ grasses have mestome sheaths (Brown, 1975). However, in many C$_4$ grasses the mestome sheath is missing from the small veins. In others it may be incomplete, only occurring external to either the xylem or the phloem (Brown, 1975; Crookston, 1980). Among C$_4$ grasses, members of the subfamily Eragrostoideae have both cell layers. Members of the subfamily Panicoideae often have only the parenchyma sheath. Whether the bundle sheath has one or two layers, the radial and outer tangential walls of the outer sheath cells of C$_4$ grasses are usually suberized and have a Casparian strip (Edwards and Walker, 1983; Laetsch, 1974). These cells are quite resistant to degradation by acids and rumen fluid (Akin et al., 1983; Wilson et al., 1983). C$_4$ grasses also have abundant plasmodesmata and primary pit fields between the bundle sheath and mesophyll cells (Edwards and Walker, 1983; Kemp et al., 1983). The plasmodesmata probably facilitate transport of assimilates from the mesophyll to the bundle sheath cells within the symplast, while the suberized wall prevents their diffusion between the mesophyll and bundle sheath via the apoplast.

Typical cross sections of C$_3$ and C$_4$ grass leaves are shown in Fig. 2.1.

Photosynthetic Pathways

Following the early work demonstrating that C$_4$ dicarboxylic acids are the first stable products of photosynthesis in C$_4$ plants, Hatch and Slack (1966) proposed a biochemical pathway for C$_4$ photosynthesis. This provided a testable hypothesis for subsequent biochemical research. A general, currently accepted biochemical pathway of C$_4$ photosynthesis is given in Fig. 2.2. The most important aspects of this pathway are:

1. Atmospheric CO$_2$ enters the intercellular air spaces and diffuses into the cytoplasm of the mesophyll cells where it is hydrated to HCO$_3^-$ + H$^+$ by carbonic anhydrase. It is then incorporated by phosphoenolpyruvate carboxylase (PEP carboxylase) into the C-4 carboxyl of oxaloacetate in the cytoplasm of the mesophyll cells.

2. Oxaloacetate is converted into either malate (in the chloroplast) or aspartate (in the cytoplasm) and is translocated (via the plasmadesmata) to the bundle sheath cells where decarboxylation of the C-4 carbon occurs.

3. CO$_2$ released by the decarboxylation reaction is refixed by RuBP carboxylase in the bundle sheath cell chloroplast and proceeds through the Calvin–Benson cycle.

4. The three-carbon compound resulting from the decarboxylation (pyruvate or phosphoenolpyruvate, depending on the species of C$_4$ plant) is converted to either pyruvate or alanine and is translocated back to the mesophyll cells where it is converted to phosphoenolpyruvate and can again participate in the carboxylation reaction described in step 1.

TABLE 2.1. Biochemical, Physiological, and Anatomical Differences Between C_3 and C_4 Grasses Related to Photosynthetic Pathway (Sources: Murata, 1981, Imai et al., 1973; Hirose, 1973; Hattersley and Watson, 1975; Laetsch, 1974; Brown, 1974, 1975; Takeoka et al., 1979; and others).

CHARACTERISTIC	C_3	C_4
Leaf anatomy		
Mesophyll cells	Chlorenchyma cells of mesophyll are not usually arranged radially around the bundle sheath. More than four chlorenchymatous mesophyll cells between adjacent bundle sheaths. Large intercellular air spaces in mesophyll.	Chlorenchyma cells of mesophyll usually arranged radially around the bundle sheath. Two to four chlorenchymatous mesophyll cells between adjacent bundle sheaths. Reduced intercellular air spaces in mesophyll.
Bundle sheath cells	Bundle sheath usually double with inner thick-walled cells and outer larger parenchyma cells.	Bundle sheath double (Erogrostoideae and others) or single (many Panicoideae). Inner thick-walled sheath often incomplete.
Choloroplasts	Bundle sheath and mesophyll chloroplasts usually similar in size and structure.	Bundle sheath chloroplasts often have reduced granal development, are often larger than mesophyll chloroplasts, and usually accumulate large amounts of starch in the light.

Lemma anatomy	Contrasting to round, elliptical or crescent-shaped silica cells. Trichomes often absent. Stomata large. Fewer silica cells per unit area than C_4.	Dumbbell- or cross-shaped silica cells. Trichomes present. Stomata small. More silica cells per unit area than C_3.
Enzyme responsible for CO_2 fixation:		
Mesophyll cells	RuBP carboxylase	PEP carboxylase
Bundle sheath cells	RuBP carboxylase	RuBP carboxylase
Photorespiration	Present	Absent, or greatly reduced
Maximum crop growth rate	Usually less than 40 g/m² day	Usually greater than 40 g/m² day
Maximum leaf photosynthetic rate	Usually less than 50 mg CO_2/dm² hr	Usually greater than 50 mg CO_2/dm² hr
CO_2 compensation point	More than 20 μliter CO_2/liter	Less than 10 μliter CO_2/liter
Light saturation of photosynthesis	0.2–0.25 of full sunlight	Full sunlight
Optimum temperature for growth	About 25°C	About 30°C
Day length sensitivity of flowering response	Long-day plants	Qualitative short-day or day-neutral plants

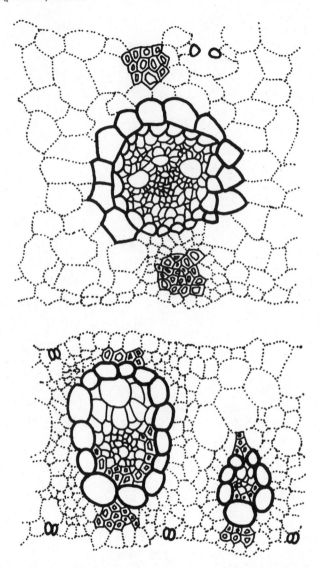

FIG. 2.1. Cross sections of typical C$_3$ and C$_4$ grass leaves. Redrawn from Gould (1968) and Art-schwager (1940).

It is important to note that RuBP carboxylase activity is restricted to the bun-dle sheath cells, and PEP carboxylase activity is restricted to the mesophyll cells. The decarboxylation of C$_4$ dicarboxylic acids in the bundle sheath cells serves as an effective mechanism of increasing the CO$_2$ concentration at the site of RuBP carboxylase. In addition, the compact radial orientation of the mesophyll cells, their proximity to the bundle sheath cells (only two to four chlorenchymatous mesophyl cells separate adjacent bundle sheaths), and the presence of numerous

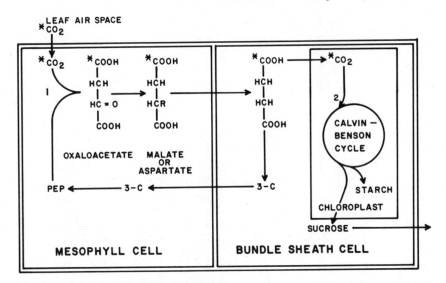

FIG. 2.2.   General pathway of carbon flow in C_4 photosynthesis. From Ray and Black (1979). Enzymes are (1) PEP carboxylase and (2) RuBP carboxylase. Reproduced by permission.

plasmadesmata between mesophyll and bundle sheath cells probably facilitate movement of metabolites between the two cell types (Kemp et al., 1983).

Photorespiration

Photorespiration is the process by which green plants take up O_2 and release CO_2 in the light. In C_3 plants normal dark respiration is replaced or supplemented by photorespiration in the light. Photorespiratory CO_2 loss often constitutes 15–20% of apparent (net) photosynthesis and its rate may be up to four times that of dark respiration. In C_4 plants photorespiration cannot be detected by the standard $^{14}CO_2$ gas exchange techniques used in C_3 plants. This section briefly describes the photorespiration in C_3 plants and explains its apparent absence in C_4 plants. More detailed reviews include Canvin (1979), Chollet and Ogren (1975), and Goldsworthy (1970, 1975).

One of the most important characteristics of RuBP carboxylase is its ability to catalyze both the carboxylation and oxygenation of RuBP. For this reason the enzyme is often referred to as RuBP carboxylase/oxygenase. Since CO_2 and O_2 compete for the active site on the enzyme, increasing the ratio of CO_2 to O_2 favors the carboxylation reaction, and increasing the concentration of O_2 relative to CO_2 promotes the oxygenation reaction. The result of RuBP carboxylation is two molecules of 3-PGA, which are the first stable products of the Calvin–Benson cycle. Oxygenation, on the other hand, results in the production of one molecule of 3-PGA and one molecule of phosphoglycolate, a two-carbon compound. The phosphoglycolate formed in the light-dependent oxygenation reaction in the chloroplasts is converted to glycolate in the chloroplasts. The glycolate moves to the peroxisomes where it is further metabolized to glycine, which is then converted to serine with the loss of CO$_2$ in the mitochondria. Figure 2.3 gives a simplified version of the reactions involved.

Unlike dark respiration, which produces ATP and reduced pyridine nucleotides, photorespiration results in the net loss of one NADH and one NADPH as well as the CO$_2$ lost in the conversion of two glycine molecules to one serine. Thus, the photorespiratory process, which is ultimately the result of the competitive RuBP carboxylase/oxygenase activity, seriously reduces the photosynthetic efficiency of C$_3$ plants.

The Calvin–Benson cycle probably evolved when atmospheric CO$_2$ concentrations were higher and O$_2$ concentrations were lower than their present ambient levels. As a result, photorespiratory CO$_2$ losses resulting from RuBP oxygenase

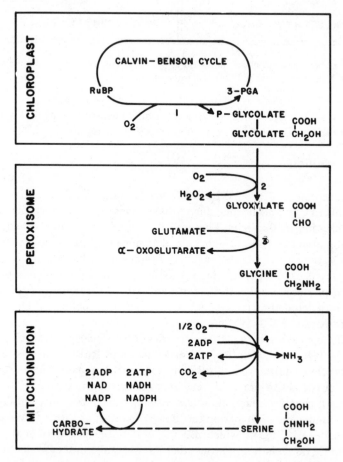

FIG. 2.3.   Photorespiratory carbon metabolism in C$_3$ plants. Enzymes are (1) RuBP carboxylase/oxygenase, (2) glycolate oxidase, (3) glutamate glyoxylate amino transferase, and (4) probably several enzymes.

activity were probably insignificant. However, C_4 photosynthesis, which evolved after CO_2 and O_2 were near their present levels, represents a means of eliminating photorespiratory CO_2 losses (Goldsworthy, 1975).

The bundle sheath cells of C_4 plants are abundantly endowed with RuBP carboxylase/oxygenase and possess peroxisomes and the enzymes responsible for glycolate metabolism. They are capable of producing CO_2 from exogenous glycolate and can even produce phosphoglycolate and glycolate under certain experimental conditions. Thus, the bundle sheath cells of C_4 grasses appear to be capable of photorespiration. However, under normal circumstances this capacity is suppressed, and little photorespiratory CO_2 efflux or O_2 uptake is detectable outside the leaf (Canvin, 1979; Furbank and Badger, 1982).

Two hypotheses have been advanced to explain this apparent absence of photorespiration: (1) inhibition of RuBP oxygenation by high CO_2 concentrations in the bundle sheath cells and (2) refixation of photorespiratory CO_2 by the mesophyll cells before it can leave the leaf. Canvin (1979) cites two types of experiments that indicate that high bundle sheath CO_2 concentrations suppress RuBP oxygenase activity and that refixation of photorespiratory CO_2 is unimportant. First, in leaves of C_4 plants glycine and serine are not labeled sequentially, as they are in leaves of C_3 plants. This suggests that the flow of carbon shown in Fig. 2.3 is not important in C_4 grasses. Second, if refixation of photorespiratory CO_2 by the mesophyll prevented its escape, increasing the O_2 concentration should increase photorespiratory CO_2 release by the bundle sheath and refixation by the mesophyll. This process would constitute a loss of photosynthetic efficiency that could be detected in the amount of apparent photosynthesis per unit light absorbed. Since photosynthetic efficiency does not change with variation in O_2 concentration in C_4 plants, photorespiratory CO_2 release from the bundle sheath cells probably does not occur under normal conditions. Thus, the apparent absence of photorespiration in C_4 plants is probably due to suppression of RuBP oxygenase activity by high bundle sheath CO_2 concentrations.

CO₂ Compensation Point

When green plants are placed in a closed atmosphere in the light, photosynthesis reduces the CO_2 concentration of the atmosphere to the point at which photosynthetic CO_2 influx equals respiratory CO_2 efflux. At this CO_2 concentration, known as the compensation point, net photosynthesis is zero. In C_3 plants the compensation point is about 40 cm³ CO_2/m³ air at 25°C and normal O_2 concentrations. Since photorespiration is the major source of CO_2 efflux in the light and photorespiration is stimulated by increasing O_2 concentrations, the compensation point of C_3 plants increases with increasing O_2 concentration. In contrast to C_3 plants, the CO_2 compensation point of C_4 plants is less than 10 cm³ CO_2/m³. In addition, it does not increase with increasing O_2 concentration. The low compensation point is undoubtedly due to the absence of photorespiration in C_4 plants, and it has been used as an easily measured indicator of the carbon fixation pathway.

^{13}C–^{12}C Ratios

Atmospheric CO_2 contains about 1.0% $^{13}CO_2$. Enzymes can partially discriminate between isotopes such as ^{12}C and ^{13}C. Thus, it might be expected that plants with different photosynthetic pathways differ in the degree to which they discriminate between $^{12}CO_2$ and $^{13}CO_2$ during photosynthetic carbon fixation.

The ^{13}C-^{12}C ratio is routinely expressed as the difference between the ^{13}C-^{12}C ratio of the sample and the ^{13}C-^{12}C ratio of a standard CO_2 gas derived from the fossil skeleton of *Belemnitella americana* from the Peedee Formation in South Carolina. This difference (δ) is expressed in parts per thousand (‰).

$$\delta^{13}C = \frac{^{13}C/^{12}C \text{ sample} - {}^{13}C/^{12}C \text{ standard}}{^{13}C/^{12}C \text{ standard}} \times 1000$$

Atmospheric CO_2 has a $\delta^{13}C$ value between -6.4 and -7.0 (‰) (Craig, 1953), meaning that it is 0.64-0.70% enriched in ^{12}C compared with the standard. Plant taxa differ in their $\delta^{13}C$ values. For example, autotrophic photosynthetic bacteria have $\delta^{13}C$ values from -28 to -39 ‰; algae, -12 to -23 ‰; C$_3$ plants, -24 to -34 ‰; C$_4$ plants, -11 to -23 ‰; and plants with crassulacean acid metabolism, -10 to -33 ‰ (Benedict, 1978; Troughton, 1971, 1979; Wong et al., 1975).

RuBP carboxylase, the primary carboxylating enzyme of C$_3$ plants, preferentially fixes $^{12}CO_2$ into PGA, that is, it discriminates against $^{13}CO_2$ (Smith and Epstein, 1971; Whelan et al., 1970). Glucose synthesized from atmospheric CO_2 by the Calvin–Benson cycle has a $\delta^{13}C$ value of approximately -25 ‰. PEP carboxylase, the primary carboxylating enzyme of C$_4$ plants, hardly discriminates against $^{13}CO_2$ at all. Malate formed by PEP carboxylase in *Sorghum* leaves has a $\delta^{13}C$ value of about -10 ‰, near the $\delta^{13}C$ value of CO_2 in the air (Whelan et al., 1970). The malate fixed in the mesophyll by PEP carboxylase is transferred to the bundle sheath cells where it is decarboxylated to CO_2 and pyruvate. This CO_2 is not in equilibrium with atmospheric CO_2; it is quantitatively fixed by RuBP carboxylase. Thus, Calvin cycle intermediates in the bundle sheaths of C$_4$ plants and the structural and metabolic compounds formed from these intermediates have the low $\delta^{13}C$ values characteristic of the first products of C$_4$ photosynthesis.

The $\delta^{13}C$ values of feces have been used to determine the relative consumption of C$_4$ and C$_3$ plants. Archaeologists have also used the $\delta^{13}C$ values of human and animal skeletons to deduce the diet of prehistoric populations. For example, human diets based on C$_3$ plants produce bone collagen with $\delta^{13}C = -21$ ‰, while those based on half C$_3$ and half C$_4$ plants produce collagen with $\delta^{13}C = -14$ ‰ (Merwe, 1982). Sudden reductions in the $\delta^{13}C$ values of American Indian skeletons probably reflect the Indians' acquisition of maize (Vogel and van der Merwe, 1977).

Other Differences

Several other anatomical and biochemical characteristics correlate with the photosynthetic pathway in grasses (Table 2.1). For example, numerous studies (see Chapter 7) have shown that the optimum and base temperatures for growth and development are higher in C_4 grasses than in C_3 grasses. In addition, C_4 grasses are almost always short day or day neutral with respect to flowering; C_3 grasses are almost always long-day plants (Purohit and Tregunna, 1974) (see Chapter 5).

Though both C_3 and C_4 plants possess RuBP carboxylase, the amino acid composition of the enzyme and its subunits is quantitatively different in the two groups (Yeon et al., 1982).

Finally, due to the great affinity of PEP carboxylase for CO_2, the light intensity at which photosynthesis becomes limited by CO_2 concentration is much higher in C_4 plants than C_3 plants (see Chapter 8).

DIFFERENCES AMONG C_4 GRASSES

As discussed above, the Kranz syndrome consists of a group of associated anatomical and biochemical characteristics in C_4 plants. The entire set of characteristics normally occurs as a group, and the discovery of one character implies the presence of the others. Thus, the presence of the Kranz syndrome can be inferred from the presence of bundle sheaths containing prominent chloroplasts and separated from adjacent bundle sheaths by two to four mesophyll cells, a low CO_2 compensation point, the absence of photorespiration, $^{13}C-^{12}C$ ratios, or the presence of four-carbon acids as the first products of photosynthesis.

Though the biochemical and anatomical similarities among C_4 grasses are remarkable, work soon after the discovery of C_4 photosynthesis demonstrated considerable biochemical and anatomical diversity within the group. For example, studies in the late 1960s showed that C_4 plants can be classified as "malate formers" or "aspartate formers" based on the major C_4 acid produced in the mesophyll cells and transferred to the bundle sheath cells (Downton, 1970). By the early 1970s, three groups of C_4 grasses could be distinguished (Gutierrez et al., 1974a,b; Hatch et al., 1975). These groups differ in the enzymes involved in the decarboxylation of C_4 dicarboxylic acids and other enzymatic, anatomical, and ultrastructural characteristics of the bundle sheath chloroplasts, the details of which were worked out by the mid-1970s (Hatch, 1976).

Table 2.2 and Fig. 2.4 give the major enzymatic and anatomical differences among the three major groups of C_4 grasses. The three groups are named for the enzyme responsible for the decarboxylation of the C_4 dicarboxylic acids transferred to the bundle sheath cells: NAD-malic enzyme species (NADme), NADP-malic enzyme species (NADPme), and PEP carboxykinase species (PEPck).

In NADme and PEPck species aspartic acid is the principal C_4 dicarboxylic

Table 2.2. Enzymatic and Anatomical Characteristics of the Three Major Groups of C$_4$ Grasses (Brown, 1977; Edwards and Walker, 1983; Edwards et al., Furbank and Badger, 1982; Gutierrez et al., 1974a; Hatch, 1976; Hattersley, 1982; Rathnam, 1978).

CHARACTERISTIC	NAD-MALIC ENZYME SPECIES	NADP-MALIC ENZYME SPECIES	PEP CARBOXYKINASE SPECIES
NAD-malic enzyme activity	70[a]	3	4
NADP-malic enzyme activity	1	30	1
PEP carboxykinase activity	1	1	>50
NADP malate dehydrogenase activity	2.5	20	3
Aspartate aminiotransferase activity	30	4	35
Alanine aminotransferase activity	25	1	20
Chlorenchymatous bundle evolved from sheath	Parenchyma sheath	Mestome sheath[e]	Parenchyma sheath
Chlorenchymatous bundle sheath cells elongated	Perpendicular to vein	Parellel to vein	Perpendicular to vein
Bundle sheath chloroplasts	Developed grana	Reduced grana	Developed grana
Bundle sheath chloroplasts	Centripetal[b,c] & centrifugal	Centrifugal	Centrifugal
Bundle sheath chloroplasts	Normal PS II	Reduced PS II	Normal PS II
Principal substrate moving from mesophyll to bundle sheath	Aspartate[d]	Malate	Aspartate
Principal substrate moving from bundle sheath to mesophyll	Alanine	Pyruvate	Alanine
δ^{13}C values	−12.7 ±0.21	−11.35 ±0.13	−11.95‰ ±0.19

Table 2.2.—*Continued*

Characteristic	NAD-Malic Enzyme Species	NADP-Malic Enzyme Species	PEP Carboxykinase Species
Photorespiratory O$_2$ uptake	Intermediate	Lowest	Highest
Ecological adaptation	Very low rainfall	Most rainfall	Intermediate
Taxonomic units	Chlorideae	Andropogoneae	Chlorideae
	Bouteloua	*Andropogon*	*Bouteloua*
	Buchloe	*Bothriochloa*	*Chloris*
	Chloris	*Heteropogon*	Eragrosteae
	Cynodon	*Saccharum*	*Eragrostis*
	Eleusine	*Sorgastrum*	*Muhlenbergia*
	Leptochloa	*Sorghum*	*Sporobolus*
	Eragrosteae	Maydeae	Zoysieae
	Eragrostis	*Euchlaena*	*Zoysia*
	Sporobolus	*Zea*	Paniceae
	Paniceae	Paniceae	*Brachiaria*
	Panicum	*Axonopus*	*Eriochloa*
		Cenchrus	*Panicum*
		Digitaria	*Urochloa*
		Echinochloa	
		Panicum	
		Paspalum	
		Pennisetum	
		Setaria	
		Aristideae	
		Aristida	

[a] Enzyme activities expressed as multiples of typical values in C$_3$ plants.
[b] Centripetal, toward inner cell wall; centrifugal, toward outer cell wall.
[c] Centrifugal in *Panicum dichotomiflorum* (Ohsugi and Murata, 1980).
[d] Relative amounts of aspartate and malate vary with leaf age and $^{14}CO_2$ exposure time (Laetsch, 1974; Rathnam, 1978).
[e] Except *Aristida* where both parenchyma and mestome sheaths are Kranz.

acid transferred to the bundle sheath cells. In NADme species decarboxylation of malate is accomplished by a NAD-dependent enzyme, NAD-malic enzyme. In PEPck species oxaloacetate is decarboxylated to PEP by PEP carboxykinase.

In NADPme species malic acid is the principal C$_4$ acid transferred to the bundle sheath cells, and it is decarboxylated to pyruvate by an NADP-dependent malic enzyme. Both NADme species and PEPck species have substantial aspartate and alanine aminotransferase activities to interconvert aspartate/oxaloacetate and alanine/pyruvate, respectively. NADPme species contain substantial

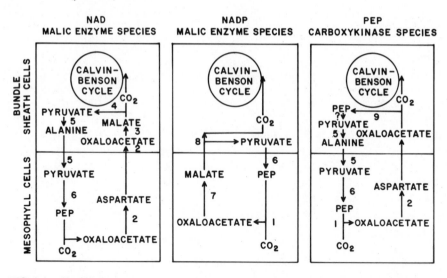

FIG. 2.4. Simplified model of net carbon flows in the three major variants of C$_4$ photosynthesis. Enzymes are (1) PEP carboxylase, (2) aspartate aminotransferase, (3) NAD malate dehydrogenase, (4) NAD malic enzyme, (5) alanine aminotransferase, (6) pyruvate P_i dikinase, (7) NAD malate dehydrogenase, (8) NADP malic enzyme, (9) PEP carboxykinase. Reprinted from Hatch (1976) by permission of University Park Press.

NADP malate dehydrogenase activity associated with the conversion of oxaloacetate to malate.

Hattersley and Watson (1976) and Brown (1974, 1975, 1977) discovered that several easily observable anatomical characteristics of the leaves correlate very well with the principal C$_4$ decarboxylating enzymes. In NADPme species the chlorenchymatous (Kranz) bundle sheath cells (chlorophyll-containing, starch-accumulating bundle sheath cells containing RuBP carboxylase) are derived from the mestome (inner) sheath (Brown, 1975). Thus, no cells separate the metaxylem vessel elements from the chlorenchymatous bundle sheath cells (Hattersley and Watson, 1976). In NADme and PEPck species the chlorenchymatous bundle sheath cells are derived from the parenchyma sheath; therefore, they are separated from the metaxylem vessels by the mestome sheath. In addition, in NADPme species the chlorenchymatous bundle sheath cells are elongated parallel to the vein, while in NADme and PEPck species they are usually wider radially than parallel to the vein (Brown 1974).

In addition to anatomical differences, Hattersley (1982) found that the three groups of C$_4$ grasses differ in their $\delta^{13}C$ values. The NADme species have a $\delta^{13}C$ mean ($\pm$ standard errors) of -11.35 ± 0.13, the PEPck species have means of -11.95 ± 0.19, and the NADme species have means of -12.7 ± 0.21. These differences could be due to differential leakage of $^{13}CO_2$ and $^{12}CO_2$ from photosynthetic tissue, to differences in intercellular CO_2 concentrations, or to unknown CO_2 fractionation processes.

Though all C_4 grasses have low rates of photorespiration, the three main groups do differ slightly in photorespiratory O_2 uptake (Furbank and Badger, 1982). The relative magnitudes of this O_2 uptake are: NADPme species < NADme species < PEPck species. However, the reasons for these differences among the groups is unclear.

Brown (1977) has proposed that the characteristics described above are useful taxonomic characters at the level of the subgenus or section, and he has postulated a sequence of genetic modifications that probably occur during the evolution of an NADPme C_4 species from a C_3 species. First, the mestome sheath cells become larger, thinner walled, and acquire chloroplasts. C_4 photosynthesis then evolves with a slight modification of the parenchyma sheath. The ratio of mesophyll to chlorenchymatous (Kranz) bundle sheath cells decreases to produce a biochemical balance between the activities of PEP carboxylase in the mesophyll cells and RuBP carboxylase in the Kranz bundle sheath cells. This reduction in mesophyll cell number and a decrease in intercellular spaces between the mesophyll cells facilitates movement of photosynthetic metabolites between the mesophyll and Kranz bundle sheath cells. Finally, the parenchyma (outer) sheath disappears or is greatly reduced, facilitating movement of metabolites from the mesophyll to the bundle sheath.

Brown (1977) has also pointed out that the NADme and PEPck types of C_4 photosynthesis seem to be adaptations to xerophytic conditions. These types occur in the subfamily Eragrostoideae and some of the more xeric species of the tribe Paniceae (e.g., section Dura of *Panicum*, *Brachiaria*, *Eriochloa*, *Urochloa*, and others).

The NADPme taxa (the tribe Andropogoneae; some species of *Panicum*; and many others of the tribe Paniceae such as *Digitaria*, *Echinochloa*, *Paspalum*, *Pennisetum*) are generally more mesophytic, though xerophytic members of the group do occur (e.g., some species of *Setaria* and *Cenchrus*). Variation in the ecological adaptation of C_4 grass subtypes has been found in South West Africa/ Namibia where NADme species are most numerous in areas with very low annual rainfall (50-200 mm). NADPme species are most abundant where rainfall is highest (400-600 mm), and PEPck species are most abundant in areas with intermediate rainfall (200-400 mm) (Ellis et al., 1980).

C_3–C_4 INTERMEDIATES

A few species have characteristics of both C_3 and C_4 species. The first to be recognized were hybrids between C_3 and C_4 species of *Atriplex* (Chenopodiaceae). *Mollugo verticillata* (Aizoaceae) also has both C_3 and C_4 characteristics. Among the grasses, *Panicum* has species that are C_3, C_4 (all three major groups), and intermediate between C_3 and C_4. *Panicum milioides* has a compensation point of 20-25 cm³ CO_2/m^3, in comparison to 40-60 for C_3 species and less than 10 for C_4 species. In addition, its leaf anatomy could be classified as Kranz or non-Kranz, depending on the criteria used. It has little O_2 inhibition of photosynthesis; however its $\delta^{13}C$ values are similar to C_3 plants. Its photosynthetic responses to temperature and to irradiance are also intermediate between C_3 and C_4 *Panicum* species (Brown and Brown, 1975; Burzynski and Lechowski, 1983; Kanai and Kashiwagi, 1975).

Neurachne is a small, endemic genus of Australian grasses with five species. Three for C_3, one is C_4, and *N. minor* may be intermediate with C_3-like values of $\delta^{13}C$ and definite Kranz anatomy (Hattersley and Roksandic, 1983).

EVOLUTION OF C_4 PHOTOSYNTHESIS

It is clear that C_4 photosynthesis evolved independently in a number of families and species (Brown, 1977; Moore, 1982). Cockburn (1983) concludes that C_4 photosynthesis is a evolutionary modification of a common, universally distributed metabolic pathway. He points out that the synthesis and consumption of four-carbon acids in the charge-balancing reactions of stomates are the likely metabolic progenitors of C_4 photosynthesis. For example, during stomatal opening, malic acid is synthesized from PEP and CO_2 (as in C_4 mesophyll cells) as a counterion to at least some of the K^+ taken up by the guard cells. During stomatal closure, malic acid is lost from the guard cells by decarboxylation (as in C_4 bundle sheath cells).

Monson et al. (1984) propose that the first step toward C_4 photosynthesis was a mechanism for reducing photorespired CO_2. This could have been an efficient recycling mechanism based on increased PEP carboxylase activity. The rest of the C_4 metabolic pathway would have evolved as a mechanism to process the C_4 acids formed in the CO_2 recycling process. At this stage photorespiration would be reduced and a limited C_4 photosynthetic pathway would be present. The final step would probably be differential compartmentation of C_3 and C_4 enzymes in the bundle sheath and mesophyll cells, respectively. Monson et al. (1984) identify C_3, C_3-C_4 intermediates, and C_4 species with characteristics similar to those described above. One of the remaining challenges for biochemists and geneticists is to describe more clearly the evolution of C_4 photosynthesis and the evolutionary advantages of the species that are intermediates in both the biochemical and evolutionary senses.

3

SYSTEMATICS AND DISTRIBUTION

Because of the ubiquity and agricultural importance of grasses, their taxonomy has been studied since the eighteenth century, and many of the "natural" groups making up today's grass tribes and subfamilies were recognized in the nineteenth century. With the increased study of grass leaf anatomy in the late nineteenth century, taxonomists recognized that the Kranz syndrome is primarily found in taxonomic groups more common to tropical than to temperate regions. In addition, Kranz species from temperate regions are nearly always species that grow and develop during the warmer parts of the year. Yet, despite scientists' long interest in grass taxonomy and evolution, the evolutionary origins of the grass family and its many tribes remain unclear.

RELATIONSHIPS WITH OTHER ANGIOSPERMS

The family Gramineae has sometimes been placed in an order with the sedges (family Cyperaceae) (order Glumiflorae, Graminales, or Poales). However, morphological and anatomical evidence suggests that the similarities between the grasses and sedges are due to convergent evolution rather than to common ancestry. The Cyperaceae may have evolved from primitive members of the class Commelinidae, which includes the spiderworts (family Commelinaceae) and yellow-eyed "grasses" (family Xyridaceae) (Stebbins, 1972). In contrast, recent treatments of grass evolution (Gould, 1968; Stebbins, 1956, 1972) emphasize the close relationship between the grasses and the primitive members of the families Liliaceae, Flagellariaceae, and Restoniaceae. Morphological differences between the Gramineae and Cyperaceae are given in Table 3.1.

GRASS SYSTEMATICS

A Historical Review

The grass family is one of the largest families of angiosperms with over 600 genera and 7500 species. Because of the economic importance of the grasses, numerous early botanists studied the group. Early taxonomic publications were

TABLE 3.1. **Morphological Differences Between the Gramineae and Cyperaceae (Modified from Gould, 1968)**

PLANT PART	GRAMINEAE	CYPERACEAE
Stems	Terete or flattened	Triangular (at least just below the inflorescence)
Leaves	2-ranked	3-ranked
Leaf sheaths	Mostly open	Mostly closed
Bract immediately subtending the flower	Usually 2-nerved	With odd number of nerves
Flowers	Terminal	Axillary
Ovaries	Probably evolved from types with parietal placentation	Probably evolved from types with free central placentation
Chromosomes	Simple centromeres	Multiple centromeres
Fruit	Usually a caryopsis	Usually an achene, never a caryopsis
Embryo	Lateral to the endosperm	Embedded in the endosperm

essentially lists of species accompanied by short descriptive phrases, and Gould (1968) cites Johann Scheuchzer's *Agrostographiae Helvetica Podromus* (1708) as the starting point in grass taxonomy. Linnaeus' *Species Planatarum* (1753) (cited in Gould, 1968) introduced the binomial nomenclature of flowering plants and included a total of 40 grass genera, classified on the basis of the number and arrangement of flower parts. This produced a highly artificial system that grouped widely different species and served primarily as an aid to identification. During the late eighteenth and nineteenth centuries the emphasis of angiosperm taxonomy shifted to the formulation of "natural systems" of classification that grouped morphologically similar species. Jussieu's *Genera Planatarum* (1789) (cited in Gould, 1968) is an example of this trend and is the origin of the accepted family name "Gramineae" from Linnaeus' "Gramina." "Gramineae" is not in strict accord with the International Code of Botanical Nomenclature, and "Poaceae" is preferred by some authors.

Early botanists had considered the spikelet to be homologous to a single flower in which the glumes corresponded to the calyx and the lemma and palea were the corolla. One of the most important contributions to grass systematics was Robert Brown's (1810) (cited in Gould, 1968) recognition that the spikelet is a reduced panicle branch rather than a single flower. Brown considered the glumes to be homologous to the bracts or involucre, the palea and lemma to be homologous to the calyx, and the lodicules to be reduced and fused corolla parts. He also correctly observed that the lower glume and the palea are the members most likely to be reduced or absent. On the basis of floral organs and geographical distribution of species, he recognized the "principal subdivisions" of the Gramineae to be the Paniceae (spikelets with two florets with lower florets imperfect and having tropical affinities) and the Poaceae (spikelets with one to many florets, imperfect florets terminal rather than basal, and having temperate affinities).

Kunth (1833) followed Brown and built upon his concepts, describing 13 tribes with no subfamilies. Bentham (1881) used Brown's principal subdivisions and divided each into tribes and subtribes based largely on those described by Kunth. Bentham's system was used with modifications by Hackel (1887, 1889), Stapf (1917–1934), Hitchcock (1935, 1951), and Bews (1929). Pilger (1954) provides much of the tribal synonymy and bibliographic data necessary for an understanding of the different treatments within this "traditional" system.

The work of Avdulov (1931) and Prat (1932 and 1936) initiated a new era in grass taxonomy and phylogeny in which microscopic characters, cytology, morphology of embryos and seedlings, and physiological characteristics contributed to a much more complete understanding of the grass family, especially at the subfamily and tribal level. The work of Avdulov (1931) was based on chromosome studies, leaf anatomy, morphology of the first seedling leaf, organization of the resting nucleus, type of starch grains in the fruit, and geographical distribution. Prat (1932, 1936) studied the leaf epidermis, leaf anatomy, cytology, morphology of seedlings, embryos, fruits, and the inflorescence. The systems produced by Avdulov and Prat were phylogenetic and captured the imagination of grass systematists. As a result, the decade from 1950 to 1960 was a period of intense activity during which new subfamily groupings were proposed by Pilger (1954), Jacques-Felix (1955), Beetle (1955), Stebbins (1956), and Tateoka (1957). The general trend has been to elevate more tribes or groups of tribes to the subfamily level, from the two subfamilies proposed by Avdulov to the four to six generally recognized today. The number of tribes recognized by the "new taxonomy" has also increased to 26 in the case of Stebbins and 42 in the case of Tateoka. This work follows the system of Gould (1968), which consists of six subfamilies composed of 25 tribes. Table 3.2 summarizes relevant morphological, cytological, and physiological characteristics of the six subfamilies.

The Fossil Record

The ancestry of the grasses is obscure. The earliest grass fossils are fruits from mid- to late-Tertiary deposits (10–40 million years old) and are similar to living species that have been referred to living genera such as *Phragmites* and *Arundo*, of the ancient and somewhat heterogeneous subfamily Arundinoideae; *Panicum* and *Setaria*, of the subfamily Panicoideae; *Stipa*, *Piptochaetium* and *Phalaris*, of the subfamily Festucoideae; *Chusquea*, a member of the Bambusoideae; and *Archaeolersia*, probably a member of the Oryzoideae (Bowden, 1965; Gould, 1968; Thomasson, 1978, 1980). Definite anatomical evidence of Kranz anatomy has been found in Pliocene (2–13 million years old) deposits in California (Nambudiri et al., 1978).

Possible grass pollen has been found in much earlier strata (Cretaceous) than the megafossils cited above, but because of the similarities between grass pollen and pollen of the Flagellariaceae and Restoniaceae it is difficult definitely to assign such pollen to the Gramineae (Daghlian, 1981).

A recent line of evidence based on current distribution of grass subfamilies and continental drift suggests that grasses arose much earlier than available megafossils would suggest, probably during the Jurassic period of the Mesozoic era (130–180 million years ago). This evidence is discussed later in the chapter.

Primitive Characters

Since the earliest fossil grasses were already highly specialized, and the relationships of the Gramineae to other angiosperms is highly speculative, reconstruction of the most "primitive" grass is difficult. Essentially, it consists of examining the variation found in individual traits and applying Occam's razor to determine the parental "type" from which all the observed variation could most easily evolve. The exercise is frankly speculative but is necessary for an "understanding" of the different groups within the Gramineae.

Taxonomists generally agree that the structure of the ancestral grass inflorescence was much more complex than that of grasses today. The early progenitors of today's grasses were probably small leafy herbs. The ends of terminal branches bore small axillary branches, each subtended by a leafy bract and

TABLE 3.2. A Taxonomic Comparison of the Principal Subfamilies in the Gramineae (Adapted from Bowden, 1965; Gould, 1968)

CHARACTERISTIC	BAMBUSOIDEAE	PANICOIDEAE	ERAGROSTIDEAE	FESTUCOIDEAE	ARUNDINOIDEAE	ORYZOIDEAE
Climatic specialization	Equatorial, moist	Tropical, monsoonal	Tropical arid	Temperate, moist	Temperate, moist	Tropical, moist
Vegetative parts	Stems woody, large; leaves petiolate	Herbaceous	Herbaceous	Herbaceous	Herbaceous	Herbaceous
Inflorescence						
Florets/spikelet	Many	Few	Few	Few	Few	Single
Glume length	Short	Long	Long	Long	Long	Long
Spikelets						
Dehiscence	Various	Shed as a whole	Breaking up at maturity	Breaking up at maturity	Breaking up at maturity	Shed as a whole
Reduction	Little, from base or apex or both	Much, lowermost reduced	Much, uppermost reduced	Much, uppermost reduced	Much, uppermost and/or lowermost reduced	Much, lowermost reduced
Compression	Little	Much, dorsal	Much, lateral	Much, lateral	Much, lateral	Much, lateral
Florets						
Components	3 stigmas, 6 stamens, 3 lodicules	2 stigmas, 3 stamens, 2 lodicules	2 stigmas, 3 stamens, 2 lodicules	2 stigmas, 3 stamens, 2 lodicules	2 stigmas, 3 stamens, 2 lodicules	1 or 2 stigmas; 6, 3 or 1 stamens; 2 lodicules
Chromosomes						
Size	Small	Small	Small	Large	Small	Small
Basic number (n)	12	5, 9, 10	9, 10	7	6, 12	12

TABLE 3.2. (*Continued*)

CHARACTERISTIC	BAMBUSOIDEAE	PANICOIDEAE	ERAGROSTIDEAE	FESTUCOIDEAE	ARUNDINOIDEAE	ORYZOIDEAE
Leaf anatomy						
Vascular sheath	Parenchyma sheath with chloroplasts; mestome sheath present	Parenchyma sheath with amyloplasts only; mestome sheath absent	Parenchyma sheath with chloroplasts; mestome sheath may be present	Parenchyma sheath with chloroplasts; mestome sheath present	Parenchyma sheath with amyloplasts only; mestome sheath present	Parenchyma sheath with chloroplasts; mestome sheath present
Mesophyll	Nonradiate	Radiate	Radiate, few chloroplasts	Nonradiate	Radiate	Nonradiate
Leaf epidermis						
Type	Complex-primitive	Complex-primitive	Complex-primitive	Complex-specialized	Both complex and simple	Complex-primitive
Microhairs	Filiform or inflated bicellular	Pointed bicellular	Inflated bicellular	Absent	Bicellular	Cylindrical bicellular
Silica bodies	Dumbbells or saddles	Dumbbells or crosses	Saddles or crescents	Round, rods or crenate	Rectangular, crescents, crosses, saddles, dumbbells	Oblongs, crosses, saddles, or dumbbells
Root hairs						
Direction	Perpendicular	Perpendicular	Perpendicular	Oblique, pointing to apex	Perpendicular or oblique	—

Characteristic						
Origin	Cell center	Cell center	Cell center	Apical end of cell	Variable	—
Caryopsis-embryo Size	Large	Large	Large	Small	Small to moderately large	Small
Epiblast	Present	Absent	Present	Present	Present	Present or absent
Scutellar cleft	Present	Present	Present	Absent	Overlapping or meeting	Overlapping
First leaf margins	Overlapping	Overlapping	Meeting	Meeting	Separated from plume	Not separated from plume
Storage carbohydrates	—	Starch	Starch	Fructosans with starch	Starch	Starch
Inhibition of germination by IPC (iso-propyl-N-phenyl carbamate)	—	Little	Little	Strong	—	—
Susceptibility to low oxygen tensions during germination	—	Little	Little	Strong	—	Little
Heat production of caryposes when moistened	—	Strongly exothermic	Strongly exothermic	Weakly exothermic	—	—
Photosynthetic pathway	C_3	C_3 and C_4	C_4	C_3	C_3	C_3

bearing a terminal flower subtended by a bracteole. The flowers were trimerous with two rings of perianth segments and two rings of three stamens. The ovary was probably already reduced from the tricarpellate condition, having a single carpel and ovule. The styles and stigmas were still trimerous, and the plant was already wind pollinated, though it had probably evolved from insect-pollinated ancestors (Stebbins, 1956; Clifford, 1961; Bowden, 1965).

During the course of evolution, the ancestral flower was further reduced and simplified. The perianth was reduced to small vestigal organs, the lodicules. The axis of the flowering stem was shortened, and the leafy bract and bracteole were reduced to the lemma and palea, respectively, and were modified to protect the flower. Thus, the floret, consisting of the lemma, palea, and reproductive organs, was formed. The terminal flowering branches were reduced so that the florets were concentrated in compact groups. Two leaves at the base of each branch were also modified to form a second pair of valvelike protective organs, the glumes, which partially or wholly enclose several florets to form the spikelet.

Stebbins (1972) cites anatomical and developmental evidence that the grass panicle is a complex structure without homology in the Monocotyledoneae. For example, the fasciculate arrangement of the lower panicle branches and the strongly developed intercalary meristems and extreme cell elongation in the panicle branches are unique to the grasses. For these reasons, Stebbins (1972) concludes that the open paniculate inflorescence common to many species was derived from the more primitive racemiform inflorescence.

Primitive grasses were probably small, tufted, herbaceous perennials with short leaves (Stebbins, 1972). The epidermis probably produced many specialized cells such as those found in the genus *Joinvillea* of the Flagellariaceae and most subfamilies other than the Festucoideae. Their leaf chlorophyll-bearing parenchyma was probably evenly distributed as it is in the Festucoideae, not with highly specialized bundle sheath cells. Carbon fixation was via the C_3 rather than the more complex C_4 pathway (Brown and Smith, 1972). The embryo was probably relatively small compared to the endosperm, and it probably possessed an epiblast and cotyledonary node. Cytologically, the ancestral grasses had six or seven gametic chromosomes that were relatively small (Stebbins, 1972).

For a number of reasons, it may be assumed that primitive grasses arose in the wet tropics, and either they or their ancestors occupied forest or near-forest habitats. First, their closest relatives, the Flagellariaceae and Restoniaceae, are distributed principally in the tropics (Stebbins, 1972). Second, most grass genera with the full complement of three lodicules are confined to warm areas. This includes *Streptochaeta*, which also has the best developed of all grass perianths (attaining a length of 1 cm or more) (Clifford, 1961).

The reduction of the perianth and development of wind pollination were probably adaptations to drier habitats. The reduction in the number of stamens from six to three is not so easily rationalized because it would appear to increase the risks associated with wind pollination. However, the development of the paniculate inflorescence with its numerous flowers, and the production of numerous male flowers, especially in the subfamily Panicoideae, seem to be developments that tend to minimize the risks of wind pollination (Clifford, 1961).

Probable primitive and advanced characters compiled from Stebbins (1956, 1972), Clifford (1961), and Gould (1968) are summarized in Table 3.3.

Biological Recognition

There is good evidence that grass subfamilies and tribes recognized by taxonomists can also be distinguished by other forms of life (Savile, 1979). For example, smuts differ in the grass taxa they infect (Watson, 1972). The smut genera and species attacking the Festucoideae are quite different than those of the Panicoideae and Eragrostoideae. The smut family Tilletiaceae most frequently infects festucoid grasses, while the family Ustilaginaceae (with the exception of the genus (Ustilago) prefers panicoid and eragrostoid grasses.

DISTRIBUTION OF C$_4$ GRASSES

The clearer understanding of grass systematics that resulted from detailed anatomical studies since the 1930s and the recent discovery of the C$_4$ photosynthetic pathway have stimulated research on the distribution of grass subfamilies and tribes. One subject of investigation has been the worldwide distribution of the tribes with special emphasis on evolutionary relationships. A second type of investigation has dealt with the occurrence of C$_3$ and C$_4$ grasses along climatic gradients and the general climatic adaptations of the two types.

Continental Distribution

The three most important groups of C$_4$ grasses in terms of their numbers of species, ecological dominance, and economic importance are the subfamily Eragrostoideae and the panicoid tribes Andropogoneae and Paniceae. Hartley, in his classical studies of the distribution of grasses (Hartley, 1950, 1958a,b; Hartley and Slater, 1960) clearly demonstrated that these groups are unequally distributed throughout the world. Though the three groups have similar temperature requirements, the Andropogoneae have attained their greatest species diversity in Africa and southern Asia, while the Paniceae are most prevalent in Central America and northern South America. In contrast, species of the subfamily Eragrostoideae are most abundant in warm arid regions such as the Chihuahuan desert of North America, the interior of southern Africa, near the Red Sea, and in parts of India and Australia. Figures 3.1–3.3, from the work of Hartley, illustrate the relative abundance of these three groups.

This curious asymmetric distribution of the Paniceae and Andropogoneae has recently been explained in terms of continental drift (Brown and Smith, 1972). According to the theory of continental drift, a single ancient world conti-

TABLE 3.3. Primitive and Advanced Characters in the Gramineae

CHARACTER	PRIMITIVE	ADVANCED
Leaf	Short as in many shade-tolerant species	Long
Leaf sheath	Open	Closed
Leaf epidermis	Numerous specialized cells including bicellular microhairs	Fewer specialized epidermal cells as in the Festucoideae
Stem internodes	Continuous pith with scattered vascular bundles	Hollow pith with concentric ring or rings of vascular bundles as in the Festucoideae
Leaf mesophyll cells	Loosely, irregularly arranged chlorenchymatous cells	Regularly arranged chlorenchymatous tissue radiating from vascular bundles
Inflorescence	Racemiform	Paniculate
Spikelet	Large, many-flowered	Small, few or 1-flowered
Glumes	Large, leaflike in texture, several-nerved, awnless or short-awned	Absent, reduced, or highly developed for flower protection or seed dispersal
Lemma	Like the glumes	Conspicuously different from the glumes
Palea	Present, 2 to many-nerved	2 or 1, fleshy or membranous, few-nerved
Stamens	6, in 2 whorls	3, 2, or 1, in 1 whorl
Stigmas	3	2 or 1
Embryo	Small compared to endosperm	Large
Epiblast	Present	Absent
Cotyledonary node	Present	Absent
Chromosomes	Small	Large
Type of photosynthesis	C_3	C_4

nent (Pangaea) existed during the late Palaeozoic and early Mesozoic eras (180–270 millions years before present). The grasses probably appeared during the Jurassic period of the Mesozoic era (130–180 million years before present), and some subfamilial and tribal diversification occurred prior to the breakup of Pangeae. The Eragrostoideae were probably the first to acquire C_4 photosynthesis, even earlier than the Panicoideae; therefore, they were well distributed through-

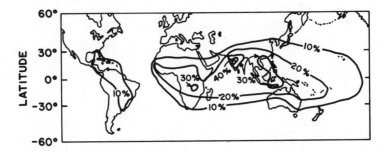

FIG. 3.1. Map of world distribution of the Andropogoneae as a percentage of species in the total gross flora. Redrawn from Hartley (1958a).

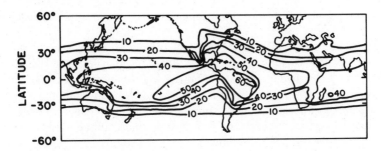

FIG. 3.2. Map of world distribution of the Paniceae as a percentage of species in the total grass flora. Redrawn from Hartley (1958b).

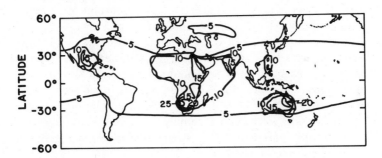

FIG. 3.3. Map of world distribution of the Eragrosteae as a percentage of species in the total grass flora. Redrawn from Hartley and Slater (1960).

out Pangeae prior to its breakup. Shortly before South America separated from Africa during the late Cretaceous, C$_4$ photosynthesis and its attendant morphological characteristics probably evolved in the tribe Paniceae. Thus, C$_4$ members of the Paniceae were well distributed in Africa and South America. After the separation of these two continents, the Andropogoneae evolved from some C$_4$ panicoid type in southern Asia. They rapidly occupied Africa and eventually reached North America, probably by way of southern Europe before North America and Europe separated during the Tertiary. However, their greatest species diversity is still found in southern Asia.

Effects of Climate

The rather recent recognition that the photosynthetic pathway of C_4 grasses is fundamentally different from that of C_3 grasses has stimulated research on the comparative physiological ecology of C_3 and C_4 grasses.

Hartley's (1950) classic study of the distribution of the six largest and most widely distributed grass tribes (viz., Agrosteae, Andropogoneae, Aveneae, Eragrosteae, Festuceae, and Paniceae) clearly revealed fundamental differences in the physiological ecology and distribution of the Agrosteae, Aveneae, and Festuceae on one hand, and the Andropogoneae, Eragrosteae, and Paniceae on the other.

Hartley obtained 70 floras and floristic lists from around the world, classified the grasses listed according to tribe, and studied the distribution of the six major tribes. He very clearly demonstrated that the Andropogoneae, Eragrosteae, and Paniceae are relatively more abundant in warm climates while the Agrosteae, Aveneae, and Festuceae are more abundant in cooler climates. In general, the Andropogoneae, Eragrosteae, and Paniceae are more abundant in areas where the mean temperature of the midwinter month exceeds 10°C. Conversely, the Agrosteae, Aveneae, and Festuceae are more abundant where the mean midwinter temperature is less than 10°C.

Hartley's work, which predated the discovery of C_4 photosynthesis, has been followed by a number of floristic studies that have shown the overriding effect of temperature on the distribution of C_4 and C_3 grasses.

In one of the earlier studies of C_4 grass distribution, Teeri and Stowe (1976) studied the occurrence of C_4 grasses in 32 regions of the United States using the published floras of each region. They found that the mean July (summer) minimum temperature explains over 94% of the variation in the percentage of grass species with the C_4 photosynthesis pathway. The mean minimum July temperature at which C_3 and C_4 grasses are equally numerous is approximately 20°C (Fig. 3.4). Above this temperature C_4 grasses are relatively more numerous, and below it C_3 grasses are more prevalent.

Similar results have been found with C_3 and C_4 species of the family Cyperaceae (Teeri et al., 1980); however, the distribution of C_3 and C_4 dicot species is apparently affected more by aridity than by temperature (Stowe and Teeri, 1978). Several studies indicate that the distribution of C_3 and C_4 grasses relative to temperature is different in the tropics than in temperate areas. Chazdon (1978), Tieszen et al. (1979), and Rundel (1980) found that in Costa Rica, Kenya, and Hawaii a floristic balance between C_3 and C_4 species is found where the mean minimum temperature of the coolest month is 8–9°C and the mean maximum temperature of the warmest month is 21–23°C. Thus, the floristic balance between C_3 and C_4 grasses occurs at cooler temperatures in the tropics

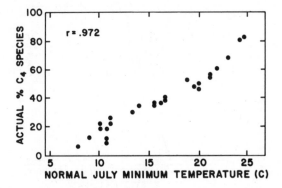

FIG. 3.4. Effect of normal July minimum temperature on percent C_4 grass species in 27 regions in North America. Redrawn from Teeri and Stowe (1976).

than in temperate, continental climates. Rundel (1980) speculates that the relatively small seasonal fluctuation in temperatures in tropical regions allows C_4 grasses to dominate at lower mean temperatures in tropical regions than in temperate continental climates. Rundel (1980) also shows that the relative percentage of C_4 and C_3 grass species along an altitudinal transect in Hawaii changes gradually with increasing altitude (Fig. 3.5); however, the relative coverage of C_4 versus C_3 species changes much more abruptly (Fig. 3.6). This suggests that physiological factors strongly affect the competitive ability of C_3 and C_4 grasses, and changes in climatic conditions that have small effects on floristic composition can dramatically affect the relative dominance of C_3 and C_4 species.

In temperate and subtropical climates, cool temperatures in the spring and fall often favor the growth of C_3 grasses, while warm summer temperatures favor C_4 grasses. For example, Ode et al. (1980) used the $\delta^{13}C$ values of biomass (see

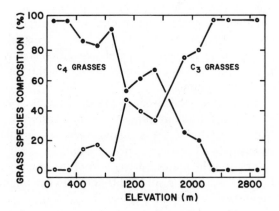

FIG. 3.5. Effect of elevation on percentages of C_3 and C_4 grass species in Hawaii (redrawn from Rundel, 1980).

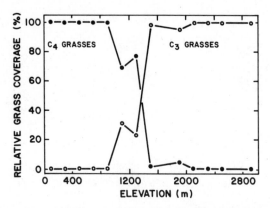

FIG. 3.6. Effect of elevation on relative cover of C_3 and C_4 grasses in Hawaii. Redrawn from Rundel (1980).

Chapter 2) in upland and lowland communities of a mixed prairie in the Northern Great Plains (North Dakota) to determine the seasonal contribution of C_3 and C_4 groups (primarily grasses) to total community production. The biomass of both lowland and upland communities possessed low $\delta^{13}C$ values in spring, indicating predominately C_3 production. In midsummer the $\delta^{13}C$ values of the upland community increased, indicating a shift to predominantly C_4 production. The increase was much less dramatic in the lowland community. In the fall, C_3 production dominated in both sites. Water availability was greater in the lowland community, and this apparently was a factor in the greater C_3 production of this community in midsummer. These results are consistent with those of Boutton et al. (1980) and Doliner and Jolliffe (1979), which suggest that dry habitats favor C_4 species.

Doliner and Jolliffe (1979) used several statistical techniques to analyze the occurrence of C_3 and C_4 species at locations differing in environmental conditions in central Europe and California. Four-carbon plants are relatively more abundant where summer or winter temperatures are relatively high and moisture availability is relatively low, suggesting that warm, dry conditions confer a competitive advantage to C_4 species. No significant differences were found between the C_3 and C_4 groups in response to variations in light, soil nitrogen, soil salinity, or continentality of climate.

Boutton et al. (1980) studied the distribution of biomass of C_3 and C_4 (primarily grass) species along an altitudinal transect in a southeastern Wyoming grassland (Fig. 3.7). They found that C_4 grass biomass increases with both increasing mean annual temperature and decreasing precipitation. In addition, C_4 biomass increases in importance in the hottest, driest part of the growing season.

Not only do C_3 and C_4 grasses respond differently to environmental factors, but the relative abundance of the different groups of C_4 grasses is affected by environment. Hartley (1958a) showed that the percentage of C_4 species belong-

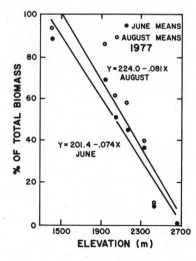

FIG. 3.7.   Effect of elevation on percentage of total community biomass as C_4 grasses during two months in 1978. Redrawn from Boutton et al. (1980).

ing to the tribe Andropogoneae is relatively greater in tropical and subtropical areas with monsoonal rainfall patterns (high midsummer rainfall) and mean midwinter temperatures greater than about 5°C. Regions within the tropics in which the frequency of species of the Andropogoneae are low are either those with high altitudes or low midsummer temperatures. Species of the tribe Paniceae are relatively more abundant where both temperature and rainfall are adequate over a long period during the year, producing a long growing season (Hartley, 1958b).

Boutton et al. (1980) noted that in a southeastern Wyoming grassland all the C_4 grasses on relatively dry soils are members of the Eragrostoideae, while in moist floodplain soils 44% of the C_4 grass species and 48% of the C_4 grass biomass belong to the subfamily Panicoideae, suggesting that members of the Eragrostoideae are generally more drought resistant than members of the Panicoideae.

The highest concentrations of species of the subfamily Eragrostoideae are usually found in arid regions with predominantly summer rainfall in which the mean temperature of the coldest month is above 10°C. The primary exception to this rule occurs in the southcentral United States (Kansas, Oklahoma, New Mexico), a region rich in Eragrostoid species with an arid or semiarid climate, predominantly summer rainfall, but with mean midwinter temperatures below 10°C. The relatively high frequency of the subfamily in this region is largely due to many species of *Muhlenbergia*. Subsequent studies (Kawanabe, 1968; Chazdon, 1978; Boutton et al., 1980) have shown that species of *Muhlenbergia* (all C_4) are especially tolerant of low temperatures and are found at higher altitudes and latitudes than other species of the Eragrostoideae.

Many species of the Eragrostoideae are especially adapted for growth under conditions of erratic rainfall. Semiephimeral annuals and perennials with special adaptations to resist prolonged drought are common. Tieszen et al. (1979) found that along an altitudinal and moisture gradient in Kenya the C_4 tribes Eragrostideae, Chlorideae, and Aristideae (all members of the Eragrostoideae) are most common at warm, arid, low-altitude sites. The tribes Paniceae and Andropogoneae are more common in higher, cooler, less arid areas. Similar observations have been made by Vogel et al. (1978), Brown (1977), and Boutton et al. (1980).

4

VEGETATIVE GROWTH

Within the grasses, C_4 physiology is a recent, highly successful specialization that evolved after the evolution of the basic reproductive and vegetative structure of the grasses. As a result, C_3 and C_4 grasses, while physiologically and ecologically distinct, share many morphological and developmental characteristics. This chapter summarizes the growth and development of the grass shoot with examples drawn from the C_4 grasses. Important references include Artschwager (1925), Barnard (1964), Clements (1980), Gould (1968), and Metcalfe (1960).

GERMINATION

Seed play two critical roles in the life of the plant: dispersal of new individuals and their maintenance during periods of adverse environmental conditions. Seed dormancy has evolved to prevent seeds from germinating at the wrong time or place. The degree of dormancy is determined by a number of genetic and environmental factors.

During the final maturation and drying of the caryopsis, cell division and synthesis of DNA, RNA, and protein cease, in that order (Osborne, 1981). Respiration may continue until water content reaches 20–25%. As long as caryopsis water content remains below about 20% respiration is not detectable. However, slow enzymatic and chemical degradation occur, resulting in a gradual loss of membrane integrity, mitochondrial function, and enzyme, DNA, and RNA integrity (Osborne, 1981).

Dormancy

When the caryopsis is rehydrated, it may or may not germinate. In fact, it is common for caryopses in the soil to undergo several wetting and drying cycles prior to germination. A rehydrated caryopsis that does not germinate under favorable temperature conditions is "dormant."

If a caryopsis is not dormant, germination begins as soon as rehydration occurs. Active metabolism is restored in approximately the following sequence: ATP production, RNA and protein synthesis, DNA repair, DNA replication, renewed cell division, and embryo growth. In dormant caryopses, rehydration does not lead to germination and growth of the embryo, but metabolism is restored in the following sequence: ATP production, RNA and protein synthesis, and DNA repair. DNA replication does not occur. Instead, a state of metabolic homeostasis is established with continuous turnover of nucleic acids, lipids, and proteins—but without DNA replication, cell division, and embryo growth. This turnover of cellular components in the dormant rehydrated caryopsis results in the repair of cellular lesions that occurred prior to rehydration (Osborne, 1981). When dormancy is "broken," DNA replication, cell division, and embryo growth occur in the normal sequence.

The percentage of dormant caryopses in a seed lot at any particular time is determined by a number of factors, including parental and caryopsis genotype and environmental conditions both during maturation and after rehydration. In most tropical grasses these factors are not well understood. The major factors are discussed below.

After-Ripening Period.   When tropical grass caryopses mature, the percentage of dormant caryopses is often very high initially. They may be induced to germinate immediately by the proper combination of light and temperature. However, this combination of environmental conditions is often extremely specific. If these dormant caryopses are stored in a dry environment, dormancy gradually decreases with time (after-ripening period), and the environmental conditions needed to break dormancy become less stringent. For example, Wright (1973) found that a 2-week moist prechill (5–10°C) treatment has little effect on germination of lehmann lovegrass (*Eragrostis lehmanniana*) just after harvest, but it strongly stimulates germination 30–50 weeks after harvest (Fig. 4.1). Fujii and Isikawa (1962) found that red light strongly inhibits germination of *Eragrostis ferruginea* for about 8 months after harvest, but this effect virtually disappears by 12 months after harvest.

The amount of after-ripening required to break dormancy varies with genotype and with the storage temperature during after-ripening. For example, postharvest dormancy of lehmann lovegrass may last for two years, while that of sideoats grama (*Bouteloua curtipendula*) disappears after four to six months (Major and Wright, 1974; Wright, 1973). In fall panicum (*Panicum dichotomiflorum*), postharvest dormancy disappears in about 13 weeks at 22°C, in about 40 weeks at 0°C, and it is delayed even longer at −15°C (Brecke and Duke, 1980) (Fig. 4.2).

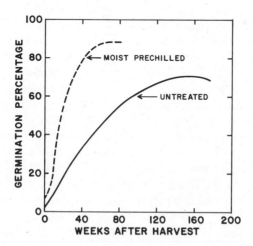

FIG. 4.1.   Effect of after-ripening period and moist prechilling treatment on *Eragrostis lehmanniana* germination. From Wright (1973). Reproduced from *Crop Science*, by permission of the Crop Science Society of America.

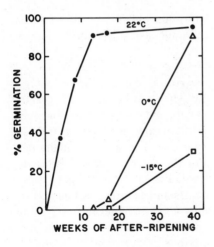

FIG. 4.2. Effect of temperature during the after-ripening period on percent germination in *Pani-cum dichotomiflorum*. From Brecke and Duke (1980). Reproduced from *Weed Science*, by permission of the Weed Science Society of America.

Light.   In the natural environment, seedlings must intercept enough radiation to become autotrophic by the time caryopsis reserves are exhausted. Therefore, when dormancy is broken, they should be close enough to the soil surface for leaf emergence to occur, and light intensity should be adequate for autotrophic growth. Thus, the dormancy mechanism should be sensitive to the light environment of the caryopsis. Many studies have shown that exposure to white light can promote germination of tropical grasses (Cole, 1977; Toole, 1973; Voigt, 1973). In caryopses the principal pigment affecting dormancy is phytochrome. Far red light (730 nm) inhibits germination by maintaining phytochrome in the P_r form. Red light (660 nm) promotes germination of dormant caryopses by converting the P_r form to the P_{fr} form. After conversion to the P_{fr} form, phytochrome reverts to the inhibitory P_r form by a thermal, nonphotochemical process with a half-life of about 5 hr at 25°C (Frankland, 1981). Thus, day length, the ratio of red to far red radiation, and the intensity of that radiation all affect the relative amounts of P_r and P_{fr}.

The mechanism by which P_{fr} breaks dormancy is unknown. However, it, like other factors affecting dormancy, must ultimately influence the activities of enzymes involved in DNA replication (Osborne, 1981).

In the natural environment the presence of a vegetation canopy and/or soil and plant litter above the caryopsis can affect the ratio of red to far red radiation reaching the caryopsis. For example, during most of the daylight hours, the ratio of red to far red radiation is about 1.2. However, because of preferential absorption of red radiation by soil and green vegetation, the red–far red ratio is about 0.55 below 3 mm of fine soil, and it is about 0.1 below a dense vegetation canopy (Frankland, 1981). Thus, the presence of a vegetation canopy or soil cover shifts the P_r–P_{fr} ratio toward the P_r form, which tends to maintain dormancy. The absence of this cover usually promotes germination by converting more phytochrome to the P_{fr} form.

White light, which normally converts much of the P_r to P_{fr}. can inhibit as well as promote caryopsis germination (Frankland, 1981; Toole, 1973). The inhibitory effects of white light normally occur under prolonged exposure to relatively high light intensities. The most effective wavelength is about 720 nm, and considerable evidence suggests that this response, termed the "high irradiance reaction," is also mediated by phytochrome. The dual (promotive and inhibitory) effects of white light on phytochrome and on germination can be explained in terms of a dual mechanism. First, white light with a normal ratio of red to far red converts sufficient P_r to P_{fr} to break the developmental block and stimulate germination. The high irradiance reaction probably inhibits a metabolic step subsequent to that promoted by P_{fr} (Frankland, 1981).

Among C_4 grasses the high irradiance reaction may be responsible for the inhibition of germination caused by white light in *Cenchrus* and *Lasiurus* (Lahiri and Kharabanda, 1964), *Panicum* (Cole, 1977), and *Eragrostis* (Toole, 1973; Toole and Borthwick, 1968a,b). One ecological explanation for the inhibitory effect of the high irradiance reaction is that caryopses germinating on the soil or litter surface in strong sunlight are likely to desiccate before roots can penetrate to moist soil layers. The high irradiance reaction may inhibit germination under these conditions.

The relative importance of the two opposing light effects is highly dependent on the physiological status of the seed. It varies among species (Lahiri and Kharabanda, 1964), ecotypes (Cole, 1977), and with duration of the after-ripening period and inhibition temperature (Toole and Borthwick, 1968a,b; Voigt, 1973).

Depth of Sowing.   When dormant caryopses are buried below the depth of light penetration, dark reversion of phytochrome to the P_r form can maintain germination at very low levels (Brecke and Duke, 1980; Harradine, 1980; Lahiri and Kharabanda, 1964). However, in field studies it may be difficult to determine whether sowing depth affects germination or the subsequent emergence of the seedling.

Germination Inhibitors.   Chemical germination inhibitors within the glumes or the spines surrounding groups of spikelets can have important effects on germination. This mechanism may be especially important in *Buchloe* (Ahring and Todd, 1977) and *Cenchrus* (Lahiri and Kharabanda, 1963) in which several caryopses (spikelets) are enclosed in thickened glumes and fall as a group (bur). Ahring and Todd (1977) found that oils present in the burs of buffalograss (*Buchloe dactyloides*) inhibit caryopsis germination. Lahiri and Kharabanda (1963) found that water and ether extracts of *Cenchrus* burs inhibit caryopsis germination.

Merrill and Young (1962) demonstrated that caryopses removed from the glumes of curly mesquitegrass (*Hilaria belangeri*) germinate more rapidly than those left within the glumes, possibly due to the presence of germination inhibitors. Similar results have been found in sideoats grama (*Bouteloua curtipendula*) (Major and Wright, 1974). In this case coumarinlike inhibitory compounds were implicated in maintaining dormancy.

Other Factors.   In addition to the effects mentioned above, numerous other treatments can reduce dormancy in C_4 grass caryopses. These include placing caryopses in a cool moist environment for several days (stratification) (Brecke and Duke, 1980; Groves et al., 1982; Wright, 1973); alternating rather than constant temperatures after imbibition (Brecke and Duke, 1980; Groves et al., 1982; Harty and Butler, 1975); chemical (concentrated acid) or mechanical scarification (Wright, 1973; Lahiri and Kharabanda, 1963); soaking in a gibberellic acid solution (Evans and Stickler, 1961; Groves et al., 1982; Mondrus-Engle, 1981); and exposure to enriched oxygen atmosphere and sodium hypochlorite (Major and Wright, 1974).

For largely unknown reasons the percentage of dormant caryopses can also vary dramatically among years, among sites from which caryopses are collected, and among ecotypes (Groves et al., 1982; Major and Wright, 1974; Wright, 1973). This variation is probably due to genotype of the parents, environmental conditions during seed maturation, and environmental conditions during the after-ripening period.

Germination of Nondormant Caryopses

After the developmental block (dormancy) preventing germination has been broken, both temperature and water availability affect germination.

Temperature.   Numerous studies have addressed the effects of temperature on germination of tropical grasses. Some of these are summarized below.

Groves et al. (1982) found little effect of temperature between 20 and 40°C on percentage germination of nondormant kangaroograss (*Themeda australis*) caryopses two weeks after sowing. However, no germination was found at 45°C, and 15°C temperatures reduced germination in some genotypes.

At near-optimum soil water contents germination and emergence of black grama (*Bouteloua eriopoda*) and *Eragrostis chloromelas* were inhibited when maximum soil temperatures exceeded about 50°C (Herbel and Sosebee, 1969).

Grain sorghum germination was over 80% at 10°C, but it was delayed and reduced at 8°C (Plinthus and Rosenblum, 1961). Germination was slower at 17 than 27°C, but it finally reached nearly 100% in both cases (Evans and Stickler, 1961).

Maximum (70–100%) germination of *Lasiurus sindicus* and *Cenchrus setigerus* occurred at about 30°C with almost no germination at 20°C and less than 20% germination above 45°C. In contrast, germination of buffelgrass (*Cenchrus ciliaris*) was above 50% at temperatures ranging from 25 to 50°C, but almost no germination occurred at 20 or 55°C (Lahiri and Kharabanda, 1964).

Sharma (1976) found that germination of *Danthonia caespitosa* was greater than 60% at temperatures between 15 and 30°C; much lower percentages were found above and below these limits.

From these studies it is evident that some species have more stringent temperature requirements than others for germination. However, in most species germination is greatly reduced at mean daily temperatures below 10°C or above 45°C.

Water Potential.   Several studies have demonstrated the effects of soil water potential on germination of C_4 grass caryopses. Evans and Stickler (1961) germinated grain sorghum caryopses in mannitol solutions of differing strengths to simulate drought stress (Fig. 4.3). They found dramatic decreases in percent germination at osmotic potentials lower than −0.5 MPa. They also found significant differences in germination response to stress among grain sorghum cultivars and among seed lots of a single cultivar. These results are consistent with those of Groves et al. (1982), who germinated kangaroograss (*Themeda australis*) caryopses in soil at varying water potentials. They found almost no germination at soil water potentials lower than −1.2 MPa. They also found large differences in germination among genotypes at higher soil water potentials.

Sharma (1976) germinated *Danthonia caespitosa* caryopses in solutions of polyethylene glycol at varying temperatures. He found that at solution water potentials above −0.4 MPa the caryopses germinated quite well at temperatures ranging from 15 to 30°C. However, the temperature requirements for germination were much more stringent at −1.5 MPa, where significant germination occurred only at 25°C. These results demonstrate that soil water and temperature interact to inhibit germination when the probability of seedling survival is low due either to suboptimal soil water or to extreme temperature.

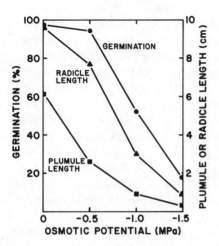

FIG. 4.3.   Effect of osmotic potential on mean percent germination, radicle length, and plumule length of four seed sources of RS610 grain sorghum. From Evans and Stickler (1961). Reproduced from *Agronomy Journal*, by permission of the American Society of Agronomy.

Reimposition of Dormancy

Nondormant caryopses can be defined as those that will germinate when allowed to imbibe under favorable temperature, moisture, and light environments. However, under some conditions nondormant caryopses will regain a certain degree of dormancy if they are sown into an unfavorable environment. Voigt (1973) found that allowing nondormant weeping lovegrass (*Eragrostis curvula*) caryopses to imbibe in cool (18–24°C, 13–18°C) environments in the dark or in alternating light and dark will reinduce dormancy in a large percentage of these caryopses. These newly dormant caryopses will not germinate when they are transferred to favorable germination environments (24–29°C in the light). Caryopsis age, temperature, and light regime during the imbibition period all affect the percentage of seeds in which dormancy is reinduced. This is consistent with field observations in which nondormant seed do not germinate right away, but germinate later when environmental conditions are again favorable.

SEEDLING GROWTH

Germination of the caryopsis results in the growth and development of the embryo. This early development can be divided into three phases: heterotrophic, transitional, and autotrophic. The energy for growth during the heterotrophic phase is derived from the remobilization of endosperm carbohydrates. During the transitional stage the endosperm continues to contribute energy to the seedling, but current photosynthate from newly expanded leaves also begins to contribute energy. The autotrophic stage begins when remobilization of endosperm carbohydrates ceases and concurrent photosynthesis provides all the energy requirements for growth (Whalley et al., 1966).

The contribution of the endosperm and scutellum to early seedling growth has been studied more carefully in maize than in other C_4 grasses (Cooper and MacDonald, 1970; Dure, 1960; Ingle et al., 1964). At 22°C the heterotrophic phase of maize seedling growth lasts for approximately 9–10 days after germination. The transitional stage begins about the time the first leaf is fully expanded (around day 9) and ends at the three- or four-leaf stage (about day 16), when the autotrophic stage begins (Cooper and MacDonald, 1970) (Fig. 4.4). During the heterotrophic stage, about two-thirds of the decrease in endosperm dry weight is converted into an increase in shoot and root dry weight. During this phase, there is almost no difference between light- and dark-grown seedlings in the conversion of endosperm dry weight to shoot and root dry weight. The mobilization of endosperm reserves has a lower base temperature than does photosynthesis. Therefore, at temperatures between 10 and 15°C the heterotrophic stage may be completed but the autotrophic stage may not begin due to low-temperature stress. This effect is discussed further in Chapter 7.

Two forms of grass seedling growth have long been recognized. Members of the subfamily Festucoideae produce an elongated coleoptile that pushes upward

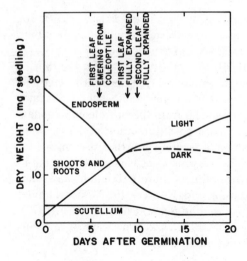

FIG. 4.4. Change in maize endosperm, shoot and root, and scutellum weights after germination. From Cooper and MacDonald (1970). Reproduced from *Crop Science*, by permission of the Crop Science Society of America.

to the soil surface. In this type of seedling the first adventitious roots are produced at the base of the coleoptile, which is at the depth of the seed. The second type of seedling growth occurs in the subfamilies Panicoideae and Eragrostoideae (Fig. 4.5). These seedlings produce an elongated subcoleoptile internode with a narrow band of meristematic tissue just below the node at the base of the coleoptile. The elongated subcoleoptile internode pushes the coleoptile node upward to near the soil surface, and adventitious roots arise from coleoptile node (Boyd and Avery, 1936; Sargent and Arber, 1915). However, if the seedling encounters dry soil conditions at the coleoptile node near the soil surface, adventitious roots will not develop. If adventitious roots do not develop, the seedling eventually dies because the seminal root system is unable to support the growing

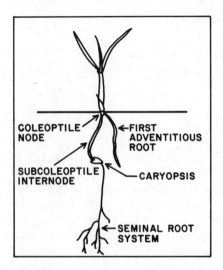

FIG. 4.5. Morphology of a typical panicoid or eragrostoid seedling.

top (Hyder et al., 1971; Sluijs and Hyder, 1974; Wilson and Briske, 1979). For example, Wilson and Briske (1979) found that consistent establishment of blue grama (*Bouteloua gracilis*) requires (1) average soil temperature above 15°C, (2) a soil water potential of about -0.03 MPa in the 0–40-cm zone at the time of emergence, and (3) two properly spaced two- to four-day periods with a continuously moist soil surface (one for emergence and another two to eight weeks later for initial growth of the adventitious root system from the subcoleoptile node).

The reserve carbohydrates of the endosperm constitutes most of the weight of the caryopsis, and the heterotrophic phase of seedling growth is responsible for early root and leaf area production. Thus, caryopsis size is positively related to germination and/or seedling growth in grain sorghum (Abdullahi and Vanderlip, 1972; Suh et al., 1974; Swanson and Hunter, 1937), pearl millet (Gardner, 1980), blue panicgrass (*Panicum antidotale*) (Wright, 1976, 1977), buffelgrass (*Cenchrus ciliaris*) (Kathju et al., 1978), and other C_4 grasses (Lahiri and Kharabanda, 1961; Kneebone and Cremer, 1955).

VEGETATIVE SHOOT GROWTH

The vegetative shoots of grasses, C_4 and C_3, are characterized by a peculiar combination of structural features:

1. Cylindrical jointed stems with short basal internodes.
2. Long narrow leaves with parallel veins and sheathing bases that arise from the stem in two distinct ranks (distichous phyllotaxy) (Fig. 4.6).
3. A primary adventitious root system arising from nodes of the stem (Fig. 4.7).
4. Leaves and axillary buds on lateral branches inserted at right angles to those of the stem from which the branch arose (Fig. 4.8).

This combination of characteristics is common to almost all grasses. Differences in structure among species are simply variations within the basic structure (Barnard, 1964).

Stems

The morphology of grass stems is quite variable and plays an important role in the adaptation of species to fire, grazing, temperature extremes, and competition for light. Culms are erect or decumbent stems. They elevate the leaves and apical meristem above the ground surface, facilitating wind pollination and allowing the plant to compete more successfully for light but exposing the apical meristem to damage by grazing and fire. Stolons are horizontal aboveground stems, and rhizomes are horizontal below-ground stems. The apical meristems

FIG. 4.6.   A well-developed maize plant illustrating the long narrow leaves in two distinct ranks opposite each other (distichous phyllotaxy).

of stolons and rhizomes are usually near the soil surface. This offers some protection from damage by grazing, fire, and frost.

Changes in the orientation of the stem with respect to gravity are sensed by statolith-containing cells. In the weed jungle rice (*Echinochloa colonum*), these cells are associated with vascular bundles within the leaf sheath bases (Parker, 1979). These cells contain large central vacuoles with up to 50 spherical starch

FIG. 4.7. A bermudagrass (*Cynodon dactylon*) stolon illustrating adventitious roots and axillary shoots arising from the nodes. The figure also shows that the first several leaves on the left tiller lack well-developed laminae.

statoliths that sediment readily in response to gravity. We have known for some time that auxin promotes stem curvature in sugarcane and other grasses (Dillewijn, 1952). For example, application of auxin to one half of the cut surface of a sugarcane stalk segment causes elongation of that side of the stem just above the node. This results in curvature of the stem. Bandurski et al. (1984) have recently demonstrated that gravistimulation causes free indole-3-acetic acid (IAA) to accumulate on the lower side of maize mesocotyls within 15 min of their being placed in a horizontal position. Curvature of the mesocotyl follows, suggesting that asymmetric distribution of endogenous IAA causes the curvature.

Culms. The basic unit of the culm is the phytomere, defined as an internode together with a leaf at its upper end and a bud at its lower end (Dillewijn, 1952; Weatherwax, 1923). Each phytomere matures basipetally. The first tissues to mature are those of the leaf lamina. The sheath then matures and reaches its final size about the time that the internode begins to elongate rapidly. At this time, the entire internode is essentially a single intercalary meristem. As it elongates, the most acropetal cells mature, and the intercalary meristem becomes restricted to the basipetal portion of the internode (Fig. 4.9). These basal cells usually retain their ability to divide for some time, and when erect culms are

FIG. 4.8.   A grain sorghum shoot with one tiller. Note that the leaves on the tillers form a plane at a right angle to the plane of leaves on the main shoot.

blown or forced down, they have the ability to renew growth on the lower side of the horizontal culm and force its apex to a more vertical position (Fig. 4.10).

Just basipetal to or within the root band is the axillary bud. This bud is the last portion of the phytomere to mature, and it, like adventitious root primordia, usually remains dormant until favorable conditions occur, when it begins to develop into an axillary shoot or tiller. The axillary bud is usually clasped tightly by the leaf sheath of the phytomere below it. When it begins to grow, it escapes from the sheath either by growing upward between the internode and sheath (intravaginal branching) (Fig. 4.10) or it may rupture the sheath and grow through the opening (extravaginal branching) (Fig. 4.11).

One characteristic of the subfamily Panicoideae is the development of culm

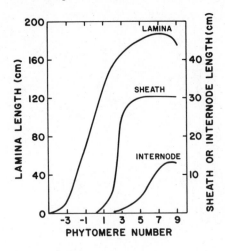

FIG. 4.9. Lengths of leaf laminae, leaf sheaths, and stem internodes of 14 phytomeres from a typical sugarcane culm. The phytomere with the most recently emerged leaf tip is designated number 1.

pulvini, meristematic swellings at the base of the internode (Fig. 4.10). In contrast, members of the subfamily Festucoideae often possess sheath pulvini, enlargements of the sheath base.

The basipetal end of the phytomere may also contain root primordia that can be stimulated to grow by a combination of hormonal and environmental factors such as a reduction in apical dominance and favorable temperature and moisture conditions. These root primordia are arranged in whorls in the "root band" area of the basipetal phytomeres (Fig. 4.7).

In transverse section the internodes are typical of monocotyledonous stems. In C_4 grasses they are normally solid with vascular bundles either scattered throughout the pith or arranged in two to several rings at its periphery. In contrast to C_4 grasses, most C_3 grasses of the subfamilies Festucoideae, Oryzoideae, and some Bambusoideae and Arundinoideae have hollow internodes with vascular bundles in concentric rings near the periphery.

The epidermis of the culm is similar to that of the leaf blade, having rows of long and short cells, stomata, and hairs of various types. Beneath the epidermis of the culm is a cortex made of sclerenchyma and chloroenchymatous parenchyma cells. The sclerenchyma often continues to develop and lignify as the stem develops, increasing the strength of the culm. The strength of this sclerenchyma is genetically controlled and contributes to lodging resistance (Thompson, 1982). The cortical cells often produce anthocyanins that give the stem a red or purple coloration, which contributes to the "colorful" names of traditional sugarcane varieties such as "Louisiana Purple" (also called "Black Cheribon") and "Rose Bamboo." The anthocyanin content of the cortical cells is often stimulated by high sugar and starch concentrations in the internodes. This leads to the typical purple coloration of maize culms when the developing ears are removed and nonstructural carbohydrates accumulate in the internodes, and the purple color typical of low-temperature stress and P deficiency.

The nodes are important organs from which leaf sheaths, adventitious roots,

FIG. 4.10. Johnsongrass (*Sorghum halepense*) culm showing curvature at the nodes, intraval branching, and culm pulvini.

and axillary buds originate. Vascular bundles from the leaves and axillary buds pass into the nodes where they anastomize with bundles from the internodes above and below the node. The nodes are important because they are the region in which the vascular tissues from successive phytomeres join. In each phytomere the vascular system, like the other tissues, matures basipetally. Thus, the bundles in the leaf blade, leaf sheath, and internode are well developed before

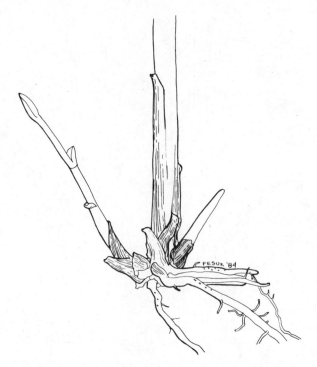

FIG. 4.11. Johnsongrass (*Sorghum halepense*) primary tiller with secondary tillers emerging through ruptured leaf sheaths (extravaginal tillering). Note lack of well-developed laminae on basal leaves.

they are connected to the vascular system of the phytomere below at the node. This is probably one reason that expanding leaf blades and sheaths are so well protected from the drying effects of the atmosphere within the enveloping blades and sheaths of older leaves.

Adventitious root primordia arise from parenchymatous tissue of the node just basipetal to the intercalary meristem of the internode (Fig. 4.7). Their steles anastomize with bundles from the internodes, leaves, and axillary buds in the node from which the root originates.

Stolons and Rhizomes.   Stolons and rhizomes of grasses are morphologically analogous to culms. They differ in being horizontal structures adapted to conditions near or below the soil surface. Several studies have shown that stolons and rhizomes are anatomically similar (Rogers et al., 1976; Stiff and Powell, 1974). They consist of an epidermis of thick-walled cells covered by a heavy cuticle. Below the epidermis is an outer cortex of large thin-walled parenchyma. The outer cortex is separated from the pith by a ring of thick-walled sclerenchyma known as the fiber band (Esau, 1953), fiber ring (Metcalfe, 1960), or fibrond (Stiff and Powell, 1974). Vascular bundles similar to those of the culm are located in the fibrond or just inside the ring in the pith. The number of bundles in the stolon or rhizome is more a function of genotype than of the size of the organ (Stiff and Powell, 1974). As in the culm, when the vascular bundles pass from the internode into the node, they lose their longitudinal orientation and anastomize with bundles from axillary buds and the steles of adventitious roots.

Stiff and Powell (1974) studied the stem (primarily stolon) anatomy of 20 turf-grass cultivars of the C_4 genera *Cynodon*, *Zoysia*, *Stenotaphrum*, and *Digitaria* and the C_3 genera *Agrostis*, *Poa*, and *Festuca*. In all genera the fibrond separates the internode into an outer cortex and inner pith region. Three types of morphology are found: (1) a single ring of vascular bundles embedded in the fibrond (*Agrostis*, *Poa*, and *Festuca*), (2) multiple rings in which a ring of small bundles is located on the outer edge of the fibrond, a ring of large bundles is embedded in or connected to the fibrond, and a third ring of large bundles lies in the pith (*Cynodon*, *Digitaria*, *Zoysia*), and (3) complex rings in which small bundles lie in the cortex and large bundles are embedded in the fibrond and also occur in the pith (*Stenotaphrum*). Thus, the C_4 species differ from the C_3 species in having multiple rings of vascular bundles. Two types of rhizome nodes were also observed: simple nodes with one leaf and axillary bud (C_3 species and *Digitaria*) and compound nodes with two to four (usually three) leaves and axillary buds (*Cynodon*, *Zoysia*, *Stenotaphrum*).

Thus, *Cynodon*, *Stenotaphrum*, and *Zoysia* have a long internode followed by a compound node composed of several short internodes and nodes from which axillary shoots and roots arise. These secondary shoots often form erect tufts of several tillers before giving rise to another horizontal stolon with alternating long and short internodes. In most cases the leaves of secondary shoots form a plane at right angles to the leaves of the stolon. One exception to this rule is *Stenotaphrum*, in which the leaves of the stolon and those of secondary shoots are in the same plane. This results in fanlike structures consisting of the stolon, the leaves of two or three short and one long internodes, and two axillary shoots arising from the nodes of the short internodes (Fig. 4.12).

Rhizomes normally have short internodes and bear scale leaves rather than the normal leaves borne by stolons and culms. If the internodes of the rhizome are short, the axillary buds produce tillers arising close to one another, and the plant has a tussock or tufted habit. This habit is typical of sugarcane and many tufted forage grasses. When rhizome internodes are longer, as in the case of weed species such as johnsongrass (*Sorghum halepense*) and African feather-grass (*Pennisetum macrourum*), the rhizome becomes an important organ of reproduction (Fig. 4.13). When undisturbed, the plant spreads by growth of the rhizomes beneath the soil surface. When plowing is used to eradicate the infestation, the rhizomes are cut into fragments that can regrow from axillary buds and may even increase the infestation (Harradine, 1980; Horowitz, 1972).

Vegetative Propagation.   A few species such as bermudagrass (*Cynodon dactylon*) have both rhizomes and stolons. Hybrid bermudagrass cultivars such as

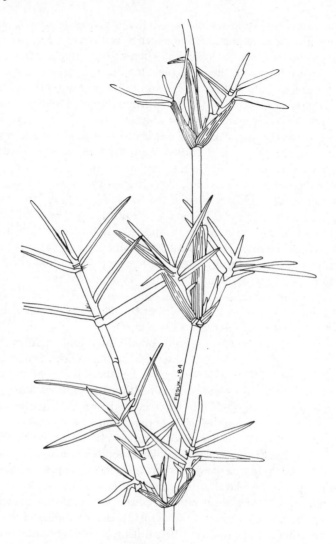

FIG. 4.12. Stolon morphology of St. Augustinegrass (*Stenotaphrum secundatum*). Note that a long internode is followed by several short internodes and nodes from which adventitious roots and axillary shoots arise. Note the fanlike axillary shoots with leaves in the same plane as those on the main stolon. This is an exception to the normal case in which leaves of the axillary shoots are in a plane perpendicular to that of the leaves on the primary shoot.

Coastal, Midland, and Coastcross, as well as other important forage grasses such as pangolagrass (*Digitaria decumbens*) and several species of *Brachiaria*, are highly sterile or produce few seeds and are normally propagated by planting stolon and/or rhizome cuttings. Sugarcane (*Saccharum* spp. hybrid) is also reproduced from stem cuttings, though in this case sections of erect culms rather than rhizomes or stolons are used (Fig. 4.14).

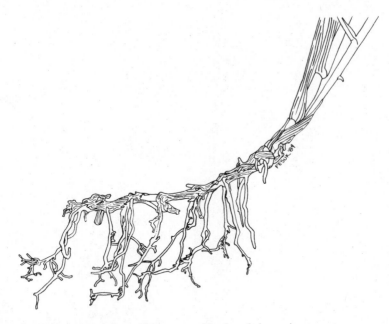

FIG. 4.13.   Rhizome morphology of johnsongrass (*Sorghum halepense*).

The primary reason that stem cuttings can be used for vegetative propagation is their ability to store nonstructural carbohydrates and then remobilize them to support the growth of axillary buds and adventitious roots. Sugarcane culms can accumulate more than 50% of their dry weight as sucrose (Clements, 1980), and in the rhizomes and stolons of C_4 forage grasses and weed species starch and other nonstructural carbohydrates can constitute 12-22% of the dry weight of the organ (Dunn and Nelson, 1974; Rogers et al., 1976; Harradine, 1980). Most of this reserve carbohydrate is stored in the pith parenchyma (Rogers et al., 1976; Clements, 1980). It usually reaches its highest concentrations in autumn when vegetative growth and flowering are at a minimum. Its lowest levels normally occur in late winter or early spring when it is utilized in the initiation of new growth (Harradine, 1980; Rapp, 1947; Weinmann, 1961).

Internode sucrose concentration is genetically as well as environmentally controlled, and sugarcane breeding programs have dramatically increased sugar yield per hectare by developing cultivars with higher percentages of sucrose in the harvested cane. For example, in Louisiana sugar contents of clean cane have increased from about 9% in the 1930s and 1940s to about 13% in the 1970s (Breaux, 1984).

Tillage is often used to control rhizomatous weed species. Four principal factors affect its success: the reserve carbohydrate level of the rhizome, the average number of nodes per rhizome fragment, the depth of burying, and the moisture content of the soil and rhizome. Harradine (1980) concludes that cultivation of African feathergrass (*Pennisetum macrourum*) should be most beneficial if

FIG. 4.14. Growth of axillary buds of sugarcane setts. Note that buds placed toward the soil surface have more rapid shoot growth but delayed root growth. Both root and shoot growth are adequate when buds are in a plane parallel to the soil surface. Reproduced from Clements (1980) with permission of University of Hawaii Press.

done in the summer when reserve carbohydrate levels are relatively low and water stress is most likely to occur. Maximizing the depth of burial and minimizing the number of nodes per fragment should also reduce regeneration.

Herbicides are also used to control rhizomatous grass weeds, but the most favorable time of application may vary among species and among herbicides. In Tasmania glyphosate controls African feathergrass best when it is applied in autumn or late spring when reserve carbohydrates are increasing. Harradine (1980) suggests that this facilitates basipetal translocation of the herbicide. However, Schirman and Buchholtz (1966) conclude that atrazine application is most effective when it is timed to coincide with minimum carbohydrate reserves.

Because of its economic importance, a great deal is known about the growth of sugarcane from stem cuttings (Clements, 1980; Dillewijn, 1952). Several factors affect the early growth of sugarcane buds. In general, all undamaged buds on a stalk are capable of germination. However, early growth is more rapid and vigorous when soil temperature and moisture are near optimum. The minimum temperature for growth of the buds is 8-9°C, but much more rapid growth occurs at warmer temperatures. The optimum temperature for early growth is genotype dependent and ranges from about 30 to 38°C. The section of culm that is planted, called the seed piece or sett, may have one to several axillary buds, depending on its length. In commercial sugarcane fields the setts normally have two or more buds to assure the survival of at least one shoot. Slight apical dominance is often found in these setts, and the apical bud grows more rapidly than more basal buds. Buds on the upper side of the sett grow more rapidly than those on the lower side. However, they produce few or no roots initially, and their subsequent development is delayed. The most favorable growth occurs when the buds are placed horizontally with respect to the soil surface. These shoots emerge more slowly than those on the upper side of the sett, but they develop adventitious roots more rapidly and soon surpass shoots growing from the upper side (Fig. 4.14).

Sugarcane setts are often treated with fungicides and hot water (52°C) prior to planting. The hot water treatment destroys certain pathogens, reduces apical dominance in long setts, and increases early vigor. However, precise control of the time and temperature of the treatment is important for maximum effectiveness.

Leaves

The internal anatomy of the lamina of C_4 grasses is discussed in Chapter 2. Further discussion and excellent line drawings and photographs are found in Artschwager (1925), Clements (1980), Esau (1953), Dillewijn (1952), and Weatherwax (1923).

Leaf Shape.   The first leaves produced by tillers and seedlings are normally different than subsequent leaves. The first several leaves on developing axillary buds of rhizomes and sugarcane setts are scalelike. These leaves, as well as the first leaf appearing above the soil surface, consist of sheaths without prominent laminae (Fig. 4.7). Likewise, the first leaf produced during the development of an aerial axillary bud is the prophyll, a leaf without a lamina. The function of these atypical leaves is protection of the immature tiller and its apical meristem.

The first leaf of seedlings is also atypical in many cases. For example, the lamina of the first leaf of the maize seedlings is broad and has a rounded tip. Subsequent laminae are typically long and narrow.

Leaf Size.   As a seedling or a tiller develops, the areas of successive leaf laminae increase until an environment- and genotype-dependent maximum size is attained. The areas of subsequent laminae may be relatively constant if the tiller remains vegetative (as in sugarcane and some forage grasses). However, after panicle initiation lamina areas of the last several leaves may decrease dramatically. This developmental effect on lamina area is especially noticeable in maize, where the areas of successive laminae increase up to a maximum then decrease (Fig. 4.15). Early work (Abbe et al., 1941; Rosler, 1928) demonstrated that the increase in lamina width is correlated with the circumference of the apical meristem at the time of leaf initiation. The change in meristem circumference is primarily due to a change in the number of cells in the apex. In some species (e.g., maize and sugarcane), the areas of the last several laminae below the inflorescence are greatly reduced. In others (e.g., grain sorghum and small grains), the flag leaf lamina (the leaf directly below the panicle) is large and is an important photosynthetic organ.

Leaf Initiation and Emergence. Leaf primordia are initiated in the apical meristem, and the lamina usually grows almost to its final size before the leaf tip emerges from the whorl of older laminae. The number of unemerged leaves and primordia within the whorl varies among species. In large grasses such as maize and sugarcane, eight or more unemerged leaves are normally found (Clements, 1980). Fewer are found in smaller species.

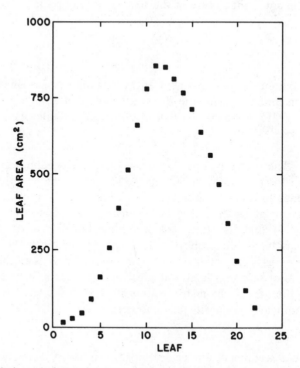

FIG. 4.15. Maximum leaf area of all leaves on a maize plant. Author's unpublished data.

The phyllochron or plastochron interval (the interval between emergence of successive leaf tips from the whorl) is strongly affected by temperature (Fig. 4.16). The degree-day (thermal time) concept can be used to quantify the response of the phyllochron interval to temperature. In vegetative tillers (such as those of sugarcane and vegetative forage grasses), the interval between initiation of successive primordia is approximately equal to the interval between the emergence of successive leaf tips from the whorl. Thus, the number of unemerged leaves is nearly constant. In sugarcane and (probably) in other species, this interval is genotype specific (Dillewijn, 1952). In the absence of water or nutrient stresses, the rate of leaf tip emergence is a linear function of mean air temperature.

In crops (such as maize, grain sorghum, and millets) in which panicle initiation occurs early in the life of a plant or tiller, the rates of leaf primordia initiation and leaf emergence are not equal. Recent work (J. R. Kiniry, personal communication) has shown that in maize the rate of primordia initiation is a linear function of thermal time above a base temperature of 8°C. A new primordium is initiated at the apical meristem each 21 degree-days. In contrast to primordia initiation, maize leaves emerge from the whorl each 42 degree-days (base 8°C) (Tollenaar et al., 1979). Thus, the number of unemerged leaves in the whorl of a typical maize plant increase from approximately 6 primordia in the seed to a maximum of approximately 10 to 12 at panicle initiation.

Leaf Expansion.   As discussed earlier, the elongation of the phytomere (an internode with a leaf at its upper end and a bud at its lower end) begins with elongation of the leaf lamina. When the lamina has almost attained its final length, the leaf sheath begins to elongate, and when the sheath has almost ceased elongation, the internode begins to elongate (Weatherwax, 1923; Dillewijn, 1952) (Fig. 4.9). In sugarcane and other large grasses the total rate of expansion of all

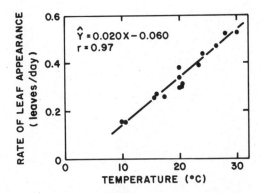

FIG. 4.16.   Effect of mean temperature on rate of maize leaf tip appearance. From Tollenaar et al. (1970). Reproduced from *Crop Science*, by permission of the Crop Science Society of America.

unemerged leaves is approximately equal, whether the lamina, the sheath, or both are elongating. Since these leaves are packed together very tightly in the whorl (Fig. 4.17), any difference in growth among leaves would result in distortion and folding or crimping of their tender tissues (Dillewijn, 1952). Thus, the unemerged leaves grow as a unit, and their tips move away from the apical meristem at the same rate.

Leaf sheath elongation ceases at about the time the leaf lamina emerges from the whorl and unrolls, thus detaching itself from the inner, tightly rolled leaves. At about this time the stem internode begins to elongate, and the leaf exhibits "gliding growth" as its sheath moves upward out of the enveloping sheath of the next older leaf (whose elongation has slowed) (Clements, 1980).

The leaf lamina of crops such as maize and sugarcane is almost fully expanded by the time its tip emerges from the whorl; therefore, the time needed for the lamina to be pushed out of the whorl by the expanding sheath and internode is highly correlated with the size of the lamina (Fig. 4.18).

The rate of leaf tip emergence is not dependent on leaf size, but the time interval from leaf tip to leaf collar emergence is. Therefore, the number of leaves expanding at any time is small (two or three) if the leaves are small, but it is large (six or eight) if the leaves are large.

Leaf blade and sheath elongation (as well as culm elongation) are often stimulated by application of exogenous gibberellic acid (see Chapter 7). Rood et al. (1983a,b) have found that at least a portion of the hybrid vigor of maize is probably due to higher levels of endogenous gibberellinlike substances in hybrids than

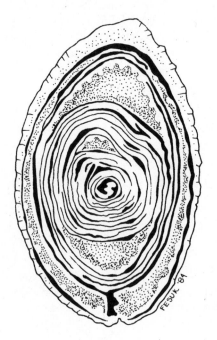

FIG. 4.17. Diagramatic drawing of a cross section of the whorl or spindle of a maize plant. Note the tight packing and overlapping margins of leaves in the whorl.

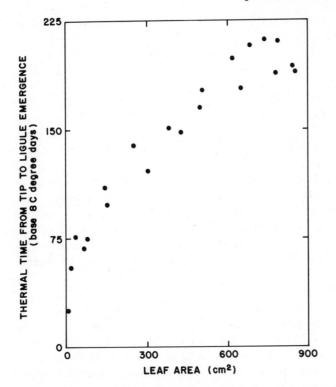

FIG. 4.18.   Relationship between the area of the lamina and the thermal time required for its emergence from the whorl. Author's unpublished data.

in inbreds. Application of exogenous gibberellin causes significant elongation of the leaf blades and sheaths of inbreds, but it has little or no effect on hybrids.

Leaf Senescence.   During the early growth of a tiller or seedling, leaves emerge at a far more rapid rate than they senesce. In maize and grain sorghum seedlings and in sugarcane tillers, the senescence of the first two or three leaves is hastened by mechanical damage to the sheath caused by the normal increase in stem diameter. In vegetative tillers such as those of forage grasses or sugarcane, the rate of leaf death soon approaches the rate of leaf emergence, and the number of live, fully expanded leaves becomes relatively constant. In sugarcane it is common to have 12–16 live, fully expanded leaves at any time. Of course, stresses have opposite effects on the rates of leaf emergence and leaf senescence. They increase the rate of senescence, and they may decrease the rate of emergence. This reduces the number of live leaves on the tiller and (along with the decrease in the size of expanding leaves) effectively reduces leaf area per tiller.

Imbalances in the source–sink ratio also affect leaf senescence. Senescence is accelerated when the source–sink ratio is excessively low due to defoliation or shading or when it is excessively high due to ear removal (Tollenaar and Daynard, 1982).

Tillers

Grass tillers are axillary shoots that grow, produce adventitious roots, and may eventually become completely independent of the culm, stolon, or rhizome from which they arose. Just as a grass tiller is made up of multiple phytomeres, a grass plant may be made up of multiple tillers, each consisting of a shoot and its adventitious roots.

Individual tillers may be short or long lived. They are determinant structures because they senesce and die after their reproductive growth ceases. However, grass plants with multiple tillers are potentially perennial by virtue of their ability to produce tillers indefinitely under favorable environmental conditions. In perennial grasses the production, maturation, and/or death of tillers usually has an annual rhythmic pattern that is synchronized with local climate and flowering.

Several phases of tiller development can be distinguished. First, axillary buds are formed. In almost all cases one bud is formed for each phytomere produced on a stem, and more buds normally form than ever develop. Second, hormonal and nutritional factors cause these buds to begin to grow using photosynthates and mineral nutrients from the parent tiller. Third, the tiller begins to produce sufficient photosynthate and to extract sufficient water and mineral nutrients from the soil to be independent of the parent tiller. Finally, the new tiller may produce small tillers and may undergo panicle initiation. Hormone balance is the main factor affecting the very early growth of the tiller. Competition for light, water, and nutrients largely control its subsequent vegetative growth.

Hormones.   Early studies on the control of tillering showed that hormones affect the release of axillary buds from the dormant condition. In grasses, as in dicots, auxin from the apical bud of the main stem (or exogenous auxin) inhibits the growth of axillary buds on that stem (Leopold, 1949). In contrast to auxin, cytokinin from the root tips (or exogenous cytokinin) promotes the growth of axillary buds (Harrison and Kaufman, 1980; Sharif and Dale, 1980). The balance of auxins and cytokinins in a bud largely determines whether the bud remains dormant or begins to grow; however, gibberellic acid (GA) and abscisic acid also have important affects. For example, exogenous GA in concentrations too low to promote stem elongation prevents bud development prior to anthesis, but it stimulates tillering following anthesis (Isbell and Morgan, 1982). Isbell and Morgan (1982) conclude that for grain sorghum application (at anthesis) of GA_3, cytokinins, auxin transport inhibitors, or a combination of these substances could be used to promote postanthesis tillering.

Genotypes.   Numerous studies have shown that tillering varies widely among cultivars of rice and wheat. In general, modern semidwarf cultivars produce more tillers than taller traditional cultivars. Similarly, grain sorghum (Escalada and Plucknett, 1975a; Isbell and Morgan, 1982) and maize (Rood and Major, 1981; Stevenson and Goodman, 1972) cultivars vary in their tendency to produce tillers.

Since all grasses produce tillers and the close relatives of maize tiller freely, it has been assumed that primitive maize also tillered freely (Mangelsdorf, 1958). However, more recent work with a large number of exotic races of maize (Stevenson and Goodman, 1972) suggests that tillering of primitive races is quite variable and is correlated with the environment where the race developed. In harsh environments with short growing seasons, early maturing genotypes have been selected. In these areas nontillering races predominate, presumably because production of sterile tillers diverts energy from the growth of the main stem. In more favorable environments tillering races were often selected because the tillers as well as the main stem can produce grain, especially at the low population densities used by primitive agriculturalists.

Environment.   In general, environmental conditions that favor growth of the main stem also favor tillering. Thus, reduced plant competition for light, nutrients, and water favors tiller production. In addition, cool temperatures sometimes stimulate tillering, presumably by reducing the photosynthate demand of the main stem and allowing abundant carbohydrate reserves to accumulate. However, the effects of temperature and day length on tillering are complex. In maize, low temperatures and short day lengths increase tillering (Stevenson and Goodman, 1972), but in grain sorghum similar conditions can result in fewer tillers (Escalada and Plucknett, 1975b).

The production of tillers in perennial forage grasses usually follows a rhythmic pattern synchronized both with climate and with reproductive growth. Tillers are produced most rapidly when climatic conditions favor growth but when most tillers are still vegetative. During the period of rapid stem elongation following panicle initiation, tillering usually decreases, probably due to competition for photosynthate. If environmental conditions remain favorable after seed maturation, tiller production increases again. Often, however, seed production precedes the dry or cold part of the year, and rapid tiller production resumes only when environmental conditions are again favorable (Singh and Chatterjee, 1965). Like seed maturation and death of reproductive tillers, defoliation reduces apical dominance and permits rapid tiller production (Escalada and Plucknett, 1975a).

Sugarcane (Clements, 1980; Dillewijn, 1952), grain sorghum (Escalada and Plucknett, 1975a), and C_3 cereals such as wheat and barley (Thorne, 1962) produce many more tillers than actually contribute to the final economic yield of the crop. However, this is not true in all crops under all conditions. For example, Krishnamurthy et al. (1973) found that in finger millet (*Eleusine coracana*) a high (80-90%) percentage of all tillers produce heads when the crop is grown under irrigation at wide row spacings. Similarly, Cable (1982) found that the forage grass Arizona cottontop (*Trichachne californica*) produces a higher percentage of reproductive tillers than other perennial grasses. In addition, it produces new tillers throughout the year.

Competition for light, nutrients, and water usually prevents most small tillers from developing. These tillers usually remain small or die (Dillewijn, 1952; Escalada and Plucknett, 1975a) due to lack of photosynthetic assimilate and/or mineral nutrients. Dayan et al. (1981) have produced a simulation model of tiller dynamics and regrowth after defoliation (TILDYN) for rhodesgrass (*Chloris gayana*). This model incorporates the nutrient diversion theory (Woolley and Wareing, 1972), in which the growth of tillers is subject to availability of assimilates and mineral nutrients. Large tillers are stronger sinks for assimilates and nutrients than the small tillers, and these substances are translocated to small tillers only at a very early stage of growth (Loomis, 1945). Thus, the growth of small tillers is often limited by nutrients or photosynthate.

5

REPRODUCTIVE GROWTH

In the life of a grass tiller no developmental event is more important than the conversion of the vegetative apical meristem to a reproductive meristem. This normally irreversible change in the apex results in the initiation of the inflorescence. Inflorescence development soon follows, including the determination of the number of caryopses in the inflorescence. Finally, caryopsis growth, development, and maturation occur. It is during the latter period that caryopsis size is determined.

PANICLE INITIATION

Grasses are among the most adaptable and most successful plant families. Ecotypes of some species are found over wide areas across which climate varies dramatically. A flexible reproductive process has evolved that allows ecotypes to maintain synchrony with the environment. In most grass species photoperiod plays an important role in the control of panicle initiation and flowering (Evans, 1964b).

Photoperiod

W. W. Garner and H. A. Allard, plant physiologists with the U.S. Department of Agriculture, are generally credited with the discovery of photoperiodism (Garner and Allard, 1920, 1923). They and numerous subsequent investigators have shown that "short-day" plants are induced to flower when the length of the night exceeds some minimum value. "Long-day" plants are induced to flower when the length of the night is less than some maximum value. Flowering of "day-length-neutral" plants is unaffected by day length.

Short-day plants and long-day plants may be either obligate or quantitative. Obligate short-day plants have an absolute requirement for appropriate long nights in order to flower. Flowering of quantitative short-day plants is hastened by long nights but does not absolutely require long nights.

All members of the Festucoideae that have been carefully studied are either day-length-neutral or long-day plants. In contrast, the vast majority of members of the Panicoideae, Eragrostoideae, and Oryzoideae are either indifferent to day length, are short-day plants, or flower only at intermediate day lengths (Evans, 1964b; Purohit and Tregunna, 1974).

Phytochrome

One of the well-known characteristics of the short-day (long-night) response is that induction of flowering can be prevented or delayed by interrupting the dark period for a few minutes with relatively low-intensity light. Red light is more effective than other wavelengths in preventing or delaying flowering, and its effects can be partially or wholly reversed by a short period of far red light following the red light. Thus, the response of short- (and long-) day plants to photoperiod appears to be mediated by a classical phytochrome response.

One puzzling aspect of the short-day response in C_4 grasses is their tendency to flower in the late summer or autumn but not in the spring when the photoperiod is similar. "Ripeness to flower" may play a role in this response; however, some evidence suggests that an unusual phytochrome response may also play a role. For example, Gorski (1980) has pointed out that the ratio of far red to red radiation at dusk is higher in spring than in autumn, and far red light at the beginning of the dark period may inhibit flowering in short-day plants (Kadman-Zahavi, 1977). The effects of day length on flower induction and development in several C_4 grasses are described below.

Maize

Garner and Allard (1923) first reported that short days hasten the flowering of maize. Much subsequent work (Francis, 1969, 1973; Palmer, 1973; Rood and Major, 1980; Kiniry et al., 1983a) has shown that maize genotypes differ in their response to photoperiod. When many tropical maize cultivars (which were selected under relatively short day lengths) are grown in higher latitudes, their flowering is delayed. They remain vegetative for a longer time and produce taller plants with more leaves. Conversely, many cultivars selected under the long-day conditions of the temperate summer mature too rapidly in the tropics. They do not attain their normal plant height or number of leaves. Despite the differences between normal temperate and tropical varieties, some maize cultivars are insensitive to day length. For example, in Colombia, Francis (1973) used incandescent lighting to reduce the night period from about 12 to 7 hr. The time from plant emergence to tassel initiation was almost unaffected in some cultivars but was reduced by more than half in others.

Since the time of tassel initiation cannot be determined without dissecting the apical meristem, a tedious and destructive procedure, Major (1980) has described the sensitivity of flowering to photoperiod in terms of the relationship between photoperiod and time from emergence to anthesis (Fig. 5.1). The "basic vegetative phase" is the minimum time from emergence to anthesis under very short days. The "maximum optimum photoperiod" is the longest photoperiod at which no photoperiod delay of anthesis is observed. "Photoperiod sensitivity" is

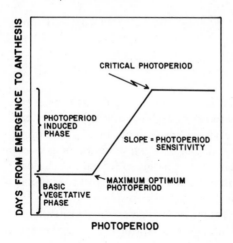

FIG. 5.1. General relationship between photoperiod and period from plant emergence to anthesis in short-day plants. From Major (1980). Reproduced from *Crop Science*, by permission of the Crop Science Society of America.

the delay in flowering per unit increase of photoperiod above the maximum optimum photoperiod (days/hr). The "critical photoperiod" is the photoperiod above which no further increase in the time from emergence to anthesis is observed. The "photoperiod-induced phase" is the difference between the basic vegetative phase and the time from emergence to anthesis at some photoperiod. Rood and Major (1980) found that in some photoperiod-sensitive maize cultivars, varying the photoperiod from 14 to 24 hr does not affect the period from anthesis to silking (Fig. 5.2). In some extremely photoperiod-sensitive cultivars the maximum optimum photoperiod is less than 14 hr and the critical photoperiod is 24 hr. In these varieties the time from emergence to anthesis steadily increases as photoperiod increases from 14 to 24 hr. In intermediate varieties the

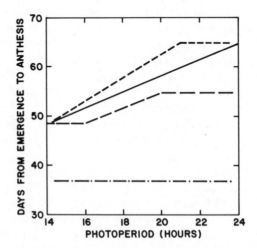

FIG. 5.2. Variation in the response of four maize cultivars to photoperiod. From Rood and Major (1980). Reproduced from *Crop Science*, by permission of the Crop Science Society of America.

maximum optimum photoperiod is greater than 14 hr and the critical photoperiod is less than 24 hr. In these varieties photoperiod sensitivity ranges from 0.0 to 2.4 days/hr increase in photoperiod. There is no correlation between photoperiod sensitivity, the basic vegetative phase, maximum optimum photoperiod, and critical photoperiod suggesting that these characteristics are under separate genetic control (Major, 1980; Rood and Major, 1980).

The intervals from plant emergence to tassel initiation and from tassel initiation to anthesis are strongly affected by temperature as well as by photoperiod. At constant photoperiod, the times from emergence to tassel initiation and from tassel initiation to anthesis are inversely related to temperature (see Chapter 7). Thus, photoperiod sensitivity should be expressed in degree-days rather than calendar days.

Kiniry et al. (1983a) modified Major's (1980) phases by assuming that the vegetative phase ends at tassel initiation instead of anthesis. This system defines the basic vegetative phase (BVP) as the interval from seedling emergence to tassel initiation under short photoperiods. The threshold photoperiod (TP) is the minimum photoperiod at which day length begins to cause a delay in tassel initiation, and photoperiod sensitivity (PS) is the additional thermal time needed for tassel initiation per hour increase in photoperiod (degree-days/hr). Kiniry et al. (1983a) found that BVP, TP, and PS are all genotype-specific constants. In early-maturing genotypes BVP ranges from 139 to 229 degree-days, TP varies from less than 10.0 to 13.5 hr, and PS ranges from 0 to 16 degree-days/hr. In a late-maturing group BVP ranges from 251 to 344, TP varies from less than 10.0 to 12.4, and PS ranges from 9 to 36.

Kiniry et al. (1983b) further divided the BVP into a juvenile phase in which the plant is insensitive to photoperiod and an inductive phase in which the plant still produces leaf primordia but is sensitive to photoperiod and becomes capable of tassel initiation. They found that maize cultivars remain insensitive to photoperiod until 4–8 days prior to tassel initiation; therefore, the juvenile phase is normally 4–8 days shorter than the BVP. Tollenaar and Hunter (1983) obtained similar results.

Grain Sorghum

Numerous studies (Gerik and Miller, 1984; Major, 1980; Quinby, 1966, 1967, 1972) have demonstrated that grain sorghum is a typical short-day plant in which the time from emergence to anthesis is affected by both photoperiod and temperature. However, one of the most important achievements of grain sorghum breeders has been the reduction of photoperiod sensitivity in cultivars adapted to temperate regions with long days during the growing season. This allows these cultivars to be planted over a wide range of latitudes and planting dates without strongly affecting their phenological development.

When photoperiod-sensitive cultivars are grown at a mean temperature of about 27°C, the basic vegetative phase ends at about 15 days or when the plant has about five leaves (Caddel and Weibel, 1972). Five short days are sufficient to induce floral initiation. Four maturity genes control the sensitivity of grain sorghum to photoperiod (Quinby, 1966; Quinby and Karper, 1945; Sorrells and Myers, 1982), and these genes apparently control the different characteristics described by Major (1980).

One of the interesting aspects of grain sorghum development is that between about 28 and 22°C leaf number is controlled almost entirely by photoperiod. However, at warmer (31°C) and cooler (14°C) temperatures floral initiation is inhibited to a greater extent than leaf initiation. As a result, at these extreme temperatures grain sorghum produces more leaves than under intermediate temperatures, even at the same photoperiod (Gerik and Miller, 1984; Quinby et al., 1973).

Pearl Millet

Pearl millet (*Pennisetum americanum*) is another typical short-day plant (Begg and Burton, 1973; Burton, 1965; Hellmers and Burton, 1972). Burton (1965) found that over 80% of the introductions from near its center of origin in West Africa are quantitative short-day plants. The remainder are obligate short-day plants. At the four-leaf stage as few as 4 short days reduce the time to floral initiation and anthesis, but 12 short days are needed to minimize the time to anthesis and the number of leaves per plant (Hellmers and Burton, 1972; Ong and Everard, 1979).

Sugarcane

In contrast to maize, grain sorghum, and pearl millet, flowering of commercial sugarcane is detrimental. Flowering tillers either die and do not contribute to sugar yield or their juice quality is poor. However, advances in sugarcane breeding depend on producing synchronous flowering in diverse genotypes so that crosses can be made.

In some respects the flowering of sugarcane resembles that of typical short-day C_4 grasses. For example, flowering can be prevented or delayed by interrupting or extending normally inductive night lengths with artificial light (Julien and Soopramanien, 1975; Midmore, 1980; Moore and Heinz, 1971). In addition, the delay in flowering caused by artificial light is proportional to the distance from the light source (Francis, 1973; Moore and Heinz, 1971), and red light is effective while far red light is ineffective in delaying or preventing flowering (Julien and Soopramanien, 1975).

However, environmental control of sugarcane flowering is more complex than normal short-day plants. Julien and Soopramanien (1975) divided the flowering of sugarcane into the following stages: panicle induction, initiation of the inflorescence axis primordium, initiation of spikelet primordia, and elongation of the inflorescence. In contrast to typical short-day plants, day length and other environmental factors continue to affect flowering after panicle induction. In fact, during the initiation of axis and branch primordia day length extension and night interruptions cause delays in inflorescence emergence and can even cause primordia to revert to a vegetative state (Julien and Soopramanien, 1975). This causes abnormalities in the vegetative apex that result in "zigzag nodes" in which several deformed internodes without axillary buds are formed but the original apex reverts to the vegetative condition. "Witches broom" or "multiple top" are abnormalities in which the original apex does not produce a panicle but dies, and axillary buds assume dominance (Clements and Awada, 1967; Dillewijn, 1952). After initiation of spikelet primordia day length extension or night interruption can delay flowering and result in abnormal panicle development (Clements and Awada, 1967; Julien and Soopramanien, 1975).

In contrast to typical short-day plants, in sugarcane far red light does not reverse the effects of night interruption by red light (Coleman, 1969; Julien and Soopramanien, 1975). In addition, induction of flowering in sugarcane normally occurs only within a narrow range of day lengths. In Hawaii, night periods of 11.5–12.0 hr are required (Clements, 1968; Moore and Heinz, 1971). However, induction only occurs in the autumn, suggesting that lower spring temperatures, increasing rather than decreasing day length, or a difference in the ratio of far red to red radiation at dusk inhibits panicle induction and development in the spring. Midmore (1980) has explained the apparent requirement for shortening day lengths by proposing that the day length requirement for induction is approximately 12 hr, while inflorescence differentiation and inflorescence elongation are stimulated by day lengths of about 11.75 and 11.5 hr, respectively. Therefore, in the spring induction may occur but subsequent differentiation of the inflorescence is inhibited by increasing day lengths.

From a practical standpoint sugarcane breeders can produce wide variation in flowering time by using night interruption with a gradient of light intensity. Thus, early-flowering genotypes will be delayed near the light source and their time of anthesis will overlap with that of late-flowering genotypes far from the light source.

Other C₄ Grasses

As previously mentioned, the vast majority of C_4 grasses are short-day, intermediate-day, or day-length-indifferent plants (Evans, 1964b; Purohit and Tregunna, 1974). Many species with wide geographical distributions have developed ecotypes with different photoperiod requirements. For example, Olmsted (1944) found that when ecotypes of sideoats grama (*Bouteloua curtipendula*) are grown at the same location, ecotypes from lower latitudes in North America (29°–32° N) flower under shorter day lengths than those from intermediate or high (42°–46° N) latitudes. Similarly, ecotypes of speargrass (tanglehead) (*Heteropogon contortus*) from Australia and Africa vary in their response to day length (Tothill and Knox, 1968), and races of kangaroograss (*Themeda australis*) originating from 6° S in New Guinea to 43° S in Tasmania differ markedly in their adaptation to photoperiod. Races from lower latitudes are strict short-day plants; races from higher latitudes flower under longer photoperiods, though they may simply be short-day plants with a long minimum photoperiod (Evans and Knox, 1969). Races of *T. australis* from dry inland areas of Australia tend to be indifferent to day length, as might be expected in areas of unreliable rainfall (Evans and Knox, 1969).

In several C_4 grasses photoperiod also affects the percentages of sexual and apomictic seed produced by the plant. Short days increase the percentage of aposporous embryo sacs in angleton bluestem (*Dichanthium aristatum*) (Knox and Heslop-Harrison, 1963) and in kangaroograss (*Themeda australis*) (Evans and Knox, 1969). As days become shorter in autumn, the percentage of apomictic seed increases. This probably represents a compromise between the advantages of sexual reproduction in maintaining genetic variation in the population and the more certain apomictic seed production as environmental conditions deteriorate under short days.

PANICLE DEVELOPMENT

The development and structure of the inflorescence of C_4 grasses has been described in numerous studies (Damptey and Aspinall, 1976; Metcalfe et al., 1975; Moncur, 1981). During the vegetative phase, the apical meristem is usually small, ranging from about 0.05 to 0.1 mm in many C_4 forage grasses (Fig. 5.3) (Moncur, 1981). The transition from vegetative to reproductive growth is signaled by a rapid enlargement and elongation of the apical meristem. Swellings

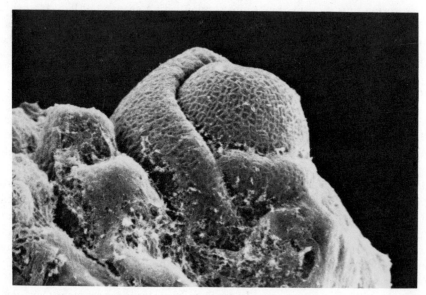

FIG. 5.3. Vegetative apical meristem of *Sorghum bicolor* (grain sorghum). Reproduced from Moncur (1981), by permission.

soon appear at the base of the meristem (Fig. 5.4). These are the primordia of the primary lateral branches. The primary branches initiate acropetally in a spiral pattern and soon cover the apical meristem (Fig. 5.5). If the species has secondary panicle branches (e.g., grain sorghum), these appear acropetally on the sides (distichously) of the primary branch primordia, and spikelets appear as bumps on the secondary branch primordia (Figs. 5.6 and 5.7). If the inflorescence is a false spike or contracted panicle (e.g., pearl millet) or an inflorescence without secondary branches [e.g., finger millet, (*Elusine coracana*)], spikelets appear as bumps on the primary branch primordia (Figs. 5.8 and 5.9). In species such as grain sorghum, tertiary and quarternary branch primordia are commonly formed on the lower inflorescence branches, but they seldom occur on the upper branches. This suggests that higher-order branches also develop acropetally.

Spikelets are paired in the tribe Andropogoneae, but they occur singly in the tribe Paniceae and in the subfamily Eragrostoideae. In the subfamily Panicoideae (tribes Paniceae and Andropogoneae), spikelets have two florets, one fertile floret above and one reduced (neuter or staminate) floret below. In the C_4 subfamily Eragrostoideae (like the C_3 subfamily Festucoideae), spikelets have one to several florets. Reduced florets, when present, are usually above the perfect florets. On an inflorescence branch or axis, spikelets may develop basipe-

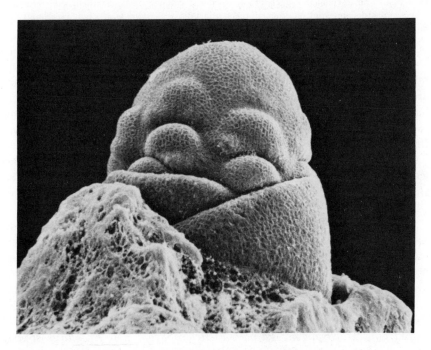

FIG. 5.4. Initiation of primary panicle brach primordia at the base of the apical meristem of *Sorghum bicolor* (grain sorghum). Reproduced from Moncur (1981), by permission.

FIG. 5.5. Primary panicle branch primordia almost covering apical meristem with secondary branch primordia appearing on primary branch primordia of *Sorghum bicolor* (grain sorghum). Reproduced from Moncur (1981), by permission.

tally, acropetally, or from the center. The first parts of the spikelets to develop are the glumes, which usually appear as two ridges, one slightly above the other, half encircling the spikelet. The lemma and palea are the first parts of the floret to develop. They are usually followed by the lodicules, then the stamens, and finally the pistil.

After florets are initiated, the panicle branches elongate rapidly in species with open panicles (Fig. 5.10). This, combined with the rapid elongation of the stem internodes below the panicle, pushes the inflorescence out of the protective whorl of leaves.

Gould (1968) points out that grass inflorescences are actually cymes because they are determinate with the terminal spikelet maturing first. However, they are usually described as panicles, racemes, or spikes. Panicles are all inflorescences in which the spikelets are not borne on the main axis but occur on branches. *Eragrostis*, *Panicum*, *Saccharum*, and *Sorghum* species have paniculate inflorescences (Fig. 5.11). Racemes are inflorescences in which some or all of the spikelets are borne on pedicels attached directly to the main inflorescence axis. *Tripsacum*, *Trachypogon*, *Paspalum*, *Heteropogon*, and *Elyonurus* species have spikelike racemes (Fig. 5.12). Spikes are inflorescences in which all spikelets are sessile on the main axis. Primary inflorescence branches are sometimes

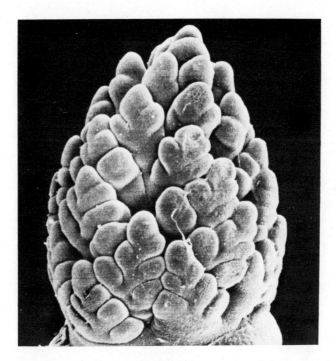

FIG. 5.6. Further development of secondary panicle branch primordia in *Sorghum bicolor* (grain sorghum). Reproduced from Moncur (1981), by permission.

FIG. 5.7. Spikelet primordia developing in *Sorghum bicolor* (grain sorghum). Reproduced from Moncur (1981), by permission.

FIG. 5.8. Spikelet primordia and bristle primordia developing on primary branches in *Pennisetum americanum* (pearl millet). Reproduced from Moncur (1981), by permission.

FIG. 5.9. Spikelet and glume primordia on the digitate inflorescence of *Elusine coracana* (finger millet). Reproduced from Moncur (1981), by permission.

FIG. 5.10. Initial phase of panicle branch elongation showing undifferentiated florets and stamens of *Eragrostis tef* (tef). Reproduced from Moncur (1981), by permission.

spicate. *Chloris* and *Hilaria* species have spicate inflorescences or inflorescence branches (Fig. 5.13) (Allred, 1982).

Differences in the form of grass inflorescences are primarily due to the number of times branching occurs before spikelet differentiation and to the relative lengths of internodes of the various branches. For example, the maize ear has resulted from the telescoping of the primary inflorescence axis plus adnation of prophylls on secondary axes to the primary axis (Nickerson, 1954).

BREEDING SYSTEMS

Connor (1979) has recently reviewed breeding systems in grasses, and Burson (1980) has reviewed breeding systems of C_4 grasses. This section gives examples of some of the variation in C_4 grass breeding systems.

The majority of C_4 grasses are sexual, hermaphroditic (male and female organs in one floret), and predominantly cross-pollinated by the wind. Several mechanisms, including self-incompatibility, monoecy, dioecy, and protogyny, promote cross-pollination.

Self-incompatibility

Self-incompatibility, the complete or near failure to set seed after self-fertilization, occurs in many genera. Many species are not completely self-incompatible, but they have poor self-fertility, and obvious inbreeding depression occurs. Self-

FIG. 5.11. Plant and spikelet of *Panicum coloratum*. From Gould (1978). Reproduced courtesy of Mrs. F. W. Gould and Texas A&M University Press.

FIG. 5.12.   Raceme and pedicellate spikelet of dallisgrass (*Paspalum dilatatum*). From Gould and Box (1965). Reproduced courtesy of Mrs. F. W. Gould.

incompatibility has been reported in numerous C_4 genera, including *Andropogon*, *Bouteloua*, *Brachiaria*, *Chloris*, *Cynodon*, *Echinochloa*, *Eragrostis*, *Paspalum*, *Pennisetum*, *Saccharum*, *Setaria*, *Sorghastrum*, *Sorghum*, and others.

Monoecism

In monoecious plants the male and female organs occur in different inflorescences (e.g., *Tripsacum* and *Coix*), and in different florets of the same spikelet (e.g., *Lophopogon*). Maize bears staminate spikelets in the apical inflorescence (tassel) and pistillate spikelets in the axillary inflorescences (earshoots). The inflorescence of eastern gamagrass (*Tripsacum dactyloides*) has one or several spicate racemes with staminate spikelets above and pistillate spikelets below.

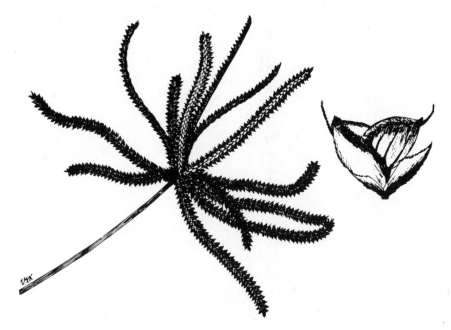

FIG. 5.13.   Spikate inflorescence of *Chloris cuculata*. From Gould and Box (1965). Reproduced courtesy of Mrs. F. W. Gould.

Dioecism

Dioecious species are those in which male and female florets are borne on different plants. Among C_4 grasses it is reported in only a few genera of the tribes Chlorideae and Paniceae. Male buffalograss (*Buchloe dactyloides*) plants bear staminate spikelets on slender, erect inflorescences that rise above the leafy portion of the plant. Female plants have three to five pistillate spikelets in a burlike cluster hidden among the leaves. Similar dimorphism occurs in *Buchlomimus* and *Jouvea*. *Spinifex* and *Zygochloa*, members of the Paniceae, have differently shaped male and female panicles.

Protogyny

Protogyny (exertion of the styles one or more days prior to exertion of the anthers) discourages self-pollination by separating the pollen from the receptive stigma in time rather than space. Among C_4 grasses, protogyny occurs in pearl millet, sugarcane, kikuyugrass (*Pennisetum clandestinum*), *Zoysia* species, and the normally apomictic buffelgrass (*Cenchrus ciliaris*) and *C. setigerus*.

Though most C_4 grasses are cross-pollinated, self-pollination is quite common. Self-compatibility may be advantageous in wind-pollinated species where pollen viability is often measured in minutes and the probability of cross-pollination declines rapidly with increasing distance between plants. Thus, even predominantly cross-pollinated species are usually capable of producing a small percentage of selfed seed. In addition, cleistogamous species find it advantageous to minimize cross-pollination.

Cleistogamy

Cleistogamous grasses are those in which the florets never open and anthers are not exerted, which guarantees self-fertilization within the unopened spikelet. Faulty exertion of anthers can be due to nonfunctional lodicules or to confinement of the floret by glumes or vegetative sheaths. In addition, anthers and other floral parts may be reduced in cleistogamous species.

Most cleistogamous species also contain chasogamous florets (florets which open and exert the anthers) (Connor, 1979), and the percentage of cleistogamous florets can vary with photoperiod or soil moisture (Campbell, 1982; Connor, 1979).

Self-fertilization imposed by cleistogamy may be advantageous in weedy species. First, self-compatibility permits establishment of new populations following long-distance dispersal of only one plant. Second, selfing increases homozygosity, which may improve adaptation to the local environment (Jain, 1976). Both these advantages appear to function in the broomsedge bluestem (*Andropogon virginicus*) complex, where taxa with the highest percentage of cleistogamous flowers are the most successful colonizers of disturbed habitats (Campbell, 1982).

Though selfing can enhance the seed set of isolated plants, it can also cause inbreeding depression. One means of avoiding this is to produce highly heterozygous seed by asexual means.

Apomixis

Asexual seed development (apomixis) is an important method of reproduction in several important C_4 forage grasses (Bashaw, 1975, 1980). Obligate apomicts are plants that reproduce only by apomixis. Normally, guineagrass (*Panicum maximum*), buffelgrass (*Cenchrus ciliaris*), weeping lovegrass (*Eragrostis curvula*), tetraploid bahiagrass (*Paspalum notatum*), and some varieties of dallisgrass (*Paspalum dilatatum*) are obligate apomicts, although some sexual plants have been found in all these species. Facultative apomicts are plants capable of both sexual and asexual reproduction. For example, angleton bluestem (*Dichanthium aristatum*) and kangaroograss (*Themeda australis*) are facultative apomicts in which the percentage of apomictic seed varies with day length (Evans and Knox, 1969; Knox and Heslop-Harrison, 1963). In guineagrass different levels of apomixis occur in different populations, which vary from nearly obligate apomicts to populations producing over 50% sexual seed (Usberti and Jain, 1978a).

Manipulation of apomixis has been used to produce improved cultivars of buffelgrass (Bashaw, 1975, 1980). In this species a rare sexual plant has been used as the female in crosses with obligate apomicts. Since apomixis is genetically controlled, obligate apomictic progeny with desirable forage characteristics could be recovered. This led to "instant" apomictic cultivars that could be immediately released to ranchers. This process of obtaining new grass cultivars is currently being applied to other species in which sexual and apomictic plants are capable of hybridization.

FLOWERING AND FERTILIZATION

The following descriptions of flowering and fertilization illustrate the wide range of sexual behavior found in important C_4 grasses.

Grain Sorghum

Sorghum bicolor is primarily self-fertilized with cross-pollination averaging about 6% (Schertz and Dalton, 1980). The inflorescence is a compact-to-open panicle with spikelets borne in pairs, the lower being sessile and fertile and the upper being pedicellate and either male or sterile. Anthesis begins near the tip of the panicle 0–3 days after its emergence from the boot. Flowering proceeds basipetally for 4–7 days. Stigmas are receptive up to 2 days before anthesis and if unfertilized remain receptive for 5–16 days after anthesis. Pollen is usually shed in the morning, but cool and wet conditions delay flowering. In the field, pollen viability is normally high for about 30 min then declines to near zero after 4 hr. Pollen germinates when it contacts the receptive stigma, and fertilization usually occurs in about 2 hr (Schertz and Dalton, 1980).

Pearl Millet

Pennisetum americanum is primarily cross-pollinated. The inflorescence (head) is a false spike (tightly contracted panicle) ranging from 5 to more than 150 cm in length and from 1 to 5 cm in diameter. Spikelets occur in pairs, each with a sessile male floret and a pedicellate perfect floret. Pearl millet is protogynous (i.e., styles are exerted several days before the anthers). Exertion of styles usually occurs soon after the inflorescence appears from the whorl. Exertion is basipetal and occurs over about 3 days (more in very long inflorescences). Anthesis begins at least 1 day after most styles are exerted, and it proceeds basipetally. Perfect florets exert stamens 2 or 3 days prior to staminate florets. Pollen is usually shed from an inflorescence for 4–6 days. Most pollen is shed in the morning, but it is delayed by cool, moist conditions. Fertilization occurs within a few hours after pollination.

Protogyny facilitates natural cross-pollination of pearl millet. However, cross-pollination is not normally complete because tillers on the same plant flower in succession, and self-pollination can occur between inflorescences of the same plant. In the absence of cross-pollination stigmas remain receptive until anthesis and self-fertilization can occur (Burton, 1980).

Maize

Zea mays is monoecious. Male florets are located in the apical inflorescence or tassel, and female florets are found in the axillary earshoots. Each male spikelet has two florets. During anthesis the anthers of the upper floret are exerted first; those of the lower floret are exerted later the same day or on following days. Pollen is shed from the tassel from one day to more than a week. In the midwestern United States pollen dispersal usually occurs between sunrise and midday, though cool temperatures and high humidity may cause a delay until afternoon. Pollen remains viable for only a few minutes after it is shed (Russell and Hallauer, 1980).

Maize plants produce several earshoots. The top earshoot of commercial cultivars grown in the United States is at the sixth or seventh internode below the tassel. Though axillary buds are present at each node below the top earshoot, only the top two or three earshoots elongate and produce functional stigmas or silks. One silk is produced for each potential kernel. They usually emerge first from the top earshoot; they emerge from the second earshoot a few hours or days later. The first silks to appear are from the basal portion of the ear and up to 6 days may be required for all silks to emerge. Silking usually occurs 1–3 days after anthesis, though environmental stresses may delay silking even more. Silks are receptive to pollen as soon as they emerge from the husk. If unfertilized, they remain receptive for about 10 days, though high temperature and low humidity may reduce this interval (Russell and Hallauer, 1980).

Prolific cultivars are those with a tendency to produce more than one ear per plant. These cultivars often have poorer synchronization between anthesis and silking than nonprolific cultivars (Russell and Hallauer, 1980).

Sugarcane

Commercial sugarcane (*Saccharum* spp. hybrid) is a vegetatively propagated hybrid of several species, including *S. officinarum*, *S. spontaneum*, and *S. robustum*. Under natural field conditions survival of seedlings is very rare, but sugarcane breeders have produced extraordinary systems to synchronize flowering and produce hybrids of widely divergent genotypes (James, 1980).

The inflorescence (arrow or tassel) of sugarcane is an open panicle with a main axis and first-, second-, and third-order laterals. Branching is greatest at the base and decreases toward the apex. Spikelets occur in pairs with the lower spikelet sessile and the upper pedicellate. Long, silky callus hairs are attached at the base of each spikelet. Anthesis usually begins 0–3 days after panicle emergence near the apex of the panicle. It proceeds basipetally for 3–14 days. Stigmas are receptive as soon as the florets open. Anther dehiscence and pollen viability range from 0 to 100%, depending on the genotype. To avoid self-pollination,

breeders normally use male-sterile genotypes as the female parent. Pollen germinates immediately upon contact with the receptive stigma and fertilization occurs within 8 hr (James, 1980).

C₄ Forage Grasses

Burson (1980) reviewed the literature on flowering of warm-season forage
grasses. He found that in most species anthesis occurs prior to midday; however,
in a few species [e.g., guineagrass (*Panicum maximum*)], it occurs during the
night. In general, cool temperatures and high humidity delay pollen shed. The
stigmas are usually considered receptive as soon as they are exerted; but if they
are not fertilized, they often remain receptive for several days. Under natural
conditions pollen remains viable for only a few minutes or at most a few hours.
The time from pollination to fertilization ranges from about 30 min to 18 hr.

CARYOPSIS DEVELOPMENT

The reproductive structures of grasses are extremely variable; however, this variation is based on a common structure, the caryopsis. The caryopsis is a dry,
hard, indehiscent, one-seeded fruit (mature ovary) in which the pericarp (ovary
wall) adheres to the seed coat (inner integument of the ovule). In only a few
genera (e.g., *Sporobolus* and *Elusine*) do the seed coat and pericarp remain separate. In these the structure may be classified as an utricle, a small bladdery one-
seeded fruit (Gould, 1968).

The rate of grass seed development after fertilization is a strong function of
temperature, and degree-days can be used to estimate the date of physiological
maturity. The time from fertilization to maturity is highly species and genotype
dependent. Small-seeded forage grasses such as weeping lovegrass (*Eragrostis
curvula*), blue panic (*Panicum antidotale*), and dallisgrass (*Paspalum dilatatum*) may mature in 8–14 days under normal conditions (Burson, 1980). Sugarcane seeds normally require a minimum of 22 days (James, 1980). Grain sorghum normally matures 35–45 days after pollination, and maize often requires
more than 60 days (Russell and Hallauer, 1980; Schertz and Dalton, 1980).

The reproductive structure that falls from the inflorescence also varies greatly
among grasses. In the C₄ subfamily Eragrostoideae and C₃ subfamilies Festucoideae, Bambusoideae, Oryzoideae, and Arundinoideae, spikelets normally
disarticulate above the glumes. In the C₄ subfamily Panicoideae disarticulation
is usually below the glumes. In the Panicoideae spikelets may fall separately, as
in *Panicum*, *Paspalum*, and most other genera. In contrast, they may be borne
singly or in clusters of several spikelets subtended by an involucre of bristles or
spines as in *Cenchrus* (Fig. 5.14) and *Pennisetum*. In this case disarticulation is
below the involucre, and the entire structure falls as a whole. In some species of
Eragrostis and *Aristida* and in tumblegrass (*Schedonnardus paniculatus*), the
entire inflorescence breaks off at the base and can act as a tumbleweed for more

FIG. 5.14. Plant, bur, and spikelet of *Cenchrus incertus*. From Gould and Box (1965). Reproduced courtesy of Mrs. F. W. Gould.

effective seed dispersal. Finally, selection by man has suppressed disarticulation in C$_4$ grasses such as maize, grain sorghum, and the millets (Gould, 1968).

Seed shattering is a major problem for commercial production of grass seed. For example, bulk-harvested seed of some C$_4$ grasses seldom exceeds 20% of the seed set (Holt and Bashaw, 1963; Humphreys, 1975). Disarticulation of the seed generally occurs at an abscission layer. In bahiagrass (*Paspalum notatum*) and dallisgrass (*P. dilatatum*), it is a layer of thick-walled cells several cells thick in the pedicel below the glumes. Several days after anthesis these cells elongate, then collapse. Spikelets remain attached to the pedicels by the epidermal cells until abscission after physiological maturity (Burson et al., 1978). In some cases spraying panicles with auxin can reduce shattering without reducing germination. This could greatly aid seed production in species such as guineagrass (*Panicum maximum*) (Weiser et al., 1979).

Details of spikelet and floret morphology in the most important subfamilies and tribes are given in Chapter 3. The following sections summarize the growth and development of maize and grain sorghum caryopses.

Maize

Zea mays fruit (grain) development is atypical because maize is a monoecious species with staminate florets in a terminal panicle (tassel) and with pistillate florets in axillary spikelike inflorescences (earshoots). However, because of its economic importance, the growth of the maize grain has been studied in greater detail than that of other C_4 grasses, and certain principles that have been established for the development of maize inflorescences probably apply to those of other more typical species. Much of this discussion is taken from Tollenaar's (1977) review.

Tassel initiation in maize is similar to panicle initiation in other grasses. In addition, the first stages of spikelet development are similar in the tassel and the earshoot. However, as the florets begin to develop, the pistils of the staminate florets (on the tassel), and the stamens of the pistillate florets (on the ear) cease differentiation (Bonnett, 1966; Fuchs, 1968). The topmost earshoot begins to elongate and differentiate into a reproductive structure at the same time that the floret primordia begin to differentiate on the tassel. Other earshoots begin to develop in a basipetal sequence (Bonnett, 1966; Fuchs, 1968).

After earshoot development begins, spikelet primordia are initiated in an acropetal direction, and primordia initiation continues until 1–2 weeks prior to silking. The cessation of spikelet initiation occurs first in the lowest earshoots, then proceeds acropetally. Thus, the longest duration of spikelet initiation occurs in the top earshoot. Spikelet initiation is less sensitive to environmental factors than are other processes such as plant growth rate and leaf expansion (see Tollenaar, 1977). This may be due to the proximity of earshoots to large pools of carbohydrate in the stem internodes and their protected location in the leaf axil, far from the desiccating influence of the atmosphere.

Elongation of the styles (silks) of the basal florets normally begins 10–15 days before silking (Sass and Loeffel, 1959). The silks of the apical florets begin to grow later than those of the basal florets, and the basal florets are the first to be fertilized (Peterson, 1942).

Under normal environmental conditions, silks are extruded from the ear at the same time that anthesis occurs. However, under adverse environmental conditions (drought or very low light intensities) or when high populations reduce light interception per plant, silk extrusion is often delayed. This can result in incomplete fertilization and even ear abortion.

Although six to eight axillary buds are transformed to earshoots at about the time of tassel initiation, all but the top two or three cease development prior to silking. Genotypes vary in their tendency to produce more than one ear per stalk. Prolific genotypes often produce two or three ears, while nonprolific genotypes rarely do so under normal cultural conditions. In addition to genetic factors, environmental conditions just prior to and just following anthesis are critical in determining the number of earshoots that become "functional" and produce mature grain (Prine, 1971; Earley et al., 1974). Earshoots become functional in a basipetal sequence, and when upper earshoots become functional, they inhibit the development of lower earshoots, probably hormonally.

Functional earshoots of prolific plants usually silk within 1 or 2 days of each other while any nonfunctional earshoots that silk do so 2–4 days later than the upper earshoot. However, earlier pollination of the functional earshoots is not the reason that lower nonfunctional earshoots fail to develop. Simultaneous pollination of all silking earshoots, functional and nonfunctional, results in no more ears per plant than natural pollination (Earley et al., 1974). However, when the upper earshoot of prolific hybrids is fertilized after the lower earshoot, a higher percentage of two-eared plants is produced (Gallais et al., 1982).

Nonfunctional earshoots can be converted to functional earshoots by either removing upper functional earshoots or covering them to prevent pollination. Unfavorable environmental conditions can also delay silking of the functional second earshoot, cause incomplete fertilization, and cause the second earshoot to become nonfunctional (Earley et al., 1974; Sass, 1960).

The first nuclear divisions of the fertilized egg occur about 30 hr after fertilization (Peterson, 1942). Nuclei in the coenocytic embryo sac begin to divide. Three or four days later cell walls begin to form, and by about 15 days after pollination starch synthesis begins. The beginning of rapid starch synthesis in the grains marks the beginning of linear grain fill, the period during which grain weight increases rapidly and linearly with time. At about this time, 2–3 weeks after anthesis, an important regulatory event occurs: Final grain number is determined. Under normal conditions more florets are fertilized than ever develop into mature grains. At approximately the beginning of linear grain fill, a number of apical florets cease development. Several studies have shown that the number of grains that continues to develop is positively correlated with assimilate supply (Barnett and Pearce, 1983). Assimilate supply in the 2- to 3-week period after silking is correlated with kernel number (Frey, 1981). Sugar content of the stalk around silking is strongly correlated with the percentage of ears that have shriveled kernels near the apex (Iwata, 1975). Finally, the number of mature grains per plant is correlated with the rate of plant growth during the preflowering period (Hawkins and Cooper, 1981). All these studies suggest that carbohydrate availability around silking is an important factor affecting final kernel number; however, the mechanism involved is unclear. For example, Tollenaar and Daynard (1978) found only small differences in sugar content between growing and nongrowing kernels at the tips of ears. In addition, kernels at the tip of the ear lag 4–5 days behind kernels at the base of the ear from the start of silk growth to linear grain fill. Tip kernels also lag behind other kernels in the rate of dry matter accumulation, possibly due to the lower volume of tip kernels at the beginning of linear grain fill.

During the 2–3 weeks between silking and the beginning of linear grain fill, no large sinks for photosynthate exist. The leaves have stopped growing, and the kernels are still growing very slowly; therefore, during this period nonstructural carbohydrates usually accumulate in the stem and other organs (Daynard et al., 1969; Hume and Campbell, 1972).

The plant now begins the period of linear kernel dry weight accumulation. During this period more than 90% of kernel dry matter accumulates, most of it due to concurrent photosynthesis. The final weight per kernel is determined by the rate and duration of this accumulation, both of which are affected by genotype. During linear grain fill, the rate of kernel dry weight accumulation is not strongly affected by short-term (day-to-day) changes in photosynthetic rate, and many studies indicate that the stem serves as a major source of carbohydrates to overcome short-term decreases in photosynthesis. For example, when the supply of photosynthate is reduced by shading (Daynard et al., 1969), drought (Jurgens et al., 1978), or leaf removal (Hanway, 1969; Jones and Simmons, 1983), kernel growth continues as the result of translocation of carbohydrate from the stem. Conversely, if ears are removed or if fertilization is prevented, stem dry weight and internode carbohydrate content increase during the period that would normally correspond to linear grain fill (Hume and Campbell, 1972; Jones and Simmons, 1983). In addition to the stem, the cob, husks, and the base of the earshoot can also store nonstructural carbohydrates (Jain, 1971; Palmer et al., 1973).

Final kernel weight is approximately the product of the rate of grain fill during linear grain fill and the duration of linear grain fill. Both these parameters are under strong genetic control (Daynard and Kannenberg, 1976; Ottaviano and Camussi, 1981; Poneleit and Egli, 1979). Temperature also has an important role in regulating the duration of linear grain fill. The growing degree-days (thermal time) required to complete linear grain fill is genotype specific. For example, Carter and Poneleit (1973) found that the growing degree-days (base 10) during grain fill ranged from 512 to 821 among 20 inbred lines.

Recent work suggests that, other factors being equal, the duration of grain filling is an important determinant of grain yield. Castleberry et al. (1984) reported that since the 1930s the genetic yield potential of maize cultivars growing in the U.S. Corn Belt has been increasing about 1% per year. This is almost exactly the rate of increase in growing degree-days required during the grain filling period for maize grown in Indiana between 1951 and 1980 (McGarrahan and Dale, 1984). This suggests that the increase in grain filling period is an important factor contributing to progress in maize breeding.

The effect of temperature on rate of grain fill is less clear, though recent work (D. P. Knievel, personal communication) indicates that when ear temperature is varied independently of whole-plant temperature, the rate of grain fill is a linear function of ear temperature. The base temperature for grain filling appears to be near 0°C, and the optimum temperature is probably between 25 and 30°C. Similar results were obtained in a study of *in vitro* kernel development of maize (Jones et al., 1981), although the rate of kernel development at low (18°C) temperatures was much reduced in this study.

From the preceding discussion, it is clear that the sink size (number of kernels) of maize is determined shortly after silking. This sink is then filled almost completely during linear grain fill. In most cases at least 80% of the final kernel weight is composed of photosynthate fixed during grain fill (Evans, 1975b). An important question for plant breeders attempting to increase maize yields then becomes: Which limits maize yield, the size of the sink or of the source? Several studies in which the sizes of the source and the sink have been manipulated independently suggest that at most latitudes the sink limits final yield (Allison et al., 1975a; Earley et al., 1967; Prine, 1973). For example, Jones and Simmons (1983) found that removal of 20–40% of the kernels at 12 or 24 days after silking (thus increasing the ratio of source to sink) has no effect on the rate of filling of the remaining kernels.

Goldsworthy and Colegrove (1974) and Goldsworthy et al. (1974) reported that during the grain filling period of tropical maize dry matter continues to accumulate in the vegetative plant parts. This suggests that the sink is inadequate to fully utilize the available photosynthate.

Other studies indicate that source limitations may be more important at high latitudes near the limits of maize adaptation (Daynard et al., 1969; Hume and Campbell, 1972).

Actually, source and sink sizes may limit grain growth at different times during grain filling. Short-term source limitations do not usually affect the rate of kernel growth during the early part of grain fill because translocation of carbohydrates from the stem and cob compensates for any reduction in concurrent photosynthesis. This suggests that sink size normally limits the total rate of grain growth (per unit land area) early in the grain filling period. However, leaf disease, drought stress, or defoliation can severely limit photosynthesis during grain filling. This increases the translocation of dry matter from the vegetative parts to the grain (Jurgens et al., 1978; McPherson and Boyer, 1977), and ultimately grain growth is limited by the size of the source.

The source–sink ratio also affects maize photosynthesis and dry matter accumulation after silking (Allison and Watson, 1966; Barnett and Pearce, 1983; Tollenaar and Daynard, 1982). Decreasing the source by removal of leaves can increase the rate of photosynthesis per unit leaf area. Decreasing the sink by reducing fertilization of the ovules or by removing the ears reduces postanthesis dry matter accumulation and leaf photosynthesis. In addition, leaf senescence can be accelerated by either assimilate starvation (low source–sink ratio) or by excessive assimilate accumulation (high source–sink ratio). This suggests that during grain filling a delicate source–sink balance exists. Disturbing it affects photosynthesis, dry matter accumulation, and leaf senescence.

In general, grain yield of maize is greater at lower (18–25°C) than at higher mean temperatures (25–30°C) because lower temperatures increase the length of the grain filling period proportionately more than they reduce the kernel growth rate (Hunter et al., 1977). This is consistent with the data cited above, where the base temperature for the rate of kernel maturation (during linear grain fill) is about 10°C, while the base temperature for the rate of kernel growth is about 0°C (Fig. 5.15). Thus at 18°C, the relative rate of kernel maturation is about 0.4 while the relative rate of kernel growth is about 0.6.

The relative importance of different leaves during grain filling has been studied by tracing the translocation of ¹⁴C fixed in various parts of the plant. In young maize plants all leaves export carbon to all other organs of the plant (Hofstra and Nelson, 1969). However, after silking the position of the leaf blade relative to the earshoot strongly influences translocation (Tripathy et al., 1972). At silking the ear leaf and those above it translocate ¹⁴C primarily to the earshoot. Leaves below the ear leaf translocate most ¹⁴C to the lower shoot and roots. Three weeks after silking (during linear grain fill) all leaves translocate ¹⁴C primarily to the ear (Eastin, 1969, 1970). However, even 2–3 weeks after silking, the translocation of ¹⁴C from the lower leaves to the ear is slower and less complete than from the upper leaves (Palmer et al., 1973). In tillered maize plants significant amounts of ¹⁴C are translocated from the tiller to the main stalk during grain fill when the main stalk has an ear but the tiller does not. Otherwise, little ¹⁴C is translocated from the tiller to the main stalk either before or after silking (Alofe and Schrader, 1975).

Physiological maturity of maize and other crops is defined as the time when maximum seed dry weight is attained (Aldrich, 1943; Shaw and Loomis, 1950). However, Daynard and Duncan (1969) have shown that a black layer (Johann, 1935; Kiesselbach and Walker, 1952) forms at the base of the kernel at approximately the time that maximum kernel dry weight is attained. The formation of

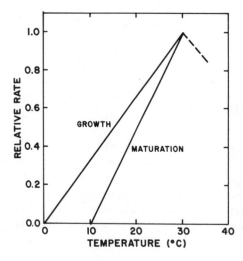

FIG. 5.15.   Effect of mean temperature on the relative rates of maize kernel growth and maturation.

this black layer prevents the translocation of further photosynthate into the kernel, and it provides a convenient indicator of physiological maturity.

The time of black layer formation is affected by both genotype and environment. As mentioned earlier, the duration of grain fill (and, therefore, the date of black layer formation) is a genotype-specific function of thermal time. Cultivars adapted to high latitudes generally require fewer growing degree-days from the beginning of linear grain fill to black layer formation than cultivars adapted to lower latitudes and longer growing seasons. Normally, black layer occurs at about the time that the endosperm completes its solidification and the "milk line" or boundary between the liquid and solid endosperm reaches the bottom of the kernel (Afuakwa and Crookston, 1984). However, stresses that inhibit photosynthesis or translocation of sucrose to the kernel can cause precocious black layer formation. For example, premature defoliation, cool temperatures, and drought stress can cause premature black layer formation, probably by inhibiting sucrose and hormone translocation to the grains (Afuakwa et al., 1984; Daynard, 1972; Tollenaar and Daynard, 1978).

Because of the variety of factors that can affect black layer formation and the cessation of grain growth, maximum dry matter accumulation can occur at a wide range of grain moisture contents (Daynard and Duncan, 1969; Rench and Shaw, 1971; Shaw and Thom, 1951), and black layer formation rather than grain moisture content is now universally used as an indicator of physiological maturity.

Grain Sorghum

Panicle and caryopsis development is probably more typical in grain sorghum than in maize. Grain sorghum, like other members of the Andropogoneae, produces paired spikelets. The lower spikelet is sessile and perfect and the upper is pedicellate and sterile. Each perfect spikelet contains two florets. The upper floret is fertile, and the lower is sterile and represented only by a lemma.

Three developmental events determine potential grain number (Lee et al., 1974). These are (1) the number of primary branch primordia that are initiated, (2) the number of higher-order branches initiated on each primary branch, and (3) the timing of spikelet differentiation, which causes branch growth to cease. If spikelet differentiation is delayed, branch growth can continue, which results in an increase in spikelet number. Both genetic and environmental factors influence final spikelet number, and stresses during panicle differentiation can severely reduce spikelet number.

As discussed previously, the initiation of branch primordia in the grain sorghum panicle is acropetal, but differentiation of spikelet primordia is basipetal. Thus, upper panicle branches have developed for a shorter time than lower branches when spikelet differentiation occurs and the branches cease developing. As a result, upper branches have fewer secondary branches and produce fewer spikelets than lower branches (Lee et al., 1974).

Under favorable field conditions spikelet abortion after anthesis is insignificant and is usually found on only the lowest branches. However, under unfavorable environmental conditions spikelet abortion can severely reduce grain number.

As in maize, concurrent photosynthesis is responsible for most of the carbon translocated to grain sorghum seeds. Some studies have found no significant retranslocation of dry matter from the stem to the grain (Krishnamurthy et al., 1973; Roy and Wright, 1973). Others have found that retranslocation is significant (Warsi and Wright, 1973; Jacques et al., 1975a; Fischer and Wilson, 1971a), and it often accounts for about 10-12% of the final grain weight (Chamberlin and Wilson, 1982; Fischer and Wilson, 1971a).

In grain sorghum the panicle branches and spikelets contain chlorophyll and make a significant contribution to photosynthesis. In several studies Fischer and his collaborators have shown that panicle photosynthesis accounts for 14-18% of grain yield, and each of the four top leaves contributes 13-25% (Fischer and Wilson, 1971b; Fischer et al., 1976).

In contrast to maize, grain sorghum grain yield is often source-limited (Fischer and Wilson, 1975; Munchow and Wilson, 1976; Pepper and Prine, 1972). However, grain yield of some tall genotypes is probably sink limited since stem weight continues to increase during grain fill (Shinde and Joshi, 1980).

Dry weight that accumulates in the grain sorghum kernel must pass through the placental area where the kernel is attached to the panicle branch. As in maize the placental area becomes darker at the same time that it becomes impermeable to photosynthate from the vegetative part of the plant (Eastin et al., 1973). Formation of this dark or black layer can be used as an indicator of physiological maturity (Eastin et al., 1973; Giles et al., 1975).

Black layer formation, like anthesis, progresses basipetally in the panicle. Several days are required for anthesis and black layer formation to progress from the apex to the base of the panicle. Therefore, when the dates of each are observed to determine the period of grain filling, the same portion of the panicle should always be used to eliminate sampling bias.

Grain sorghum kernel dry weight often declines slightly after reaching a maximum at physiological maturity (Collier, 1963; Kersting et al., 1961a). This decline continues until the moisture content of the kernel reaches 20-25%. The total decrease in kernel dry weight appears to be influenced both by genotype and by the time interval between physiological maturity and when the kernels reach 20-25% moisture (Fig. 5.16). If kernel drying is delayed, over 10% of the dry weight of the kernels can be lost (Kersting et al., 1961a,b).

During grain filling the carbohydrate composition of grain sorghum caryopses changes. Total sugars decrease from more than 5% shortly after anthesis to about 2% at maturity. Starch increases to a maximum of over 60% when maximum caryopsis dry weight is reached, then it often declines due to respiration loss until caryopsis drying inhibits respiration (Fig. 5.17) (Kersting et al., 1961b).

In "senescent" grain sorghum genotypes, senescence of the leaves on a tiller soon follows physiological maturity of the grain on that tiller. However, "non-

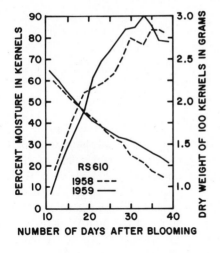

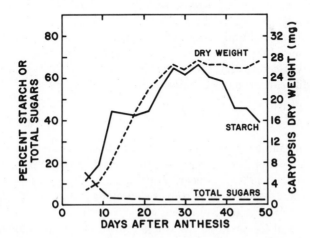

FIG. 5.16.   Decrease in kernel water content and increase in kernel dry weight during grain fill. Note that kernel dry weight decreases in 1959 from the time when maximum kernel dry weight is attained until kernel water content reaches about 25%. From Collier (1963). Reproduced from *Crop Science*, by permission of the Crop Science Society of America.

senescent" genotypes maintain green leaf area for some time after the grain matures. Nonsenescent genotypes can also accumulate starch and sucrose in the culms after physiological maturity of the grain, McBee et al. (1983) suggest that nonsenescent genotypes may be desirable since the carbohydrates accumulated after black layer formation might have other uses (e.g., as forage or biomass).

The tannin content of grain sorghum seed is an important factor affecting the relative susceptibility of cultivars to bird damage. High tannin contents make cultivars less attractive to birds; however, they also decrease digestibility. Thus, the ideal cultivars would have high tannin concentration in the early stages of grain growth when bird damage is most likely and low tannin at maturity. In

FIG. 5.17.   Caryopsis dry weight and percent total sugars and starch in grain sorghum caryopses. From Kersting et al. (1961a,b). Reproduced from *Agronomy Journal*, by permission of the American Society of Agronomy.

fact, seed tannin concentrations and contents do decrease following anthesis, and the pattern is affected by both genotype and environment (Hoshino and Duncan, 1982).

GRAIN QUALITY

Grain quality is an important factor in breeding C_4 cereals such as maize and grain sorghum. However, quality varies with the intended use of the grain. For example, amino acid composition is important for feeding nonruminants, where low grain lysine concentration often limits nutritional quality. For popcorn, the ratio of volume after popping to that before is of primary importance. For some industrial uses, the relative amounts of various starches are important.

Grain quality has been studied more carefully in maize than in other C_4 grasses, and the following discussion is taken from reviews of Alexander and Creech (1977) and Creech and Alexander (1978). Data for grain sorghum and pearl millet are from Jambunathan (1980).

Dry kernels of common dent maize contain abut 80% carbohydrate, 10% protein, 4.5% oil, 3.5% fiber, and 2% minerals. Grain sorghum caryopses consist of about 72% carbohydrates, 14% protein, 3% oil, 2% fiber, 2% minerals, and 1% tannin. However, large genotypic variation is found in all components.

Carbohydrates

In normal maize kernels starch is about 75% amylopectin, a high-molecular-weight branched molecule, and about 25% amylose, a lower-molecular-weight nonbranched molecule. However, genotypes containing from 0% (waxy maize) to over 50% (amylomaize) amylose have been identified, and both waxy and amylomaize are now well-established specialty crops.

Normal maize and grain sorghum caryopses contain about 1–4% sucrose. However, almost 20% of the caryopsis dry weight of some sweet corn cultivars is sugar. Genes affecting caryopsis carbohydrates are well known and have been used to modify carbohydrate composition for special purposes.

Proteins

Protein quality and amino acid composition of C_4 cereals are also under genetic control. Jumbunathan (1980) reported wide variation in the protein content of grain sorghum (11–19%) and pearl millet (8–23%) genotypes. In Illinois, over 70 generations of selection for extremes of protein content has produced maize genotypes with more than 25% or less than 5% protein.

The protein of C_4 cereal grains is less nutritious for nonruminants than is protein from legumes, primarily because of deficiencies in lysine and tryptophan. In 1964 a known maize mutant (opaque-2) was found to have higher concentrations of lysine and higher nutritional value than normal maize. However, the opaque-2 gene also led to more fragile grains and lower yields. Since then, breeders have gradually reduced the problems associated with the opaque-2 gene, and hybrids combining high-yield, good physical grain qualities, and high lysine content are available. Commercial production of these hybrids is expected to rise, especially in countries unable to produce adequate protein supplements.

Similar high-lysine mutants have been found in grain sorghum, and considerable genotypic variation in lysine concentration is found in pearl millet (Jambunathan, 1980).

Oil

Oil is an important by-product of maize grain processing, and breeders have developed both high-oil (18-20% of dry weight) and low-oil (0.5%) genotypes. Though hybrids with adequate yield and high oil contents (6.5-8.0%) have been developed, commercial production of high-oil maize is very limited. Relatively wide variation is also found in the oil content of grain sorghum (2-7%) and pearl millet (3-8%) genotypes.

ROOT GROWTH AND FUNCTION

The growth and function of C_4 grass roots are not well understood. This is largely due to the difficulty of separating roots from their natural medium, the soil. Grass roots systems often extend more than 2 m into the soil, and they are exceedingly fragile. Thus, it is very difficult to remove the entire root system from the soil, much less obtain an accurate estimate of its total weight. An additional complication involves the rapid turnover of the root system. Roots continually grow, die, and decompose, and only recently have techniques been developed to estimate accurately the rates of both growth and death. In addition, root function is extremely difficult to measure under natural conditions. Roots absorb both water and nutrients and translocate them to the shoot. Despite decades of research, it is very difficult to obtain accurate estimates of root activity in a particular portion of the soil. Finally, the root system is extremely plastic and dynamic. Root growth and function are very sensitive to drought, temperature, soil strength, aeration, toxic ions, and nutrient deficiencies. The root system is quick to respond to transient stresses in one part of the soil profile and to shift its growth and activity to another less stressful part.

Because of the difficulty in studying plant root systems in their natural environment, quantitative generalizations about their growth and function are difficult to make. The objective of this chapter is to describe root system morphology, development, and function in largely qualitative terms. Root response to environmental factors will be discussed in greater detail in chapters dealing with those factors.

MORPHOLOGY

Since grass roots are interdeterminate and freely branching organs, they grow most rapidly where environmental conditions are most favorable. For this reason, sweeping generalizations concerning the rate of elongation of roots in the soil and the distribution of the root system within the soil profile are impossible. Nevertheless, general descriptions of "typical" root growth and development provide a point of reference for subsequent discussions of environmental effects. The following sections describe typical root growth in sugarcane, maize, and grain sorghum.

Sugarcane

The early growth of sugarcane roots has been described by Clements (1980) and Dillewijn (1952). Under favorable environmental conditions, the axillary buds on sugarcane "seed pieces" or "setts" begin to swell and within 3 days the meristem becomes active. Within a week of planting the bud has enlarged considerably and "sett roots" begin to grow from the root band of the sett where their dormant tips could previously be seen. "Shoot roots" begin to appear from the short basal internodes of the bud about the time that the young shoot emerges from the soil (Fig. 6.1). The first shoot roots are much thicker than the sett

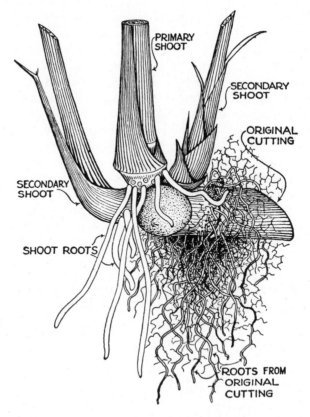

FIG. 6.1. Young sugarcane plant showing primary and secondary shoots arising from sett and two kinds of roots, sett roots arising from the sett and shoot roots arising from the shoots. From Martin (1938). Reprinted courtesy of the Hawaiian Sugar Planters' Association.

roots; their rate of growth is more rapid than sett roots; they tend to grow steeply into the soil; and they produce few branches initially. Shoot roots produced later are finer and branch more freely than thicker, earlier shoot roots.

Bud position at planting has an important influence on sett root development, undoubtedly because of hormone balance within the bud. Sett roots grow more vigorously from nodes in which the bud is on the side of the sett rather than on the top. In setts with multiple nodes, sett roots grow more vigorously from the basal nodes, while shoots grow more vigorously from the apical nodes.

Glover (1967) described the early growth of sugarcane roots in several soils as observed through the glass wall of a rhizotron in South Africa. In this study, sugarcane setts were planted against the glass at the top of the rhizotron window. The soils were well aerated, warm, and moist. Under these conditions the first sett roots began to grow from the root bands of the sett within 24 h of planting. The sett roots were initiated over a period of about 5 days. Initially, they grew only a few millimeters per day, but by the fifth day the rate of elongation was about 20 mm/day. Maximum growth rates reached 24 mm/day. A very fibrous root mass consisting of fine sett roots with extensive branching was produced. However, sett root growth stopped after about 11 days when they were 15–25 cm long. Suberization began to discolor the roots by the fifth day, and the whole system was reddish-brown by the eleventh day. The sett root system turned dark brown and decomposed rapidly, disappearing within 2 months after planting.

The primary shoot roots began to develop 5–7 days after planting. By the time sett root growth had ceased (11 days after planting), the primary shoot roots were 2–5 cm long. After a few days this rate of growth increased. Individual roots grew very rapidly (75 mm/day) for periods of 1 or 2 days, but over periods of a week or more their growth averaged about 40 mm/day. Unlike sett roots, the growth rate of primary shoot roots varied among cultivars and soils. Heavy soils slowed root elongation. In most cases, thick, early shoot roots grew downward into the soil forming a roughly conical shape.

These thick shoot roots are the "buttress" roots described by Evans (1935). Late shoot roots are finer and more freely branched "superficial absorbing" roots. Evans (1935) also described "rope" roots, which are actually several roots and their intertwined branches which descend a common pore or crack. These rope roots can reach depths of 4 m or more.

Glover (1967) found that in sandy soils the primary shoot roots were usually less than 2 mm in diameter; however, in undisturbed clayey soils roots were less numerous, were thicker, and developed fewer branches. When these thickened roots grew into layers of sandier subsoils, they reverted to the thinner, more freely branching type. Branches were always initiated at right angles to primary roots. Similar results are reported by Venkatraman and Thomas (1922).

Despite the ability for roots to grow vertically at rates averaging 40 mm/day, the bulk of the root system normally remains concentrated in the top 20 or 30 cm. For example, Evans (1937) found that 70% of the total root length was concentrated in the upper 30 cm, and 91% was found in the upper 61 cm, even though roots extended to more than 183 cm (Fig. 6.2). Of course, a higher per-

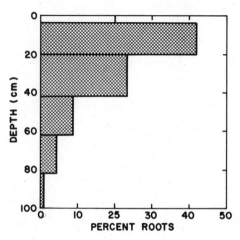

FIG. 6.2. Percentages of roots of 21-month-old furrow-irrigated H109 sugarcane in various soil layers. From data in Lee (1926a).

centage will normally occur in the deeper soil layers if environmental conditions (especially water) are more favorable there. A lower percentage will occur in the deeper layers if poor aeration, compaction, aluminum toxicity or other factors inhibit their growth at depth.

Sugarcane roots, especially those growing deep in the soil profile, have the capacity to remain alive and metabolically active for long periods after being separated from the shoot. Glover (1967) observed one at a depth of 100 cm that had been cut by an insect 160 days earlier. In the spring when a flush of new root growth occurred in the rest of the root system, it produced three branches covered with turgid root hairs.

Despite their ability to remain alive after harvest, the roots of the previous crop gradually die and ratoon shoots develop their own root systems.

Maize and Grain Sorghum

Maize and grain sorghum, like other C_4 grasses grown from seed, produce seminal roots from the seed and nodal or adventitious roots from stem nodes (Fig. 6.3). The seminal root system is important during the first week to 10 days after emergence. During this period, the number and length of seminal root apices first increase rapidly then become stable. For example, Maizlish et al. (1980) found that in maize the number of seminal root apices per plant increased from about 50 at emergence to about 170 at 8 days after emergence. During this period, the length of seminal roots increased from 0.6 to 4.4m/plant. During the

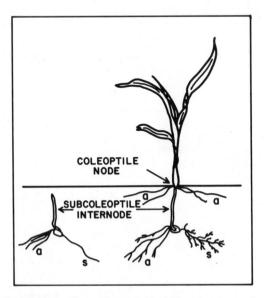

FIG. 6.3. Germinating maize seedlings showing seminal (s) root system and adventitious (a) root system growing from the base of the subcoleoptile internode as well as from the coleoptile node.

same period, the adventitious system increased from about 100 to 900 apices/ plant while the length of the system increased from about 2 to 22 m/plant. By 29 days after emergence, the total length of the adventitious system was over four times that of the seminal system, which was no longer increasing.

During the first week after emergence of maize and grain sorghum plants, the root system contains 20–30% of the dry matter of the whole plant. During the next few days, the root system grows rapidly and by the eight-leaf stage about 40% of the total plant dry matter is in the root system (Maizlish et al., 1980). Subsequently, the shoot grows at a relatively more rapid rate than the root system, and the percentage of total plant dry weight in the root system decreases gradually to about 10% maturity (Fig. 6.4 and 6.5).

Many studies of maize and sorghum root growth suggest that after anthesis total root weight either remains relatively constant or decreases slightly (Kaigama et al., 1977; Foth, 1962, Zartman and Woyewodzic, 1979); however, in other studies or in other years of the same studies (Mengel and Barber, 1974; Myers, 1980), root weight continues to increase after anthesis. This suggests that the balance between supply and demand for carbohydrate affects root growth after anthesis. During the early stages of maize and grain sorghum growth, vertical root growth is more rapid than horizontal root growth. Rates of vertical root growth ranging from 1.3 to 6 cm/day have been reported (Allmaras et al., 1975; Myers, 1980; Nakayama and van Bavel, 1963).

As in sugarcane, the greatest maize and sorghum root length densities are normally found in the upper soil layers (Foth, 1962; Lal and Maurya, 1982; Myers, 1980). Maximum root length densities vary widely among studies. For example, Allmaras et al. (1975) reported maximum densities of 0.43 cm root/ cm^3 for maize grown on a sandy clay soil. Barber (1971) reported maximum densities ranging from 0.7 to 3.0 cm/cm^3 for maize on a silt loam. Taylor and Klepper (1973) found up to 10 cm/cm^3 for maize on a sandy loam in a rhizotron, and Myers (1980) found up to 12 cm/cm^3 for grain sorghum grown on a clay

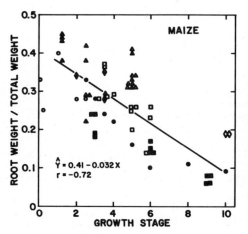

FIG. 6.4. Relationship between maize growth stage and the ratio of root weight to total plant weight. Growth stages: 0 = emergence, 2 = eight fully expanded leaves, 4 = 16 fully expanded leaves, 6 = blister stage, 8 = early dent stage, 10 = physiological maturity. ● = Foth (1962), ○ = Maizlish et al. (1980), ■ = Chakravarty and Karmakar (1980), □ = Cal and Obendorf (1972), ▲ = Paddick (1947), △ = Allmaras and Nelson (1977), ◆ = Shank (1943), ◇ = Fehrenbacher and Alexander (1955).

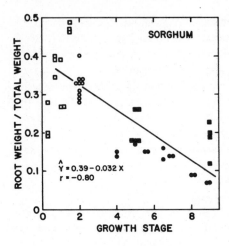

FIG. 6.5.   Relationship between grain sorghum growth stage and the ratio of root weight to total plant weight. Growth stages: 0 = emergence, 2 = five fully expanded leaves, 4 = final leaf visible in whorl, 6 = half bloom, 8 = hard dough, 9 = physiological maturity. ● = Kaigama et al. (1977), ○ = Nour and Weibel (1978), ■ = Meyer (1980), □ = Arkin and Monk (1979).

loam. Commonly reported ratios of maize root length to root weight are from 1 to 6 cm/mg (Allmaras et al., 1975; Barber, 1971).

When soils are well drained but uniformly moist and subsoils offer no chemical or physical impediments, root length density normally decreases exponentially with depth (Fig. 6.6). However, the soil environment plays a very large role in the growth and distribution of the root system. For example, Follett et al.

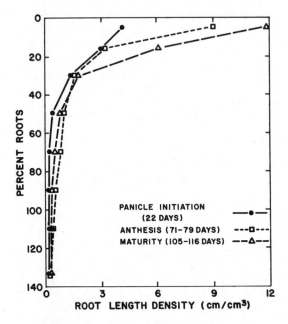

FIG. 6.6.   Root length density of TX 610 grain sorghum at different depths in a clay loam soil. From Myers (1980). Reproduced with permission of the author and *Field Crops Research*.

(1974) studied maize root distribution on 10 sandy soils with declining water tables and found the percentage of total root weight in the 0–30 cm depth varied between 69 and 97%, depending on water table depth. When soil water contents decline in the surface soil, root length density may actually decrease in these layers and increase in the deeper, moister layers (Fig. 6.7). More detailed discussions of the effects of soil water, temperature, aeration, compaction, and nutrients on root growth can be found in other chapters.

GENOTYPIC VARIATION IN ROOT DEVELOPMENT

Agronomists and plant breeders have long recognized that genotypic variation in root systems affects plant adaptation to environment. Though the difficulty in observing root systems has limited research in this area, a number of studies illustrate the existing genotypic variation.

Sugarcane

Genotypic variation in sugarcane root systems is well documented (Evans, 1935, 1936; Lee, 1962a,b; Raheja, 1959). Stevenson and McIntosh (1935) found a correlation between varietal differences in tillering and adventitious root growth from these tillers. Cultivars that produce a second flush of tillers also produce a second flush of root growth associated with those tillers. Another common observation is that the degree of root branching and the predominant direction of root growth vary among sugarcane cultivars. Stevenson and McIntosh (1935) noted that some cultivars have a higher degree of branching than others. Stevenson (1936) found that crosses of noble canes (*Saccharum officinarum*) with wild *Saccharum* spp. have more profuse root branching than the noble canes. Evans (1936), Mukerji and Alan (1959), Stevenson and McIntosh (1935), and Venkatraman (1930) observed that in some cultivars root growth is predominantly horizontal with few roots penetrating deep into the soil profile. Other cultivars have more vertical and fewer horizontal roots. Cultivars with weakly gravitropic (more horizontal) root orientation are more resistant to lodging in high winds

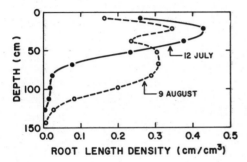

FIG. 6.7. Maize root length densities at different depths on two dates. The decrease in root length density above 50 cm is due to dry topsoil. From Allmaras et al. (1975b). Reproduced from *Soil Science Society of America Proceedings*, by permission of the Soil Science Society of America.

than cultivars with strongly gravitropic root systems (Mukerji and Alan, 1959; Stevenson and McIntosh, 1935).

Maize

As in sugarcane, maize cultivars exhibit genotypic variation in root characteristics (Chakravarty and Karmakar, 1980). Weihing (1935) found consistent differences among maturity groups in virtually all root characteristics. Late-maturing cultivers had larger shoots and root systems at maturity, but the shoot–root ratio of late cultivars was greater than that of early cultivars. Shank (1934) found consistent differences in shoot–root ratios among inbreds and among hybrids. The shoot–root ratios of the hybrids were similar to those of the parent with the lower shoot–root ratio, suggesting that vigorous root system development is dominant.

Maize breeders have developed rapid techniques to select for desirable rooting characteristics. Jenison et al. (1981) and Penny (1981) found consistent differences among maize genotypes in their resistance to uprooting. These differences were consistent across environments and were positively correlated with root dry weight and the lateral spread of the adventitious root system. In addition, there was a negative correlation between the decrease in resistance to uprooting during grain fill and the incidence of root rot. This suggests that root rot is responsible for root system deterioration during grain filling.

Grain Sorghum

Several recent studies have revealed considerable genotypic variation in grain sorghum root system development. For example, Nour and Weibel (1978) found significant variation among 10 cultivars in root weight, length, volume, and shoot–root ratio. In general, the more drought-resistant cultivars had greater root weight and volume and had smaller shoot–root ratios. As in maize, maturity genotype affects root development. Blum et al. (1977a) found that early-flowering genotypes started producing adventitious roots slightly earlier than late genotypes. At 38 days after planting, the total number of adventitious roots was greater in early than in late genotypes; however, late genotypes had greater root volume due to greater lateral branching. Zartman and Woyewodzic (1979) reported that during grain filling root length density decreases more rapidly in senescent than nonsenescent genotypes.

Hybrid vigor also affects root development in grain sorghum. Blum et al. (1977b) found that hybrids have greater seminal root length, a greater rate of adventitious root growth, greater lateral root growth, and greater root volume than inbreds. However, heterosis had little effect on the number of adventitious roots or the ratio of root length to leaf area. These results are similar to those of Smith (1934) and Shank (1945), who found that hybrid maize has a higher ratio of lateral to primary roots than inbreds. They attributed the hybrid's relative insensitivity to P deficiency to their greater root surface area and root volume.

HORMONES, GROWTH, AND GRAVITROPISM

Early studies with sugarcane (Evans, 1936; Stevenson, 1936; Stevenson and McIntosh, 1935; Venkatraman, 1930) revealed that the mean angle of root pene- tration into the soil varies among genotypes. Some genotypes produce predomi- nantly shallow root systems with many roots growing almost horizontal to the soil surface. Other genotypes produce root systems in which more roots grow at a steep angle into the soil. Thus, it seems probable that root response to gravity (gravitropism) as well as root growth varies among genotypes.

By the beginning of the twentieth century plant physiologists had recognized that root gravitropism is normally correlated with the presence of starch gran- ules (statoliths) in the root cap (or in other root cells near the extending zone) (Nemec, 1900; Haberlandt, 1900). Much subsequent work (see Wilkins, 1977 and 1979) has confirmed the close correlation between the presence of statoliths and the capacity for gravitropism in roots.

Soil temperature has a major effect on gravitropism. Mosher and Miller (1972) found that maize roots grow more vertically in warm than in cool soils. Light also affects root gravitropism. The roots of some maize cultivars having statoliths in the root cap do not respond to gravity unless exposed to light for a brief period. Suzuki et al. (1980) have shown that an unknown photoirreversible pigment (not phytochrome) with an absorbance peak at about 640 mm is respon- sible for the induction. The site of the photoreaction is in the apical 2 mm of the root, and the reaction triggers several biochemical changes in the root apex, in- cluding an increase in the energy charge and concentration of NADPH in the root tips (Steck and Pilet, 1979; Suzuki et al., 1981b) and the appearance of abscisic acid (ABA) in the root cap (Wilkins and Wain, 1974). In roots of maize cultivars that do not require light for gravitropism the energy charge of the roots is high without exposure to light (Steck and Pilet, 1979).

Thus starch statoliths and (in some cultivars) light are needed for the induc- tion of gravitropism. After induction, gravitropism is controlled by hormones. Ethylene and cytokinins generally inhibit root growth, while auxins may either stimulate or inhibit it. All these hormones affect gravitropism as well as growth (Jackson and Barlow, 1981; Mulkey et al., 1982; Pilet and Chanson, 1981; Stenlid, 1982). ABA and (perhaps) other root growth inhibitors are produced in the root cap (Pilet and Rivier, 1980), accumulate in the lower part of the root apex, and are translocated basipetally from the tip to the zone of cell elongation. Georeaction is enhanced by the availability of indole acetic acid (IAA), which moves acropetally to the zone of elongation, presumably from the caryopsis in seedlings (Ney and Pilet, 1980) and from the shoot in older plants (Hirota and Watanabe, 1980; Pilet and Elliott, 1981). Gravitropism resulting from the concomitant actions of ABA and IAA is due to a reduction in growth on the lower side of the root and a slight enhancement of growth on the upper side (Pilet and Ney, 1981). Differences among maize cultivars in gravitropism may be due to differences in ABA concentrations in the apical 1 mm of the root tip (Rivier and Pilet, 1981; Chanson and Pilet, 1981; Hirota and Watanabe, 1980) or to differences among genotypes in their sensitivity to ABA (Pilet and Chanson, 1981).

Moloney et al. (1981) found that root growth is always associated with H^+ efflux from the root tissue, and auxin enhances this efflux (Cleland et al., 1977; Moloney et al., 1981). In georeacting roots, H^+ efflux always increases on the upper side and decreases on the lower side of the root (Mulkey and Evans, 1981; Mulkey et al., 1981). This suggests that a common hormone-mediated mechanism is involved in H^+ efflux, root growth, and root georeaction.

WATER EXTRACTION

For plants growing in moist soils, the dominant resistance to water movement from the soil to the atmosphere is in the root system (Boyer, 1971; Hansen, 1974a,b; Newman, 1969a,b). Clipping the root tips opens the xylem directly to the solution surrounding the roots and results in a rapid, short-term influx of water (Kramer, 1938). Thus, the major resistance to water flow probably lies between the root surface and the xylem (Taylor and Klepper, 1975). Newman (1976) suggests that this is primarily resistance to water movement through the symplasm of the cortex and stele.

If resistance to water movement through the symplasm is the dominant resistance for plants growing in moist soils, water uptake from different layers of a uniformly moist soil profile should be roughly proportional to rooting density within that profile. There is ample evidence that this is often the case in C_4 grasses and other plants (Burch et al., 1978; Taylor and Klepper, 1971, 1973, 1975) When adverse soil conditions (such as compaction or aluminum toxicity) reduce rooting density in a particular soil layer, the rate of water extraction from that layer is reduced. For example, Grimes et al. (1975) found that water extraction by maize from the different layers of a moist soil with compacted zones is proportional to the rooting density in those layers (Fig. 6.8).

The major exception to the proportionality between water uptake and root length occurs when some soil layers are drier than others. When soil water potentials fall below about -0.1 MPa in sandy layers or about -0.8 to -1.0 MPa in clayey layers, the resistance to water movement from the bulk soil to the root surface becomes the dominant resistance (Reicosky and Ritchie, 1976). Uptake of water per unit root length then decreases in the drier layers.

Another major exception to the proportionality between water uptake and root length occurs when large numbers of old suberized roots occur in the top part of the profile. These old roots are less permeable than the young roots deeper in the profile, and uptake of water per unit length of root is greater where roots are younger (Burch et al., 1978; Taylor and Klepper, 1973).

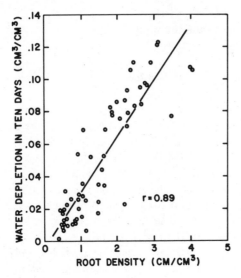

FIG. 6.8. Relationship between rooting density in a layer and water extraction from that layer in a uniformly moist soil. From Grimes et al. (1975). Reproduced from *Agronomy Journal*, by permission of the American Society of Agronomy.

NUTRIENT UPTAKE

A detailed review of the mechanisms of ion movement in the soil and the regulation of ion uptake by plant demand and nutrient supply at the root surface is beyond the scope of this book. However, a few important concepts and relationships should be understood to appreciate the general effects of transpiration rate, metabolic energy supply, compaction, and aeration on nutrient uptake by the plant. Recent books with more comprehensive discussions of these topics include Torrey and Clarkson (1975), Lüttge and Pitman (1976), Russell (1977), Nye and Tinker (1977), and Lüttge and Higinbotham (1979).

Movement to Roots

The two principal mechanisms of nutrient movement to plant roots are (1) diffusion down a concentration gradient and (2) mass flow in soil water moving toward the root in response to transpiration.

Nye (1977) has described the combined effects of mass flow and diffusion of nutrients toward the plant root with Eq. 1:

$$I = 2\pi RDMF \frac{dC}{dR} + 2\pi RVC \qquad (1)$$

where I = flux of nutrient crossing a radial boundary outside the root (mol/cm s)

R = radius of the boundary (cm)

D = diffusion coefficient of the nutrient in a solution (cm^2/s)

M = volumetric water content of the soil (cm^3/cm^3)

F = diffusion impedance factor
C = concentration of the nutrient in the soil solution (mol/cm^3)
V = water flux toward the root (cm^3/cm^2 s)

The left term describes the effect of diffusion; the right term describes the effect of mass flow.

Nutrients such as NO_3^- and Cl^- interact weakly with soil particles and are highly soluble in the soil solution. Under many conditions mass flow of these nutrients in the transpiration stream is sufficient to supply the NO_3^- requirements of the plant (Barber, 1962; Barber et al., 1963). Ca^{2+} and Mg^{2+} are more strongly attracted to the cation exchange sites of the soil than are anions such as NO_3^- and Cl^-. However, in most fertile soils their concentrations in the soil solution are great enough to supply the growing plant's requirement through mass flow (Barber et al., 1963), and they may accumulate near the plant root when more Ca^{2+} and Mg^{2+} move to the root surface than are absorbed.

Nutrients such as HPO_4^{2-} and K^+ are tightly bound to soil clay particles, and their concentrations in the soil solution are normally too small for mass flow to supply the needs of the growing plant (Barber, 1962; Barber et al., 1963). These ions tend to move to the root over very small distances by diffusion in response to the concentration gradient caused by ion absorption by the root.

A more controversial and less important mechanism of nutrient supply is "contact exchange." Roots growing into soil probably "intercept" some nutrients as they displace soil, and hydrogen ions released by root tissue may be exchanged directly for cations on clay particles (Jenny, 1966). However, with the exception of Ca^{2+} and Mg^{2+} only a small percentage of the nutrients needed by the crop could be obtained by contact exchange (Barber et al., 1963). In addition, movement of nutrients from the point of contact across the root mucigel and root apoplast would require movement in the solution phase (Olsen and Kemper, 1968).

Table 6.1 summarizes the principal means by which major and minor nutrients are thought to move from the soil to the root. It should be recognized that both mass flow and diffusion are normally involved and that the principal mechanism can vary with differences in transpiration rate, root length density, and soil solution concentration of the nutrient (Nye, 1977).

Absorption by Roots

One of the most important functions of the root system is to supply the shoot with the proper amount and balance of mineral nutrients needed for growth. Since the mineral composition and balance of the soil solution is very different than that needed by the shoot, the root system must be highly selective in its absorption of nutrients. To accomplish this, it expends energy to absorb some nutrients and to exclude or secrete others. Ions that are often actively absorbed include NO_3^-, NH_4^+, HPO_4^{2-}, and K^+. Na^+, Ca^{2+}, and Mg^{2+} are often excluded or secreted. However, nutrients that are excluded under some conditions are

TABLE 6.1. Principal Mechanisms of Nutrient Movement to the Roots (Compiled in part from Olsen and Kemper, 1968)

NUTRIENT	ROOT INTERCEPTION	DIFFUSION	MASS FLOW	REFERENCES
Nitrate and Cl		+	+	Barber (1962), Nye (1977)
P, S		+		Barber (1962), Nye (1977), Olsen et al. (1961), Lai and Mortland (1961)
K, Rb		+		Barber et al. (1963), Barber (1962), Nye (1977), Place and Barber (1964)
Ca, Mg			+	Barber (1962), Barber et al. (1963), Oliver and Barber (1966a)
Mo		+[a]	+[b]	Lavy and Barber (1964)
Fe, Mn, Zn		+		Oliver and Barber (1966b)
Al, B, Cu, Sr	+		+	Oliver and Barber (1966a)
Si			+?	Jones and Handreck (1965)

[a] In young root tissue.

[b] In older root tissue.

actively absorbed in others. For example, in fertile agricultural soils Ca^{2+} may be excluded from the roots and build up to high concentrations near the root surface. However, in Ca-deficient acid soils it may be actively absorbed.

Many factors affect uptake of a particular nutrient by the root system. These include plant demand, the concentration of the ion at the root surface, transpiration rate, the "leakiness" of the root system, and numerous environmental factors such as temperature, soil water potential, soil aeration, soil strength, and the presence of toxic elements. The first four factors are discussed below; environmental effects are discussed in other chapters. The reviews of Anderson (1972, 1976), Bowling (1981), Higinbotham (1973, 1974), Hodges (1973), Lüttge and Higinbotham (1979), Nye (1977), Pitman (1977), and Racusen (1979) can be consulted for further details.

Plant Demand. Even when nutrients are abundantly supplied to the root system, plants usually do not accumulate nutrients above some upper limit or "optimum" concentration. This optimum nutrient concentration in the plant varies with growth stage and is normally higher in young than in old plants. For example, Jones (1983) described the effect of growth stage on the optimum shoot N and P concentrations for maize and grain sorghum (Fig. 6.9, 6.10). Cowie (1973) and Jones (1983) have demonstrated that maize and grain sorghum N concentrations vary between the optimum concentration (when N is abundant) and about half the optimum (when N is very deficient). Thus, at any time the whole plant

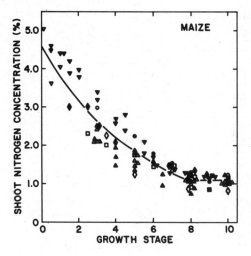

FIG. 6.9. Maize shoot nitrogen concentrations at different growth stages and at near-maximum dry matter yields in several field experiments. From Jones (1983). Reproduced with permission of the author and *Field Crops Research*.

demand for a nutrient is the amount of that nutrient needed to bring that nutrient to the optimum concentration in the plant.

This principal is illustrated in the results of an experiment conducted by Cowie (1973). In this experiment grain sorghum was grown in solution cultures of either high or low N concentration. The growth period was divided into three parts (early, medium, and late), and plants could be grown in different nutrient

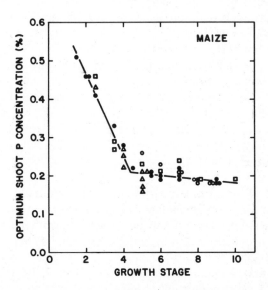

FIG. 6.10. Relationship between growth stage and maize shoot phosphorus concentration in treatments with adequate phosphorus nutrition. From Jones (1983). Reproduced with permission of the author and *Field Crops Research*.

solutions during the different parts of the season (Fig. 6.11). Plants that were always maintained in the solution with high N concentration (HHH) had plant N concentrations that decreased from about 4.5% at 10 days to about 2.5% at 100 days. Plants that were always grown in solutions with low N concentrations (LLL) had tissue N concentrations about half this high. However, plants that were in solutions of different N concentration during different periods (HLL, HHL, LLH, and LHH) had tissue N concentrations that increased or decreased to the concentrations found in the HHH and LLL treatments. Whenever plants were moved from a solution with low N concentration to one with high N concentration, its rate of nutrient uptake per unit dry matter accumulation increased until its nutrient concentration approached that of HHH treatment.

In contrast to plant demand for a nutrient, Nye (1977) has described root demand in terms of the root's nutrient conductance or absorbing power (P):

$$U = 2\pi APK \qquad (2)$$

where U = rate of nutrient uptake per length of root (mol/cm s)

A = root radius (cm)

K = concentration of the nutrient in the soil solution at the root surface (mol/cm^3)

P = root absorbing power (cm/s)

The nutrient status of the root has an important effect on its absorbing power (P). This is clearly demonstrated in numerous experiments with excised "low-salt roots."

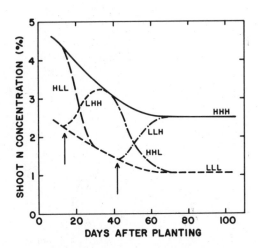

FIG. 6.11.   Changes in grain sorghum shoot nitrogen concentration as a result of changes (at arrows) in nutrient solution nitrogen concentration from high (H) to low (L) or from L to H. For example, the treated LHH was grown at L solution concentration until the first arrow, then it was grown at H concentration until the end of the experiment. Redrawn data from Cowie (1973).

When plants are grown in solutions of low nutrient availability, they become nutrient deficient and "low-salt roots" are produced. These roots can then be excised, placed in solutions with adequate nutrient concentrations, and their absorption of these nutrients studied. After an initial rapid influx into the apparent free space, a slower and gradually diminishing rate of ion uptake occurs for several hours. During this period of slow absorption, ion uptake exceeds ion efflux from the roots. After a few hours, influx decreases to the point at which it is balanced by efflux, and the roots are said to be in the salt-saturated state.

Thus, the nutrient status of the root governs its rate of nutrient absorption. In intact plants the nutrient concentration of the shoot is normally in equilibrium with that of the root. When excess ions begin to accumulate in the shoot, they are retranslocated to the root in the phloem, thereby increasing root nutrient concentrations and slowing ion uptake.

Solution Concentration. As implied by Eq. 2, the soil solution nutrient concentration at the root surface affects nutrient uptake. These effects have been studied by numerous workers. The results of Epstein and his collaborators (Epstein, 1966, 1972; Epstein and Hagen, 1972; Nissen, 1974) suggest that accumulation of ions by plant roots is governed by two systems. System I operates at external ion concentrations below about 0.5 mM, concentrations similar to the actual concentrations of most important mineral nutrients in the soil solution (see Barber et al., 1963). Michaelis-Menten kinetics describe the effects of concentration on the rate of ion uptake by system I.

System II operates between about 1.0 and 50 mM, above most soil nutrient solution concentrations. Ion uptake by this system does not exhibit true Michaelis-Menten saturation kinetics but appears to be a collection of separate Michaelis-Menten-like curves described as "multiphasic kinetics" (Fig. 6.12) (see Nissen, 1974). Systems I and II also differ qualitatively. For example, K$^+$ and Na$^+$ do not compete with each other at low concentrations corresponding to system I, but they do compete at the high concentrations corresponding to system II. The same behavor is exhibited with Cl$^-$ and NO$_3^-$, where the two anions compete at high but not at low (<0.5 mM) concentrations. As a result of this

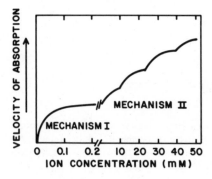

FIG. 6.12. Diagrammatic representation of the velocity of ion influx into roots as a function of the external ion concentration. Redrawn from Hodges (1973). Reproduced with permission of the author and *Advances in Agronomy.*

competition, roots are often less selective at high than at low solution concentrations (Pitman, 1965a; Epstein, 1972).

External solution concentration also affects root membrane ATPase activity. In maize and several other species the effects of solution K^+ concentration on K^+ uptake and ATPase activity are almost identical (Fisher et al., 1970; Beffagna et al., 1979). The results strongly suggest that the plasma membrane fraction of roots contains ATPase activity that is responsive to external K^+ concentration and that is highly correlated with K^+ uptake at concentrations corresponding to both systems I and II (Fig. 6.13) (from Leonard and Hodges, 1973).

The quantitative and qualitative differences in systems I and II suggest that different uptake mechanisms are involved in the two systems. A number of hypotheses have been advanced to explain the kinetic data available. These hypotheses involve varying numbers of hypothetical carriers; whether these carriers transport anions, cations, or both; the location of the carriers at the plasmalemma or both the plasmalemma and tonoplast; the source of energy driving ion movement; and the subunit structure and detailed kinetics of the carriers (see Hodges, 1973 and Lüttge and Higinbotham, 1979 for discussions). However, despite voluminous research and speculation, no clear consensus has emerged.

Transpiration.   During the 1950s considerable controversy existed concerning effects of transpiration on ion uptake and translocation. Hylmo (1953, 1955, 1958) took the position that a significant proportion of the ions translocated to the shoot are passively drawn into the root by mass flow in the transpiration stream. However, Brouwer (1954, 1956a,b, 1965) contended that translocation, while stimulating ion uptake, does so by increasing the rates of metabolic processes affecting ion uptake. Subsequent work (Pitman, 1965b) suggests that while large increases in transpiration often cause only small increases in ion uptake, they reduce ion selectivity. For example, Pitman (1956b) found that tran-

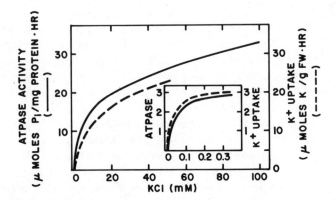

FIG. 6.13.   Effect of solution KCl concentration on ATPase activity and K^+ uptake by mechanism I (inset) and mechanism II in oat roots. Redrawn from Leonard and Hodges (1973). Reproduced with permission of *Plant Physiology*.

spiration has little effect on the total uptake of the monovalent cations (K^+ and Na^+); however, high transpiration rates increase the ratio of Na^+ uptake to K^+ uptake. The effect of transpiration on selectivity is similar to that of solution concentration (Pitman, 1965a), suggesting that the effect of transpiration is at least partially due to an increase in ion concentration at the plasmalemma of the cortical cells due to mass flow of ions into the apoplast with the transpiration stream.

Root Hairs. Root hairs are involved in nutrient and water uptake, mucilage secretion, and anchorage of the growing root tip (Barley, 1970). In most C_4 grasses they develop from the center of root epidermal cells a few millimeters behind the root apex. However, in sugarcane (Clements, 1980), as in festucoid grasses, they develop from the apical end of the epidermal cells. In well-aerated, uncompacted soils root hairs are long and undeformed; however, root hairs that develop in compacted soils are often malformed, branched, and covered with soil particles. Roots in poorly aerated soils often have no root hairs (Clements, 1980).

Root hairs of grasses are long lived and often persist for several months. They grow very close together, can penetrate small soil pores, and adhere to soil particles by secretion of mucilage. Evans (1937) found that the surface area of root hairs is 41% of the total surface area of the root system. Since they grow very close together, the distance from any point within their radius around the root to the nearest root hair is very small. This reduces the distance that immoble ions such as $H_2PO_4^-$ must diffuse to reach the root surface. Lewis and Quirk (1967) found that in an acid soil with high P fixation the radius of P uptake around the root axis was equal to length of the root hairs.

Mycorrhizae. Mycorrhizal fungi occur in close symbiotic associations with plant roots. These fungi utilize carbohydrate from the host plant and facilitate uptake of water and nutrients by the plant. In the grasses and many other families, the most important group of mycorrhizae are the vesicular-arbuscular mycorrhizae (VAM). These fungi are Phycomycetes of the family Endogonaceae. They are named for the vesicles (bladderlike structures) and arbuscules (shrublike structures) they form within the root cortex. Their structure consists primarily of an internal mycelium in the cortex connected with an external mycelium in the soil.

The importance of VAM has been reviewed (Abbott and Robson, 1982; Bowen and Smith, 1981; Tinker and Sanders, 1975), and it is clear that VAM can facilitate plant uptake of relatively immobile nutrients such as $H_2PO_4^-$, Zn^{2+}, and NH_4^+. Most studies have focused on the effects of VAM on P nutrition. P uptake and/or plant growth are stimulated by VAM, especially at low levels of soil P. This had been documented for C_4 species such as maize (Khan, 1972), grain sorghum (Krishna and Bagyaraj, 1981), finger millet (*Eleusine coracana*) (Krishna et al., 1981), wheat (Azcon and Ocampo, 1981; Khan, 1975), barley

(Clarke and Mosse, 1981; Owusu-Bennoah and Mosse, 1979; Saif and Khan, 1977), and other crops (Yost and Fox, 1979).

As with root hairs, the effect of VAM is to reduce the distance that nutrients must diffuse through the soil to either the roots or the fungal hyphae. Thus, the response to VAM is most dramatic for relatively immobile nutrients under conditions of low nutrient concentrations and/or low soil water—conditions under which diffusion is most likely to limit uptake. In addition, organic as well as inorganic P sources can be used (Allen et al., 1981a).

In addition to increased P uptake and growth, VAM infection of blue grama (*Bouteloua gracilis*) can increase leaf conductivity to water vapor diffusion, reduce whole-plant resistance to water flow, increase photosynthetic rates (Allen et al., 1981b), and increase root and shoot cytokinin activity (Allen et al., 1980). Increase in leaf thickness, cell sizes, plastid numbers, starch grain size, and protein content have resulted from VAM infection in finger millet (*Eleusine coracana*) (Krishna et al., 1981).

Even prior to infection of the root, the fungal mycelium is dependent on exuded mucilage and other compounds for its energy supply. Plant mucilages are complex acidic or neutral polysaccharides. In grasses they are secreted primarily by the root system. In *Sorghum* roots, golgi bodies, endoplasmic reticulum, and mitochondria participate in mucilage secretion (Werker and Kislev, 1978). Dictyosomes and golgi bodies are involved in its secretion by maize root caps (Juniper and Pask, 1973; Morre et al., 1967; Paul and Jones, 1976). Root exudates of weeping lovegrass (*Eragrostis curvula*) contain a number of amino acids, sugars, and organic acids (Visser, 1965), and hydrolysis of mucilage from maize root caps produce glucose, galactose, galacturonic acid, arabinose, xylose, and fucose. These compounds serve as a rich energy supply for microorganisms, including VAM.

Vietor (1982) studied the proportion of photosynthate released into nutrient solution from the root systems of kleingrass (*Panicum coloratum*), pearl millet (*Pennisetum americanum*), and bahiagrass (*Paspalum notatum*). Bahiagrass translocated 7% of ^{14}C-labeled photosynthate to the roots, and 1.1% was released from the roots. In contrast, pearl millet and kleingrass translocated 12% to the roots, but only 0.4% was released.

In sudangrass (*Sorghum sudanense*), low P nutrition causes increased membrane permeability and loss of metabolites from the root system. The sugars and amino acids lost from the roots probably stimulate mycelial growth and infection of the host. Infection and establishment of the VAM symbiosis improves P nutrition causing P uptake to increase and both membrane permeability and metabolite loss to decrease (Graham et al., 1981; Menge et al., 1978; Ratnayaka et al., 1978). Wheat cultivars differ in their susceptibilty to mycorrhizal infection, and absence of infection is sometimes associated with lack of sugar exudation from the roots (Azcon and Ocampo, 1981).

Electrochemical Potential.   The chemical concentration of root tissues and xylem fluid are clearly different than those in the soil solution. In addition, root tissues normally maintain electropotentials 60–200 mV more negative than the soil solution. Since nutrients such as K^+, Na^+, and Cl^- are ionized and are soluble in both the plant and the soil, they respond to both chemical and electropotential gradients between the protoplasm and the external solution. The combined effects of the chemical and electropotentials are described by Eq. 3:

$$P = C + RT \ln A + ZFE \tag{3}$$

where P = electrochemical potential of an ion in a cell relative to its chemical
$\quad\quad$ potential at a standard state outside the cell
$\quad C$ = chemical potential at the standard state
$\quad R$ = universal gas constant
$\quad T$ = absolute temperature
$\quad A$ = ionic activity of the ion
$\quad Z$ = valence of the ion
$\quad F$ = faraday constant
$\quad E$ = electropotential difference between the cell and the standard state

If the standard state described above is the solution surrounding the root, the electrochemical potential of an ion within the root (P) could either be greater or less than the chemical potential (C) outside the root.

Active, Passive, and Energy-dependent Transport.   Hodges (1973) has described several types of ion transport. "Passive" transport refers to movement of ions down an electrochemical potential gradient. "Energy-dependent" transport depends directly or indirectly on metabolism. "Active" transport is a type of energy-dependent transport against an electrochemical potential gradient, and it is directly coupled to exothermic reactions. Transport down an electrochemical potential gradient may be energy dependent if the gradient is dependent on exothermic reactions.

When an ion is distributed across a semipermeable membrane, net movement toward the side of lower electrochemical potential will occur until the electrochemical potential outside (o) the membrane equals that inside (i). Thus, at equilibrium (see Eq. 3),

$$(RT \ln A + ZFE)_o = (RT \ln A + ZFE)_i \tag{4}$$

Rearranging and substituting concentration (C) for activity,

$$E_o - E_i = \frac{\ln(C_i/C_o)}{RT/ZF} \tag{5}$$

Equation 5, the Nernst equation, can be used to predict the ratio of the concentrations of an ion inside and outside the root symplasm that would be produced at equilibrium as a result of maintenance of a particular electropotential difference between the symplasm and the external solution. For example, the Nernst equation predicts that at 20°C and an external solution concentration of 1 mM K$^+$, a root with an electropotential of -166 mV with respect to the external solution will have an equilibrium protoplasm K$^+$ concentration of 100 mM.

In practice, the Nernst equation has been used to determine whether electropotential differences measured across membranes in roots (and other tissues) are sufficient to explain observed ion concentrations in those tissues. Most studies with higher plants (Higinbotham et al., 1964, 1967) have shown that anions (Cl$^-$, NO$_3^-$, H$_2$PO$_4^-$, and SO$_4^-$) are present at higher concentrations in the tissues than would be predicted by the Nernst equation, suggesting that they are actively accumulated against an electrochemical potential gradient. Cations such as Na$^+$, Ca^{2+}, and Mg^{2+} are present at lower concentrations than those predicted by the Nerst equation, indicating that they are actively secreted or excluded as a result of low membrane permeability. When external salt concentrations are low, K$^+$ may be actively accumulated (Bowling and Ansari, 1971, 1972; Dunlop and Bowling, 1971a,b,c) or it may be near electropotential equilibrium and be passively distributed (Higinbotham et al., 1964, 1967). However, when salt concentrations are high, it, like other cations, is probably excluded or actively secreted.

Similar results have been obtained by using radioisotopes to analyze the rates of flux from the solution to the cytoplasm and from the cytoplasm to the solution. In general, Na$^+$ is actively secreted, Cl$^-$ is actively accumulated, and results with K$^+$ are variable [see Hodges (1973) for discussion].

Electrogenic pumps, which move ions from one side of the membrane to the other, are probably responsible for much of the membrane electropotential difference. These pumps presumably consist of carrier proteins or "porters." "Primary porters" are carrier proteins driven by ATP that generate electropotential differences by moving H$^+$, K$^+$, Cl$^-$, or other ions from one side of the membrane to the other (Beffagna et al., 1979; Higinbotham and Anderson, 1974; Hodges, 1976, Spanswick, 1974). "Secondary porters" are carrier proteins that facilitate ion movement down an electrochemical potential gradient created by the primary porters. These secondary porters may facilitate movement of a single ion (uniporters) or (in order to maintain electrical neutrality) they may facilitate movement of two ions of like charges in opposite directions or two ions of different charges in the same direction (symport). A specialized form of symport may operate by coupling sugar or amino acid transport to H$^+$ transport (Etherton and Rubinstein, 1978; Racusen and Galston, 1977).

The mechanisms of ion transport discussed above may explain some of the effects of solution concentration on ion uptake. For example, cell membrane electropotential differences decrease when solution nutrient concentrations are increased. Cheeseman and Hanson (1979b) conclude that below about 0.5 mM K$^+$ in the external solution, this decrease in electropotential difference is due to increased active uptake of K$^+$. Above this concentration, the electrochemical potential gradient favors passive K$^+$ movement into the tissue, and depolarization results from this passive uptake.

Axial Variation.   Root tissue varies axially in age, maturation, suberization, and metabolic activity. Many studies have shown that axial nonuniformity is an important factor in ion absorption and movement to the xylem. When transpiration is inhibited most ion uptake occurs in the apical 10 cm of the root with the apical 3–10 mm contributing little to translocation (Eshel and Waisel, 1972, 1973; Hodges and Vaadia, 1964; Smith and Majeed, 1981). In addition, the rate of ion translocation to the xylem at a particular location along the root is positively correlated with the number of xylem vessels still possessing cytoplasm (Anderson and House, 1967). This cytoplasm persists well after the perforation plates between xylem elements are formed and functional (Higinbotham et al., 1973), suggesting that ion movement from the parenchyma of the stele to the lumen of the xylem is mediated by the cytoplasm of the xylem cells, and the rate of xylem maturation is at least partly responsible for axial nonuniformity in ion uptake.

When transpiration is rapid, as is often the case in plants growing in the field, ion uptake and translocation to the xylem increase dramatically in the basal portions of the root. Thus, recent experiments with intact plants suggest that translocation of K^+, Na^+, HPO_4^{2-}, and NH_4^+ to the xylem can occur throughout the entire root system of both monocotyledons and dicotyledons (see discussion in Russell, 1977). This apparently does not occur with Ca^{2+}, which is preferentially translocated to the xylem from the apical region of the root of intact plants. These results clearly illustrate that relationships obtained with excised roots are often inadequate to explain results with intact transpiring plants.

Radial Movement.   From a teleological point of view, the role of the root in plant nutrition is to transfer a balanced array of nutrient ions from the soil solution to the xylem vessels. In order to accomplish this, the root can expend metabolic energy to both accumulate and excrete ions against electrochemical potential gradients. After nutrient ions have been absorbed by root cells they must be transferred to the xylem vessels in order to be translocated to the shoot. During translocation within the xylem, the root must prevent excessive leakage of ions out of the xylem back into the soil solution.

Most authorities now agree that at least two sites of active ion transport are needed to perform the functions described above (Bowling, 1981). One of these sites mediates movement of ions from the soil solution and apoplast of the cortex across the plasmalemma and into the cortical cells. These mechanisms were discussed above. The cells of the cortex are interconnected by plasmodesmata, and radial movement of ions across the cortex and endodermis into the stele is primarily by diffusion via the plasmodesmata (Anderson, 1972b, and 1976; Dunlop, 1974; Pitman, 1977). The mechanism of ion movement out of the symplasm of the stele and into the xylem vessels is still unclear; however, three distinct hypotheses are available.

One of the oldest hypotheses is that nutrient ions in the stele leak passively from the symplasm into the stelar apoplast due to the partially anaerobic conditions inside the suberized endodermis (Crafts and Broyer, 1938; Arisz, 1956).

This hypothesis assumes that the endodermis constitutes a significant barrier to O_2 diffusion, an assumption that Anderson (1972) and others have challenged. Dunlop (1974) has modified the "leaky stele" hypothesis by postulating that energy-dependent transport between the stelar parenchyma and the xylem regulates the solution concentration of the xylem fluid. This hypothesis predicts that low carbohydrate supply to the roots would inhibit uptake of ions from the xylem fluid to the stelar parenchyma and explains the transient increase in the concentration of ions into the xylem fluid under these conditions (Pitman, 1971).

A similar hypothesis is that of the "stelar pump"(Lauchli et al., 1971; Pitman, 1971), which suggests that stelar parenchyma cells have specialized mechanisms to pump ions into the stelar apoplast and/or the xylem vessels. This hypothesis is based on studies with metabolic inhibitors (Lauchli et al., 1971; Pitman, 1971; Jarvis and House, 1970) and the observation that ions sometimes appear to move against an electrochemical gradient into the xylem (Ginsburg and Ginzburg, 1974; Davis and Higinbotham, 1976).

A third hypothesis is based on work of Higinbotham et al. (1973) who showed that the outer metaxylem vessels of maize roots contain plasma membranes at distances up to 10 cm from the root apex. These membranes may mediate movement of ions into the xylem (Davis and Higinbotham, 1976). However, this hypothesis does not explain radial ion movement into the xylem in older portions of the root system where xylem membranes have disintegrated. It also fails to explain ion uptake in species in which xylem vessel membranes extend only a few millimeters from the root tip.

All the hypotheses mentioned above assume that the cortical parenchyma is the primary nutrient accumulatory system, that ions move inward from the cortical symplasm to the symplasm of the stele via plasmodesmata in the endodermis, and that the suberized walls of the endodermis are a barrier preventing significant efflux of ions and water from the stelar apoplast to the cortical apoplast.

With regard to the mechanism of ion translocation from the stelar parenchyma to the xylem, "researchers are well provided with theories to explain any contingency that may arise" (Dunlop, 1974). The radial and axial variation in root anatomy and ion uptake as well as the effects of transpiration, nutrient concentration, and root metabolism make the system difficult to study, and it is doubtful that an integrated model of ion uptake will be accepted by all authorities in the near future.

Leakage into Roots.   Active ion uptake is regulated by membranes bounding the symplast. This regulation could be avoided if ions moved to the xylem wholly within the apoplast. However, the Casparian strip of the endodermis, secondary and tertiary thickenings in endodermal cell walls (Clarkson and Robards, 1975; Harrison-Murray and Clarkson, 1973; Ferguson and Clarkson, 1975), and the suberization of the hypodermis (Ferguson and Clarkson, 1976a,b) clearly restrict apoplastic movement of some ions into stele, and their concentrations are often much higher outside these layers than inside. Important nutrients whose uptake can be restricted by these layers are Ca^{2+}, Mg^{2+}, and SO_4^{2-} (Lüttge and

Weigl, 1962; Weigl and Lüttge, 1962; Ferguson and Clarkson, 1976a,b). Heavy metals such as lanthanum and lead, Al^{3+} and several dies are also blocked by the endodermis (Robards and Robb, 1974; Nagahashi et al., 1974; Peterson et al., 1981).

Lateral roots are initiated in the stele and rupture the endodermal cell layer when they grow outward through the cortex and epidermis. Recent evidence indicates that the ions and dies discussed above can enter the stele via the apoplast through the disrupted endodermal layer (Peterson et al., 1981; Dumbroff and Peirson, 1971; Ferguson, 1979). In addition, some of these ions and dyes may enter the stele before the Casparian strip has developed (Peterson et al., 1981; Karas and McCulley, 1973; Robards and Jackson, 1976). These breaks in the endodermal barrier may explain why plant root systems are not completely effective in excluding toxic ions and high-molecular-weight organic dyes that move apoplastically. In addition, it may explain why high transpiration rates and high solution concentrations of nutrients can cause reduced selectivity among nutrients.

7

TEMPERATURE

Temperature is one of the most important factors affecting the growth of C_4 grasses. In statistical analyses of environmental factors correlated with growth and yield, temperature is often among the most important variables (Chudleigh et al., 1977; Nelson and Dale, 1978; Vong and Murata, 1977).

GENERAL EFFECTS OF TEMPERATURE

Ambient temperatures affect species distribution, photosynthesis and other biochemical processes, dry matter accumulation and partitioning, expansion growth, phenological development, and other aspects of crop growth.

Species Distribution

The distribution of C_4 grasses was discussed in detail in Chapter 3. On a worldwide basis the most important factor affecting the relative distributions of C_3 and C_4 grasses is temperature. C_3 grasses are best adapted to the cooler temperatures found in temperate climates and at high elevations in the tropics. C_4 grasses are best adapted to warmer temperatures prevailing during the summer in temperate and subtropical regions and to warm lowland areas of the tropics. Species, ecotypes, and cultivars of C_4 grasses differ in their temperature responses (Cooper and Tainton, 1968; Sosebee and Herbel, 1969; Sweeney and Hopkinson, 1975). These genotypic differences are evident in tolerance to freezing (Hacker et al., 1974; Ivory and Whiteman, 1978b; Ludlow and Taylor, 1974), chilling (Lee and Estes, 1982; Mock and McNeill, 1979; Stamp et al., 1983), and high temperatures (Ivory and Whiteman, 1978c; Sosebee and Herbel, 1969; Sweeney and Hopkinson, 1975). Such genotypic differences are very important in determining species and ecotype distribution across altitudinal and longitudinal gradients. In addition, differences in temperature responses among ecotypes and cultivars normally reflect differences in the thermal environment of their native environments (McWilliams, 1978) or the areas in which they were selected (Lee and Estes, 1982; Stamp et al., 1983). For example, frost tolerance of kangaroograss (*Themeda australis*) ranges from very good in high-altitude races from southern Australia to very poor in low-altitude races from New

Guinea (Mison, 1972, cited in McWilliam, 1978). Similarly, inbred maize lines selected in cool regions are more tolerant of cool temperatures at the seedling stage than are lines selected in warm regions (Stamp et al., 1983).

Photosynthesis

C_4 grasses have higher optimum temperatures for photosynthesis than festucoid C_3 grasses such as wheat, barley, oats, and most temperate forage grasses. For example, Vong and Murata (1977) found that the optimum temperature for wheat photosynthesis is 18°C while that of maize, grain sorghum, and two millets is 30°C (Fig. 7.1).

For several tropical C_4 pasture grasses grown at 30°C, the minimum temperature for net photosynthesis is 5–10°C, the optimum temperature is 35–40°C, and the maximum temperature at which net photosynthesis is positive is 50–60°C (Ludlow and Wilson, 1971a) (Table 7.1).

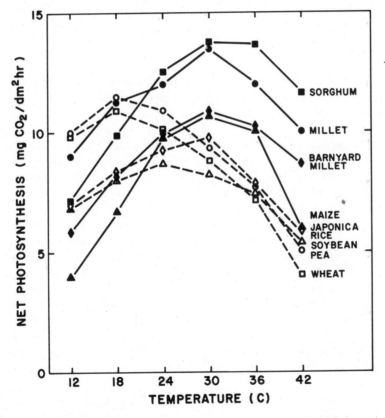

FIG. 7.1. Response of net photosynthesis to air temperature in C_3 (———) and C_4 (———) plants. From Vong and Murata (1977). Reproduced by permission of the authors and the *Japanese Journal of Crop Science*.

TABLE 7.1.   Minimum, Optimum, and Maximum Leaf Temperatures for Positive Net Photosynthesis (modified from Ludlow and Wilson, 1971a)

	TEMPERATURE (°C)		
SPECIES	MINIMUM	OPTIMUM	MAXIMUM
Buffelgrass (*Cenchrus ciliaris*)	6.0	39.0	61.0
Napiergrass (*Pennisetum purpureum*)	6.7	36.7	58.8
Guineagrass (*Panicum maximum*)	9.7	37.5	58.3
Molassesgrass (*Melinis minutiflora*)	6.2	39.2	58.0
Congograss (*Brachiaria ruziziensis*)	8.5	37.8	55.8
Columbusgrass (*Sorghum almum*)	5.0	39.5	52.0
Mean	7.0	38.3	57.3

Factors such as CO_2 concentration, light intensity, and vapor pressure deficit of the air also affect net photosynthesis and presumably affect the optimum temperature for net photosynthesis.

Dry Matter Accumulation

The general effects of temperature on dry matter accumulation have been studied in a number of experiments conducted in growth chambers and temperature-controlled greenhouses (McWilliam, 1978; Sweeney and Hopkinson, 1975; Whiteman, 1968). In most of these studies plants were grown in pots and no attempt was made to control soil temperatures. In general, they indicate that the optimum mean daily air temperature for C_4 grass dry matter accumulation is about 30–35°C. Growth is severely reduced above about 40°C and below about 20°C (Fig. 7.2 and Table 7.1). In contrast, festucoid C_3 grasses grow most rapidly at 20–25°C and grow poorly below 10°C and above 30°C.

The effect of temperature on dry matter accumulation is similar, though not identical, to its effect on photosynthesis. At suboptimal temperatures the relative photosynthetic rate is usually higher than the relative rate of dry matter accumulation (Fig 7.3). For example, the minimum temperature for maize and C_4 grass net photosynthesis is usually less than 10°C (Table 7.1), but the minimum temperature for dry matter accumulation is about 10°C or slightly higher (Duncan and Hesketh, 1968; Miedema and Sinnaeva, 1980; Singh et al., 1976). As a result, plants tend to accumulate nonstructural carbohydrates in the leaves, stems, and roots at suboptimal temperatures. TNC is predominantly starch in most C_4 grasses, but sucrose is the most important TNC in sugarcane and "sweet" sorghum.

TNC is usually highest at the end of the day and lowest at the end of the night. Low temperatures decrease the magnitude of this daily fluctuation, probably due to a greater reduction in dark respiration than in gross photosynthesis (Burns, 1972; Burris et al., 1967; McKell et al., 1969). In sugarcane, maximum

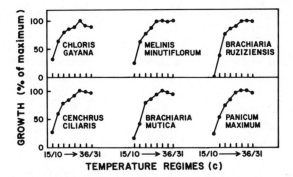

FIG. 7.2. Response of grass dry matter accumulation to day/night temperatures of 15/10, 18/13, 21/16, 24/19, 27/22, 30/25, 33/28, and 36/31°C. From Sweeney and Hopkinson (1975). Reprinted with permission of *Tropical Grasslands*.

stalk production is obtained at a mean temperature of 30 C° (Glasziou et al., 1965); however, maximum sucrose storage occurs at lower mean temperatures, especially when diurnal variation is wide (Brodie et al., 1969; Clements, 1980). Conversely, high mean temperatures combined with high night temperatures severely reduce the sucrose content and sugar yield of sugarcane.

Increase in both tiller number and tiller weight contribute to the growth of most grasses. For festucoid grasses the optimum temperature for tiller number is generally much lower than that for tiller size. This is also true for buffelgrass (*Cenchrus ciliaris*), rhodesgrass (*Chloris gayana*), and kikuyugrass (*Pennisetum clandestinum*); however, there is little difference in the optimum temperatures for tiller number and tiller size in slender guineagrass (*Panicum maximum* var. *trichoglume*) and kleingrass (*Panicum coloratum*) (Ivory and Whiteman, 1978c). The lower optimum temperature for tillering may reflect the greater

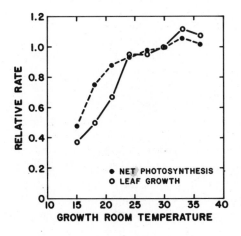

FIG. 7.3. Relative rates of maize net photosynthesis and leaf growth at different temperatures. From Duncan and Hesketh (1968). Reproduced from *Crop Science*, by permission of the Crop Science Society of America.

availability of nonstructural carbohydrates in the parent tiller at lower temperatures.

These results are consistent with studies on *Sorghum* growth (Escalada and Plucknett, 1975b; Major et al., 1982). For example, Downes (1968) found that axillary buds do not expand into tillers when the mean daily temperature is greater than 18°C. Plants tiller only when they spend a period (8–16 days) below 18°C at the four- or six-leaf stage. Major et al. (1982) found that most grain sorghum tillers appear from 10 to 24 days after emergence. Reducing ambient temperatures from 23/18 to 13/8°C during this period dramatically increases tiller number.

High temperatures reduce the duration of grain filling; cool temperatures lengthen it (Fussell et al., 1980; Hunter et al., 1977; Pearson and Shah; 1981). Few data are available on the effect of temperature on the rate of grain filling in C_4 grasses. A few studies indicate that warm temperatures increase the rate of kernel filling (D. P. Knievel, personal communication; Fussell et al., 1980); others suggest that temperatures between 35/25 and 25/10°C have little affect (Badu-Apraku et al., 1983). However, in the latter study, source rather than sink appeared to limit growth at high temperatures. Thus, while warm temperatures may increase the potential rate of kernel filling, high-temperature inhibition of photosynthesis may limit the amount of photosynthate available.

Expansion Growth

Leaf and stem elongation are very sensitive to temperature. The effects of temperature on sugarcane stalk growth and maize leaf growth have been carefully studied. When water and nutrients are adequately supplied, sugarcane stalk elongation is highly correlated with air temperature (Das, 1933a, 1936; Halais, 1935; Dillewijn, 1952). However, the relationship between elongation and temperature is affected by other factors. For example, cane often grows more rapidly at a given temperature in spring than in autumn (Fig. 7.4), probably due to differences in mineral nutrition, soil moisture, and crop age (Clements, 1980; Stender, 1924; Sun and Chow, 1949).

Grass leaf elongation is also very sensitive to temperature (Kleinendorst and Brouwer, 1970). However, in many grasses the meristematic region of the leaf is at or below the soil surface during much of the growth of the tiller. Until recently is was unclear whether the temperature of the meristem in the soil or the leaf in the air was most important in regulating leaf elongation. Watts (1971, 1972, 1973, 1974) studied the independent effects of root, meristem, and leaf temperatures on maize leaf elongation. Soil solution and air temperatures were regulated independently, and leaf meristem temperature was controlled with a thermostatic collar fitted around the base of the shoot. Relative humidity of the air was maintained at 100% to eliminate transpiration and ensure that root temperatures had minimal effect on leaf water potential. When the temperature of the meristematic region was held at 25°C, air and root zone temperatures had little short-term effect on leaf elongation. In contrast, when air and root temperatures

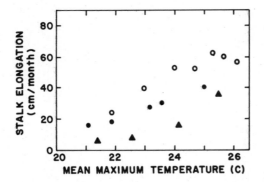

FIG. 7.4. Relationships between mean monthly maximum temperature and sugarcane stalk elongation during three periods of the year in Hawaii (Stender, 1924). ● = Oct. 1922–Feb. 1923 (decreasing temperatures); ○ = Mar. 1923–Sept. 1923 (increasing temperatures); △ = Oct. 1923–Jan. 1924 (decreasing temperatures and ripening). Courtesy of Hawaiian Sugar Planters' Association.

were maintained at a constant 25°C but the temperature of the meristematic region was varied, leaf elongation was strongly affected (Fig. 7.5). When root and air temperatures were varied independently and meristem temperature was allowed to fluctuate normally, leaf elongation was a linear function of measured meristem temperature (Fig. 7.5). Other studies have confirmed the strong effect of meristem temperature on leaf elongation in maize (Barlow et al., 1977) and wheat (Kemp and Blacklow, 1982). The optimum meristem temperature for leaf elongation is about 35°C for maize and 28°C for wheat.

Phenological Development

The phenological development of plants is accelerated by warm temperatures and delayed by cool temperatures. For example, for the African C_4 bunchgrass *Andropogon gayanus*, the time to anthesis decreases to less than half when mean temperature is raised from 17.5 to 27.5°C (Tompsett, 1976). High temperatures after anthesis shorten the grain filling period of grain sorghum and can reduce both seed weight and number (Balasko and Smith, 1971; Peters et al., 1971).

The effects of temperature on the duration of the various growth stages has been studied most intensively in maize. Several indices have been used to quantify these effects. The "growing degree-day" (GDD) is the most common index. The value of GDD for a given day is the difference between the daily mean temperature and a threshold temperature, which for maize is usually assumed to be about 10°C. A GDD index is obtained by summing daily GDD from planting to some stage of development, usually silking or maturity. A better index, the "effective degree" index, is obtained when the GDD is modified by assuming that (1) any daily minimum temperature below 10°C is equal to 10°C and (2) any daily maximum temperature above 30°C is corrected by subtracting from the daily mean temperature the number of degrees by which the daily maximum temperature exceeded 30°C (Gilmore and Rogers, 1958). Another index, the

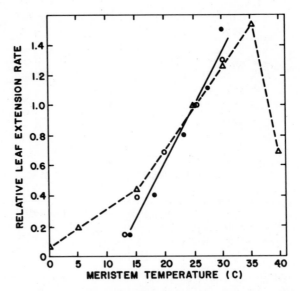

FIG. 7.5. Effect of meristem temperature on relative leaf elongation rate. Redrawn from Watts (1972). ——●—— Measured meristem temperature at various air temperatures; root temperature 25°C. ——○—— Measured meristem temperatures at various root temperatures; air temperature 25°C. --△--Meristem temperatures controlled by thermostatic collar; air and root temperatures 25°C. Reproduced by permission of Oxford University Press from *Journal of Experimental Botany*.

"modified GDD" (MGDD), sets any daily maximum temperature greater than 30°C equal to 30°C before computing the daily mean temperature (Barger, 1969). The "temperature function" (FT) method (Coelho and Dale, 1980) uses a complex function relating relative growth rate to temperature. The value of FT increases from 0.0 at 6°C to 1.0 at 28°C; it then decreases linearly from 1.0 at 32°C to 0.0 at 44°C.

Similar degree-day methods have been used to predict the time from sowing to emergence in field crops. For example, Angus et al. (1981) found that a simple linear degree-day system using screen air temperature could account for over 90% of the variation in thermal time from planting (3 cm) to emergence in seven C_4 cereals (Table 7.2).

Das (1933a,b), Halais (1935) and others have used modified degree-day calculations to predict sugarcane growth and yields. They could also be used to predict sugarcane emergence, which requires about 150 degree-days (base 10°C) (Clements and Nakata, 1967).

For grasses, the number of leaves on a tiller is fixed when the apical meristem differentiates to form the panicle. High temperatures reduce the interval between emergence of successive leaves, thereby reducing the interval between panicle initiation and panicle emergence. For example, Downes (1968) found that grain sorghum leaves appear at a rate of 0.19 leaves/day at a mean daily temperature of 13°C. The rate of leaf appearance increases linearly with increas-

TABLE 7.2. Base Temperatures (t_b) and Degree-Days (DD) Required for Emergence of Seven C_4 and Three C_3 Cereals when Planted at 3 cm at Monthly Intervals at Three Sites in Eastern Australia (modified from Angus et al., 1981)

SPECIES	t_b (°C)	DD	ADJ. R^2
C_4 Species			
Maize (*Zea mays*)	9.8	60.8	91
Grain sorghum (*Sorghum bicolor*)	10.6	47.9	96
Pearl millet (*Pennisetum typhoides*)	11.8	39.5	97
Proso millet (*Panicum miliaceum*)	10.4	44.7	94
Foxtail millet (*Setaria italica*)	10.9	42.4	90
Japanese millet (*Echinochloa frumentaceae*)	7.9	63.6	90
Finger millet (*Eleusine coracana*)	13.5	40.1	99
C_3 Species			
Wheat (*Triticum aestivum*)	2.6	77.9	46
Barley (*Hordeum vulgare*)	2.6	79.3	39
Oats (*Avena sativa*)	2.2	91.1	32

ing mean temperature to 0.29 leaves/day at a mean temperature of 22°C. Tollenaar et al. (1979) have shown that several temperature-based methods can successfully predict the rate of maize leaf appearance in environments with fluctuating temperatures.

Rates and durations of growth are not affected equally by temperature. For example, the maximum rate of leaf elongation occurs at about 30°C, but maximum final leaf length occurs at about 20°C because the duration of leaf extension is much longer at 20°C than at higher temperatures (Grobellaar, 1963) (Fig. 7.6). Similarly, maximum tiller and leaf numbers are often found at temperatures several degrees cooler than those at which maximum leaf area and tiller weight occur (Downes, 1968; Escalada and Plucknett, 1975b; Ivory and Whiteman, 1978c).

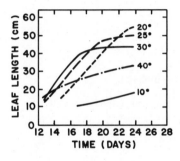

FIG. 7.6. Change in the length of maize leaf 5 when the shoots were at 20°C and the roots were at the temperatures shown. Redrawn from Monteith (1979) and used by permission of John Wiley & Sons.

There is also an inverse relationship between temperature and *in vitro* digestibility of C$_4$ grasses (Dirven and Deinum, 1977; Ivory et al., 1974; Wilson et al., 1976). For example, Henderson and Robinson (1982) found that digestibility of four C$_4$ grasses decreases about 1% for each degree increase in daytime temperature (between 26 and 35°C at 14 or 21 days regrowth in controlled environment). Pitman and Holt (1982) found up to 3.9% decrease in digestibility per degree increase in maximum temperature (for four C$_4$ grasses grown in the field).

Digestibility of C$_4$ grasses by ruminant animals generally decreases as the percentage of stem and/or structural carbohydrates increases. More rapid physiological development at high temperatures (with the attendant increase in the percentage of stems) is undoubtedly responsible for some of the decrease in digestibility. However, even leaf digestibility is reduced at high temperatures (Pitman and Holt, 1982).

Thermal Adaptation

Several studies have shown that the photosynthetic rate of C$_4$ grasses is able to adapt to different temperature regimes (Ludlow and Wilson, 1971b; Monson et al., 1983; Phillips and McWilliam, 1971). Plants grown at high temperatures generally have higher temperature optima for photosynthesis than plants grown at cool temperatures. For example, Bennett et al. (1982) reported that maize plants grown at 16, 25, and 33°C had optimum photosynthetic temperatures of 31, 33, and 38°C, respectively (Fig. 7.7). However, a single day at a higher or lower temperature was sufficient for considerable photosynthetic adaptation to the new temperature regime (Bennett et al., 1982c; Ludlow and Wilson, 1971b).

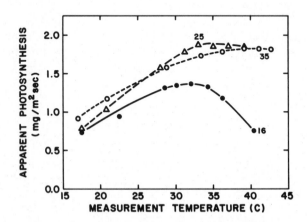

FIG. 7.7.   Effect of growth temperature (16, 25, or 35°C) on temperature response of apparent photosynthesis in maize leaves. From Bennett et al. (1982). Reproduced by permission of *Australian Journal of Plant Physiology* and CSIRO Editorial and Publications Service.

Diurnal Temperature Variation

Went (1944) proposed that maximum plant growth rates are obtained when days are warmer than nights, presumably because high day temperatures favor rapid photosynthesis while low night temperatures reduce dark respiration. However, low night temperatures can also cause reduced photosynthesis on the following day, possibly due to reduced stomatal conductance (Ku et al., 1978; Pasternak and Wilson, 1972; West, 1970). Glasziou et al. (1965) reported that for sugarcane grown at equal mean temperatures, diurnal temperature variation resulted in higher growth rates than constant temperatures. Ivory and Whiteman (1978a) studied different combinations of day and night temperatures using several C_4 forage grasses. At suboptimal mean temperatures equal day/night temperatures give the highest growth rates. However, at optimum or supraoptimum mean temperatures, the highest growth rates occur when day temperatures are about $10°C$ higher than night temperatures (Fig. 7.8). They conclude that at suboptimal temperatures a reduction in night temperature causes more reduction in net photosynthesis than in dark respiration. At optimum and supraoptimum temperatures the reverse is true.

FREEZING TEMPERATURES

Most temperate grasses are physiologically adapted to withstand some frost. Though cold temperatures severely reduce their growth, tissue damage is often minor. In contrast, C_4 grasses are more susceptible to frost. Freezing temperatures severely damage tissues and frequently kill the entire plant.

Environmental and Plant Factors

Environmental factors such as the temperature and duration of the frost period, the relative humidity during the period, and the temperature during the "hardening" period prior to freezing all affect the plant's response to frosting. For example, Ivory and Whiteman (1978b) showed that in several varieties of buffelgrass (*Cenchrus ciliaris*) a frosting temperature of $-1.75°C$ caused less than 50% foliage death, but a frosting temperature of $-2.5°C$, a difference of less than one degree, caused over 90% foliage death. Increasing the duration of freezing from 0.5 to 4.0 hr increased frost damage of setariagrass (*Setaria anceps*) and green panic (*Panicum maximum* var. *trichoglume*) by 35–58% (Ivory and Whiteman, 1978a). Reducing the rate of cooling increased frost damage (presumably by increasing the size of ice crystals formed in the tissue), but changing the reheating rate had no effect (Ivory and Whiteman, 1978a). At low relative humidities a temperature of $-5°C$ caused almost no frost damage to buffelgrass, but at near 100% relative humidity the same temperature killed more than 90% of the tillers (Ludlow and Taylor, 1974).

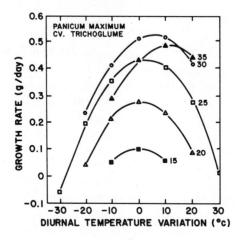

FIG. 7.8. Effect of diurnal temperature variation (day minus night) on daily dry matter accumulation of green panic (*Panicum maximum* var. *trichoglume*) at mean temperatures of 15, 20, 25, 30, and 35°C. From Ivory and Whiteman (1978c). Reproduced by permission of *Australian Journal of Plant Physiology* and CSIRO Editorial and Publications Service.

Plant nutrient and moisture status also affect frost tolerance (Adams and Twersky, 1960; Reeves et al., 1970). For example, Ivory and Whiteman (1978a) found that low plant water potential increased frost tolerance of setariagrass and green panic (Table 7.3). In other studies the balance of N, P, and K appears to be important (Davis and Gilbert, 1970). High tissue N may increase top kill due to frost (Beaty et al., 1973) but also hastens recovery in the spring (Reeves and McBee, 1972).

Hardening

Winter hardening is a well-known phenomenon in festucoid grasses, which modify their physiological responses to withstand low winter temperatures (Dexter, 1956). Four-carbon grasses harden much less than that of their C_3 relatives, but recent work has clearly shown that numerous C_4 grasses can harden and,

TABLE 7.3. **Effect of Leaf Water Potential on Foliar Frost Damage (Percent Leaf Area Killed) in Setariagrass (*Setaria anceps*) and Green Panic (*Panicum maximum* var. *trichoglume*) (modified from Ivory and Whiteman, 1978a)**

SPECIES	FROSTING TEMPERATURE (°C)	LEAF WATER STATUS		
		−1.5 MPA	−0.8 MPA	−0.6 MPA
Setariagrass	−1.5	7	0	0
	−2.5	9	6	15
	−3.5	16	35	40
Green panic	−1.5	6	1	2
	−2.5	63	64	92
	−3.5	89	98	100

thereby, limit the damage caused by frost (Hacker et al., 1974; Rogers et al., 1977; Rowley, 1976).

The degree of hardening depends on the species and the length and temperature of preconditioning. For most C_4 forage grasses, maximum hardening occurs when night temperatures are in the range of 5–10°C for about one week. More frost-resistant species may require lower temperatures for maximum hardening, and the least frost-resistant species apparently cannot be hardened (Ivory and Whiteman, 1978a; Ludlow and Taylor, 1974).

If a tropical grass plant has not been hardened by exposure to cold temperatures, a single cold night (1–10°C) can dramatically reduce the plant's stomatal conductance and net photosynthesis rate on the following day, regardless of the day temperature (Ivory and Whiteman, 1978d; Pasternak and Wilson, 1972). However, plants harden after exposure to several successive cool nights.

Cold hardiness of tropical grasses is normally greatest early in the winter and gradually declines until spring (Davis and Gilbert, 1970; Dunn and Nelson, 1974), presumably due to decreases in soluble carbohydrate and protein reserves late in winter. Dehardening (deacclimation) is the process by which warm temperatures cause a reduction in cold tolerance. C_4 lawn grasses such as St. Augustinegrass (*Stenotaphrum secundatum*), centipedegrass (*Eremochloa ophiuroides*), and bermudagrass (*Cynodon dactylon*) normally lose much of their cold tolerance with a few days of exposure to warm temperatures (Chalmers and Schmidt, 1979; Johnson and Dickens, 1976; Reeves and McBee, 1972), but deacclimation of axillary buds may be slower (Chalmers and Schmidt, 1979).

Genotype Variation

Though hardened C_4 grasses rarely attain the frost tolerance of festucoid C_3 grasses, they often have sufficient tolerance to resist mild frosts. In addition, considerable ecotypic variation in frost tolerance occurs among C_4 grass species and ecotypes. For example, Ivory and Whiteman (1978b) found that in controlled freezing studies the temperature at which 50% of the leaves were killed (LT_{50}) varies. In five *Cenchrus ciliaris*, four *Setaria anceps*, and six *Chloris gayana* ecotypes the ranges of LT_{50} were −1.8 to −2.1°C, −2.6 to −3.5°C, and −2.1 to −4.3°C, respectively.

In general, ecotypic variation in frost tolerance relates to the altitude or latitude of origin of the ecotype. Greater frost tolerance is normally found in ecotypes from cooler climates (Hacker et al., 1974). For example, frost tolerance of kangaroograss (*Themeda australis*) ranges from very good in high-altitude races from southern Australia to very poor in low-altitude races from New Guinea (Mison, 1972, cited in McWilliam, 1978).

The search for frost-tolerant genotypes is important in extending the range of C_4 forage grasses farther into the temperate zone. A good example is the development of Midland bermudagrass (*Cynodon dactylon*), a cultivar combining the frost tolerance of northern European bermudagrasses with vigor and digestibility similar to the relatively intolerant Coastal bermudagrass (Beaty et al., 1973).

Screening germplasm for frost tolerance is possible either by measuring tissue death in controlled environments (Beard et al., 1981; Hacker et al., 1974; Ivory and Whiteman, 1978b) or by measuring the electrical conductivity of the cell sap (Calder et al., 1966). These techniques give results similar to those obtained in the field, and they permit rapid screening of large numbers of genotypes.

CHILLING TEMPERATURES

Many plants of tropical origin are severely damaged by prolonged exposure to chilling temperatures (0–15°C) (Lyons et al., 1979). In C_4 grasses chilling temperatures cause retarded growth, stomatal malfunction, chlorosis, apparent phosphorus deficiency, chlorotic bands (Faris bands) on the leaves, and poor seed germination.

Seedling Growth

Seedling growth of C_4 grasses is quite susceptible to chilling temperatures (Harper, 1955; Herbel and Sosebee, 1969). Because maize is frequently planted in cool soils in temperate climates, the effect of chilling on maize seedlings is well understood. In addition, laboratory tests have been developed to select genotypes with superior seedling growth at chilling temperatures (Burris and Navratil, 1979; Isley, 1950).

Seedling growth of maize is divided into three stages: a heterotrophic stage when seedlings are wholly dependent on seed reserves, a transition stage during which both heterotrophic and autotrophic growth can occur, and an autotrophic stage during which growth is dependent on photosynthesis (Cooper and Mac-Donald, 1970; Hardacre and Eagles, 1980).

The heterotrophic and autotrophic growth of maize seedlings have different temperature minima. Heterotrophic growth can proceed at temperatures below 13°C (Blacklow, 1972; Eagles and Hardacre, 1979); however, photosynthesis occurs slowly or not at all below 15°C (McWilliam and Naylor, 1967). Thus, under periods of prolonged low temperatures, the heterotrophic phase may be completed, but at the beginning of the autotrophic phase growth may not occur due to low temperatures. Under these conditions plants become chlorotic and may even die.

Hardacre and Eagles (1980) showed that considerable genotypic variation exists in the ability of maize to grow autotrophically at low temperatures. For example, three U.S. Corn Belt dent varieties did not grow autotrophically at 13°C; however, a number of high-altitude races from Peru and Mexico and some crosses of these high-altitude races with Corn Belt dents did grow. Genes present in these populations should enable breeders to develop hybrids with lower minimum temperatures for autotrophic growth. It is possible that lower minimum temperatures will be associated with lower optimum temperatures. For example, Duncan and Hesketh (1968) found that high-altitude Mexican races grow faster than low-altitude races at temperatures below 24°C, but the low-altitude races grew faster at temperatures above 24°C. This suggests that the entire temperature response of maize can be adjusted to local conditions.

Stomatal Conductance

One symptom of chilling injury is rapid wilting. Maize is quite susceptible during the seedling stage, when it is often exposed to cool soil and air temperatures during early spring in temperate climates (Gupta and Kovacs, 1974).

Under chilling soil and air temperatures, maize stomates remain partially open despite water stress caused by reduced hydraulic conductivity of the root system. The resultant low plant water potential produces wilting. The loss of stomatal transpiration control probably results from cold-induced changes in guard cell permeability and inactivation of the K^+-ATPase pump involved in stomatal regulation (Garber, 1977; Nobel, 1974).

Chilling-tolerant maize cultivars maintain lower stomatal conductance in chilling temperatures than do sensitive cultivars (Vigh et al., 1981). Their photosynthetic rate also recovers more rapidly when warmer temperatures return (Mustardy et al., 1982).

Chlorosis

Chilling temperatures can induce chlorosis or browning of leaves in maize (Millerd and McWilliam, 1968; Mock and McNeill, 1979; Stamp, 1981a,b), bermudagrass (*Cynodon dactylon*), japanese lawngrass (*Zoysia japonica*) (Sachs et al., 1971), and other C_4 grasses (Gallopin and Jolliffe, 1973). This is partially due to reduced synthesis and esterification of a precursor of protochlorophyll (McWilliam, 1978; McWilliam and Naylor, 1967). Inhibition of carotene synthesis by cold temperatures probably aggravates the photodestruction of existing chlorophyll. The number of photosynthetic units per unit leaf area may also decrease (Teeri et al., 1977). Ultrastructural abnormalities of chloroplasts may also occur (Millerd et al., 1969; Pearson et al., 1977), though they are not necessarily associated with chlorosis (Taylor and Craig, 1971).

Faris Bands

Transverse chlorotic lesions called Faris bands develop on the leaves of some C_4 grasses after exposure to low night temperatures (Faris, 1927; Taylor et al., 1975). Bands do not appear immediately after treatment, but they become visible at the point of leaf emergence about 24 hr after the chilling stress. The bands are irreversibly chlorotic due to structural abnormalities in the mesophyll chloroplasts. The bundle sheath chloroplasts usually contain chlorophyll and have normal chloroplast lamellar structure. The abnormal development of the mesophyll plastids may be at least partly due to a chilling-induced failure of plastid ribosome production (Slack et al., 1974).

Generally, Faris band production is greatest when night temperatures are cool and day temperatures are warm. Plants normally adapt to the low-temperature regime within a few days and cease to produce the bands. Though there is wide genotypic variation in banding, plants that produce Faris bands are not always among the most chilling-sensitive genotypes. For these reasons absence of Faris bands is probably a poor criterion for the selection of low-temperature hardiness (Taylor et al., 1975).

Phosphorus Deficiency Symptoms

Both chilling temperatures and P deficiency cause anthocyanins to accumulate in the leaf sheaths and internodes of C_4 grasses, and there is good evidence that in both cases the purplish hue is associated with P deficiency in the shoot. The effects of chilling temperatures on P metabolism has been studied in sugarcane.

Under normal conditions the internode tissue of sugarcane contains a pool of labile P that is translocated to the leaves and incorporated into organic compounds in the event of P deficiency (Hartt, 1958). Cool air temperatures can increase the concentration of P in this pool, presumably by reducing the P demand of photosynthetic tissues; however, cool root temperatures can also reduce P uptake by the root system and its translocation to the top (Hartt, 1958). For example, decreasing root temperature from 22.2 to 16.7°C caused a sevenfold decrease in ^{32}P uptake; however, after 24 hr at the lower root temperature, the ^{32}P content of the root system was 2.5 times that at 22.2°C due to inhibition of translocation to the shoot (Hartt and Kortschak, 1967).

Membrane Fluidity

Chlorosis, reduced P uptake, and poor seed germination are probably all symptoms of membrane disfunction caused by a chilling-induced phase change in membrane lipids. The phase change is apparently a gradual change in the membrane that occurs over a temperature range of several degrees. The liquid–crystalline membrane is transformed to a gel phase as the temperature is lowered (Raison, 1974). This may alter the configurations and, thereby, modify the activities of membrane-bound enzymes.

Wolfe (1978) proposed that at warm temperatures the lipid bilayer of the membrane is completely fluid, allowing free movement of proteins attached to or embedded in it. As the temperature falls, the lipids become less fluid, restricting the movement of proteins and perhaps reducing diffusion through membrane pores. Ordering of water molecules in the vicinity of the membrane increases. This affects cation concentrations and electric charge near the hydrophilic portions of the proteins. The changes combine to affect the energy of activated enzyme complexes, and conformational changes may even destroy enzyme function.

Changes in membrane fluidity are commonly estimated from the paramagnetic resonance spectra of membranes into which spin-labeled aliphatic carbon compounds (compounds containing a stable nitroxide group) have been infused. A change in the ordering of the spin-labeled compounds within the membrane can be represented graphically as a change in the slope of the Arrhenius plot (the log of spin-label motion against the reciprocal of absolute temperature).

Raison et al. (1979) found that the phase change usually occurs around 0°C in festucoid grasses but between 6 and 12°C in C_4 grasses. Genotypic variation in the critical temperature at which chilling injury occurs is related to the lipid composition of their membranes. Plants with a greater percentage of unsaturated, short-chained lipids are more chilling resistant (Raison, 1974). The effects of temperature on the motion of a spin label infused into chloroplasts of two ecotypes of kangaroograss (*Themeda australis*) of different susceptibilities to chilling are shown in Fig. 7.9. The ecotype from Tasmania is more resistant to chilling than that from New Guinea, and this is reflected in the lower temperature associated with the phase change of the chloroplast membrane.

Enzyme Activity

The temperature-induced change in the fluidity of membrane lipids affects the activation energy of enzyme complexes embedded in the membrane.

Arrhenius plots which employ rates of enzyme activity instead of the motion of a spin-label can be used to detect the temperature at which changes in activation energy of enzymes occur.

A change in slope (Fig. 7.10), corresponding to a change in activation energy of PEP carboxylase, occurs at about 12°C in the chilling-susceptible *Sorghum sudanense*. However, *S. halepense*, a chilling-tolerant species adapted to subtropical regions, shows no increase in activation energy (McWilliam and Ferrar, 1974).

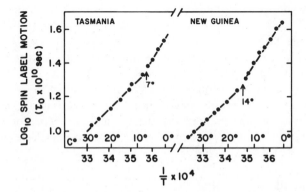

FIG. 7.9. Temperature-dependent changes in the motion of a spin label infused into chloroplasts from two kangaroograss (*Themeda australis*) cultivars. From Raison (1974). With permission of the Royal Society of New Zealand.

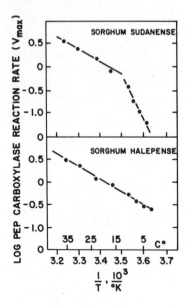

FIG. 7.10. Arrhenius plots of PEP carboxylase temperature response in sudangrass (*Sorghum sudanense*) and johnsongrass (*S. halepense*). From McWilliam and Ferrar (1974). With permission of the Royal Society of New Zealand.

Cool temperatures affect activities of maize enzymes in both the C_4 and Calvin–Benson photosynthetic pathways (Taylor et al., 1974). RuBP carboxylase and NADP-malate dehydrogenase activities decreased rapidly when maize seedlings were transferred from warm to chilling temperatures (Stamp, 1980, 1981a). In contrast, PEP carboxylase and NADP-malic enzyme activities were relatively insensitive to chilling temperatures (Stamp, 1980).

Levitt (1980) stresses the role of energy-dependent ion pumps in maintaining normal ionic concentrations within the cells. According to his theory, the activity of these pumps decreases abruptly below the temperature at which the phase change occurs in the membranes.

The simultaneous changes in membrane fluidity, activation energy of enzymes involved in carbon fixation and photoreduction of NADP, and leakage of solutes from chilling-sensitive tissues is strong evidence that low temperatures dramatically affect lipid–protein interactions of chilling-sensitive plants. Downton and Hawker (1975) have even suggested that the lipids involved need not be associated with membranes. They propose that lysolecithin, a nonmembrane lipid that makes up 85% of the lipid component of starch, is involved in the activation of starch synthetase in a variety of plants. Since the activation energies of both soluble and starch-bound maize starch synthetase show a temperature-dependent change at about 12°C, they suggest that a complex between lysolecithin and starch synthetase is affected by low temperatures.

Hormone Effects

Chilling temperatures reduce the rate of sugarcane internode elongation more than the rate of leaf appearance. As a result, internodes elongating during cool periods are shorter than those elongating during warm periods. The effect of temperature on internode length can be largely overcome by applying gibberellins to the foliage during cool periods (Bull, 1964; Mongelard and Mimura, 1972; Yates, 1972). The gibberellin is absorbed and translocated to the meristem, where it stimulates elongation. Serial applications at 15–30-day intervals during the cool winter months result in 2–5% increases in sugar yields in Hawaii (Buren et al., 1979; Moore and Buren, 1978; Moore et al., 1982). Maximum response to gibberellin application occurs when cool temperatures limit growth. Under these conditions, gibberellin production by or translocation from the roots may be inhibited, which would explain the effects of exogenous gibberellin (Mongelard and Mimura, 1972). Genotype, age, the amount applied, and the frequency of application all affect response to gibberellin and application.

Karnok and Beard (1983) found that cool (7/5°C) temperatures reduce photosynthesis dramatically in St. Augustinegrass (*Stenotaphrum secundatum*) and bermudagrass (*Cynodon dactylon*). Gibberellin application reverses these effects in bermudagrass, but not in St. Augustinegrass. In similar work, Karbassi et al. (1971a) demonstrated that gibberellins can reverse many of the effects of cool (10°C) nights on pangolagrass (*Digitaria decumbens*).

Chloroplast Starch Accumulation

When environmental conditions favor active growth, most of the carbon fixed by C_4 grasses during the day is exported within a few hours. That which is not exported is stored (primarily as starch) in the chloroplasts and is exported during the following night, when it is used in dark respiration and growth. Warm temperatures decrease the amount of starch present in the leaves at the end of the day and increase its translocation from the leaves at night. For example, in dallisgrass (*Paspalum dilatatum*) plants grown at temperatures between 28/20 and 15/10°C, leaf starch varied from 3.0 to 5.3% of dry weight at the end of the day and from 0.8 to 5.3% at the end of the night (Forde et al., 1975). The same general pattern of accumulation and mobilization of leaf starch has been found in pangolagrass (*Digitaria decumbens*) and pearl millet (*Pennisetum americanum*) (Chatterton et al., 1972; Hilliard and West, 1970; West, 1970, 1973). West and his colleagues noted that the severe growth depression resulting from chilling night temperatures are associated with accumulation of starch granules in both the mesophyll and bundle sheath chloroplasts. Cultivars of pangolagrass and weeping lovegrass (*Eragrostis curvula*) whose growth is least affected by chilling temperatures are those that retain the least starch in their chloroplasts after chilling nights and that have the highest amylolytic enzyme activity at chilling temperatures (Carter et al., 1972; Hilliard, 1975; Karbassi et al., 1971a,b).

In contrast to lovegrass and pangolagrass, pearl millet accumulates sucrose rather than starch under chilling conditions (Pearson and Derrick, 1977). In fact, the most chilling-sensitive genotype had the highest photosynthetic rate, but also, it had the highest leaf sucrose content at chilling (18/13°C) temperatures.

The negative correlation between growth and carbohydrate retention in the leaf at chilling temperatures suggests that either photosynthate demand by the meristem or translocation to the meristem limits growth under chilling conditions. Low leaf temperatures drastically reduce starch translocation, even when nearby sinks are maintained at high temperatures (to increase their demand for photosynthate) (Garrard and West, 1972). When cooling is restricted to a band above but not including the leaf meristem, it has little effect on meristem cell division or on leaf water potential. However, it can dramatically reduce cell elongation, probably by reducing carbohydrate available for cell wall formation in the region of cell elongation (Kleinendorst and Brouwer, 1972). Thus, chilling temperatures can limit both translocation of photosynthate to meristematic tissue and the ability of meristems to utilize photosynthate.

The reduction in net photosynthesis associated with low night temperatures, though sometimes temporary (Ludlow and Wilson, 1971a), has been attributed to physical damage of the chloroplasts by the large starch grains (West, 1973), feedback inhibition of photosynthesis (Forde et al., 1975; Glasziou and Bull, 1971; but see Pearson and Derrick, 1977), and disruption of photophosphorylation and electron transport (Pearson et al., 1977). The swelling and starch accumulation commonly observed in mesophyll chloroplasts do not always occur in bundle sheath chloroplasts (Forde et al., 1975; Hilliard, 1975; Millerd et al., 1969). This is consistent with chloroplast damage in Faris bands, which is most severe in the mesophyll cells (Slack, et al., 1974).

Root Zone Temperatures

The optimum soil temperatures for both shoot and root growth are normally higher for C_4 than for festucoid C_3 grasses (Fig. 7.11) (Morrow and Power,

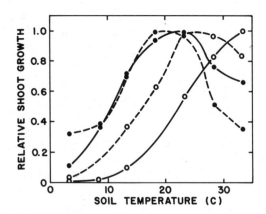

FIG. 7.11. Response of shoot (———) and root (— — —) growth to soil temperature at constant (23°C) air temperature in C_3 crested wheatgrass (●) (*Agropyron desertorum*) and C_4 blue grama (○) (*Bouteloua gracilis*) (Morrow and Power, 1979). Reproduced from *Agronomy Journal*, by permission of the American Society of Agronomy.

1979). Soil temperature is one of the most important factors in the spring growth of C_4 forage grasses in subtropical regions (Chudleigh et al., 1977; Morrow and Power, 1979). It is also very important in the germination and early growth of annual crops such as maize and grain sorghum in temperate and subtropical regions (Beauchamp and Lathwell, 1966; Mock and McNeill, 1979; Rykbost et al., 1975), and in high altitudes in the tropics (Arkel, 1980).

The important effects of soil temperature on leaf elongation have been discussed above. Other effects of chilling root zone temperatures are reduced nutrient absorption, reduced incorporation of inorganic N and P into organic compounds in the root, decreased organic N and P translocation to the shoot (Hartt and Kortschak, 1967; Theodorides and Pearson, 1982), increased anion leakage from the roots (Holobrada et al., 1981), and reduced hydraulic conductivity of root cell membranes (Kuiper, 1964; Markhart et al., 1979; McWilliam et al., 1982). For example, pearl millet (*Pennisetum americanum*) took up 20% less NO_3^- at 18/13°C than at 30/25°C, but the reduction in amino acid movement into the xylem was even greater (40%). At 30/25°C, 50% of the absorbed NO_3^- was reduced in the roots, but only 20% was reduced in the roots at 18/13°C. At 18/13°C, NO_3^- reduction in the roots was not adequate to supply even the root system with reduced N, and reduced N was imported from the shoot (Theodorides and Pearson, 1982). Theodorides and Pearson (1982) suggest that low NO_3^- reduction in the roots is due to inadequate translocation of photosynthate to the root system at chilling temperatures.

Genotypic Variability

In maize, cold tolerance has been found in populations from all latitudes. It is genetically controlled, and recurrent selection has been used to improve cold tolerance of inbreds (Mock and McNeill, 1979; Mock and Skrdla, 1978; Vallejos, 1979), which are generally more sensitive to chilling than are hybrids (McWilliam and Griffing, 1965; Pesev, 1970; Pinnell, 1949).

Similar genetic variation in cold tolerance has been found among species of *Sorghum* (Bagnall, 1979; McWilliam et al., 1979) and among grain sorghum (*S. bicolor*) cultivars (Stickler et al., 1962). For example, at chilling temperatures grain sorghum has lower net photosynthetic rates, lower relative growth rates, and a greater reduction in leaf chlorophyll than do the more chilling-tolerant species johnsongrass (*S. halepense*) and *S. leiocladum* (Bagnall, 1979). However, germination, respiration, and mesocotyl extension rate are less sensitive to chilling temperatures in grain sorghum than in *S. leiocladum* (McWilliam et al., 1979). Thus, a single enzymatic or growth parameter should not be used as an overall index of genetic tolerance of chilling temperatures.

HIGH TEMPERATURES

The general effects of high temperatures on the growth of tropical grasses have been discussed earlier in this chapter. In general, C_4 plants, including tropical grasses, have higher temperature optima than C_3 plants, though considerable variation exists in each group.

Plant Biochemistry

Heat stress affects a number of biochemical processes in plants. Membrane properties are affected by high as well as by chilling temperatures (Raison et al., 1980). Cytoplasmic, mitochondrial, and nucleolar ultrastructure of maize roots are altered and RNA synthesis is inhibited by heat shock (46°C) (Fransolet et al., 1979). Photophosphorylation (Lawanson, 1976), photosynthetic enzyme activities (Bjorkman et al., 1980), and net photosynthesis (Sullivan et al., 1977; Bjorkman et al., 1980) are also inhibited by heat stress. As a result, the CO_2 compensation point of maize increases at temperatures above 40°C (Bykov et al., 1981). Finally, when the temperature is raised from 25 to 40 or 45°C, maize roots produce "heat shock proteins," whose importance and role in plant biochemistry remains unclear (Cooper and Ho, 1983).

There is considerable evidence that plants, including C_4 grasses, can be hardened to high-temperature stresses (Raison et al., 1980; Sullivan et al., 1977; Teeri, 1980). The thermal responses of enzymes, membranes, and whole plants adapt to ambient temperatures; and high-temperature hardening may last for several weeks, though genotypes differ in their ability to harden (Sullivan et al., 1977).

Germination and Seedling Growth

One of the most severe effects of high temperature on C_4 grass growth is its inhibition of germination and early growth. The high midday light intensities found in the tropics and subtropics often cause temperatures of bare surface soils to exceed 50°C for several hours during the day (Sosebee and Herbel, 1969). Though C_4 grasses have higher optimum germination temperatures than most temperate species, their germination and early growth are often reduced above 40 or 45°C (Harrison-Murray and Lal, 1979; McWilliam et al., 1970).

Low surface soil moisture and high soil temperature dramatically reduce forage grass seedling survival and growth (Herbel and Sosebee, 1969); however, surface soil mulch or shrub cover can improve the microclimate for seedling survival. For example, shrub cover can reduce the maximum soil temperature at 13 mm by 20°C in semiarid subtropical regions (Sosebee and Herbel, 1969), and straw mulch (6 tons/ha) can reduce soil temperatures at 5 cm by over 5°C (Harrison-Murry and Lal, 1979). The reduction in surface soil temperatures is usually associated with increased plant growth.

There is considerable genetic variability among C_4 grasses in susceptibility to high soil temperatures. For example, maximum soil temperatures of 53°C at 13 mm depress shoot and root growth of boer lovegrass (*Eragrostis chloromelas*) but hardly affect those of black grama (*Bouteloua eriopoda*) (Herbel and Sosebee, 1969). Usberti and Jain (1978b) found that mortality rates of guineagrass

(*Panicum maximum*) seedlings varied from 13 to 77% among six populations when plants were exposed to air temperatures of 51.5°C for 5 hr. Survival was greatest in populations with greatest leaf and sheath pubescence, a character associated with tolerance to drought and heat stress in other species (Ehleringer, 1980).

Reproduction

In most temperate and subtropical maize-growing regions, high air temperatures are associated with drought stress. It has long been known that maize is most susceptible to drought stress during pollination (Robbins and Domingo, 1953), due partly to pollen sterility (Denmead and Shaw, 1960; Lonnquist and Jugenheimer, 1943). However, Herrero and Johnson (1980) found that even when tassels are excised and placed in water, maximum air temperatures of 38°C reduce pollen germination to levels below those found at 27 and 32°C.

The potential number of florets per panicle is determined during early panicle development, and Downes (1972) showed that heat stress at panicle initiation reduces the number of grain sorghum florets that eventually develop. High temperatures during the boot stage reduce grain sorghum seed set (Pasternak and Wilson, 1969; Sullivan, 1972). Heat tolerance increases as panicles mature and emerge from the boot. Finally, high temperatures reduce the grain filling period, thereby reducing the total amount of light interception and photosynthesis occurring during this critical period (see Chapter 5).

8

LIGHT, CO$_2$, PHOTOSYNTHESIS, AND GROWTH

When temperature, soil water, and mineral nutrition are adequate for rapid growth, light interception and CO$_2$ concentration are usually the most important growth-limiting factors. This chapter provides an overview of the effects of light intensity, CO$_2$ concentration, light interception, leaf angle, shading, competition for light, and assimilate demand on leaf and canopy photosynthesis and crop growth.

The solar spectrum can be divided into three main regions: the ultraviolet (300–400 nm), the visible (400–700 nm), and the infrared (700–3000 nm). Solar radiation (SR) or short-wave radiation refers to radiation originating from the sun with wavelengths between 300 and 3000 nm. Short-wave irradiance varies from 0 at night to over 900 and 1200 mW/cm^2 at solar noon on clear days in temperate and tropical regions, respectively. Photosynthetically active radiation (PAR) refers to radiation in the visible (400–700 nm) region. About 45% of direct SR is PAR, but when both diffuse and direct components of SR are considered, PAR is about 50% of total SR (Monteith, 1973). Other useful definitions and conversion factors are given in Table 8.1.

LIGHT INTENSITY AND CO_2 CONCENTRATION

Light intensity has a profound effect on plant growth and development. Whether one discusses a square centimeter of leaf blade exposed to different irradiances in a growth chamber or an entire maize or sugarcane field, the light intensity to which that leaf or field is exposed can greatly affect its rates of photosynthesis and growth.

TABLE 8.1.   Definitions and Conversion Factors Used in Chapter 8 (Monteith, 1973; Shibles, 1976)

Definitions

Solar radiation (SR): Radiation in the 300–3000-nm wave band.

Photosynthetically active radiation (PAR): Radiation in the 400–700-nm wave band. PAR is about 50% of SR.

Photosynthetic photon flux density (PPFD): Photon flux density of PAR. The number of photons (400–700 nm) incident per unit time on a unit surface. [Suggested units: nE/cm^2 s (E = Einsteins)]

Photosynthetic irradiance (PI): Radiant energy flux density of PAR. The radiant energy (400–700 nm) incident per unit time on a unit surface. [Suggested units: mW/cm^2 (W = Watts)]

Intercepted PAR (IPAR): Photosynthetically active radiation intercepted over some time period. [Suggested units: MJ/m^2 (J = joules)].

CO_2 exchange rate (CER): Net rate of CO_2 diffusion from air to a tissue or plant. [Suggested units: nmol/cm^2 s (mol = moles)].

Conversion Factors

1 nE/cm^2 s (PPFD) = 0.22 mW/cm^2 (PI)

1 MJ/m^2 (IPAR) = 23.9 cal/cm^2 (Langleys)

1 cal/cm^2 min = 69.7 mW/cm^2

1 nmol CO_2/cm^2 s = 0.063 mg CO_2/dm^2 hr

PAR available for crop growth varies from month to month, from day to day, and from minute to minute depending on the time of year, the angle of the sun, day length, and cloud cover (Monteith, 1978). In addition, the percentage of incident PAR that is intercepted and absorbed by the crop canopy varies with the size of the canopy and its structure. PAR that is absorbed by photosynthetic tissue is converted to chemical energy, which is used to fix atmospheric CO_2 into a wide variety of chemical compounds. The efficiency of this conversion depends on numerous environmental and plant factors, including temperature, soil water, plant nutrition, age and health of the leaves, demand for assimilate, and availability of atmospheric CO_2. The effects of temperature, soil water, and plant nutrition are discussed in other chapters. The capture and utilization of light and CO_2 by C_4 grasses are discussed in this chapter.

Since 1958, atmospheric CO_2 concentrations have risen from about 315 cm^3/m^3 to slightly over 330 cm^3/m^3. If present trends of fossil fuel use continue, concentrations will double in about 50 years. The impact of this change on global temperatures and crop production has been widely debated (Allen, 1980; Hansen et al., 1981; Idso, 1980, 1981, and 1983a; Kimball, 1983; Kramer, 1981). The principal concern is that use of fossil fuels will increase atmospheric CO_2 concentrations, increase the absorption and reradiation of infrared radiation by the atmosphere, and increase global temperatures. Current estimates for the increase in global temperatures due to a doubling of atmospheric CO_2 concentrations range from about 0.25 to 4°C. Whatever the future effects of increased CO_2 concentrations on global temperatures, it seems clear that CO_2 concentrations will continue to increase. This will probably have a significant effect on crop productivity in the next 50 years (Rosenberg, 1981).

Leaf Photosynthesis

Light intensity has a marked effect on the photosynthetic rate of plant leaves. At normal ambient CO_2 concentrations, the net photosynthetic rate of individual C_3 leaves rarely exceeds about 1.3 nmol CO_2/cm^2 s, and C_3 leaves are usually light saturated between 35 and 70 mW/cm^2 (7.7 and 15.4 nE/cm^2 s) (400–700 nm) (Bull, 1969; Hesketh and Baker, 1967). However, at normal CO_2 concentrations the photosynthetic rate of C_4 grass leaves is not light saturated at maximum clear-day solar irradiance (105–140 mW/cm^2), where net photosynthetic rates of 3.2–3.8 nmol CO_2/cm^2 s are often found (Bull, 1969; Hesketh and Baker, 1967; Hesketh and Moss, 1963) (Fig. 8.1).

CO_2 concentration interacts with light intensity to affect leaf photosynthesis. Increasing CO_2 concentrations above ambient has little effect on photosynthetic rate at low light intensities; however, at high light intensities increasing the CO_2 concentration can cause a significant increase in photosynthesis (Fig. 8.2).

At near-optimum temperatures and light intensities, the CO_2 concentration at which net photosynthesis approaches its maximum rate is generally lower in C_4 than in C_3 leaves (Akita and Moss, 1973); Morison and Gifford, 1983). For example, Morison and Gifford (1983) found that the net CO_2 assimilation rate of

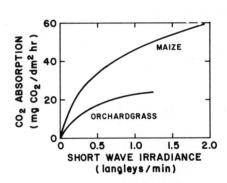

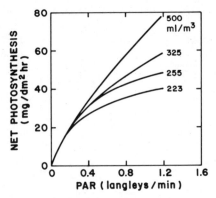

FIG. 8.1. Effect of short-wave irradiance on CO_2 uptake by maize and the C_3 species orchardgrass (*Dactylis glomerata*). From Hesketh and Baker (1967). Reproduced from *Crop Science*, by permission of the Crop Science Society of America.

FIG. 8.2. Effect of photosynthetically active radiation (PAR) and CO_2 concentration on net photosynthesis of maize leaves. From Hesketh and Moss (1963). Reproduced from *Crop Science*, by permission of the Crop Science Society of America.

single attached maize and brownseed paspalum (*Paspalum plicatulum*) leaves reached a maximum at CO_2 concentrations of 300–350 cm^3/m^3. However, in the C_3 species, rice and *Phalaris aquatica*, leaf photosynthesis continued to increase to CO_2 concentrations of 600 cm^3/m^3. These results indicate that photosynthesis of C_4 plants is less sensitive to above-ambient CO_2 concentrations than is C_3 photosynthesis. It suggests that yield increases resulting from increasing ambient CO_2 concentrations will be greater in C_3 than in C_4 plants.

This suggestion is borne out by several studies in which long-term (5–11 weeks) growth of C_3 and C_4 species have been compared at several CO_2 concentrations in growth chambers and in the field (Carlson and Bazzaz, 1980; Carter and Peterson, 1983; Patterson and Flint, 1980; Rogers et al., 1983). In virtually all cases, the increase in growth due to elevated CO_2 concentrations was greater in C_3 than in C_4 plants. In several cases growth of C_4 grasses was actually depressed by CO_2 concentrations of about 1000 cm^3/m^3 (Fig. 8.3), probably due to stomatal closure at these high external CO_2 concentrations.

Canopy Photosynthesis

Under field conditions, solar radiation, temperature, and CO_2 concentration in the crop vary seasonally, daily, diurnally, hourly, and vertically within the crop (Allen, 1971). Since all these variables affect leaf photosynthesis, estimation of canopy photosynthesis from meteorological information would appear to be a formidable task. Fortunately, under most field situations light intensity is the predominant factor limiting canopy photosynthesis, and it often accounts for most of the variation in crop photosynthesis. For example, Moss et al. (1961) measured net canopy photosynthesis of maize over 347 one-hour periods be-

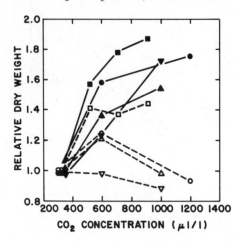

FIG. 8.3. Effect of CO_2 concentration on dry weight accumulation in C_3 (closed symbols) and C_4 (open symbols) plants. $\bigcirc$ = maize, $\bullet$ = soybean (35 days). (Carlson and Bazzaz, 1980). ∇ = maize, $\triangle$ = *Rottboellia exalta*, $\blacktriangledown$ = soybean, $\blacktriangle$ = *Abutilon theophrasti* (45 days) (Patterson and Flint, 1980). $\square$ = maize, $\blacksquare$ = soybean (77 days). (Rogers et al., 1983).

tween July 15 and October 1 in New York. The found a 0.95 correlation between incident solar radiation and net photosynthesis during this 2.5-month period. In another study Hesketh and Baker (1967) found a good correlation between net canopy photosynthesis and solar radiation intercepted by the maize canopy (Fig. 8.4). It is now generally recognized that in the absence of water, temperature, or nutrient stress, there is a linear relationship between canopy light interception and net canopy photosynthesis in maize and probably in other C_4 crops (Hatfield, 1977; Hesketh and Baker, 1967).

LIGHT INTERCEPTION

Light that is not intercepted by the crop canopy cannot be used for crop growth. The interception and absorption of light by crops has been studied intensively since the early 1950s (Monsi and Saeki, 1953). These studies have produced several geometrical and statistical models of light interception by individual plants and by crop canopies. These have been reviewed by Lemeur and Blad (1974) and Thornley (1976). This discussion is limited to a simple, though useful, application of Beer's law to light interception by crop canopies, first attempted by Monsi and Saeki (1953).

As applied to crop canopies, Beer's law simply states that the fraction of incident photosynthetically active radiation (PAR) passing through the canopy decreases exponentially as leaf area index increases. Thus,

$$PI_g = PI_o e^{-kL}$$

where PI_o and PI_g are photosynthetic irradiance above the canopy and at ground level, respectively; k is the canopy extinction coefficient; and L is leaf area index. The value of k varies with the inclination and orientation of the foliage, light

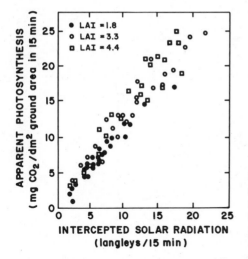

FIG. 8.4. Relationship between intercepted solar radation and apparent photosynthesis of a maize canopy. From Hesketh and Baker (1967). Reproduced from *Crop Science*, by permission of the Crop Science Society of America.

scattering and transmission by the foliage, sun angle, row spacing, and row direction, and latitude (Thornley, 1976). For example, Uchijima et al. (1968) found that in maize k decreases from about 1.7 when the sun is only 15° above the horizon, to about 0.9 at 30°, to 0.5–0.7 at sun angles above 60°. Muchow et al. (1982) found little effect of plant population on midday k in grain sorghum, but isolateral plant spacing produced higher values of $k(0.63)$ than normal 75-cm rows ($k = 0.45$).

PI$_o$ perpendicular to the crop surface is highest near midday when solar angle is high. Since most of the energy available for photosynthesis is received when solar angle is greater than 60°, and k is quite stable at solar angles between 60° and 90°, many researchers have used an "average" value of k obtained near-solar noon to describe daily canopy light interception. Under these conditions, crops with horizontal leaf orientations such as clover, bermudagrass, and pangolagrass have "average" values of k of 0.7–1.0 (Brougham, 1960; Brown et al., 1966; Ludlow et al., 1982). Crops such as maize, grain sorghum, and pearl millet with a mixture of erect upper leaves and more horizontal lower leaves have values of k ranging from 0.4 to 0.7 (Begg et al., 1964; Bonhomme et al., 1982; Brougham, 1960). Grasses with extremely erect leaves have values of k as low as 0.3 (Ludlow et al., 1982; Monteith, 1965). The effects of LAI and k on the fraction of PI passing through the canopy are given in Fig. 8.5.

LEAF ANGLE

Dewit (1965) divides plant canopies into several types based on the distribution of leaf angles within them. The two extreme forms are the planophile canopy in which most leaves are horizontal and the erectophile canopy in which most leaves are erect. In the planophile canopy the upper leaves intercept most of the

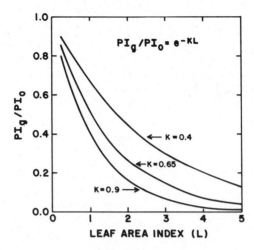

FIG. 8.5. Effect of leaf area index (L) and the crop canopy extinction coefficient (K) on the fraction of solar radiation reaching the soil surface (PI_g/PI_o).

light, and the lower leaves are severely shaded. In the erectophile canopy direct sunlight penetrates more deeply into the canopy. When the solar angle is high, the upper leaves of planophile canopies are more nearly perpendicular to the sun's rays than are the leaves of erectophile canopies. The leaves of planophile C$_3$ species (in which net photosynthesis is light saturated at about one-third full sunlight) cannot make efficient use of this direct sunlight. However, the leaves of erectophile species are not perpendicular to the rays of the sun. The same amount of light energy is spread over a larger leaf area. Thus, if the same amount of light is intercepted by two otherwise identical canopies, a C$_3$ crop with an erectophile canopy should fix more CO$_2$ than one with a planophile canopy because the light is distributed over a larger leaf area and is utilized more efficiently. Simulation models based on light interception theory (Duncan, 1971; Monsi et al., 1973) predict that C$_3$ crops with erectophile canopies should be more productive than planophile canopies at high plant populations and leaf area indexes; however, C$_3$ crops with planophile canopies should be more productive at low populations because they intercept more light than erectophile canopies. In view of this analysis the ideal C$_3$ crop canopy should have erect upper leaves to make efficient use of sunlight at the top of the canopy and horizontal lower leaves to intercept light not absorbed by the upper leaves (Mock and Pearce, 1975).

In contrast to C$_3$ plants, the photosynthetic response of C$_4$ leaves to irradiance is not saturated when full sunlight falls perpendicularly on the leaf (Bull, 1971). If the C$_4$ photosynthetic response to irradiance were linear, no advantage of erectophile canopies would be predicted because the same amount of light energy would be equally effective whether it fell on a small leaf area in a planophile canopy or was distributed over a larger leaf area in an erectophile canopy. How-

ever, there is sufficient curvilinearity (Fig. 8.1) in the C_4 photosynthetic response
to irradiance to predict a slightly greater photosynthetic rate in erectophile canopies than planophile canopies if almost all radiation is intercepted (Duncan,
1971).

Considerable experimental evidence suggests that the preceding analysis is
correct. When maize leaves are mechanically manipulated to produce more
erectophile canopies, they usually produce higher grain yields at high population
densities (where light interception is nearly complete) and lower yields at low
population densities (where light interception by the erectophile canopy is lower
than the normal canopy) (Pendleton et al., 1968; Winter and Ohlrogge, 1973;
Fakorede and Mock, 1978). Similar results have been obtained with isogenic
hybrids differing only in leaf angle (Pendleton et al., 1968; Lambert and Johnson, 1978).

The amount of yield increase due to the erectophile canopy structure undoubtedly varies with environmental conditions, leaf area index, plant population, and the susceptibility of the hybrid to barrenness under high population
density (Trenbath and Angus, 1975). The effects of plant population on the ratio
of grain yield in erectophile canopies to that in normal canopies is shown for two
studies (Fig. 8.6), one involving mechanical manipulation of leaves (Winter and
Ohlrogge, 1973) and the other involving isogenic hybrids varying in leaf angle
(Lambert and Johnson, 1978).

Similar results have been found with mechanical manipulation of leaf angle
in C_4 salt marsh grasses (*Spartina patens* and *Distichlis spicata*) (Turitzin and
Drake, 1981). Whether erectophile canopies were mechanically manipulated to
produce planophile canopies or the reverse, erectophile canopies had higher
photosynthetic rates.

Other studies with isogenic and nonisogenic hybrids differing in leaf angle are
not consistent with the results discussed above (Ariyanayagam et al., 1974;
Hicks and Stucker, 1972; Pommer et al., 1981; Russel, 1972). In most of these
studies the hybrids with more horizontal leaf orientation intercepted more radia-

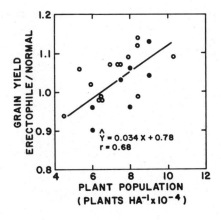

FIG. 8.6.   Effect of plant population on the ratio
of grain yield of erectophile maize to that of normal
maize. ● = Lambert and Johnson (1978), ○ =
Winter and Ohlrogge (1973).

tion than those with more erect leaves. Thus, the possible advantage of erect leaves (more efficient use of intercepted radiation) was probably offset by the greater interception of radiation by the hybrids with more horizontal leaves. These results clearly illustrate that in C$_4$ grasses erectophile canopies are advantageous only under conditions of high leaf area index and almost complete light interception (Duncan, 1971).

SHADING AND COMPETITION

The amount of light available to an individual plant is reduced when it is shaded by taller plants and when it is grown at high population densities. For example, the growth of weeds germinating below the crop canopy is often limited by shading, and individual plants grown at high population densities are usually smaller than those grown at lower populations. Plants adapt to shading and competition for light in several ways. The gross structure of the plant as well as leaf morphology and biochemistry adjust to changes in the light regime. This adaptation to suboptimal irradiance has recently been reviewed by Patterson (1980b).

One of the most common observations in shading studies is that shaded plants become elongated, the weight per unit leaf area (specific leaf weight, SLW) decreases, and the leaf area per unit shoot weight (leaf area ratio, LAR) increases (Earley et al., 1966; Ludlow et al., 1974; Eriksen and Whitney, 1981). Another common observation is that shading reduces root and rhizome growth proportionately more than shoot growth. This results in an increase in the ratio of shoot weight to root or rhizome weight (Burton et al., 1959; Patterson, 1980a; Eriksen and Whitney, 1981) (Table 8.2). Shading may also severely reduce the number of leaves, tillers, and rhizomes produced by the plant (Patterson, 1980a), though under certain conditions moderate shade may increase tillering (Inosaka et al., 1977) and even shoot growth (Eriksen and Whitney, 1981; Singh et al., 1974; Wong and Wilson, 1980).

TABLE 8.2. Effects of Shading on Components of Plant Dry Matter in Cogongrass *Imperata cylindrica* **(from Patterson, 1980a)**[a]

AVAILABLE LIGHT (%)	LEAF WEIGHT / TOTAL WEIGHT (g/g)	STEM WEIGHT / TOTAL WEIGHT (g/g)	ROOT WEIGHT / TOTAL WEIGHT (g/g)	RHIZOME WEIGHT / TOTAL WEIGHT (g/g)	LEAF AREA / LEAF WEIGHT (dm^2/g)	LEAF AREA / TOTAL WEIGHT (dm^2/g)
100	0.36	0.13	0.093	0.41	1.91	0.69
56	0.51	0.14	0.063	0.39	2.31	0.94
11	0.50	0.29	0.055	0.16	3.31	1.64

[a] Reprinted by permission of the Weed Science Society of America.

Because of reduced photosynthesis, nonstructural carbohydrate concentration usually decreases dramatically under shade (Burton et al., 1959), and in forage grasses heading may either be hastened or delayed (Inosaka et al., 1977).

A strong interaction between light intensity and nitrogen nutrition is often observed in shading experiments. When nitrogen fertility is low, shade can stimulate shoot dry matter yields and increase the N concentration of the shoot (Eriksen and Whitney, 1981; Navarro-Chavira and McKersie, 1983; Wong and Wilson, 1980); however, when nitrogen is abundant, increasing shade results in an almost linear decrease in shoot dry matter yield (Burton et al., 1959).

The response of grain crops to shading is more complex than that of forage crops, and a number of studies on the effects of shading have been conducted with maize and grain sorghum. In addition, a few studies have been conducted on the effects of light-enriched environments on maize growth and yield.

In grain sorghum two stages of plant development are particularly sensitive to shading. If shading occurs during the 2-week period prior to 50% anthesis, the number of seeds per panicle can be severely reduced. If shading occurs during the milk and dough stages, seed size may be reduced. However, when the number of seed per panicle is reduced by shading prior to anthesis, increased seed weight may partially or completely compensate for the reduced seed number (Pepper and Prine, 1972).

In maize, severe shading (60–90% shade) from plant emergence to just prior to tasseling or from just prior to tasseling to just after silking can severely reduce the number of kernels per ear. Shading after silking reduces final kernel weight (Earley et al., 1967).

Increasing the light intensity within the crop with reflectors or flourescent light can increase maize growth and grain yield per plant and per unit ground area (Graham et al., 1972; Pendleton et al., 1967). These studies suggest that light is a primary factor limiting maize yield in the field.

Maize plants growing under low levels of solar radiation acclimate by producing greater leaf area per unit leaf and shoot weight. This results in lower RuBP carboxylase activity and fewer stomata per unit leaf area than leaves developing in full sun (Singh et al., 1974; Gaskel and Pearce, 1980). In addition, maize leaves that develop in shade have lower rates of net photosynthesis at both low and high light intensities than leaves that develop in full sun (Singh et al., 1974; Gaskel and Pearce, 1980). Acclimation to shade conditions does not continue after leaves are fully expanded; therefore, leaves that are shaded by younger leaves maintain their potentially high rate of net photosynthesis after being shaded (Gaskel and Pearce, 1980). This explains why plants with the upper leaves removed produce as much grain per unit leaf area as those with the lower leaves removed (Schmidt and Colville, 1967).

The effects of population density and row spacing on maize yields have been studied extensively throughout the world. Bunting (1973), Downey (1977), and Major and Daynard (1972) reviewed a number of studies, and their conclusions are summarized below.

Increasing plant density from suboptimal (less than 4 plants/m^2) to supraoptimal (more than 8 plants/m^2) levels generally results in greater plant height, greater leaf area and biomass per unit ground area (but less leaf area and biomass per plant), thinner more erect leaves, fewer ears per plant, fewer grains per ear, smaller ears, lower grain quality, lower stalk strength (and more lodging), fewer tillers, delayed ear initiation and silking and a greater interval between tasseling and silking (resulting in increased barrenness) (Bunting, 1973; Iremiren and Milbourne, 1980; Wilson and Allison, 1978), and a decrease in the grain filling period (Poneleit and Egli, 1979).

Because of barrenness, maximum grain yield occurs at lower plant densities than does maximum dry matter production. For example, at low and mid latitudes, optimum grain yields are often obtained at 4-6 plants/m^2, but total dry matter production continues to increase up to 9 or 10 plants/m^2. At higher latitudes, such as in Great Britain and southern Canada, optimum grain yields are normally obtained at higher populations, 7-9 plants/m^2. This results from the shorter, cooler growing season in these regions. Smaller plants are produced, and more plants are needed to obtain maximum yields (Bunting, 1973; Downey, 1971).

Row spacing, fertility, and irrigation management also affect the optimum plant density. Narrower rows allow more rapid ground cover and better light interception during early growth. Better fertilizer and irrigation management reduce other sources of interplant competition. Thus, as farmers gradually reduce row spacing and improve plant nutrition and irrigation, optimum plant populations will increase. Production of smaller hybrids with more erect leaves and less barrenness will also promote this trend.

LEAF AGE

C_4 grass leaves normally lose part of their photosynthetic efficiency as they age (Ludlow and Wilson, 1971c; McPherson and Slatyer, 1973). In most studies, photosynthetic rates just prior to leaf senescence are about 50% those of newly expanded leaves (Thiagarajah et al., 1981; Vietor et al., 1977) (Fig. 8.7). Both stomatal and mesophyll resistance to CO_2 uptake increase with leaf age. In some studies increased stomatal resistance seems most important (Ludlow and Wilson, 1971c). In others, increased mesophyll resistance is the culprit (Thiagarajah et al., 1981).

McCree (1972) found that as maize leaves age the photosynthetic quantum yield decreases at all wavelengths between 400 and 700 nm, but the general shape of the action spectrum hardly changes. These results suggest a general decline in potential photosynthetic activity as C_4 grass leaves age. It is usually associated with a gradual decrease in leaf chlorophyll and nitrogen content. The photosynthetic metabolism of older leaves may change quite drastically in C_4 grasses. For example, Bull (1969) found evidence of photorespiration in an attached leaf of an 18-month-old sugarcane plant grown at low (17°C) temperatures. The leaf had a lower photosynthetic rate and lower chlorophyll content

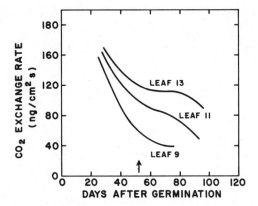

FIG. 8.7.   Effect of maize leaf age on CO_2 exchange rate. From Thiagarajah et al. (1981). Arrow indicates silking date. Reproduced by permission of the National Research Council of Canada from the *Canadian Journal of Botany*.

than others in the study. Like C_3 plants, its photosynthetic rate increased in an O_2-depleted atmosphere. Bull suggests that high levels of leaf assimilates in the leaves of old plants may produce measurable photorespiration in sugarcane.

ASSIMILATE DEMAND

The hypothesis that a buildup of starch and/or sugars in leaves can reduce photosynthesis of those leaves was first proposed in the mid-1800s, and it is still actively debated (Herold, 1980; Neales and Incoll, 1968; Pinto, 1980). A number of studies with C_4 grasses have revealed that the ratio of source to sink can affect leaf photosynthetic rate and plant dry matter accumulation. For example, reducing the sink size in maize by removing the ear or preventing pollination can cause a reduction in leaf photosynthesis (Barnett and Pearce, 1983; Moss, 1962) and shoot dry matter accumulation (Loomis, 1934–1935; Kiesselbach, 1948; Tollenaar and Daynard, 1982; but also see Allison and Watson, 1966; Verduin and Loomis, 1944). Similarly, removal of half the spikelets of grain sorghum shortly after anthesis can cause a reduction in total shoot weight at harvest (Fischer and Wilson, 1975). This suggests that reduction of the sink relative to the source can cause a decrease in photosynthesis and shoot dry weight accumulation.

Treatments that prevent assimilates from moving to or being utilized at the sink can also reduce photosynthetic rates. For example, physically breaking or killing the phloem at the base of a sugarcane leaf results in an increase in leaf sucrose content and a decrease in leaf photosynthetic rate (Hartt, 1963; Waldron et al., 1967).

As discussed in Chapter 7, chilling night temperatures can reduce C_4 grass growth even when day temperatures are warm. Hilliard and West (1970) suggest that low night temperatures prevent translocation of carbohydrates out of the chloroplasts, and the remaining chloroplast starch inhibits photosynthesis the following day. However, it is also clear that these effects of cool nights can be partially or completely negated by either the presence of active young tillers (Chatterton et al., 1972) or application of gibberellic acid (Karbassi et al., 1971a). Thus, increasing sink strength can override the inhibitory effects of cool night temperatures on photosynthesis.

Finally, decreasing the source–sink ratio may actually stimulate photosynthesis. For example, Barnett and Pearce (1983) showed that partially defoliating maize plants during ear growth causes the CO$_2$ exchange rate of the remaining leaves to increase. In addition, when maize plants are exposed to very low irradiance for one or two days (to reduce nonstructural carbohydrates) then are returned to a normal light regime, posttreatment photosynthetic rates are usually higher than pretreatment rates (Andre et al., 1982). These studies suggest that photosynthesis responds positively to a decrease in the source–sink ratio.

Though alternative interpretations have been offered for many of the studies sited above, both the mass and the diversity of evidence suggests that photosynthetic rates are sensitive to the ratio of source to sink. Possible biochemical control mechanisms sensitive to source and sink strengths are described in Herold (1980) and Pinto (1980).

PHOTOSYNTHESIS AND CROP YIELD

When rapid methods of measuring leaf photosynthesis were developed in the 1960s, plant physiologists and breeders were eager to exploit the new tool to select genotypes with more efficient photosynthesis, more rapid growth, and higher economic yields. Though they knew that growth and yield are determined by many interacting processes, they assumed that increasing the leaf's CO$_2$ exchange rate (CER) per unit leaf area was a key to increasing crop productivity.

Researchers began to search for genotypic differences in leaf CER. These were soon found in maize (Edmeades and Daynard, 1979; Gaskel and Pearce, 1981; Hanson, 1973; Vietor et al., 1977) and sugarcane (Irvine, 1967; Rosario, 1972). As expected, significant heterosis for maize leaf CER was also found (Crosbie et al., 1978; Heichel and Musgrave, 1969; Monma and Tsunoda, 1979). Several studies also revealed that CER is positively related to leaf chlorophyll, nitrogen, and DNA contents as well as to dry weight per unit area (Bhatt et al., 1981; Fleming and Palmer, 1975; Irvine, 1967; Monma and Tsunoda, 1979; Rao et al., 1978).

Within a few years it became evident that breeders could successfully select for genotypes with high leaf CER (Crosbie et al., 1977, 1978, 1981a). However, it also became evident that short-term measurements of leaf CER made once or twice during the season were not correlated with high growth rates (Duncan and Hesketh, 1968). For example, Hanson (1971, 1973 found that maize selected for high dry matter production during the vegetative stage had lower CER and lower chlorophyll and DNA contents than genotypes selected for low dry matter production.

Elmore (1980) and Evans (1975b) have cited several reasons why one would not expect short-term standardized measurements of photosynthesis to reflect long-term changes in dry matter and yield. After all, crop growth and yield occur over a wide range of environmental conditions, not only those conditions used to measure short-term leaf photosynthesis. Zelitch (1982) has pointed out that numerous measurements of canopy photosynthesis made frequently throughout the season are needed to predict dry matter accumulation accurately. For example, Kanemasu and Hiebsch (1975) and Vietor and Musgrave (1979) found relatively good agreement between canopy photosynthesis and crop dry matter accumulation in grain sorghum and maize, respectively.

MAXIMUM GROWTH RATES

At optimum temperatures, high levels of solar radiation, and normal CO_2 concentrations, the CER of C_4 leaves is normally higher than that of C_3 leaves. This suggests that under these near-ideal conditions the crop growth rate (CGR) of C_4 plants would be higher than that of C_3 plants when light interception by the canopies is equal. Gifford (1974) and Evans (1975a,b) have disputed this conclusion. They reviewed the literature on maximum crop growth rates observed in experiments and found that maximum growth rates by crops of the two photosynthetic pathways are similar. However, Monteith (1978) reviewed the same studies and concluded that experimental or numerical errors were responsible for the highest crop growth rates reported for C_3 plants. He concluded that, indeed, C_4 plants do have higher maximum short-term growth rates in the field than do C_3 plants (Table 8.3). All maximum short-term daily rates reported for C_3 plants were below 40 g/m^2 day while all those of C_4 plants were greater than 50 g/m^2 day.

Monteith (1978) also demonstrated that the higher short-term crop growth rates of C_4 plants can be translated into higher long-term growth rates (Fig. 8.8). Over the course of the growing season the mean CGR of the C_4 plants was 22.0 $\pm$ 3.6 g/m^2 day[1]. The mean CGR of the C_3 group was 13.0 $\pm$ 1.6 g/m^2 day[1]. These values are approximately 0.4 of the maximum short-term values of C_4 and C_3 plants in Table 8.2, probably reflecting the combined effects of incomplete interception of light at the beginning of the crop, senescence at the end, and suboptimal weather conditions at various periods during the growth of the crops.

LIGHT INTERCEPTION, CROP GROWTH, AND PHOTOSYNTHESIS

Monteith (1977) proposed that for several C_3 crops growing in Great Britain, dry matter accumulation is approximately a linear function of cumulative intercepted solar radiation. The same relationship can be estimated for several C_4

TABLE 8.3.   Maximum Short-Term Growth Rates of C$_3$ and C$_4$ Plants Grown in the Field[a]

PHOTO-SYNTHETIC PATHWAY	SPECIES	CROP GROWTH RATE (g/m^2 day)
C$_4$	Pearl millet (*Pennisetum americanum*)	54
	Maize (*Zea mays*)	52
	Grain sorghum (*Sorghum bicolor*)	51
C$_3$	Potato (*Solanum tuberosum*)	37
	Rice (*Oryza sativa*)	36
	Cattail (*Typha latifolia*)	34

[a] Reprinted by permission from J. L. Monteith. 1978. Reassessment of maximum growth rates for C$_3$ and C$_4$ crops. *Experimental Agriculture*, Cambridge University Press.

grasses (Bonhomme et al., 1982; Marshall and Willey, 1983; Muchow et al., 1982; Williams et al., 1968). These estimates are given in Fig. 8.9. From these relationships it is apparent that during vegetative growth C$_4$ grasses can produce 3.5–4.0 g shoot dry matter per MJ intercepted PAR. During grain filling, the rate of dry matter accumulation declines to 2.0–3.0 g shoot dry matter per MJ intercepted PAR in maize (Bonhomme et al., 1982; Williams et al., 1968) and grain sorghum (Muchow et al., 1982). Marshall and Willey (1983) reported rates of about 1.5 g/MJ intercepted PAR in pearl millet during the latter part of grain filling. The reduction in the rate dry matter accumulation per unit intercepted PAR during grain filling is probably due both to declining leaf photosynthetic capacity as the crop approaches maturity and the higher caloric value of a given weight of grain in comparison with the same weight of vegetative material (McDermitt and Loomis, 1981).

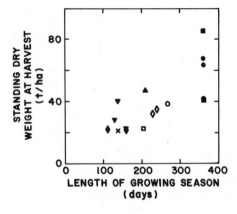

FIG. 8.8. Relationship between length of growing season and standing dry weight at harvest. C$_4$ grasses: ◆ = pearl millet, ▼ = maize, ▲ = grain sorghum, ● = sugarcane, ■ = napiergrass. C$_3$ crops: × = kale, ▽ = potatoes, ◇ = sugar beet, □ = rice, ○ = cassava, △ = oil palm. Reproduced from Monteith (1978).

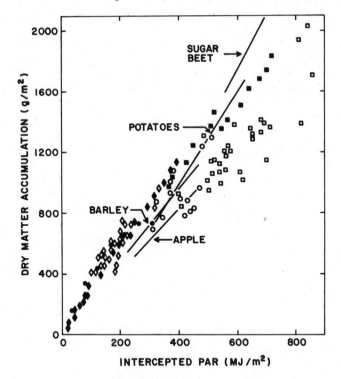

FIG. 8.9. Relationship between cumulative intercepted photosynthetically active radiation (PAR) and dry matter production of C_4 grasses. ■ = Maize, U.S., several plant populations, all growth stages (Williams et al., 1968). ◇ = Maize, France, several sites, prior to flowering (Bonhomme et al., 1982). □ = Maize, France, several sites, flowering to beginning of leaf senescence (Bonhomme et al., 1982). ◕ = Pearl millet, India, several growth stages (Marshall and Willey, 1983). ○ = Grain sorghum, Australia, several sites, grain filling period (Muchow et al., 1982). —— = Several C_3 crops, Great Britain (Monteith, 1977).

Over the entire growing season the C_3 and C_4 crops shown in Fig. 8.9 had similar rates of dry matter production per unit PAR absorbed. If this is true, why were both maximum short-term and long-term growth rates of C_4 grasses higher than those of C_3 crops (Table 8.2, Fig. 8.8)? Monteith (1978) points out that C_4 grasses are normally grown under higher solar radiation than C_3 crops; however, no definitive field studies have examined the efficiency of conversion of inter-cepted PAR in several C_3 and C_4 crops under near-optimum (25–30°C) tempera-ture regimes.

The problem of efficiency of light utilization has been addressed from a dif-ferent point of view. Several laboratory studies have examined the quantum yields of C_3 and C_4 plants at different temperatures, CO_2 concentrations, and O_2 concentrations. Both Ehleringer and Pearcy (1983) and Monson et al. (1982) found that at ambient CO_2 concentrations the quantum yield (moles CO_2/mol photons absorbed) of C_3 plants decreased with increasing temperatures. However, the quantum yield of C_4 grasses remains constant between 15 and 40°C. These two studies suggest that the quantum yields are approximately equal (0.05–0.06 mol CO_2/mol photons) at 25°C. Above 25°C, C_4 grasses are more efficient; below 25°C, C_3 plants have an advantage.

The shoot growth data shown in Fig. 8.9 are consistent with the quantum yield data cited above for C_4 grasses. If we assume a maximum conversion of intercepted PAR to shoot dry matter of 4.0 g/MJ, assume that 40% of the dry matter is carbon, assume 25% of fixed carbon is lost in dark respiration, and 25% is diverted to the root system, applying equations in Table 8.1 we obtain a quantum efficiency of 0.059 mol CO_2/mol photons. This is very similar to the values reported by Ehleringer and Pearcy (1983) and Monson et al. (1982).

9

DEFOLIATION

Defoliation plays an integral role in the management of most C_4 grass crops. For example, harvest of most range and forage grasses consists of defoliation either by animals or by machines. Burning is often used to remove unpalatable residues and stimulate grass regrowth. Sugarcane and grain sorghum can be ratooned after harvest to produce more than one crop from a single seeding.

Proper defoliation is a key to good rangeland and pasture management. If defoliation is too frequent and severe, desirable species may be reduced in importance, and herbaceous and woody weeds and unpalatable grasses may increase. If no defoliation occurs, large amounts of tall, unpalatable reproductive stems may accumulate. They exert strong apical dominance and shade young tillers. The result can be a pasture with large amounts of low-quality biomass but with low productivity. Improper management of sugarcane and grain sorghum defoliation results in stand reduction, slow regrowth of tillers, and poor sugar or grain yields of the ratoon crop.

Thus, defoliation is an important factor in the management of most C_4 grass crops, and understanding the plant's response to defoliation is crucial to good management.

TYPES OF DEFOLIATION

The three major types of defoliation—clipping, grazing, and burning—have different effects on grass growth and development. Clipping is often used to harvest hay and silage. It has the advantage of efficient, rapid, and uniform removal of regrowth; however, its effects on pastures are different than those produced by grazing. Large herbivores trample plants, select the most palatable species and plant parts, and return most plant nutrients to the pasture in the urine and feces. Thus, highly palatable species and species susceptible to trampling may maintain themselves under clipping but may completely disappear under selective grazing pressure.

Fire has been used since prehistoric times to remove dead, unpalatable residues, kill or burn back undesirable woody species, stimulate the growth of palatable leaves, and stimulate tillering and seed germination. It is an important management tool where grazing pressure cannot be controlled to prevent residue accumulation. It is also used to facilitate the harvest of sugarcane by removing dead leaves.

PHYSICAL EFFECTS OF DEFOLIATION

When the crop canopy is relatively complete and the water content of the soil profile is adequate, most solar radiation is absorbed by the canopy and converted to latent heat through transpiration. However, when severe defoliation occurs, especially due to burning, coverage can be reduced to the point that most of this radiation reaches the soil surface. The surface soil layers then tend to dry quickly, and absorbed radiation is dissipated advectively and as long-wave reradiation rather than as latent heat. This results in increased soil surface temperatures and increased vapor pressure deficits near the soil surface.

The changes in the pasture's physical environment are often accompanied by decreased water infiltration due to removal of foliage and mulch and the impact of raindrops on the unprotected soil surface. Runoff may increase, and severe interplant erosion can occur (Daubenmire, 1972; Savage, 1980; Wright and Bailey, 1982).

Burning also causes volatilization of N and S from plants and litter. However, soil organic matter transformations often increase as a result of high soil temperatures, oxidation of nitrification inhibitors, and increased cation availability. The resultant increase in soil nitrate and ammonium concentrations can stimulate plant growth (Wright and Bailey, 1982).

BIOLOGICAL EFFECTS OF DEFOLIATION

Defoliation has multiple biological effects. If leaf areas are severely reduced, crop photosynthesis and dry matter production decrease until leaf area recovers. Root growth and reserve carbohydrate concentrations of the roots and stem bases often decrease, but tillering and leaf growth may increase due to reduction of apical dominance.

The reduced food supply and changes in the canopy microclimate can dramatically affect insect and pathogen populations. In fact, heavy grazing or burning is sometimes prescribed as an economical means of dealing with insect attacks on pastures.

Shoot Apex Removal

During the development of most grass tillers, internodes are initially very short, and the apical meristem remains close to the soil surface. However, in many C_4 grasses panicle initiation marks the beginning of rapid internode elongation and the elevation of the apical meristem above the soil surface. In wind-pollinated species the elevation of the inflorescence prior to anthesis is needed to increase the efficiency of cross-pollination. However, it also makes the apical meristems and much of the leaf area susceptible to removal by grazers. Researchers have long recognized the importance of meristem height in defoliation tolerance (Aitken, 1962; Booysen et al., 1963; Branson, 1953). For example, Branson (1953) found that many of the apical meristems of switchgrass (*Panicum virgatum*) plants are elevated well above the soil surface early in the growing season. Big bluestem (*Andropogon gerardii*), on the other hand, has growing points that remain near the soil surface until much later in the season, and it is generally more tolerant of grazing than is switchgrass.

Shoot apex removal can severely limit dry matter production if the plant does not have the capacity to produce new tillers following defoliation. However, some species (e.g., Arizona cottontop, *Trichachne californica*) respond to defoliation by rapidly producing new tillers, regardless of their stage of development. These species can tolerate severe defoliation even though a large percentage of their tillers are normally reproductive and have elevated apical meristems (Cable, 1982).

Nonstructural Carbohydrates

Graber et al. (1927) first pointed out that carbohydrate reserves decrease in response to defoliation and that they play an important role in the plant's subsequent recovery. Numerous studies and literature reviews have since documented these effects (May, 1960; McIlroy, 1967; Smith, 1972c).

After defoliation, total nonstructural carbohydrates (TNC) are used both for maintenance respiration and for the synthesis of new tissue (Ehara et al., 1967). For this reason the TNC concentrations of stem bases and roots usually decrease during the first week after defoliation. The degree of TNC reduction depends on the severity of defoliation and the TNC concentration prior to defoliation. The higher the initial TNC concentration and the more severe the defoliation, the greater the subsequent decrease in TNC (Bartholomew and Booysen, 1969; Steinke and Booysen, 1968).

If TNC reserves are reduced to different levels before plants are defoliated, plants with higher initial TNC concentrations sometimes produce more regrowth than those with lower TNC concentrations (Humphreys and Robinson, 1966; Ward and Blaser, 1961). This effect of TNC is clearest when regrowth occurs in the dark and concurrent photosynthates are not available for regrowth. For example, Adegbola (1966) has shown that regrowth in the dark is positively related to the amount of TNC in giant stargrass (*Cynodon plectostachyus*) and in guineagrass (*Panicum maximum*). However, TNC losses cannot explain all regrowth in the dark (Davidson and Milthorpe, 1966; Humphreys and Robinson, 1966). Other substances also contribute, and Burton and Jackson (1962a) suggest that dark regrowth is a direct measure of maximum energy reserves for regrowth.

In both C_3 and C_4 grasses, regrowth after complete defoliation depends on reserve carbohydrates for about one week. Thereafter, reserves have usually been depleted, and concurrent photosynthesis from new leaves must support continued regrowth. The original levels of reserves are often restored within three to five weeks (Bartholomew and Booysen, 1969; Ehara et al., 1967; Steinke and Booysen, 1968).

Leaf Area

Defoliation often reduces leaf area below the level needed for near-complete interception of solar radiation. When this occurs in C_4 grasses, the rate of regrowth in the light is usually more dependent on residual leaf area than on TNC concentration (Jones and Carabaly, 1981; Ludlow and Charles-Edwards, 1980; Morgan and Brown, 1983). For example, Humphreys and Robinson (1966) studied the regrowth of buffelgrass (*Cenchrus ciliaris*) and green panic (*Panicum maximum* var. *trichoglume*). They used combinations of defoliation height and frequency to produce treatments in which residual leaf area varied from 0.0 to 0.75 and TNC varied by a factor of 2.7. When moisture and nutrients were adequate, growth was more dependent on LAI than on TNC. Plants having the lowest TNC but some residual leaf area regrew faster than those with higher TNC but little residual leaf area.

When almost all leaf area and many tiller meristems are removed, regrowth is often determined by the number of regrowing tillers (Dovrat et al., 1980; Ferraris and Sinclair, 1980). Since LAI and light interception are directly proportional to the number of regrowing tillers, regrowth is also proportional to total leaf area, light interception, and concurrent photosynthesis.

Tillering

Defoliation can either stimulate or reduce tillering. It often stimulates the growth of auxillary buds, especially when the apical meristem is removed. For example, clipping grain sorghum below the apical meristem at any growth stage (from 10 cm high to the dough stage) stimulates tillering (Singh and Colville, 1962). Similarly, removal of the apical meristem of Arizona cottontop (*Trichachne californica*) at any time during the growing season stimulates production of basal tillers (Cable, 1982).

However, severe and frequent defoliation can reduce leaf area, deplete carbohydrate reserves, and ultimately reduce tillering (Evers and Holt, 1972). For example, severe defoliation of maize above the apical meristem at the four-leaf stage can reduce the production of tillers, probably due to a reduction in TNC during the critical period for tiller initiation (Crockett and Crookston, 1980, 1981).

Root Growth

Resource allocation changes dramatically after severe defoliation. Photosynthesis decreases due to leaf area removal, and respiratory reserves normally used for root growth are diverted to shoot growth. Root initiation and elongation may cease within 24 hr (Crider, 1955; Hodgkinson and Baas Becking, 1977; Oswalt et al., 1959), and for several days or weeks root growth is more severely affected than shoot growth (Branson, 1956; Davis et al., 1959; Harradine and Whalley, 1981). Respiratory reserves are preferentially used to expand leaf area as quickly as possible. For example, Carman and Briske (1982) found that severe defoliation of all tillers on little bluestem (*Schizachyrium scoparium*) plants actually increases the rate of leaf area expansion on young tillers while decreasing the rate of root growth.

Defoliation decreases live root weight by reducing the growth of existing roots and causing them to die. Root death is often greater in plants with a high proportion of reproductive tillers, presumably because those tillers are incapable of regrowth after defoliation (Dovrat et al., 1980). Severe defoliation causes greater reduction in root growth than does moderate defoliation. Removal of 40% of the shoot may temporarily stop root elongation, but repeated removal of 70% causes long-term cessation of root growth (Bernardon et al., 1967; Crider, 1955).

Sugarcane is often ratooned after harvest, and the effects of harvest on subsequent root growth have been studied in detail (Evans, 1935; Glover, 1968). When the sugarcane stalk is cut near the soil surface, root elongation ceases within two to five days (Glover, 1968). However, these roots remain metabolically active for over one month. They are capable of water and nutrient absorption and support the initial regrowth of ratoon tillers until these develop their own root systems (Clements, 1980; Glover, 1968).

The ability of the old root system to support ratoon tiller growth is sensitive to soil disturbance at and after harvest. Mechanized harvesting can cause severe compaction. Various methods of subsoiling have been used to correct this compaction; however, subsoiling near the harvested plants severely damages their root systems and delays ratoon regrowth (Clements, 1980).

TOLERANCE OF DEFOLIATION

A plant's ability to tolerate defoliation depends on a number of factors, including the severity and type of defoliation, the stage of growth and morphology of the plant at the time of defoliation, its reserve carbohydrate status, and environmental conditions during regrowth.

Defoliation Height and Frequency

The height and frequency of defoliation can dramatically affect the growth and nutritive value of C_4 forage grasses (Evers and Holt, 1972; Middleton, 1982; Vicente-Chandler et al., 1959). Dry matter production, protein concentration, and digestibility are all affected. In general, frequency of defoliation is more important than height of defoliation (Fig. 9.1). Short intervals between defolia-

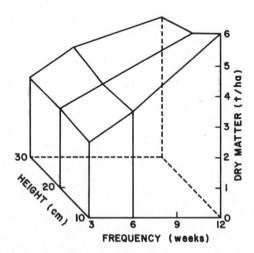

FIG. 9.1.   Effects of cutting height and frequency on cumulative dry matter production of three tufted C_4 grasses over 48 weeks in northeastern Queensland, Australia. From Middleton (1982). Reprinted by permission of *Tropical Grasslands*.

tion usually reduce cumulative dry matter yield; however, they also produce forage with high crude protein concentration and high digestibility (Fig. 9.2). The decrease in dry matter production is primarily due to the reduction in leaf area index and light interception each time the plants are defoliated (Morgan and Brown, 1983). The higher crude protein concentration and digestibility result from a high percentage of leaves with lower lignin contents in frequently defoliated plants (Fig. 9.3–9.5).

Cutting heights below about 10 cm can also severely reduce dry matter production in C_4 grasses. The reduction is usually due to excessive leaf area removal, severe depletion of carbohydrate reserves (Steinke, 1975), and reduction in the number of live tillers.

Frequent and severe defoliation can also change the morphology of the plant. For example, repeated and close defoliation often kills tillers near the center of the plant; however, new tillers emerge from the periphery of the crown. These tillers usually have a more horizontal orientation than the erect tillers near the center (Menke and Trlica, 1983). This produces plants with most of their leaf area concentrated near the soil surface (Ludlow and Charles-Edwards, 1980; Menke and Trlica, 1983). Plants whose morphology has adjusted to frequent close defoliation are better able to recover than those with erect tillers, probably because substantial leaf area remains near the soil surface following defoliation (Evers and Holt, 1972).

Though frequent or close defoliation can reduce dry matter production, some defoliation is often necessary to maintain high dry matter productivity. Removal of mature photosynthetically inefficient tillers stimulates the growth of new tillers by reducing apical dominance and allowing light to penetrate to the level of young basal tillers (Cable, 1982).

Owen and Wiegert (1981) suggest that many grass species have evolved in the presence of large herbivores and that a mutualistic relationship has developed. This relationship has favored species with low basal meristems out of reach of most grazers; with rapid vegetative reproduction by rhizomes, stolons, and spreading tillers; and with relatively high palatability, which encourages grazing and stimulates vegetative reproduction. This "strategy" results in well-adapted genotypes with efficient vegetative reproduction, a great deal of tolerance to

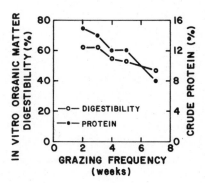

FIG. 9.2.   Effect of grazing frequency on *in vitro* organic matter digestibility and crude protein of forage. Means of 16 stoloniferous C_4 grasses harvested in June over two years in Florida. From Mislevy et al. (1982).

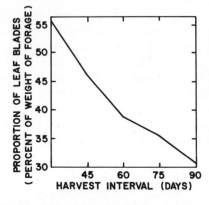

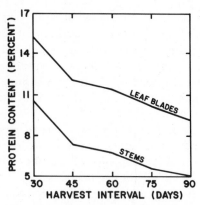

FIG. 9.3. Effect of harvest interval on mean percentage of leaf blades in pangolagrass (*Digitaria decumbens*), guineagrass (*Panicum maximum*), and napiergrass (*Pennisetum purpureum*) in Puerto Rico. From Vicente-Chandler et al. (1959). Reprinted by permission of the authors.

FIG. 9.4. Effect of harvest interval on mean protein content of leaf blades and stems of pangolagrass (*Digitaria decumbens*), guineagrass (*Panicum maximum*) and napiergrass (*Pennisetum purpureum*) in Puerto Rico. From Vicente-Chandler et al. (1959). Reprinted by permission of the authors.

grazing, and a low probability of extinction (Cook, 1979). Indeed, they may require periodic defoliation to remain vigorous and dominant (Mack and Thompson, 1982; McNaughton, 1979).

Several studies suggest that mammalian and insect grazing, and even mammalian saliva left on defoliated plants during grazing, can stimulate subsequent C_4 grass growth, though the nature of the stimulation is still unclear (Dyer, 1980; Dyer and Bokhari, 1976; Reardon and Merrill, 1978).

Burning

Wide genotypic variation is found in the rate of C_4 grass regrowth after burning (Jones and Carabaly, 1981). This factor as well as the effects of time of year, amount of litter, heat of the fire, and weather are discussed in detail by Wright and Bailey (1982).

The most significant effect of fire is often to increase forage palatability. This increases forage utilization by livestock (Klett et al., 1971; Heirman and Wright, 1973) and makes fire a preferred management tool in areas such as the llanos of Colombia and Venezuela and the Cerrado of Brazil.

Many grasslands throughout the world have developed in the presence of frequent fires caused by lightning and by man. Therefore, many of their dominant C_4 grasses are quite tolerant of burning. In the U.S. southern desert grasslands, fires occurred at intervals of approximately 10 years prior to 1880. This, combined with competition by grasses, controlled invasion by woody plants. However, overgrazing reduced the ability of the grasslands to carry fires, and by the

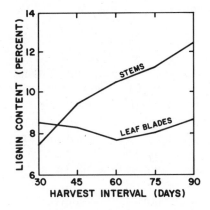

FIG. 9.5. Effect of harvest interval on mean lignin content of leaf blades and stems of pangolagrass (*Digitaria decumbens*), guineagrass (*Panicum maximum*) and napiergrass (*Pennisetum purpureum*) in Puerto Rico. From Vicente-Chandler et al. (1959). Reprinted by permission of the authors.

early 1900s shrub and tree invasion was occurring (see Wright and Bailey, 1982). Quite frequent burning is required to control invasion in more humid grasslands. For example, a three-year burning interval is needed to maintain C_4 grass dominance in the tallgrass prairie of the U.S. Central Great Plains. Burning at 3–10-year intervals is recommended for shrub suppression in the Rio Grande Plains of Texas (Wright and Bailey, 1982).

In relatively arid grasslands rainfall affects recovery of grasses after burning. Burning during dry years can reduce dry matter production for more than a year after the fire; however, burning during a wet year causes increased dry matter yields for several years (Neuenschwander et al., 1978; Wright and Bailey, 1982).

Maize and Grain Sorghum

Hail damage and insect attack can severely reduce crop leaf area. Complete or nearly complete defoliation of maize at the four- to five-leaf stage has only a small effect on grain yields. Early defoliation sometimes actually increases maize grain yields slightly (Crockett and Crookston, 1980 and 1981; Hicks and Crookston, 1976; Hicks et al., 1977), though small (0–15%) yield decreases may also occur (Camery and Weber, 1953; Eldredge, 1935; Johnson, 1978). Crookston and Hicks (1978) concluded that the response depends on timing of the defoliation. Defoliation before ear initiation usually decreases yields. Defoliation after ear initiation often increases grain yields, primarily due to increased grain numbers. Crockett and Crookston (1980, 1981) discovered that high harvest indexes of sweet corn defoliated at the four- or five-leaf stage are due to a reduction in tiller number. Decreased tillering may result from a reduction in reserve carbohydrates due to defoliation during the critical stage of tiller initiation. After defoliation leaf area quickly recovers, and the absence of tillers may actually increase photosynthate availability to the ear, resulting in less kernal abortion.

Defoliation prior to tassel initiation also delays silking by up to 7 days (Johnson, 1978; Remison, 1978), but defoliation after tassel initiation has little effect. Prior to panicle initiation, "flaming" to control weeds can also delay anthesis in maize and grain sorghum and cause significant yield reduction (Green, 1949; Vanderlip et al., 1977).

Defoliation has been used to estimate the effects of leaf area destruction by hail, insects, and diseases on grain production (Allison et al., 1975a,b: Remison, 1978; Singh and Nair, 1975). In general, the most severe effects are found when defoliation occurs at approximately silking. This is especially true when defoliation is almost complete. For example, complete defoliation at the "10-leaf," "silking," and "roasting ear" stages reduced maize grain yields to 70, 2, and 31% of the undefoliated control, respectively (Hanway, 1969). The effect on kernel number was greatest when defoliation occurred just prior to silking. The effect on final kernel weight was greatest when plants were defoliated at the "roasting ear" stage.

After silking, the effects of defoliation depend on both its severity and timing. For example, Egharevba et al. (1976) found that the reduction in maize grain yield due to defoliation decreases linearly with time after silking (Fig. 9.6).

Ratoon Crops

Sugarcane and grain sorghum produce tillers after they are harvested, and these tillers can produce a second (ratoon) crop without the need to replant.

There are three common methods of harvesting sugarcane. The cane may be harvested by hand, it may be harvested by machine while the cane is still standing, or it may be harvested by machine after it has lodged. Hand-harvested stalks are cut individually near ground level and are generally loaded onto small carts for transport to the mill. This causes a minimum of damage to the plant, and ratoon growth is usually rapid. Clements (1980) reported that up to 17 consecutive ratoon crops were harvested in Hawaii when cane was cut by hand.

Standing sugarcane is harvested mechanically in many parts of the world. Several kinds of machines have been developed. Most of them are mounted on conventional farm tractors, cut the cane near ground level with some type of knife, and harvest one row of standing cane at a time (Churchward, 1967). These

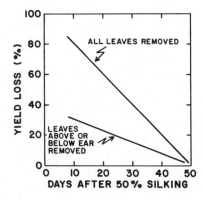

FIG. 9.6. Effect of timing and severity of maize defoliation on grain yield loss. From Egharevba et al. (1976). Reproduced from *Agronomy Journal*, by permission of the American Society of Agronomy.

machines normally cause minimal damage to the underground portion of the plant and regrowth is rapid.

In Hawaii, sugarcane is grown for more than 18 months prior to harvest. During this time the plants lodge, and a mass of tangled stalks must be harvested. Very heavy tracked machinery is used to break the stalks near ground level and push them into piles or windrows. They are then lifted into heavy vehicles for transport to the mill. The use of this heavy machinery can cause severe damage to the roots and stem bases, and poor ratoon stands often result. Partial replanting is often needed to obtain an adequate stand. Severe compaction can occur on wet soils, and interrow subsoiling is often necessary. This can further damage surviving plants and slow ratoon growth (Clements, 1980; Duncan, 1960).

In the southeastern United States grain sorghum can be ratooned to produce a second grain harvest prior to frost (Duncan, 1981a). After combine harvest of the first crop, the stubble is cut with a sickle-type mower to a height of 5–12 cm to stimulate tillering at the lower internodes. Rotary mowers should not be used because they split and uproot stalks, reducing tiller populations.

Ratooning both sugarcane and grain sorghum requires fertilization of the ratoon as well as the plant crop. Weed and insect control can be special problems in ratoon grain sorghum (Banks, 1981; Gardner, 1981). Cultivars of sugarcane and grain sorghum can also differ considerably in their ability to produce rapid ratoon regrowth. Where ratooning is practiced, management practices and cultivars must be selected to promote rapid plant regrowth after harvest.

10

DROUGHT

Ecologists and agronomists have long recognized that drought stress affects the biochemical, physiological, morphological, and developmental processes of plants. Several recent reviews have discussed different aspects of plant adaptation to drought stress (Hanson and Hitz, 1983; Jensen and Cavalieri, 1983; Jordon et al., 1983a; Krieg, 1983; Passioura, 1983; Rosenow et al., 1983). Four-carbon grasses evolved in and are important components of tropical and subtropical vegetation, where they frequently experience periodic drought stress. In addition, yields of maize, grain sorghum, sugarcane, and the millets, important food crops for much of the world's population, are often affected by drought. As a result, considerable research has been done on the response of C_4 grasses to drought stress.

GROWTH AND DEVELOPMENT

The most obvious effects of drought stress on grasses are reductions in the rates of stem and leaf growth, premature senescence in older leaves, and reduction in seed yields.

Leaf Growth and Senescence

Final leaf size is the product of two easily measured parameters: average growth rate and growth duration. In most plants drought affects growth rate more than duration (Karamanos, 1979). The reduction in final leaf size is usually due to a decrease in both cell size (Acevedo et al., 1971; Boyer, 1968, 1970) and the number of cells per leaf (McCree and Davis, 1974).

Even when soil water is abundant, hot and dry atmospheric conditions can cause reductions in leaf length, leaf width, and both total number and average area of leaf cells. For example, McCree and Davis (1974) found that decreases in daytime leaf water potential (ψ_l) as small as 0.1 MPa can cause significant decreases in leaf area and leaf cell size and number in unhardened grain sorghum plants. Hot and dry atmospheric conditions combined with soil water deficits cause a further reduction in leaf length and width, due entirely to a decrease in the number of cells per leaf.

During drought stress, partially expanded cells accumulate in the meristematic regions of leaf blades and sheaths and in stem internodes. When leaf turgor and photosynthesis are restored by watering, these cells can expand rapidly, causing rapid organ growth (Fig. 10.1), which may briefly exceed that of well-watered plants (Fig. 10.2) (Acevedo et al., 1971; Kleinendorst, 1975; Ludlow and Ng, 1977). In some cases, organs elongating after the relief of drought stress may even exceed the lengths of a well-watered control (Fig. 10.3).

The decline in leaf growth during drought stress has been attributed to decreased leaf turgor (Acevedo et al., 1979; Boyer, 1968, 1970; Sharp and Davies, 1979). However, Michelena and Boyer (1982) have shown that in moderately drought-stressed maize plants osmotic adjustment of the intercalary meristem is sufficient to maintain its turgor throughout the day, even though leaf elongation and turgor in expanded blades decline. They suggest that some factor other than meristem turgor is responsible for reduced leaf elongation during

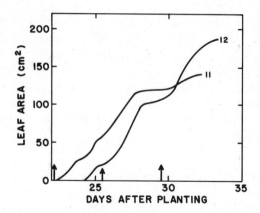

FIG. 10.1. Change in the exposed areas of leaves 11 and 12 of a container-grown maize plant during three drying periods. Dates of watering shown by arrows. From McCree and Davis (1974). Reproduced from *Crop Science*, by permission of Crop Science Society of America.

moderate drought stress. This factor may be the difference in water potential between the xylem and the expanding tissue, which governs the rate of water movement to the expanding cells (Westgate and Boyer, 1984).

In addition to slowing leaf expansion, drought stress can reduce the rate of leaf appearance and the duration of leaf growth (McCree and Davis, 1974; Stout et al., 1978). For example, after irrigation the leaf appearance rate of pangolagrass (*Digitaria decumbens*) declines linearly with increasing cumulative pan evaporation (Blunt and Jones, 1980).

When soil water is adequate, the net photosynthetic rate of a leaf is at its maximum shortly after its complete expansion. Thereafter, it declines slowly until normal senescence occurs (Ludlow and Wilson, 1971c; Ludlow, 1975). This

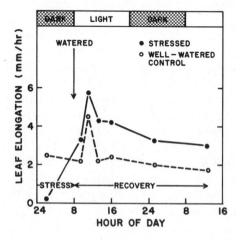

FIG. 10.2. Leaf elongation of water-stressed green panic (*Panicum maximum* var. *trichoglume*) prior to and following rewatering. From Ludlow and Ng (1977). Reproduced by permission of *Australian Journal of Plant Physiology* and CSIRO Editorial and Publications Service.

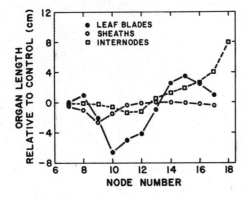

FIG. 10.3. Effect of drought stress beginning at the appearance of the ninth leaf blade (and continuing until plants no longer regained turgor at night) on relative blade, sheath, and internode lengths of grain sorghum plants grown in containers. Redrawn from data in Mirhadi and Kobayashi (1979).

normal aging is apparently controlled by an endogenous "biological clock." In young leaves water stress severe enough to cause wilting stops the clock. When stress is relieved, the net photosynthetic rate returns to its prestress level rather than to the level of unstressed leaves of the same chronological age; thus, normal senescence is delayed (Ludlow and Ng, 1974; Ng et al., 1975; Ludlow, 1975).

Contrary to its effects on young leaves, severe drought stress accelerates the senescence of older leaves in maize (Aparicio-Tejo and Boyer, 1983), grain sorghum (Stout et al., 1978), and other plants (Addicott, 1969; Gates, 1968; Hsiao, 1973). This reduces transpiring leaf area and can reduce transpiration, preventing even more severe desiccation of the remaining green leaf tissue. For example, severely drought-stressed sugarcane stalks commonly have only four to six green leaves, about one-third to one-half the normal number.

In grain sorghum and sugarcane, leaves of young tillers are more susceptible to drought stress than are those of the primary shoot. During moderate drought stress, leaves on tillers begin to roll sooner than those on the primary shoot (Clements, 1980), and during severe stress leaf death is much more pronounced on tillers (Stout et al., 1978). One reason for the greater drought susceptibility of young tillers may be their relatively poor root system development compared with that of older shoots.

Stem and Inflorescence Growth

In maize and sugarcane moderate drought stress has little effect on the rate of leaf appearance (Clements, 1980; Coelho and Dale, 1980); however, the rate of internode elongation and final internode length are dramatically reduced. In fact, in sugarcane and other grasses that produce long vegetative tillers, permanently shortened internodes serve as a convenient record of the timing, length, and severity of drought stress. The effect of drought stress on sugarcane stem elongation is shown in Fig. 10.4, where the relative rate of stem elongation is linearly and negatively related to the depletion of soil water since the last irriga-

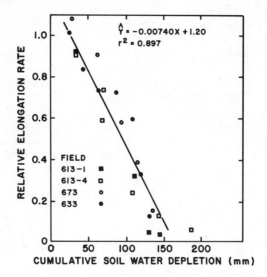

$$\hat{Y} = -0.00740X + 1.20$$
$$r^2 = 0.897$$

FIELD
613-1 ■
613-4 □
673   ○
633   ●

FIG. 10.4. Effect of drought stress (cumulative soil water depletion minus cumulative rainfall) on the relative elongation rate of sugarcane stalks. From Koehler et al. (1982). Reproduced from *Agronomy Journal*, by permission of the American Society of Agronomy.

tion (Koehler et al., 1982). Considerable genotypic variation exists in the percentage reduction of internode length during stress (T. Tew, 1980, personal communication), and this trait can be used to select for drought-resistant genotypes.

Sugarcane is most susceptible to severe drought stress during the first three to four months after planting. At this time severe stress can reduce stands and make replanting necessary. Later, sugarcane plants are more resistant to being killed by drought stress; however, severe yield losses can occur. Losses in cane production are proportional to the decrease in water use by the crop (Jones, 1980); and in Hawaii, where sugarcane is grown for two years before harvest, reductions in water use due to drought stress produce equivalent reductions in yield, whether drought stress occurs in the first or second year of growth (Fig. 10.5).

In grain sorghum and sugarcane, drought stress prior to panicle initiation delays that initiation and thereby delays the date of panicle emergence (Clements, 1980; Stout et al., 1978). In sugarcane, flowering reduces production because reproductive stalks die after seed maturity. Sugarcane panicle initiation normally requires a specific combination of day length and temperature (see Chapter 5), however, drought stress can inhibit initiation. Therefore, prior to periods conducive to floral initiation, controlled drought stress is used to inhibit panicle initiation until noninductive day lengths return (Clements, 1980).

Under some conditions frequent irrigation can cause a slight delay in grain sorghum floral initiation. For example, Done et al. (1984) found that a 7-day furrow irrigation interval delayed floral initiation by about 2 days compared with

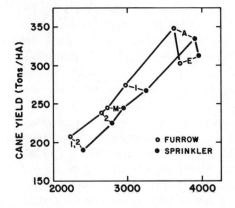

FIG. 10.5. Response of sugarcane fresh cane yields in Hawaii to furrow and sprinkler irrigation with severe drought stress imposed in the first (1), second (2), or both (1, 2) summers; moderate stress both summers (M); and adequate (A) and excessive (E) irrigation. From Jones (1980). Reprinted courtesy of Hawaiian Sugar Planters' Association.

longer (14- or 28-day) intervals. This delay was probably due to lower soil and meristem temperatures at the short irrigation interval.

Drought stress after panicle initiation normally accelerates phenological development of grain sorghum (Stout et al., 1978). For example, decreasing furrow irrigation from once weekly to once in the crop (at sowing) hastened anthesis and physiological maturity by six and eight days, respectively (Done et al., 1984). Phenological development of wheat is normally hastened by mild stress, but it can be delayed when severe stress occurs after floral initiation (Angus and Moncur, 1977). The delay in anthesis may be caused by a suspension of aging similar to that caused by drought in leaves (Ludlow and Ng, 1974).

Root Growth

The dynamic nature of root growth in perennial grasses is well documented, especially in sugarcane (Evans, 1936; Glover, 1967, 1968; Wood and Wood, 1967). Root growth and nutrient uptake are very sensitive to soil moisture, and root distribution is greatly influenced by rainfall. As surface layers of the soil profile dry, roots in these layers become less active, and root length density may actually decrease as small roots deteriorate. Nutrient uptake from the dry portions of the profile decreases, and both root growth and nutrient uptake increase in the deeper parts of the profile that still have adequate water. When rains rewet the upper layers, both root growth and nutrient uptake increase in these layers and gradually decrease in the deeper layers (Glover, 1968; Wood and Wood, 1967).

The ability of roots to proliferate in zones of adequate moisture, no matter where they occur, accounts for the very deep rooting of certain perennial tropical grasses. For example, kikuyugrass (*Pennisetum clandestinum*) growing in a deep well-drained soil during a pronounced dry season can produce roots to a depth of at least 4.9 m (Hosegood, 1963) and can extract water to depths of more than 6 m (Dagg, 1969).

The frequency and amount of irrigation also affect rooting depth. When sprinkler irrigation is applied frequently in small amounts, maize roots become concentrated close to the soil surface; however, when frequency is decreased but greater amounts are applied at each irrigation, the soil profile is wetted to a greater depth, and more roots are found deeper in the soil (Howe and Rhodes, 1955; Robertson, et al., 1980). Similarly, reducing the frequency of furrow irrigation from once in 7 days to once in 21 days dramatically increases the percentage of the sugarcane root system found below 1.0 m (Baran et al., 1974).

Some drought-resistant grain sorghum cultivars have greater root mass, greater root length, a higher root–shoot ratio, and a greater percentage of roots deep in the profile (Blum, 1974: Jordan et al., 1983a; Nour and Weibel, 1978). Other drought-resistant varieties use water slowly prior to anthesis because of slow root and/or leaf area development, but they extract more soil water during the reproductive phase. This strategy does not necessarily increase total water use, but it increases grain production per unit water use (Blum, 1972; Wright and Smith, 1983).

Several studies (Briske and Wilson, 1978; Portas and Taylor 1976; Volk, 1947) have shown that growth of C_4 grass roots effectively ceases at soil water potentials below -0.5 to -1.0 MPa, even though root tips may remain alive in even air-dry soil and may rapidly resume growth when water is available. However, Trouse (1978) suggests that soil strength (which increases as soil water potential decreases) and total plant water status are more important for root growth than the moisture content of the soil around the growing root tip.

The seminal root system of grass seedlings supports the initial growth of the top, but continued growth depends on the development of the adventitious root system. Adventitious roots arise from stem internodes and provide an important source of new roots for the seedling plant and subsequent tillers. However, surface soil moisture has an important effect on the growth of adventitious roots. Those roots arising at or above the soil surface only penetrate the surface when it is moist. Side oats grama (*Bouteloua curtipendula*), blue grama (*Bouteloua gracilis*), and other species of the tribe Chlorideae have short-lived seminal roots, and successful seedling establishment requires the development of an adequate adventitious root system. However, these species have a short coleoptile, and the apical meristem is pushed to the soil surface by the elongation of the subcoleoptile internode. This places initial adventitious root development on or very near the soil surface (Hoshikawa, 1969; Hyder et al., 1971; Olmsted, 1941). In most cases adventitious root growth begins about 10 days after germination, but if the moisture potential of the surface soil is less than about -0.5 MPa, adventitious roots cannot elongate (Briske and Wilson, 1978; Wilson and Briske, 1979). Thus, the seedling may die because the seminal root system is unable to provide sufficient water to meet the transpirational demand of the growing shoot. Wilson and Briske (1979) found that surface soil must remain moist for 2–4 days for blue grama seeds to germinate, and a second moist period of 2–4 days is required 2–8 weeks later for adventitious roots to begin growing and reach a depth sufficient to sustain the shoot.

Yield Components

In determinate grain crops such as grain sorghum and maize, the timing of drought stress plays an important role in the plant's response to the stress. A severe stress during the vegetative stage will often have less effect on final grain yield than a relatively short stress during a very critical stage such as female meiosis.

Grain Sorghum. Both the severity and timing of drought stress affect dry matter production and seed yield of grain sorghum. The growth stage found to be most susceptible to drought stress varies among studies. Lewis et al. (1974) found that the crop is most susceptible from boot through bloom. Others have suggested that the most susceptible period is the boot stage (Inuyama et al., 1976), heading to milk stage (Plant et al., 1969), and heading or the end of anthesis through grain filling (Garrity et al., 1982a,b; Musick and Dusek, 1971). Garrity et al. (1982a,b) found that with a limited amount of irrigation, water use efficiency is greatest when a small fraction of the crop's water requirement is applied each week, allowing stress to gradually increase throughout the season.

When drought stress occurs during the vegetative stage, the yield component most adversely affected is the number of panicles per plant, primarily due to a reduction in tillering. During reproductive growth, the periods of panicle initiation and floret development are very susceptible to stress (Bennett and Sullivan, 1978a,b,c; Eck and Musick, 1979a; Lewis et al., 1974). Drought during panicle initiation reduces the total number of florets per inflorescence. After panicle initiation but before floret development, the immature panicle is relatively resistant to drought stress; however, during floret development drought stress can cause abortion of developing florets (Wright et al., 1983a). Reductions in the number of florets per inflorescence or the percentage of fertile florets irreversibly reduces yield potential, and adequate water during grain filling can do little to ameliorate the loss. Drought during grain filling affects seed yields by reducing photosynthate available for storage in the seed, thereby reducing final weight per seed.

Stomatal response to drought stress also depends on growth stage. Prior to anthesis, stomata are very sensitive to reductions in leaf water potential (ψ_l) or turgor potential (ψ_p). Their closure reduces water loss and prevents or reduces leaf and tiller death at the expense of photosynthesis (Ackerson and Krieg, 1977; Bennett and Sullivan. 1978a,b,c; Garrity et al., 1984; Henson et al., 1983). During grain filling, the plant must maintain high rates of photosynthesis in the upper leaves, must efficiently translocate photosynthate to the panicle, and must avoid producing small, nonviable seeds. The plant accomplishes this by reducing stomatal sensitivity to low ψ_l. For example, Ackerson and Krieg (1977) found that after reproductive growth begins, stomates remain open even when ψ_l declines to -1.4 MPa. Fereres et al. (1978) and Garrity et al. (1984) found similar insensitivity to about -2.2 MPa. This response permits photosynthesis to continue in the upper leaves at the expense of accelerated senescence of the lower leaves (Ackerson and Krieg, 1977).

The different responses of grain sorghum to drought in the vegetative and reproductive stages are illustrated by the results of Sullivan (1972). When several sorghum genotypes at different stages of development were stressed uniformly, leaves of those in which the panicle had not emerged had little foliar damage. The leaves of those in which the panicle had emerged were severely damaged, although their panicle branches and rachises remained green and healthy and grain filling continued.

Maize.   Maize seed yields are very sensitive to the timing of drought stress (Hall et al., 1980, 1981, 1982; Moss and Downey, 1971; Stegman, 1982) (Fig. 10.6). In general, stress prior to tassel emergence reduces yields much less than stress during the reproductive stage, though it may delay the ontogenetic development of the crop. Drought stress during female gametophyte development causes the production of many abnormal embryosacs, delays silking, and reduces final grain yield even though adequate pollination is assured. Drought stress at or just before tasseling delays pollen shedding, reduces the amount of pollen produced, and can significantly reduce final yield due to incomplete pollination. One or two days of wilting during early tasseling or pollination can reduce yields by 22%; six or eight days of wilting can reduce yields by 50% (Robbins and Domingo, 1953).

Drought stress during grain filling can also cause severe yield reductions, though not usually as severe as those occurring at tasseling or silking (Barnes and Woolley, 1969; Miller and Duley, 1925; Stegman, 1982). If drought occurs

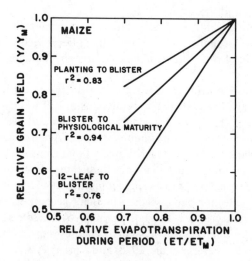

FIG. 10.6.   Effect of evapotranspiration deficits during three periods on relative maize grain yields. From Stegman (1982). Reproduced courtesy of the author and Springer-Verlag.

during the first two weeks after silking, final kernel number is reduced much more than final kernel weight. Later stress (after grain filling has begun) affects kernel weight more than kernel number (Harder et al., 1982).

PHYSIOLOGY AND BIOCHEMISTRY

Photosynthesis

Drought stress reduces photosynthesis in three ways: by reducing leaf area available to intercept solar radiation, by reducing the diffusion of CO_2 into the leaf, and by reducing the chloroplasts' ability to fix the CO_2 that reaches them. The effect on leaf area has already been discussed. The other major effects of drought stress on photosynthesis can be understood in terms of resistances to CO_2 diffusion from the air into the stroma of the chloroplasts. Gaastra (1959) expresses the rate of photosynthesis.(P) in terms of a driving force and a series of resistances:

$$P = \frac{[CO_2]_a - [CO_2]_c}{r_a + r_l + r_m}$$

where $[CO_2]_a$ and $[CO_2]_c$ are the CO_2 concentrations in the bulk air and the chloroplasts, respectively (g/cm^3), and r_a, r_l, and r_m are the resistances to CO_2 diffusion in the leaf boundary layer, from the leaf surface through the intercellular air spaces, and within the mesophyll, respectively (s/cm). The resistances have been further subdivided to include resistances to movement through the stomates and cuticle, the mesophyll cell wall, plasmalemma, cytoplasm, chloroplast membrane, and stroma (see Cooke and Rand, 1980; Nobel, 1974; Tenhunen et al., 1980). The reciprocal of stomatal resistance, stomatal conductance (c_s), is also frequently used to characterize stomatal aperture.

Stomatal control of c_s is one of the most important plant reactions to drought stress. When leaf water potential decreases and finally exceeds the threshold for stomatal closure, conductance can decrease within a few minutes from more than 0.25 to less than 0.025 cm/s. In C_4 grasses this decrease in conductance often occurs over a relatively small range (0.2–0.8 MPa) in leaf water potential (ψ_l), beginning somewhere between -1.0 and -2.0 MPa (Ackerson et al., 1980; Jones and Rawson, 1979; Klar et al., 1978) (Fig. 10.7).

The literature contains many reports of parallel changes in stomatal conductance and photosynthesis during periods of drought stress. Plant physiologists long assumed that stomates close in response to decreases in leaf water potential, and that photosynthesis decreases principally as a result of low CO_2 availability to the photosynthetic apparatus. However, recent work (reviewed in Farquahar and Sharkey, 1982: Krieg, 1983; Osmond et al., 1980; Radin and Ackerson, 1982) suggests that this view oversimplifies what is actually a more complex relationship between c_s and photosynthesis.

They point out that when water stress increases very rapidly (due to leaf exci-

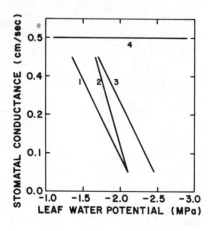

FIG. 10.7. Effect of leaf water potential on stomatal conductance of three groups (1, 2, 3) of grain sorghum genotypes prior to flowering and all genotypes after flowering (4). From Ackerson et al. (1980). Reproduced from *Crop Science*, by permission of the Crop Science Society of America.

sion, transfer of intact plants to a solution of high osmotic pressure, or rapid change in vapor pressure deficit) stomata begin to close almost immediately. This immediate closure is probably due to "peristomatal transpiration" (evaporation of water directly from guard cell surfaces), which removes water faster than it can be replaced. Radin and Ackerson (1982) consider rapid stomatal closure as a "passive" process in which guard cells lose turgor by nonmetabolic means. In contrast, slow imposition of water stress, as usually occurs in the field, results in the near-simultaneous closure of stomata, increase in abscisic acid (ABA) content, and decline in the biochemical processes of photosynthesis.

Changes in c_s associated with variations in water stress (Krieg, 1983; Osmond et al., 1980), vapor pressure deficit (Fig. 10.8) (Aho, 1980; Coyne et al., 1982; El-Sharkawy et al., 1984), irradiance (Fig. 10.9) (Idso, 1983b; Turner and Begg, 1973), leaf age (Ludlow and Wilson, 1971c; Turner and Begg, 1973), and other factors may actually be caused by variation in the biochemical component of photosynthesis, through its effect on leaf ABA content, and the effect of that ABA on stomatal aperture. Cowan et al. (1982) and Radin and Ackerson (1982) suggest that a decrease in photosynthetic rate causes chloroplast pH to decrease, allowing ABA to leak from the chloroplasts of the mesophyll cells into the apoplast. It then moves to the stomata in the transpiration stream, where it produces stomatal closure.

Stomatal sensitivity (and presumably photosynthetic sensitivity) to ψ_l and soil water potential (ψ_s) varies among C_4 grass species and genotypes (Ackerson et al. 1980; Henzell et al., 1975, 1976; Jones et al., 1980) (Fig. 10.7). In general, genotypes whose stomatal conductance is very sensitive to ψ_s and ψ_l are not well adapted to drought-prone environments. These species "avoid" low ψ_l by closing their stomates, and their leaves often roll. They thereby minimize radiation interception and water loss at a cost of drastically reduced photosynthesis. In contrast, species whose conductance declines gradually over a wide range of leaf or soil water potential generally perform better during drought stress (Beadle et al., 1973; Blum, 1974; Klar et al., 1978). Apparently, the ability to "tolerate" low ψ_l

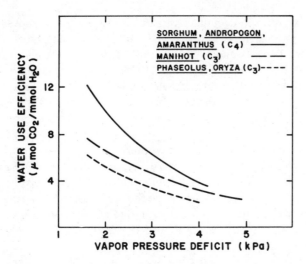

FIG. 10.8. Effect of vapor pressure deficit on water use efficiency of C_3 and C_4 plants. From El-Sharkawy et al. (1984). Reproduced from *Crop Science*, by permission of the Crop Science Society of America.

while maintaining some transpiration and photosynthesis improves genotypic adaptation to drought conditions (Hofmann et al., 1984). However, the cost of this strategy is often increased leaf senescence due to drought stress.

As discussed previously, growth stage has a large effect on grain sorghum's response to ψ_l (Fig. 10.7). Ackerson et al. (1980) found that stomates are quite sensitive to ψ_l prior to flowering, but after flowering they remain open over a wide range of ψ_l. This may permit higher rates of photosynthesis and dry matter production during grain fill under drought stress.

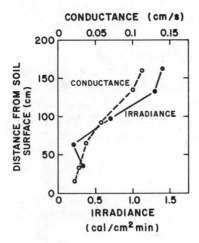

FIG. 10.9. Profiles of irradiance and leaf stomatal conductance in a maize canopy at midday. From Turner and Begg (1973). Reproduced courtesy of *Plant Physiology*.

Previous exposure to drought stress (hardening) often improves the plant's capacity for photosynthesis and growth during subsequent drought stress (Ashton, 1956; Bennett and Sullivan, 1981; Jones and Turner, 1978; Ludlow and Ng, 197ʻ). For example, Klar et al. (1978) found that hardening reduces the ψ_l of complete stomatal closure of guineagrass (*Panicum maximum*) by more than 0.5 MPa. Similar results were obtained by McCree (1974a) for grain sorghum. Hardening will be discussed in more detail in the section on osmotic adjustment.

Finally, water stress reduces photosynthesis after, as well as during, drought stress. Ludlow et al. (1980) measured ψ_l, c_s, and photosynthetic rate in green panic (*Panicum maximum* var. *trichoglume*) leaves during recovery from varying degrees of drought stress. During the first 24 hr after rewatering, net photosynthesis was determined primarily by the rate of increase in ψ_l and c_s. Leaves that had experienced ψ_l less than -4 MPa recovered more slowly than less stressed leaves.

Translocation

Studies with maize, grain sorghum, and sugarcane have demonstrated that under normal conditions more than 60% of recent photosynthate is translocated out of the leaf within 4 hr. When drought stress occurs, translocation of photosynthates decreases (Hartt, 1967). For example, Brevedon and Hodges (1973) found that the amount of ^{14}C remaining after 2 hr in an area of leaf exposed briefly to $^{14}CO_2$ was 42% in a well-watered treatment and 58% in a stressed treatment. Sung and Krieg (1979) found that 30% remained after 2 hr in a well-watered treatment, and 48% remained in a stressed treatment.

Several studies suggest that in C_4 grasses translocation is more sensitive than photosynthesis to drought stress (Hartt, 1967; Brevedon and Hodges, 1973). Others suggest the opposite (Wardlaw, 1967; Johnson and Moss, 1976). For example, drought stress severe enough to stop dry matter accumulation by the whole plant still allows movement of reserves from the stem to the grain (Jurgens et al., 1978; McPherson and Boyer, 1977). Recent work (Sung and Krieg, 1979) suggests that the relative sensitivities of the two processes vary with the degree of stress. For example, a slight reduction (0.5 MPa) in grain sorghum ψ_l causes a significant decrease in photosynthesis but almost no decrease in translocation. However, as stress becomes greater (0.5–1.2 MPa below the control), translocation is affected relatively more than photosynthesis (Fig. 10.10). Thus, moderate drought stress may even increase the percentage of photosynthate translocated from the leaf tissue, but severe drought stress reduces that percentage.

Respiratory Efficiency

Most analyses of the physiological response to drought stress concentrate on its effect on photosynthate production, though growth is equally dependent on the

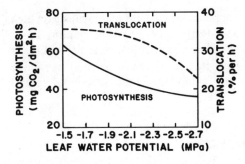

FIG. 10.10. Effect of grain sorghum leaf water potential on photosynthesis and translocation. From Sung and Krieg (1979). Similar relationships were found at panicle differentiation and grain filling stages. Reproduced courtesy of *Plant Physiology*.

utilization of that substrate for growth (Wilson et al., 1980; Bassham, 1977; Evans, 1975b).

Two parameters can be used to characterize substrate utilization: (1) growth conversion efficiency, the increase in dry matter (in carbon equivalents) per unit of substrate carbon used and (2) the maintenance coefficient, the rate at which substrate carbon is used for the maintenance of existing dry matter carbon (Thornley, 1976). Even though stomatal closure, increased mesophyll resistance, reduced leaf area expansion, and lower maintenance coefficients are associated with lower rates of dry matter production in drought-stressed tropical grasses, growth efficiency is apparently unaffected (Wilson et al., 1980). For sorghum plants grown to the eight-leaf growth stage in controlled environments then subjected to a rather wide range of drought stress, about 0.71 g of dry matter carbon are produced per gram of substrate carbon used (Wilson et al., 1980). This implies that no significant change in metabolic efficiency occurs, even when plants are in the process of adapting to drought stress. These results are consistent with others that indicate that growth efficiency is insensitive to changes in temperature (McCree, 1974b), photoperiod (Hansen and Jensen, 1977; McCree and Kresovich, 1978), and the daily light regime (Wilson et al., 1978).

Osmotic Adjustment

Plants, including C_4 grasses, have long been known to "adapt" (become hardened) to prolonged or repeated drought stress (Stocker, 1960). Adaptation involves changes in morphological characteristics such as root system development and leaf morphology as well as reductions in the sensitivity of stomatal aperture (McCree, 1974a) and photosynthesis (Ashton, 1956; Todd and Webster, 1965). For example, Ashton (1956) studied the effects of repeated cycles of water stress on photosynthesis of sugarcane grown in containers. He observed that the reduction in net photosynthesis decreases with each successive stress cycle. He also found that recovery from stress is more rapid in plants that had been previously stressed. This suggests that plants are able to adjust their metabolism and thereby reduce the effects of water deficits.

Small changes in turgor (or pressure) potential (ψ_p) are one means by which water stress affects metabolic processes (Hsiao, 1973; Zimmermann, 1978; Turner and Jones, 1980). Therefore, the maintenance of ψ_p near normal levels during drought stress should enable the plant to maintain normal metabolic rates. In leaves ψ_p depends on leaf water potential (ψ_l) and osmotic potential (ψ_o):

$$\psi_p = \psi_l - \psi_o$$

When leaves are turgid, ψ_p is positive and both ψ_l and ψ_o are small negative numbers. When drought stress occurs, ψ_l becomes more negative. If ψ_o remains constant, ψ_p must also decrease, and turgor-dependent processes such as leaf rolling are soon affected. However, plants have two mechanisms for decreasing ψ_o and thereby reducing the ψ_l required to reduce turgor beyond critical levels. These responses are osmotic adjustment and cell volume reduction.

Since ψ_o is determined by the concentration of solutes in the cell, ψ_o becomes more negative if either the amount of solutes increases or the volume of water decreases. The volume of water can be reduced by having cells with elastic walls. The volume of these cells decreases in response to decreasing ψ_p. This, in turn, causes ψ_o to become more negative since the same amount of solute is dissolved in a smaller volume of water. This mechanism may account for the decrease in volume of bulliform cells during leaf rolling. However, Wenkert (1981) estimates that in most maize leaf cells changes in cell volume account for less than 10% of the change in ψ_o over a wide range of turgor. Henson et al. (1982b) and Wright et al. (1983b) obtained similar small effects in pearl millet (*Pennisetum americanum*) and grain sorghum.

A much more important response to drought stress is osmotic adjustment, the accumulation of solutes that causes ψ_o to become more negative and helps maintain ψ_p as ψ_l decreases. Many studies suggest that leaf adaptation to drought stress is largely due to osmotic adjustment, which lowers the critical value of ψ_l at which reductions in leaf growth and photosynthesis occur (Jones and Turner, 1978; McCree, 1974a; Thomas et al., 1976; Wright et al., 1983b).

Osmotic adjustment has long been recognized as an adaptation to salinity that confers tolerance to the low soil water potentials resulting from saline soils (see Flowers et al., 1977 and Hellebust, 1976 for reviews). However, only in the 1970s was it recognized as a common plant response to drought stress. After all, plants growing in saline soils have a ready supply of inorganic ions that can be taken up to counteract those in the soil solution. These ions are not necessarily available to plants growing in nonsaline soils. When osmotic adjustment occurs in response to salinity, halophytes (plants tolerant of saline soils) accumulate large amounts of Na^+ and Cl^-. Glycophytes (plants intolerant of salinity) often accumulate K^+ when grown in saline soils, and malate may be synthesized as a counter ion to balance the charge of the K^+.

In drought-stressed plants inorganic solutes are taken up from the soil solution and organic solutes are synthesized to provide the requisite osmotic effect. In field-grown buffelgrass (*Cenchrus ciliaris*), K^+, Cl^-, sucrose, oxalate, and betaine were the most important solutes (Ford and Wilson, 1981). In the same study green panic (*Panicum maximum* var. *trichoglume*) and speargrass (*Heteropogon contortus*) behaved quite differently. For example, in green panic, Na^+, malate, *trans*-aconitate, and proline were also important. In fully expanded grain sorghum leaves, sucrose, glucose, fructose, K^+, and Cl^- were the most important solutes (Jones et al., 1980). In field-grown sugarcane, reducing sugars accounted for much of the increase in solute concentration, but in container-grown sugarcane, sucrose, reducing sugars, and soluble amino acids all increased (Koehler et al., 1982; P. H. Moore, 1980, personal communicaton). These and other results suggest that the organic and inorganic solutes responsible for osmotic adjustment vary among species and among experiments with the same species, probably due to varying soil and experimental conditions.

Osmotic adjustment can result in either full or partial maintenance of ψ_p. Total osmotic adjustments of 0.4–1.0 MPa are routinely found in grain sorghum, maize, sugarcane, and other C_4 grasses (Fig. 10.11) (Jones and Rawson, 1979; Koehler et al., 1982; Wenkert, 1981; Wilson et al., 1983). Interestingly, one of the most drought-tolerant cereals, pearl millet, may have little capacity for osmotic adjustment — 0.1–0.3 MPa (Henson et al., 1982b).

If soil water potential and leaf water potential (ψ_l) decrease very rapidly, as often occurs in plants grown in small containers, little or no osmotic adjustment may occur (Wilson et al., 1983). For example, Jones and Rawson (1979) imposed drought stress at three rates on grain sorghum grown in containers. When stress increased slowly (predawn ψ_l decreased by 0.5 or 0.7 MPa/day), a total osmotic

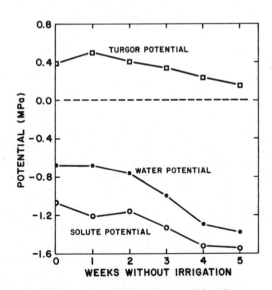

FIG. 10.11. Changes in midday water potential, solute potential, and turgor potential of sugarcane leaves during a 5-week period without irrigation. From Koehler et al. (1982). Reproduced from *Agronomy Journal*, by permission of the American Society of Agronomy.

adjustment of 0.6 MPa occurred. However, when ψ_l decreased by 1.2 MPa/day, no osmotic adjustment occurred, and photosynthetic rates were drastically affected.

In the field, osmotic adjustment normally occurs at a rate of 0.1 MPa/day or less (Hsiao et al., 1976; Turner et al., 1978a). Under these field conditions osmotic adjustment is often sufficient for complete maintenance of turgor during the initial phase of drought stress, but adjustment is incomplete when drought stress becomes more severe. For example, Jones and Turner (1978) found that for grain sorghum, ψ_o decreased by approximately 0.4 MPa as predawn ψ_l decreased from near 0.0 to -0.4 MPa. However, ψ_o decreased to only -0.7 MPa as ψ_l decreased from -0.4 to -1.6 MPa.

Several studies show that osmotic adjustment of sorghum and buffelgrass disappears within about 10 days of relief of stress (Wilson et al., 1980; Turner et al., 1978b). For example, in grain sorghum an osmotic adjustment of 0.5 MPa decreased to 0.15 MPa 3 days after relief of stress and disappeared entirely within 11 days. In addition, the hardening of prestressed plants to subsequent drought stress disappeared if high ψ_l were maintained for 10 days between drought stresses (Jones and Rawson, 1979).

During the night and early morning, maize leaf extension is often correlated with temperature (Watts, 1974). However, midday reductions in elongation and photosynthesis are often associated with periods of reduced ψ_l, ψ_p, and ψ_o (Acevedo et al., 1979: Turner, 1975). Diurnal osmotic adjustment may play a key role in maintaining photosynthetic rates and in preventing more drastic midday reductions in leaf elongation (Westgate and Boyer, 1984). This diurnal osmotic adjustment in maize (up to 0.4 MPa) is due primarily to changes in the total sugar concentration of the leaf tissue (Acevedo et al., 1979).

Drought-induced reductions in ψ_l and ψ_o have been well documented. However, even in well-watered maize and grain sorghum, ψ_l and ψ_o decline with increasing plant age (Fereres et al., 1978; Shackel et al., 1982). From 40 to 100 days after planting both midday ψ_l and ψ_o decreased by approximately 0.6 MPa in well-watered grain sorghum plants. Since both ψ_l and ψ_o decreased together, ψ_p remained at approximately $+0.5$ MPa throughout the period (Fereres et al., 1978). Wenkert (1981) obtained similar results with maize.

Genotypes and species differ in their capacity for osmotic adjustment. For example, C_4 grasses normally have greater drought-induced osmotic adjustment than legumes (Turner et al., 1978a; Wilson et al., 1980), and grain sorghum, wheat, and pearl millet genotypes differ in their capacity for osmotic adjustment (Henson et al., 1982b; Wright et al., 1983b).

Osmotic adjustment is undoubtedly a means by which plants adapt to low ψ_l. It allows the plant to maintain high stomatal conductance and photosynthesis during moderate drought stress. In addition, some studies have found good correlations between osmotic adjustment and leaf and root growth rates (Hsiao et al., 1976; Sharp and Davies, 1979). However, others have shown that most leaf osmotic adjustment occurs only after leaf expansion has slowed dramatically (Koehler et al., 1982; Michelena and Boyer, 1982; Munns et al., 1979). In addition, Michelena and Boyer (1982) found that during drought stress maize leaf elongation decreases even though osmotic adjustment is adequate to maintain meristem turgor potential. This suggests that osmotic adjustment may not be an important factor in maintaining leaf elongation during drought stress.

Nitrogen Metabolism

Drought stress affects almost all aspects of plant N metabolism. In the field, N uptake may be reduced both by reduced availability of N in dry soils and by reduced plant demand due to slower growth. Within the plant, numerous metabolic reactions and pools are affected.

Nitrate reductase, which catalyzes the reduction of nitrate to nitrite, is sensitive to both drought and heat stresses (Garg et al., 1981; Maranville and Paulsen, 1972; Teare et al., 1974). In maize, grain sorghum, and pearl millet, significant reductions in nitrate reductase activity occur by the time leaves are stressed to the point that they do not recover overnight. However, for young maize (Mattas and Pauli, 1965) and sugarcane (P. H. Moore and A. Maretzki, 1980, personal communication), nitrate reductase activity begins to decrease even earlier, before most symptoms of drought stress have appeared. Inhibition of activity may be due to reduced enzyme synthesis (Morilla et al., 1973) and/or to direct enzyme inhibition (Heur et al., 1979). In most cases the recovery of nitrate reductase activity is rapid after rewatering (Huffaker et al., 1970; P. H. Moore and A. Maretzki, 1980, personal communication).

Drought stress also inhibits protein synthesis and increases protein hydrolysis (Botha and Botha, 1979; Dhindsa and Cleland, 1975; Dungey and Davies, 1982). Protein synthesis apparently decreases due to an increase in RNAase activity, which decreases mRNA and polyribosome concentrations (Arad and Richmond, 1976; Bewley and Larsen 1982; Hsiao, 1970). Drought-induced decreases in ATP concentration (Barlow et al., 1976b) and the aggregation of chromatin in the nucleus (Nir et al., 1970) may also affect protein synthesis. Leaf protein hydrolysis probably increases during drought stress in sugarcane (P. H. Moore and A. Maretzki, 1980, personal communication) and barley (Dungey and Davies, 1982), but it can decrease in maize seedlings (Maranville and Paulsen, 1972).

Maranville and Paulsen (1972) conclude that nitrate reductase activity is an early, sensitive indicator of drought stress. RNA and protein synthesis are somewhat less sensitive, and protease activity and free amino acid accumulation are least sensitive.

Free amino acids are important components of nitrogen metabolism. During drought stress, the combined effects of reduced protein synthesis and increased protein hydrolysis can cause free amino acid concentrations to increase up to twofold (Jones et al., 1980; Koehler et al., 1982; Thakur and Rai, 1982), although larger increases can occur if stress is very severe (P. H. Moore and A. Maretzki, 1980, personal communication.).

Certain amino acids, primarily proline and betaine, usually increase dramatically during drought stress. Paleg et al. (1981) suggest that accumulation of these amino acids in the cytoplasm stabilizes protein conformation and protects enzymes from inactivation. Proline accumulation occurs in C_4 grasses such as bermudagrass (*Cynodon dactylon*) (Barnett and Naylor, 1966), sugarcane (Rao and Asokan, 1978), grain sorghum (Blum and Ebercon, 1976; Waldren and Teare, 1974), and maize (Carceller and Fraschina, 1980a,b; Oaks et al., 1970). In sorghum, 200-fold increases in proline content due to drought stress have been observed (Waldren and Teare, 1974). Sixteen- to 100-fold increases occur in sugarcane (Rao and Asokan, 1978; Koehler et al., 1982).

Proline begins to accumulate at the same threshold leaf water potential as stomatal closure and initial wilting (Blum and Ebercon, 1976; P. H. Moore and A. Maretzki, 1980, personal communication), and significant amounts of proline accumulate only after plants are severely stressed and visibly wilted (Waldren and Teare, 1974; Waldren et al., 1974). In fully turgid barley leaves, proline synthesis from glutamate occurs in the cytoplasm (Morris et al., 1969). Wilting causes a loss of end-product inhibition by proline, a decrease in proline oxidation in the mitochondria, a breakdown of intercellular compartmentation resulting in the reconversion of oxidized metabolites to proline, and a decrease in proline utilization for protein synthesis (Dungey and Davies, 1982).

Proline accumulation in barley may be triggered by loss of the ability to translocate nitrogenous compounds out of stressed leaves. For example, accumulation also begins immediately after translocation is interrupted by killing the phloem at the base of the leaf with heat, cooling it, or excising the leaf (Tully et al., 1979).

Proline accumulation varies among genotypes of C_4 grasses, but researchers disagree about its importance. Rao and Asokan (1978) found that sugarcane cultivars differ in their production of free proline during drought stress, and those known for their drought tolerance accumulate greater amounts. These results are consistent with those of Thakur and Rai (1981, 1982) for maize and Blum and Ebercon (1976) for grain sorghum. For example, upon rewatering drought-stressed sorghum there is a rapid reduction in free proline concentration and a simultaneous rapid increase in the concentration of free ammonia. Maximum free proline accumulation, maximum free ammonia concentration after relief of drought stress, and the dark respiration rate during recovery all correlate well with plant recovery from drought stress, indicating that free proline may serve as an energy source during recovery (Blum and Ebercon, 1976; Oaks et al., 1970).

In contrast, studies with maize have shown that, though proline accumulation varies among genotypes, this variation is not correlated with growth reductions due to drought stress (Carceller and Fraschina, 1980a,b; Garg et al., 1981; Ilahi and Dorffling, 1982). These results are consistent with those of Hanson et al. (1979) who concluded that free proline accumulation in barley is only a symptom of severe water deficit. Genotypes selected for high proline accumulation are less vigorous in both favorable and stress conditions than those selected for low proline accumulation. Though cultivars often vary in proline accumulation during drought stress, the variation is apparently due to differences in plant water potential during stress rather than to genotypic variation in proline accumulation at the same plant water potential (Quarrie, 1980; Hanson et al., 1979).

Betaine and other amino acids may also accumulate in C_4 grasses during drought stress (Ford and Wilson, 1981; Thakur and Rai, 1982). These amino acids are largely confined to the cytoplasm, and they may play an important role as a cytoplasmic osmoticum (Ford and Wilson, 1981).

Drought stress has an important effect on the distribution of N in the plant. It causes N to decrease in the leaves and to increase in the grain (Done et al., 1984). For example, Jenne et al. (1958) found that at harvest, maize leaf N concentration was 1.3% in well-watered plants and 1.1% in drought-stressed plants. In contrast, grain N concentration was 1.8% in nonstressed and 2.0% in stressed plants. In the southwestern United States, low-yielding, drought-stressed grain sorghum usually has grain N concentrations greater than 3%, but in high-yielding, well-watered plants it is often less than 2% (Worker and Ruckman, 1968). In another study, well-watered grain sorghum plants had lower grain N concentrations (1.2–1.8%) than stressed plants (1.8–2.4%) (Stone and Tucker, 1969). This suggests that drought has less effect on N translocation to the grain than on carbohydrate availability and/or translocation.

The effect of drought stress on leaf N is well known in sugarcane (Clements, 1980; Samuels et al., 1953). There is a high correlation between leaf sheath water content and leaf N concentration. During drought stress, the leaf N concentration decreases whether adequate soil N is available or not. The relationship is so well documented that leaf N analyses are routinely adjusted to take the effects of drought stress into account (Clements, 1980).

Nonstructural Carbohydrates

By the 1940s it was known that drought stress increases the activities of enzymes affecting carbohydrate metabolism (Spoehr and Milner, 1939), and increases the reserve carbohydrate concentrations of leaves, roots, and rhizomes of C_4 grasses (Clements and Kubota, 1943; Julander, 1945).

In sugarcane, early-morning total sugar concentration of young leaf sheaths falls as low as 5% in rapidly growing, well-watered plants; however, during drought stress it frequently increases to more than 10%. Similar results have been obtained for maize, where drought stress causes a simultaneous decrease in leaf growth and an increase in soluble carbohydrates in the leaf (Barlow and Boersma, 1976; Barlow et al., 1976a). Relief of stress causes a simultaneous rapid increase in leaf elongation and decrease in leaf carbohydrates (Acevedo et al., 1971).

Total nonstructural carbohydrates in roots and crowns of C_4 grasses increase during moderate drought stress, probably due to a greater reduction in demand for photosynthate than in their supply (Brown and Blaser, 1970; Julander, 1945; Sosobee and Wiebe, 1971). This is reflected in a change in the distribution of cells containing starch grains. During periods of drought stress, more starch grains are found in the crown. They disappear when stress is relieved and new growth commences (Nursery, 1971). Since drought-stressed plants accumulate reserve carbohydrates in the crown, they can regrow more rapidly than non-stressed plants after being defoliated, then rewatered (Kigel and Dotan, 1982).

Hormones

Drought, like other stresses, influences plant hormone levels, especially abscisic acid (ABA), ethylene, and cytokinins.

Many stress conditions cause tissue ABA concentrations to increase, and it is now clear that the production, transport, and metabolism of ABA play a vital role in the plant's response to drought stress. Application of ABA to the leaf surface or its introduction into the transpiration stream causes rapid stomatal closure (Kriedemann et al., 1972; Rajagopal and Andersen, 1978) and has long-term effects on shoot and root growth that mimic drought stress (Watts et al., 1981). In experiments in which the level of drought stress varies, higher ABA concentrations are found in treatments with more severe drought stress. Thus, ABA concentrations are often negatively correlated with plant height, ψ_l, c_s, and grain yield (Beardsell and Cohen, 1974, 1975; Kannangara et al., 1983) (Fig. 10.12).

Much recent work suggests that drought resistance is associated with low ABA accumulation (Ilahi and Dorffling, 1982; Kannangara et al., 1982; Quarrie, 1980). For example, Durley et al. (1983) found that drought-tolerant grain

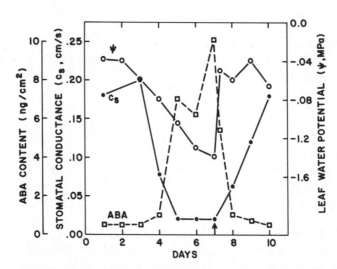

FIG. 10.12. Changes in maize leaf water potential, stomatal conductance, and ABA content during a period of drought stress and recovery after rewatering on day 7. From Beardsell and Cohen (1975). Reproduced courtesy of *Plant Physiology*.

sorghum genotypes accumulate less ABA than less tolerant genotypes. These genotypes have more metabolites of ABA than less tolerant genotypes, suggesting that they are able to metabolize ABA more effectively.

ABA concentrations, like ψ_l, c_s, and leaf solute concentration, are highly dynamic. ABA concentrations may vary two- to threefold during the day (Henson et al., 1982a; Kannangara et al., 1982). However, the dynamic behavior of these indicators differs during periods of drought stress. For example, when leaves of pearl millet were excised and allowed to dry, stomatal closure occurred within 20 min, but ABA did not increase for 25–30 min (Henson, 1981a). During recovery from drought stress, ABA concentrations normally decrease more rapidly than c_s increases. This is the case both for recovery after transient midday stresses (Henson and Quarrie, 1981; Henson et al., 1982a) and after rewatering drought-stressed plants (Beardsell and Cohen, 1975; Henson, 1981b; Ludlow et al., 1980).

Cowan et al. (1982) and Radin and Ackerson (1982) have proposed similar models to explain the effects of ABA on stomatal aperture. ABA is synthesized in the cytoplasm of mesophyll cells and diffuses to the chloroplasts as an uncharged molecule. In the light it is trapped in the chloroplasts because the high pH of the stroma converts ABA to an immobile anion. Reductions in photosynthesis due to drought cause the pH gradient between the chloroplast and cytoplasm to decrease, permitting ABA to leak out of the cytoplasm into the apoplast. It then moves in the transpiration stream to the guard cells, where it causes stomatal closure. This model explains the changes in stomatal aperture due to slow imposition of water stress. The more rapid effects observed in detached leaves (where the increase in ABA content lags behind stomatal closure) can be explained either by "passive" stomatal closure due to rapid evaporation of water from the guard cells (Radin and Ackerson, 1982) or to redistribution of existing ABA from the chloroplasts to the apoplast (Cowan et al., 1982).

Another stress hormone of C_4 grasses is the sesquiterpenoid all-*trans* farnesol, which, like ABA, builds up in maize and sorghum leaf tissue during the early stages of drought stress (Ogunkanmi et al., 1974; Wellburn et al., 1974). Exogenous all-*trans* farnesol causes reversible stomatal closure (Fenton et al., 1976, 1977). Mansfield et al. (1978) suggest that high endogenous levels alter chloroplast membrane permeability to ABA, causing its release from mesophyll chloroplasts. Wilson and Davies (1979) found farnesol-like antitranspirant activity in the leaves of grain sorghum and maize lines. The activity was similar in well-watered plants of two drought-tolerant grain sorghum lines and a drought-susceptible maize line; however, when drought stress was imposed, the increase in antitranspirant activity of the maize line was 7–10 times that of the grain sorghum lines.

Cytokinins are produced in the meristematic regions of roots (Feldman, 1975) and are translocated in the xylem to other parts of the plant. Both temperature and water stress reduce the cytokinin activity in root exudates (Skene, 1975; Vizarova, 1978), suggesting that at least part of the decrease in cell division caused by drought stress is mediated by cytokinins.

Epicuticular Wax

Epicuticular wax or "bloom" is deposited on the epidermis of many plants, including C_4 grasses. This wax may be in the form of filaments or an amorphous layer covered with waxy plates (Hull et al., 1978; Blum, 1975a,b). The thickness of the layer varies among genotypes (Ebercon et al., 1977; Hull, et al., 1978; Peterson et al., 1979), and it is modified by environmental conditions. Drought stress is especially effective in increasing the wax load of several crops, including grain sorghum (Jordan et al., 1983b).

Bloom is present on "normal" grain sorghum cultivars and is controlled by the presence of a single dominant gene, but several bloomless and sparce-bloom variants are known (Peterson et al., 1979). The presence of bloom is associated with reduced cuticular (dark) transpiration, approximately 5% greater leaf reflectance (Blum, 1975a,b), and a higher ratio of net carbon exchange to transpiration (Chatterton et al., 1975). Comparisons between pairs of bloom and bloomless isolines demonstrate a consistent yield advantage of about 15% in favor of the bloom type, especially when drought stress limits production (Ross, 1972; Webster, 1977).

Breeding for Drought Resistance

The complexities of breeding for drought resistance have been reviewed recently (Blum, 1979, 1983; Jensen and Cavalieri, 1983; Rosenow et al., 1983). Breeders and crop physiologists have identified genetic variability in almost every process and attribute associated with the water status of crops. However, simple screening methods that successfully predict overall drought resistance have eluded physiologists and breeders alike. This is undoubtedly due to the complexity of plant response to drought stress, a response involving all parts of the plant and changing with the severity of the stress and the ontogeny of the plant.

Rosenow et al. (1983) point out that two distinct types of drought stress response have been found in grain sorghum. One type is expressed when stress occurs prior to flowering. Symptoms of susceptibility include leaf rolling, uncharacteristic leaf erectness, leaf tip and margin senescence, delayed flowering, poor panicle exertion, floret abortion, and reduced panicle size. The other type of stress response occurs after flowering. Symptoms of susceptibility include premature senescence, stalk lodging, and reduction in seed size. No genotypes are known to be highly tolerant to drought stress both prior to and after flowering.

A good example of the complexity of selecting for drought resistance is given by Blum (1979). He evaluated eight sorghum genotypes for six characters thought to be associated with drought resistance: (1) drought avoidance (maintenance of high leaf water potential as soil water content decreases), (2) dehydration avoidance (maintenance of tissue water content with decreasing leaf water potential), (3) epicuticular wax deposition, (4) membrane desiccation tolerance (maintenance of membrane integrity during desiccation or heat stress), (5) stomatal desiccation tolerance (maintenance of photosynthesis during stress), and (6) recovery after drought stress. No single genotype could be classified as universally "drought resistant." Different genotypes were more "resistant" according to different parameters. Blum (1979) suggested that superior genotypes could be developed by intercrossing genotypes with different types of resistance and selecting for combinations of traits associated with drought tolerance and avoidance.

A Composite Response to Drought Stress

No studies have attempted to follow all the responses of a C_4 grass during a period of slowly increasing drought stress in the field. However, the sequence of responses can be pieced together from a number of independent studies, each of which measured a few of the many effects of stress.

The first response of C_4 grasses to drought stress is normally a reduction in leaf and stem elongation. This is accompanied or soon followed by a reduction in cell division and a decrease in the activity of certain enzymes such as nitrate reductase and RuBP carboxylase. This is followed by a decrease in plant water potential, stomatal closure resulting in lower transpiration and CO_2 assimilation, a decline in leaf cytokinin, an increase in abscisic acid concentration in the tissues, an increase in the activities of hydrolytic enzymes, and a reduction in ion transport. Increases in the activities of enzymes such as malic dehydrogenase, invertase, protease, and amylase cause increased tissue concentrations of sugar alcohols, organic acids, amino acids (especially proline), and soluble carbohydrates. These soluble compounds decrease the ψ_o of the tissue and help keep ψ_p positive.

When drought stress becomes so severe that decreases in ψ_o cannot offset decreases in ψ_l, leaves (especially those of vegetative tillers) often begin to roll longitudinally. Leaf rolling occurs because bulliform cells on the adaxial surface of the blade lose turgor and collapse. This reduces the leaf area and the interception of radiant energy by the canopy, further reducing transpiration and allowing ψ_l to remain high enough to prevent tissue death. If drought continues, older leaves, which are more susceptible to drought stress than younger leaves, begin to senesce. As stress continues, senescence advances to increasingly younger tissues until all leaf tissue is dead. Even though all leaf tissue may die, dormant buds and other apical and intercalary meristems may remain alive until rainfall causes storage carbohydrates and proteins to be remobilized,and regrowth begins.

As discussed previously, the typical response of tillers is somewhat different after anthesis. These tillers normally maintain open stomata and some photosynthesis to lower leaf water potentials than comparable vegetative tillers. After anthesis leaf rolling is typically less pronounced, and leaf senescence is more rapid.

SOIL STRENGTH AND AERATION

The plant shoot is dependent on the root system for water, nutrients, and hormones (cytokinins and gibberellins). Disruption of its ability to supply one or more of these can cause reduced or aberrant shoot growth. Numerous soil conditions can impede root growth and function, including extreme soil temperatures, nutrient deficiencies, chemical toxicities, low soil moisture, high soil strength, and poor aeration. High soil strength (resistance to penetration or compaction) often limits root growth and function in dense soil layers produced during normal pedogenesis (fragipans, gravelly, or cemented layers), in layers compacted by machinery or animal traffic, or in dry layers. Soil aeration often limits root growth in saturated or nearly saturated layers. These two constraints to normal root growth and function are discussed in the same chapter because soil compaction exacerbates both problems. Indeed, in some compact soil layers either strength or aeration limits root growth at virtually all soil water contents.

SOIL COMPACTION

Most uncompacted well-drained soils have well-connected macropores that permit rapid infiltration and drainage of water, rapid diffusion of O_2 to the roots, and root growth unimpeded by mechanical resistance. However, animal and machine traffic often compact soils to the point that either high soil strength or poor aeration limit root growth.

Compaction is a two-phase process consisting of destruction of the structural units of the soil and the compression of soil particles into existing soil macropores. Tillage is normally responsible for the destruction of structural units, and hoof, wheel, and disc pressure are primarily responsible for the compression of the disturbed soil. In general, moist soil is more easily compressed than dry soil, and traffic on wet soils is one of the major causes of severe compaction.

Not all compact zones are caused by animal or machine traffic. A fragipan is a natural subsurface layer with a bulk density higher than that of the soil above it and into which very few or no roots can penetrate. Fragipans restrict movement of both air and water and may also be strongly acid with toxic levels of aluminum. Duripans (subsurface horizons cemented by silica), some petrocalcic horizons (cemented by $CaCO_3$), and gravelly layers also form natural barriers to root penetration (Soil Survey Staff, 1975).

Zones of Compaction

Fields under mechanized agriculture are rarely compacted uniformly. Up to five zones of differing bulk densities can be found (Trouse, 1978). These result from different cultural practices and have different effects on root growth and function. The plow pan or traffic pan is formed by tire or track pressure on previously loosened soil. It begins as bands formed under the wheel or track. These bands coalesce after several years to form a continuous pan. Below the plow pan is the undisturbed soil below the tillage depth. This soil normally has resisted compaction because its structural units have not been destroyed by tillage. However, it may have layers that have been compacted or cemented during natural pedogenesis. These layers can also restrict root growth. The third zone is typically the harrow sole formed above the plow pan by the pressure of the disc edges on soil disturbed by plowing. The fourth zone is that soil loosened and pulverized by harrowing, and the fifth zone is the compact interrow formed by subsequent field operations in which wheels or tracks are confined to the interrows. Because field soils are not compacted uniformly, quantification of the compaction and its effects on root and shoot growth are difficult at best (Bowen, 1981; Trouse, 1971; Voorhees, 1977a,b).

Measures of Compaction

When soils are compacted, water and soil particles displace soil air. This causes a simultaneous increase in the resistance of the soil to root penetration and a decrease in the diffusion of O_2 within the soil. Since it is difficult to measure directly the resistance of the soil to root penetration (Eavis and Payne, 1968), the soil's bulk density or its penetrometer resistance is usually measured instead. Both bulk density and penetrometer resistance are surrogates for resistance to root penetration. However, in a particular soil, root penetration is often highly correlated with both measures. When several soils of different textures are compared, root growth usually correlates better with penetrometer resistance than with bulk density (Monteith and Banath, 1965; Taylor et al., 1964a,b; Taylor and Bruce, 1968).

Since compaction also decreases porosity and the rate of O_2 diffusion in the soil, air-filled porosity at a standard soil water content is sometimes used as an index of the effects of compaction on soil aeration (Trouse, 1964; Trouse and Baver, 1962; Trouse and Humbert, 1961).

For a particular soil, no technique is consistently better than another to estimate the effects of soil compaction, primarily because root growth is sometimes limited by soil strength and is sometimes limited by aeration. For example, when compacted soils are wet, penetrometer resistance is often low, but the limited soil pores are largely filled with water, and aeration limits root growth. As the soil dries aeration improves, but soil strength increases and mechanical impedance may limit root growth (Taylor et al., 1964b). It is not surprising, therefore, that in some studies (Trouse, 1965) root growth is better correlated with aeration porosity than with bulk density or penetrometer resistance, and in others (Monteith and Banath, 1965) the opposite is true.

SOIL STRENGTH

A number of studies with C_4 grasses have shown that root growth, morphology, and metabolism are sensitive to soil strength. The effects of high soil strength include reduced root growth, water extraction, and nutrient uptake, and thickening, deformation, and abnormal branching of roots.

Root Growth

Good negative correlations have been found between several measures of soil density and strength and either rooting densities or elongation rates (Figs. 11.1 and 11.2) for maize (Barley, 1963; Grimes et al., 1975; Henry and McKibben, 1966; Phillips and Kirkham, 1962a), sugarcane (Glover, 1967; Monteith and Banath, 1965), and C_4 forage grasses (Barton et al., 1966; Zimmerman and Kardos, 1961).

Soil strength and bulk density of layers within the soil profile affect root distribution (Fryrear and McCully, 1972; Grimes et al., 1975). Fewer roots penetrate the deeper soil layers in compacted soil than in noncompacted soil (Fig. 11.3). This often reduces water extraction by the crop and increases drought stress (Weatherly and Dane, 1979). However, even very dense, high-strength soil layers have natural fracture planes and previous root tracks that provide avenues of low resistance to root penetration. This leads to the formation of "rope roots," intertwined roots and rootlets that descend a common soil pore (Evans, 1935; Glover, 1967). Slowly permeable, cracking clay soils often produce dense peds. Roots are restricted to the faces of these peds where better aeration and low penetration resistance favor their growth (Burnett and Tackett, 1968).

When root systems penetrate a compacted soil layer and enter a deeper layer with less mechanical impedance, they often proliferate (Glover, 1967, 1968). This can reduce the effects of local compaction.

Root Deformation

Roots of grasses grown in nutrient solutions or in media with nearly ideal physical and chemical properties develop in an orderly manner. Each successive order of branching consists of younger, smaller roots. In solution culture sugarcane

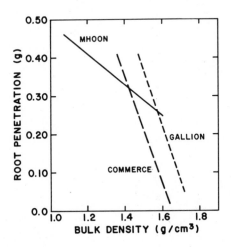

FIG. 11.1. Effect of bulk density on the penetration of sudangrass roots (*Sorghum sudanense*) into soil cores. From Meredith and Patrick (1961). Reproduced from *Agronomy Journal*, by permission of the American Society of Agronomy.

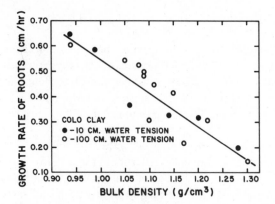

FIG. 11.2. Effect of bulk density on root growth rate at two water tensions in a Colo clay soil. From Phillips and Kirkham (1962). Reproduced from *Soil Science Society of America Proceedings*, by permission of the Soil Science Society of America.

normally produces four orders of rootlets, with each successive order having fewer xylem vessels, fewer cells in cross section, and less pith (Exner, 1971). Sugarcane roots growing in uncompacted soil normally produce at least two orders of laterals that develop at right angles to the roots from which they branch (Glover, 1967). This normal pattern of branching and development is changed by injury or mechanical impedance of the growing root tip. When the apex is damaged or impeded, root production near the tip may be stimulated (Exner, 1971; Evans, 1935; Kamerling, 1903, cited in Dillewijn, 1952). A rootlet branch-

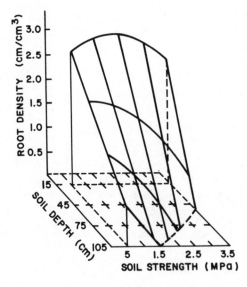

FIG. 11.3. Effects of soil strength and depth on maize root length density. From Grimes et al. (1975). Reproduced from *Agronomy Journal*, by permission of the American Society of Agronomy.

ing from the damaged or impeded root may become dominant and attain the size of the root from which it branched.

High soil strength causes an increase in root diameter as well as deformation of root hairs and the root cortex. Thick, slow-growing roots with short, deformed branches develop in layers with high soil strength. When these roots enter layers of lower soil strength, their diameters decrease (Glover, 1967, 1968; Burnett and Tackett, 1968). Root deformation is caused in part by production of ethylene in response to high soil strength. Exogenous ethylene causes root elongation to slow and root diameter to increase. Ethylene inhibitors such as DIHB (3,5-diiodo-4-hydroxybenzoic acid) stimulate root and seedling growth in high-strength soils and in the presence of exogenous ethylene (Kays et al., 1974; Wilkins et al., 1976, 1977, 1978).

Root hairs of sugarcane are potentially long lived and can remain turgid and apparently functional for up to four months (Artschwager, 1925; Glover, 1967); however, root hairs produced in compacted soils are sparse, deformed, and short lived. Soil compaction can also cause severe distortion of the root cortex and stele. In fact, Trouse (1965) described seven stages of increasing surgarcane root cortex and stele deformation associated with increasing bulk density and decreasing aeration porosity (Table 11.1).

Like compaction, soil texture affects root diameter. Roots in heavy clay soils are often thicker and branch less than those in sandy soils (Glover, 1967). Thick roots are produced even in reconstructed profiles of heavy clay soils in which the soil has been removed and replaced (Glover, 1967; Burnett and Tackett, 1968). However, root diameter decreases if these soils are completely pulverized by rototilling so that peds are broken, aeration is improved, and bulk density is reduced (Burnett and Tackett, 1968).

Sand and gravel layers occur in many soils, where they reduce volumetric water and nutrient contents. Root growth is often reduced and distorted in these layers. In studies with varying percentages of gravel mixed with soil, root growth usually varies inversely with gravel percentage and with penetrometer resistance (Babalola and Lal, 1977a,b; Vine et al., 1981).

Nutrient Uptake

High soil strength can reduce both total nutrient accumulation by the plant and the foliar concentration of mineral nutrients (Trouse and Humbert, 1961). Maize (Phillips and Kirkham, 1962b) and sugarcane (Beater, 1964) grown on compacted topsoils may have lower tissue N, P, and K concentrations than plants grown on uncompacted soils. Juang and Uehara (1971) found that while compaction reduces sugarcane P and K uptake, the effect on P uptake is much more severe than on K uptake. Potassium uptake was reduced by about 50% while P uptake was reduced by over 90% by severe compaction (Table 11.2).

This suggests that part of the effect of high soil strength on shoot growth and yield can be due to inadequate nutrition. However, deep plowing to disrupt pans may reduce fertility by mixing infertile subsoil with topsoil (Hauser and Taylor,

TABLE 11.1. Values of Bulk Density^a Associated with Various Stages of Root Deformation Caused by Soil Compaction (modified from Trouse, 1965)

STAGES OF ROOT DEGRADATION	SOIL TYPE			
	TYPIC CHROMUSTERTS	TROPEPTIC EUTRUSTOX	TYPIC TORROX	TYPIC HYDRANDEPT
A. Ideal roots, branching in all planes, no distortion.	1.14	1.03	1.03	0.66
B. Reduced proliferation, no distortion.	1.22	1.07	1.15	0.74
C. Reduced proliferation, slight flattening of rootlets, secondary roots tend to follow channels of primary roots.	1.31	1.20	1.23	0.79
D. Fair distribution, most rootlets and some roots flattened, tendency for root development in fracture zones.	1.44	1.31	1.38	0.85
E. Poor distribution, inadequate for agricultural production, all portions of all roots flattened, more roots confined to fracture zones.	1.55	1.39	1.47	0.91
F. Very few roots, most roots confined to fracture planes and badly flattened.	1.76	1.52	1.57	0.96

^{a}Measured in grams per cubic centimeter.

TABLE 11.2.   Effect of Bulk Density on Phosphorus (^{32}P) and Potassium (^{86}Rb) Uptake and Sugarcane Root Growth in a Clay Loam Soil (Juang and Uehara, 1971)

	PHOSPHORUS		POTASSIUM	
BULK DENSITY (g/ml)	ROOT DRY WEIGHT IN CORE (g)	^{32}P UPTAKE (cpm/g soil)	ROOT DRY WEIGHT IN CORE (g)	^{86}RB UPTAKE (cpm/g soil)
1.2	0.46	385	0.42	10
1.4	0.24	274	0.36	12
1.6	0.18	73	0.30	4.5
1.8	0.15	35	0.05	5.5

1964), and yields can often be improved by a combination of deep tillage and deep incorporation of fertilizer and/or soil amendments (Bradford and Blanchar, 1977; Patrick et al., 1959; Robertson et al., 1957).

Shoot Growth

A number of field and laboratory studies have clearly demonstrated that high soil density and strength can reduce emergence, shoot growth, and economic yields of maize (Gaultney et al., 1982; Phillips and Kirkham, 1962b; Smittle et al., 1981), grain sorghum (Bradford and Blanchar, 1977; Taylor et al., 1964a), sugarcane (Ricand, 1977), and C_4 forage grasses (Barton et al., 1966). Several examples illustrate these effects.

Soil crusts formed by raindrop impact or other factors can affect seedling emergence. For example, emergence of maize, grain sorghum, switchgrass (*Panicum virgatum*), and barnyardgrass (*Echinochloa cruzgalli*) is delayed and finally reduced when the strength of the soil above the seed exceeds about 0.6 MPa (Stibbe and Terpstra, 1982; Taylor et al., 1966) (Fig. 11.4).

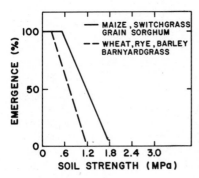

FIG. 11.4. Effect of soil strength (at −0.033 MPa matric potential) on seedling emergence of several species. From Taylor et al. (1966). Reproduced from *Agronomy Journal*, by permission of the American Society of Agronomy.

TABLE 11.3. Effects of Profile Modification and Chemical Amendments on Grain Yield of Hybrid Grain Sorghum (Bradford and Blanchar, 1977)[a]

TREATMENT	YIELD (kg/ha)
Outside trench area	1841
Nontrench area	3229
Trench	4322
Trench + lime	4906
Trench + lime + fertilizer	5145
Trench + lime + fertilizer + sawdust	5987

[a] Reproduced by permission of the Soil Science Society of America.

Bradford and Blanchar (1977) studied the effects of very deep tillage (removing then replacing soil from a trench to 150 cm) on an acid soil with a thick fragipan. They also included treatments with deep fertilizer and amendment (lime and sawdust) placement. Trenching alone improved grain sorghum yield and water extraction below the normal plow layer, and addition of amendments that reduced soil acidity and improved the water-holding capacity of the subsoil produced further yield increases (Table 11.3).

Gaultney et al. (1982) found that a simulated plow pan at 29 cm reduced maize yields by more than 50% on a poorly drained silt loam soil (Fig. 11.5). However, in this study water stress increased the yield of the treatment with severe compaction, presumably because poor aeration of the compacted zone reduced root development more than high soil strength did.

Smittle et al. (1981) found that maize root growth, plant growth, yield, and nutrient uptake efficiency were proportional to the volume of low-strength soil produced by preplant tillage. Finally, high bulk density in the top 5–10 cm re-

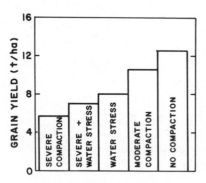

FIG. 11.5. Effects of several subsoil compaction and water stress treatments on maize grain yields on a poorly drained silt loam soil. Bulk densities of the uncompacted, moderately compacted, and severely compacted treatments were 1.69, 1.71, and 1.76 g/cm³, respectively. From Gaultney et al. (1982).

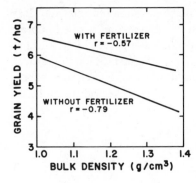

FIG. 11.6.   Effect of bulk density in the 0–7.5-cm layer and fertilizer on maize grain yield in a clay soil under sprinkler irrigation. From Phillips and Kirkham (1962b). Reproduced from *Agronomy Journal*, by permission of the American Society of Agronomy.

duces emergence, growth, and yield of many crops, including maize and grain sorghum (Phillips and Kirkham, 1962b; Taylor et al., 1964a) (Fig. 11.6, Fig. 11.7).

Taylor et al. (1964a) conclude that at least five factors affect crop response to a compact layer.

1. *Depth to the compact layer.* A soil pan near the surface will cause more frequent and more severe yield reductions than a deeper pan.
2. *Water availability.* Yields will be reduced less if water is readily available (but not excessive).
3. *Perennial vs. annual crops.* Perennial crops are less susceptible than annual crops because their roots are more likely to eventually penetrate the pan.
4. *Plant population.* High plant populations require more water and nutrients and are more susceptible to reduced rooting volume.
5. *Bulk density of the compact layer.* In a particular soil, strength increases and root growth decreases with increased bulk density.

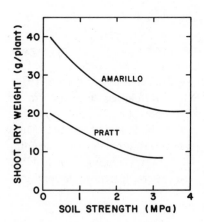

FIG. 11.7.   Relationships between grain sorghum shoot weight and soil strength (at field capacity) of the layer below the seed for two soils. From Taylor et al. (1964a). Reproduced from *Agronomy Journal*, by permission of the American Society of Agronomy.

Thus, if water and nutrients are readily available, a dense but relatively deep soil pan may hardly affect top growth, because the root system is able to obtain adequate water and nutrients from the superficial uncompacted layer. If, however, soil water becomes inadequate in the uncompacted surface soil, top growth is likely to be reduced by water stress and nutrient deficiencies (Fehrenbacher et al., 1958; Johnston and Wood, 1971; Patrick et al., 1959; Weatherly and Dane, 1979).

SOIL AERATION

Periodic flooding and poor soil aeration can dramatically affect plants. Factors influencing plant response include duration of the stress, depth of flood or water table, season and climatic conditions, the species involved, and, during flooding, the rate of water movement (Colman and Wilson, 1960).

Transient flooding is most deleterious when it occurs during active growth. Dormant or partially dormant plants have greater resistance than actively growing plants. But even if flooding occurs during a period of active growth, some species can survive indefinitely while others are killed in a few days. In addition, the tops of some species [such as kikuyugrass (*Pennisetum clandestinum*)] are readily killed, but their rhizomes remain viable for long periods (Colman and Wilson, 1960).

Depth of flooding is an important factor affecting survival. Though many caespitose *Paspalum* species are well adapted to moist conditions with higher water tables, they cannot survive when completely inundated. On the other hand, rhizomatous species such as paragrass (*Brachiaria mutica*), which is also well adapted to moist conditions, can survive deep floods, presumably because their rhizomes float to the surface.

Moderate water movement can reduce flooding damage by supplying sufficient oxygen to maintain physiological processes; however, rapid water movement can cause mechanical damage and smother plants under silt deposits (Colman and Wilson, 1960).

Tolerance to flooding and poor soil aeration, like tolerance to other environmental stresses, is quite variable among ecotypes and cultivars of tropical grass species (Table 11.4). Variation in flooding tolerance has been found among cultivars of forage species like kleingrass (*Panicum coloratum*) (Anderson, 1972a), which has good resistance, and buffelgrass (*Cenchrus ciliaris*) (Anderson, 1974), which is moderately tolerant of flooding. Among buffelgrass cultivars, greater flooding tolerance is found in tall, rhizomatous genotypes.

Sugarcane has a great deal of genotypic variation in flooding tolerance. Many varieties are very susceptible to poor drainage; however, others grow in standing water. Dillewijn (1952) cites examples of Indian varieties growing in up to 1.5 m of water for periods of up to five months. These varieties can even be combined with aquaculture.

TABLE 11.4.   Relative Tolerance of Some C$_4$ Pasture Grasses to Flooding (Whiteman, 1980)

LEVEL OF TOLERANCE	SPECIES
GOOD	*Brachiaria mutica*
	Panicum coloratum var. *makarikariense*
	P. coloratum cv. Bambatsi
	P. coloratum cv. Kabulubulu
	Paspalum plicatulum
	Paspalum dilatatum
	Digitaria decumbens
	Setaria anceps
Moderate	*Pennisetum clandestinum*
	Pennisetum purpureum
	Brachiaria decumbens
	Cenchrus ciliaris
	Chloris gayana
	Panicum maximum
Poor	*Brachiaria ruziziensis*
	Panicum maximum var. *trichoglume*
	Panicum antidotale
	Melinis minutiflora
	Urochloa mosambicensis

Root Growth

Poor soil aeration has dramatic effects on root growth, morphology, and metabolism. For example, high water tables or poorly aerated soil layers frequently restrict the depth of rooting in sugarcane (Escolar and Allison, 1976; Evans, 1964a; Webster and Eavis, 1971), maize (Baser et al., 1981; Jat et al., 1975; Lal and Taylor, 1969), and pearl millet (*Pennisetum americanum*) (Campbell and Phene, 1977; Williamson et al., 1969). The decrease in root growth is accompanied by changes in root morphology. Existing roots often stop growing, and root development from basal nodes of the stem increases. These new roots often remain white and unsuberized, develop air spaces in the root cortex, and grow upward toward the soil surface. Genotypic tolerance of poor aeration is associated with the formation of cortical air spaces and the absence of aerotropic (upward) root growth under poor aeration (Dillewijn, 1952; Evans, 1964a; Srinivasan and Batcha, 1963).

Though poor aeration reduces root growth in some parts of the soil profile, compensatory growth may occur in better-aerated zones. For example, Juang and Uehara (1971) found that root length density in well-aerated layers was greatest when poor drainage severely limited the aerated volume of soil.

It is clear that root growth of C_4 grasses is quite sensitive to the O_2 concentration of the air surrounding the roots. In most studies root growth decreases nonlinearly to about 50% of maximum as the O_2 content of the root zone air decreases from ambient (21%) to about 3–6%. Further decreases in O_2 content virtually stop root growth (Banath and Monteith, 1966; Gill and Miller, 1956; Gingrich and Russell, 1956). These studies are consistent with the effects of soil O_2 on shoot growth. For example, Campbell and Phene (1977) found that regrowth of pearl millet (*Pennisetum americanum*) decreases by about 50% as the O_2 concentration of the root zone air declines from 20 to 5%.

Linn and Doran (1984) suggest that the percentage of total soil pore space filled with water correlates well with soil microbial activity in a variety of soils. Maximum aerobic activity (measured in terms of respiration, bacterial numbers, ammonification, nitrification, and nitrogen fixation) is found at about 60% water-filled pore space (WFP). Below this value, activity is limited by low soil water. Above this value, it is limited by poor aeration. Several studies suggest that when water-filled pore space exceeds 60%, root growth is inhibited by poor aeration. For example, Grable and Seimer (1968) found that maize root growth decreases rapidly as WFP increases from 60 to 80%. Trouse (1964) reported that in several Hawaiian soils, sugarcane roots became increasingly deformed and stunted as WFP increases from 60–70% to 95%. Similarly, Silberbush et al. (1979) reported that poor aeration limits root growth below trickle irrigation emitters. Below the top 5–10 cm, maximum root development was found when WFP was about 60% or less. Very little root development occurred at WFP exceeding 75%.

One of the most important root adaptations to poor aeration is the development of continuous air spaces (lysigenous aerenchyma) in the cortex. Norris (1913) and Dunn (1921) were among the first to note the production of aerenchyma (in maize roots) as a result of poor root aeration. McPherson (1939) clearly described the disappearance of cortical protoplasm, bulging and death of the cells, and collapse of cell walls associated with the formation of the aerenchyma. The response facilitates axial movement of O_2 from the atmosphere through the cortex to the root and rhizosphere (Jat et al., 1975; Jensen et al., 1964; Lee et al., 1981). However, this diffusion of O_2 does not provide sufficient O_2 to support aerobic respiration in the root system. Luxmoore et al. (1970) estimate that it could provide only 25% of the respiratory demand, and Crawford (1978) concludes that its main effect is to oxidize toxins in the rhizosphere. Thus, even in flood-tolerant plants such as rice and the C_4 marsh grass *Spartina alterniflora* (which have well-developed aerenchyma), root metabolism is primarily anaerobic when flooded (John and Greenway, 1976; Mendelssohn et al., 1981).

Whatever the functional role of aerenchyma, it is apparently important because genotypes that produce aerenchyma are normally more flood tolerant than those that do not (Yu et al., 1969; Srinivasan and Batcha, 1963). For example, *Saccharum spontaneum*, a wild species which hybridizes readily with sugarcane, is often tolerant of poor soil aeration. Flood-tolerant hybrids of sugarcane and *S. spontaneum* produce more adventitious roots with prominent aerenchyma than do intolerant hybrids.

Ethylene is produced by plant roots as a result of several stresses, including poor aeration. Recent work (Drew et al., 1979, 1981; Jackson et al., 1981) clearly shows that development of root aerenchyma in maize is mediated by this production of ethylene. For example, exogenous ethylene stimulates aerenchyma development, and inhibitors of ethylene activity prevent it.

Though poor aeration stimulates aerenchyma production, C_4 grasses such as blue grama (*Bouteloua gracilis*), maize, and sugarcane often develop typical aerenchyma under well-aerated conditions (Beckel, 1956; Konings and Verschuren, 1980; Trouse, 1978). In these species, aerenchyma forms in the cortex of older roots prior to the degeneration and sloughing of the cortex. In addition, low nutrient solution N concentrations (NO_3^- or NH_4^+) stimulate aerenchyma formation in maize (Konings and Verschuren, 1980). It is not known whether aerenchyma production associated with natural aging and low N concentrations are also mediated by ethylene production.

Carbon Metabolism

McManmon and Crawford (1971) propose that differences in flooding tolerance are due primarily to differences in root metabolism during anaerobic conditions. Flooding prevents normal levels of O_2 in the roots and thereby prevents normal aerobic respiration. In the roots of intolerant genotypes, the activity of alcohol dehydrogenase (ADH), which catalyzes conversion of acetaldehyde to ethanol, increases, and ethanol accumulates in the roots. ADH activity does not increase in tolerant genotypes.

Schwartz and Endo (1966) found that maize has two isozymes of ADH, denoted ADH^F (fast) and ADH^S (slow). Genotypes homozygous for ADH^S are more tolerant than those homozygous for ADH^F. Heterozygous plants are intermediate (Marshall et al., 1973). This confirms that, indeed, ADH activity is sometimes associated with flooding tolerance. McManmon and Crawford (1971) suggest that ethanol is toxic to plant tissues and is responsible for flooding injury to intolerant genotypes. However, recent work indicates that (1) roots are quite tolerant of ethanol concentrations in excess of those normally measured in stressed roots (Jackson et al., 1982; Rumpho and Kennedy, 1983), (2) ethanol is a major product of anaerobic root metabolism in flood-tolerant species (Smith and apRees, 1979), and (3) across a transect of decreasing soil aeration, growth of the flood-tolerant C_4 marsh grass *Spartina alterniflora* decreases by almost 50% before ADH activity increases significantly. Jackson et al. (1982) conclude that high ADH activity and ethanol accumulation are consequences rather than causes of anaerobic injury.

Nutrient Uptake

Poor aeration of the root system reduces active ion uptake (Cram and Pitman, 1972; Shaner et al., 1975). In addition, anaerobic conditions cause rapid leakage of previously accumulated K from maize roots (Marschner et al., 1966; Mengel and Pfluger, 1972). Thus, it is not surprising that high water tables and flooding reduce plant uptake and tissue concentrations of N, P, and K in sugarcane (Gascho and Shih, 1979; Juang and Uehara, 1971), maize (Goins et al., 1966; Lal and Taylor, 1969, 1970; Schwab et al., 1966), and other species. Low nutrient solution O_2 concentrations have much the same effect (Banath and Monteith, 1966; Barley, 1962; Gill and Miller, 1956). The effects of poor aeration on Ca and Mg concentrations are variable (Gascho and Shih, 1979; Lawton, 1945; Shapiro et al., 1956). Lal and Taylor (1970) found that a high water table (15 cm) reduces maize ear leaf concentrations of N, P, K, Zn, Cu, and B (compared with well-aerated controls) while increasing the concentrations of toxic elements such as Al, Fe, Mn, and Mo.

Leaf chlorosis caused by waterlogging is similar to that caused by N deficiency, and part of the maize yield reduction caused by flooding can sometimes be overcome by high N fertility (Lal and Taylor, 1969; Ritter and Beer, 1969). Sugarcane leaf blade N and leaf sheath K concentrations are positively correlated with leaf sheath moisture content (Clements, 1980). Poor soil aeration causes all three to decrease, and much of the decrease in sheath N and K concentration can be accounted for by the decrease in sheath moisture. However, it is difficult to determine whether low nutrient concentrations cause low water contents or vice versa.

Water Uptake

Poor soil aeration reduces stomatal conductance and transpiration (Letey et al., 1962). Sojka and Stolzy (1980) showed a clear positive relationship between O_2 diffusion rate in the root zone and stomatal conductance; however, there was no relationship between leaf water potential and stomatal conductance. In fact, leaf water potential of waterlogged plants is usually equal to or greater than that of nonwaterlogged controls. This suggests that flooding does not affect stomatal conductance by reducing root system hydraulic conductivity and leaf water potential. Instead, it may cause abscisic acid contents of the shoots to increase (Hiron and Wright, 1973), thereby reducing stomatal conductance.

Shoot Growth

Though the most striking morphological adaptations to poor aeration concern the root system, grass shoots also respond. One of the most unusual adaptions is the stimulation of internode elongation in some varieties of rice and in certain flood-adapted C_4 grasses such as *Brachiaria mutica* (Bernal, 1971). The elongation response has not been carefully studied in C_4 grasses, but it is probably similar to that in deep-water rice.

Deep-water (floating) rice is grown in low-lying, riparian areas of Southeast Asia, which are flooded every year during the rainy season. The seeds are usually broadcast on dry soil and germinate after the first rains at the beginning of the rainy season. Several weeks after germination the flood waters begin to rise and flood the rice. It responds by producing elongated internodes that grow rapidly enough (up to 25 cm/day) to keep a substantial part of the plant floating on the surface. This is very important since total submergence for more than 1 week results in drastic yield decreases. In this way, tillers can reach up to 7 m length before flowering, which is controlled by photoperiod and usually occurs after water levels have stopped rising (De Datta, 1981).

Internode elongation of deep-water rice is mediated by ethylene, whose production (as in roots) is stimulated by flooding. Experimentally, exogenous ethylene stimulates internode elongation under well-aerated conditions, and inhibitors of ethylene activity prevent elongation (Metraux and Kende, 1983). Submerged shoots of deep-water rice are similar to roots in another way. They obtain oxygen by diffusion in the gas phase directly from the atmosphere. However, instead of relying on internal air spaces, they have continuous external air layers on the hydrophobic, corrugated surfaces of the leaves. In deep-water rice these air layers constitute 45% of the leaf volume, much more than is found on flooding-sensitive C_3 cereal leaves. When the air films are prevented from forming by washing the leaves with mild detergent, submerged plant parts die due to lack of aeration (Raskin and Kende, 1983). Though very little is known about the development of external air layers in submerged C_4 grasses, it may play an important role in their aeration.

Another morphological adaptation to flooding in maize is the development of enlarged stem bases with increased numbers of vascular bundles (Kuznetsova et al., 1981). However, it is unclear what importance, if any, this has in helping the plant survive anaerobic conditions.

Two types of studies have been used to quantify the effects of flooding and poor soil aeration on the growth and yield of maize and sugarcane. They involve (1) maintenance of artificial water tables over relatively long periods of time and (2) intermittent flooding for short periods.

Maize. Optimum conditions for maize shoot growth apparently occur when the volume of well-aerated soil is sufficient to supply adequate mineral nutrition and the water table is shallow enough to supply adequate water. However, the optimum water table is deeper when rainfall or irrigation supply most of the crop's demand for water. For example, Doering et al. (1976) found that the optimum static water table depth was about 100 cm when irrigation supplied only 30% of the crop water requirement. When the water table was deeper, greater irrigation rates were needed to produce maximum yields. Chaudhary et al.

(1975) obtained similar results under rain-fed conditions. In a wet year the optimum water table depth was 120 cm. In a dry year it was 60 cm.

In most soils grain and dry matter yields are greatest when the water table is more than 60 cm deep (Diseker and van Schilfgaarde, 1958; Lal and Taylor, 1969; Williamson and Kriz, 1970). However, the optimum depth often varies with soil texture, being deeper in sandy soils than in clayey soils. For example, in greenhouse experiments, optimum maize yields are obtained with water tables at about 30 cm in clays and silty clay loams (Baser et al., 1981; Goins et al., 1966) and with water tables below 75 cm in loams and sandy loams (Goins et al., 1966; Williamson and van Schilfgaarde, 1965).

Temporary flooding early in the growing season normally reduces maize yields more than do high static water tables (Lal and Taylor, 1969, 1970; Ritter and Beer, 1969; Schwab et al., 1966), probably because high-static water tables allow some root growth in the upper part of the profile. In addition, the crop is most sensitive to flooding at the seedling stage. For example, surface flooding once for only 4 days at 2 weeks of age can reduce yields by 40% (Chaudhary et al., 1975). The crop is much less sensitive to flooding after silking. Duration of flooding is also very important (Fig. 11.8); therefore, drainage systems should be designed to decrease the water table as quickly as possible to avoid severe yield reduction.

Hardjoamidjojo et al. (1982) developed a stress-day index for flooding that takes into account water table depth, duration of water tables higher than 30 cm, and the relative susceptibility of the crop at a particular growth stage. They showed that the stress-day index was highly correlated with relative maize yields in four studies from the United States and India.

One practical method for avoiding yield reductions due to poor surface drainage and intermittent soil saturation on poorly drained soils is to grow maize and sugarcane on raised beds. This permits at least a small volume of well-aerated

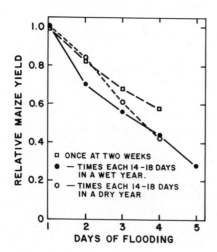

FIG. 11.8.   Effect of duration of surface flooding on maize yield. From Chaudhary et al. (1975).

soil and better root growth even when water is standing in the furrows (Norden, 1964; Smith, 1977)

Sugarcane.   Even though sugarcane can survive extremely wet conditions for extended periods, yields are severely reduced by high water tables and even short periods of flooding (Escolar and Allison, 1976; Gumbs and Simpson, 1981; Juang and Uehara, 1971). Optimum water table depth is generally below 76 cm, and high water tables are most damaging when they occur immediately after planting or during early regrowth of the ratoon crop, primarily because of their adverse effects on germination and tillering (Gosnell, 1971; Webster and Eavis, 1971).

As with maize, high water tables and slow decline of the water table after irrigation often reduce cane and sugar yields. For example, Carter (1976) found that in Louisiana maximum yields are obtained when maximum water table depth is at least 120 cm and when it reaches that depth within 7 days of irrigation. These results suggest that a stress-day index like that described for maize could be used to estimate the effects of poor aeration on sugarcane growth.

Somewhat different conditions exist in the Florida Everglades, where sugarcane is grown on organic (muck) soils, which have developed under flooded conditions and which subside (oxidize and decrease in thickness) as a result of drainage. The organisms responsible for soil organic matter oxidation are aerobic; therefore, the rate of subsidence is directly proportional to the depth of the water table. To minimize subsidence, water tables should be kept as high as possible. Since Florida sugarcane cultivars have been selected for growth with adequate to excessive soil water, they are probably less sensitive to high water tables than are most cultivars selected elsewhere (LeCroy and Orsenigo, 1964), and some cultivars grow well with water tables as high as 30–45 cm, especially if adequate fertilizer N is applied (Andreis, 1976; Gascho and Shih, 1979). This and other work with C_4 grasses (Pate and Snyder, 1979) suggests that N deficiency due to reduced mineralization limits crop production on muck soils with high water tables.

SOIL STRENGTH OR AERATION?

It is difficult to separate the effects of mechanical impedance and aeration on root growth because both are affected by soil compaction. When soils are compacted and solids displace air-filled voids, soil strength is increased, voids large enough for roots to enter are reduced in number, and the rate of O_2 diffusion in the soil is decreased.

Gill and Miller (1956) and Barley (1962) used experimental techniques that enabled them to separate the effects of rhizosphere O_2 concentration from those of mechanical impedance. They grew maize roots in a medium of very small (0.05 mm) glass beads through which nutrient solution and gases of different O_2 concentrations were passed. The glass beads were compacted by applying pressure to the medium with a diaphragm. In order to elongate, the maize roots had to displace the beads. Their ability to do so was reduced by both increased diaphragm pressure and reduced O_2 concentration. In addition, poor aeration reduced the pressure required to produce a given reduction in root growth. Thus, at a particular O_2 content and soil strength, the two stresses interact to reduce root growth more than either stress alone.

Root hormone responses to high soil strength and poor aeration are similar. Both stresses cause root ethylene production to increase (Goeschl et al., 1966; Kays et al., 1974; Robert et al., 1975). The morphological response to exogenous ethylene (reduced root elongation, swollen root tips) is similar to that caused both by poor aeration and high soil strength (Jackson et al., 1981; Kays et al., 1974; Wilkins et al., 1974, 1977, 1978). Finally, treatment of compacted soil or poorly aerated nutrient solution with the ethylene inhibitor DIHB (3,5-diiodo-4-hydroxybenzoic acid) stimulates root and seedling growth (Robert et al., 1975; Wilkins et al., 1976, 1977, 1978). The similarity of root responses to high soil strength, poor aeration, and exogenous ethylene suggests that ethylene production mediates plant response to both high soil strength and ethylene.

12

NITROGEN

The nitrogen content of C_4 grasses ranges from more than 5% to less than 1% of the dry matter. Though N makes up a relatively small percentage of plant weight, it is indispensible for growth and development, and inadequate N nutrition affects almost all aspects of plant growth and metabolism.

The nitrogen in C_4 grasses can come from a variety of sources, including mineralization of organic N; fertilizer N; N fixation by bacteria associated with the roots; absorption of volatile N-containing gases such as NH_3, NO, and N_2O; and the small amounts of N contained in rainfall. The most important sources are mineralization of organic N, fertilizer N, and N fixation.

Mineralization of organic N is greatest in soils with high organic N contents and high microbial activity. Microbial activity and mineralization of N are greatest in warm, moist, well-aerated soils with large amounts of organic matter. Tillage often increases mineralization by increasing aeration and raising soil temperatures. Natural cycling of N often provides all the N needed for growth of C_4 forage grasses in arid regions. However, most C_4 grass crops respond to N fertilization, especially when drought stress does not seriously limit growth.

Most fertilizer N used on C_4 grasses is applied as anhydrous ammonia, urea, or ammonium nitrate. In moist, warm, well-aerated soils NH_4^+ is rapidly converted (nitrified) to NO_3^-, which can be lost by leaching and denitrification. Nitrification inhibitors are sometimes used to prevent these losses.

Nitrogen fertilizers can be applied by different methods and at different times during the growth of the crop. In temperate climates preplant applications may be made in the fall or early spring, depending on soil type and rainfall pattern. In areas where significant waterlogging and/or deep percolation are common, fall-applied N may be lost through leaching and denitrification. After planting, N may be applied as side dressings of dry or liquid products or it may be mixed with irrigation water. Frequent small applications of N in irrigation water can reduce leaching losses while assuring adequate supplies of N throughout the growing season.

TISSUE NITROGEN CONCENTRATION

Four-carbon grasses typically have lower N concentrations than C_3 grasses (Henzell and Oxenham, 1973; Wilson, 1975; Wilson and Ford, 1971) and respond more to N fertilizer application (Hallock et al., 1965; Prine and Burton, 1956; Wilson and Haydock, 1971) (Table 12.1 and Fig. 12.1). Brown (1978b) concluded that the higher N use efficiency of C_4 grasses is due to the compartmentation of phosphoenolpyruvate carboxylase (PEPC) and ribulose biphosphate carboxylase (RuBPC) in the mesophyll and bundle sheath cells, respectively. This arrangement allows the plant to concentrate CO_2 in the bundle sheath cells and results in high rates of CO_2 fixation per unit RuBPC (see Chapter 2). As a result, in C_4 plants only about 20% of leaf protein is RuBPC and 10% is PEPC (Brown, 1978). In C_3 plants as much as 50% of the leaf protein is RuBPC (Bjorkman et al., 1976; Blenkinsop and Dale, 1974; Medina, 1970). The relatively lower con-

TABLE 12.1.   Dry Matter Yields and N Concentrations at Various Levels of N Fertilization for C_3 and C_4 Grasses (after Brown, 1978b)[a]

SPECIES	N APPLIED (kg/ha)	DRY MATTER YIELD C_3	DRY MATTER YIELD C_4	N CONCENTRATION (%) C_3	N CONCENTRATION (%) C_4
		(tons/ha)			
Festuca arundinaceae (C_3)	112	3.8	8.3	2.52	2.13
Cynodon dactylon (C_4)	224	5.8	11.4	2.77	2.26
	448	7.2	16.1	3.25	2.75
	896	6.9	17.5	3.50	3.00
		(g/pot)			
Digitaria macroglossa and	0	12	33	2.50	0.91
Paspalum dilatatum (C_4)	67	20	48	2.20	0.94
Lolium perenne and	134	20	60	2.61	1.17
Phalaris tuberosa (C_3)	269	31	65	2.90	1.78

[a]Reproduced from *Crop Science*, by permission of the Crop Science Society of America.

centration of photosynthetic carboxylases in C_4 plants and the higher affinity of PEPC than RuBPC for CO_2 results in more rapid rates of CO_2 fixation per unit leaf N under favorable climatic conditions. This allows C_4 plants to exploit their environment more effectively than C_3 plants when available soil N is low. Thus, high N use efficiency, adaptation to high temperature and irradiance, and high water use efficiency give C_4 plants an adaptive advantage in warm dry climates with limited available soil N.

Several factors (other than N deficiency) affect tissue N concentration, including age and plant part, temperature, and drought stress.

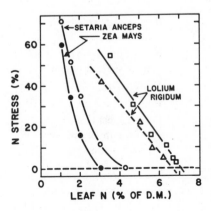

FIG. 12.1. Effect of leaf N concentration on N stress (percent of maximum growth rate) in C_3 (*Lolium rigidum*) and C_4 (*Setaria anceps* and *Zea mays*) grasses. From Brown (1978). Reproduced from *Crop Science*, by permission of the Crop Science Society of America.

Age and Plant Part

The N concentration of grass tissues varies with age and among plant parts. In all cases the same general trends are found. The highest shoot N concentrations occur in young plants or tillers, and the concentration declines with age and with increasing growth stage (Table 12.2). Figures 12.2, 12.3, 6.9, and 9.5 show typical shoot N or protein (6.25 × N) concentrations in maize, sugarcane, grain sorghum, and several C_4 forage grasses with increasing age after planting or cutting. The decrease in shoot N concentration is sometimes paralleled by decreasing N concentration in the roots (Fig. 12.4), though root N concentrations tend to be rather constant with increasing age in maize and grain sorghum (Fig. 12.5).

The decrease in sugarcane N concentration with age is due primarily to the production of internodes containing very low concentrations of N (Table 12.3). Similar effects of age and plant part are found in maize and forage grasses where rapid stem elongation following panicle initiation is associated with a rapid decrease in shoot N concentration. The N concentration of successively older leaves declines in a manner similar to its decline in successively older internodes. For example, Ayres (1936) found that the N concentration of sugarcane leaves de-

TABLE 12.2. **Definitions of Growth Stages for Maize (Hanway, 1963) and Grain Sorghum (Vanderlip and Reeves, 1972)**[a]

Growth Stage	Maize	Grain Sorghum
0	Emergence. Coleoptile visible at soil surface	Emergence. Coleoptile visible at soil surface
1	Collar of 4th leaf visible	Collar of 3rd leaf visible
2	Collar of 8th leaf visible	Collar of 5rd leaf visible
3	Collar of 12th leaf visible	Growing point differentiation. Approximately 8 leaf stage
4	Collar of 16th leaf visible. Tips of many tassels visible	Final leaf visible in whorl
5	75% of plants have silks visible	Boot. Head extended into flag leaf sheath
6	Kernels in "blister" stage	Half bloom. Half the plants at some stage of bloom
7	Very late "roasting ear" stage	Soft dough
8	Early dent stage	Hard dough
9	Full dent stage	Physiological maturity. Maximum dry matter accumulation
10	Physiological maturity	

[a] Reprinted by permission of Field Crops Research and Elsevier Scientific Publishers.

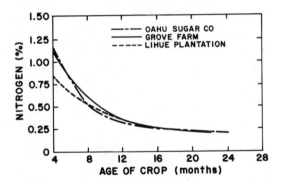

FIG. 12.2. Effect of age on optimum shoot N concentration in sugarcane at three sites in Hawaii. From Stanford and Ayres (1964). Reproduced by permission of *Soil Science*.

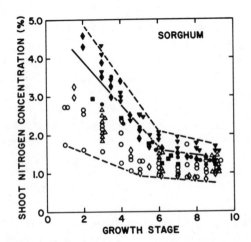

FIG. 12.3. Sorghum shoot nitrogen concentrations at different growth stages. Solid symbols, near-optimal N levels; open symbols, suboptimal N levels. From Jones (1983). Reproduced by permission of *Field Crops Research* and the author.

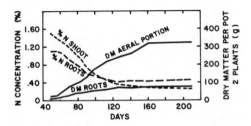

FIG. 12.4. Effect of age on sugarcane root and shoot N concentrations. From Dillewijn (1952). Reproduced courtesy of Hawaiian Sugar Planters' Association.

237

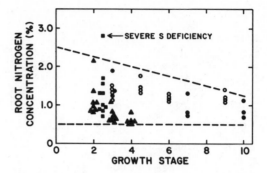

FIG. 12.5. Variation in grain sorghum (open symbols) and maize (closed symbols) root N concentrations at different growth stages. From Jones (1983). Reproduced by permission of *Field Crops Research* and the author.

clines from 1.2% in the youngest leaves to about 0.4% in the oldest green leaves and about 0.2% in senescent leaves still attached to the plant. Similarly, Hanway (1962a,b,c) and Walker and Peck (1972) found that the lower leaves of maize usually have lower N concentrations than the middle or upper leaves, especially when plants are N deficient (Fig. 12.6).

Tissue analysis is often used to diagnose N deficiency and determine whether additional N fertilizer should be applied. Since plant part and stage of development affect the optimum tissue N concentration, these factors are standardized for most systems of tissue analysis. Critical levels and optimum ranges of tissue N concentration for maize, sugarcane, and grain sorghum are given in Table 12.4.

TABLE 12.3. Tissue Nitrogen Concentration of Sugarcane in Hawaii at 12 Months of Age (modified from Clements, 1980)

PLANT PART	N CONCENTRATION (%)
Meristem	1.77
Young blades	1.18
Old blades	0.89
Young sheaths	0.45
Old sheaths	0.31
Internodes	
10–12	0.17
7–9	0.11
4–6	0.14
1–3	0.15

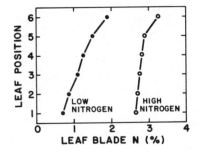

FIG. 12.6. Effects of leaf position and N fertility on maize leaf blade N concentration during early grain fill. Position 1 = lower 2 to 4 leaves, 2 = next 2 higher, 3 = leaf opposite and just below uppermost ear, 4 = next 2 higher, 5 = next 3 higher, 6 = top 2 to 4. From Hanway (1962c). Reproduced from *Agronomy Journal*, by permission of the American Society of Agronomy.

Grain N concentrations are affected by N nutrition, drought and temperature stresses, and genotype. The effects of drought stress are also discussed in Chapter 10. Table 12.5 gives published ranges of maize and grain sorghum grain and stover N concentrations. The concentration of N in the developing grain usually decreases during the first third of the grain filling period. It then remains relatively constant until physiological maturity (Frolich et al., 1980; Jones and Simmons, 1983; Kreig and Rice, 1975).

While grain N content generally increases with increasing N fertility, the concentration of lysine, which limits the nutritive value of maize, grain sorghum, and pearl millet for nonruminants, does not increase significantly (Mirhadi et al., 1980). This limits the effectiveness of N fertilization for improving nutritional quality of C_4 cereals.

Temperature

Low temperatures often increase leaf and whole-plant N concentrations. For example, in Hawaii, sugarcane grown under cool wet conditions at high elevations or on the windward side of an island typically has higher leaf N concentrations than that grown in warmer leeward locations, even when water and fertilizer N are abundant (Clements, 1980; Dillewijn, 1952). Colman and Lazenby (1970) found a similar response in two C_4 forage grasses, *Digitaria macroglossa* and *Paspalum dilatatum*, grown at two temperature regimes. When adequate N was available, plants grown at 35/23°C day/night temperatures produced more dry matter and had lower N concentrations than those grown at 24/13°C (Table 12.6). The lower N concentration found at high temperature is normally due to dilution of tissue N by greater dry matter accumulation, not reduced N uptake.

High-temperature stress (48°C) of C_4 grasses causes proteolysis, increases in amino acid, NH_3, and NO_3^- concentrations, reflecting a general increase in catabolic processes (Lahiri and Singh, 1969).

Drought Stress

The effects of drought stress on plant nitrogen uptake and metabolism are complex. In general, low soil water contents reduce nitrification of NH_4^+ (Sabey,

TABLE 12.4. Critical N Concentrations for Various Tissues of C_4 Grasses

Species	Tissue	Stage of Development	Critical Level or Range (%)	Reference
Zea mays (maize)	Ear leaf	Silking or tasseling	3.0	Melsted et al. (1969)
			2.6–4.0	Neubert et al. (1969)
			2.8–3.5	Jones (1967)
			2.9	Tyner (1946)
			2.9	Gallo et al. (1968)
			3.1	Arnon (1975)
			3.2	Hanway and Dumenil (1965)
			2.6–3.2	Escano et al. (1981a)
	Whole shoot	8-leaf	3.2	Jones (1983)
		12-leaf	2.6	
		Tip of tassel visible	2.2	
		75% plants silking	1.8	
		Early dent	1.1	
		Physiological maturity	1.1	
Sorghum bicolor (grain sorghum)	Youngest leaf	23–39 days	3.5–4.0	Lockman (1972c)
		37–56 days	3.2–4.2	
	Third leaf below head	Bloom stage	3.3–4.0	
		Dough stage	3.3–4.0	
Saccharum spp. hybrid (sugarcane)	Leaf blades 3–6	"Boom stage"	1.5–2.7	Humbert (1973)
	Internodes 8–10		0.25–6.0	
Pennisetum americanum (pearl millet)	Whole shoot	4–5 weeks regrowth	2.5–3.5	Martin and Matocha (1973)
Sorghum halepense (johnsongrass)		Boot stage	1.6–1.8	Martin and Matocha (1973)
Cynodon dactylon cv. Coastal (bermudagrass)		4–5 weeks regrowth	2.5–3.0	Martin and Matocha (1973)
Digitaria decumbens (pangolagrass)		4–5 weeks regrowth	1.7–2.0	Martin and Matocha (1973)

TABLE 12.5. **Mean Optimum, Standard Deviation of Mean Optimum, and Range of Grain and Stover N Concentrations (% of Dry Matter) in Maize and Grain Sorghum [Numbers in Parentheses are the Numbers of Treatments Used to Establish Each Value (Jones, 1983)[a]]**

| | OPTIMUM N CONCENTRATION | | |
	MEAN	STANDARD DEVIATION	RANGE
	Grain		
Maize	1.58 (82)	0.17	0.90–2.09
Grain sorghum	1.67 (53)	0.25	1.02–3.20
	Stover		
Maize	0.72 (117)	0.21	0.41–1.28
Grain sorghum	0.80 (39)	0.26	0.36–1.26

[a] Reproduced by permission of *Field Crops Research* and Elsevier Science Publishers.

1969) and mineralization of soil organic matter (Miller and Johnson, 1964; Stanford and Epstein, 1974; Reichmann, et al., 1966). It reduces the diffusion and mass flow of NO_3^- to the root system and often causes a reduction in root length in the drier surface layers of the soil (South African Sugar Association, 1967; Stevenson and McIntosh, 1935). It also decreases the demand for nitrogen since dry matter accumulation is reduced.

Despite the reduction in soil N availability and plant demand, drought stress often increases tissue N concentration. In field experiments with sudangrass

TABLE 12.6. **Effect of Growth Temperature and Fertilizer N rate on Dry Matter Yield and Forage N Concentration of *Digitaria macroglossa* and *Paspalum dilatatum* [Data are Averages of the Two Species (after Brown, 1978)[a]]**

TEMPERATURE (°C)	N APPLIED (kg/ha)	DRY MATTER YIELD (tons/ha)	FORAGE N CONCENTRATION (%)
23–35	0	33	0.91
	67	48	0.94
	134	60	1.17
	269	65	1.78
13–24	0	22	0.91
	67	35	1.18
	134	41	1.61
	269	48	2.00

[a] Reproduced from *Crop Science*, by permission of the Crop Science Society of America.

(*Sorghum sudanense*), starr millet (*Pennisetum glaucum*), and sart sorghum (*Sorghum bicolor* cv. Sart), dry matter production was more severely affected by the stress than was nitrogen accumulation (Bennett et al., 1964; Doss et al., 1964) (Fig. 12.7).

Despite the increase in the N concentration of the shoot during drought stress, the N concentration of different parts of the shoot are not affected equally. Jenne et al. (1958) found that maize leaf N concentrations were reduced by drought stress, but the N concentrations of the stalk, grain, husks and cobs increased (Table 12.7).

The strong correlation between leaf blade N concentrations and drought stress has long been recognized in sugarcane, where "normal nitrogen" concentrations are calculated from both leaf blade nitrogen concentrations and tissue moisture content. When tissue moisture drops due to drought stress, the "normal" or critical concentration of nitrogen in the leaf tissue is adjusted downward to reflect the translocation of soluble N compounds out of the blade (Clements, 1980; Clements and Moriguchi, 1942; Samuels et al., 1953).

Lahiri and Singh (1968, 1970) showed that in pearl millet (*Pennisetum americanum*) seedlings soluble N increases and protein decreases during the initial stages of drought stress. All forms of soluble N analyzed (amino acid N, ammonia, nitrate and nitrite, and amide N) increased during the period of drought stress, probably due to decreased nitrate reductase activity, decreased protein synthesis, and increased protease activity.

Genotype

Maize, grain sorghum, sugarcane, and C_4 forage grass genotypes differ in tissue N concentration and yield response to fertilizer N (Balko and Russell, 1980a,b; Dillewijn, 1952; Kamprath et al., 1982; Moll et al., 1982; Murali and Paulsen, 1981; Sharma et al., 1979; Vicente-Chandler et al., 1974). Remobilization of N

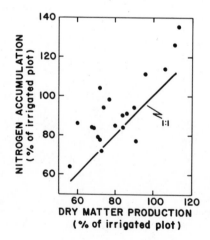

FIG. 12.7. Relationship between relative dry matter production and relative N accumulation in unirrigated plots (compared with irrigated plots). From Doss et al. (1964) and Bennett et al. (1964).

TABLE 12.7. **Nitrogen Concentrations in Several Parts of Mature Maize Plants at Harvest and in Leaves at Several Stages of Development (from Jenne et al., 1958)**[a]

| GROWTH STAGE | PLANT PART | NITROGEN CONCENTRATION (%) ||
		IRRIGATED	UNIRRIGATED
Early tasseling	Leaves	2.6	2.6
Full tassel		2.2	2.1
Milk stage		2.0	1.9
Mature		1.3	1.1
	Stalks & sheaths	0.7	1.0
	Husks	0.7	0.8
	Cobs	0.3	0.4
	Grain	1.8	2.0
	Whole shoot	1.3	1.3

[a] Reproduced from *Agronomy Journal*, by permission of the American Society of Agronomy.

from the leaves and stem varies among maize genotypes (Beauchamp et al., 1976; Below et al., 1981), and grain N concentration varies among maize, grain sorghum, and pearl millet genotypes (Frolich et al., 1980; Kumar et al., 1983; Ross et al., 1981).

NITROGEN UPTAKE

Nitrogen uptake by the plant is affected by two principal factors; the N demand of the plant and the supply of available soil N. However, what appears to be a simple case of supply and demand is complicated by the effects of environmental conditions, genotype, and stage of development.

Soil Supply

Edwards and Barber (1976) have shown that in solution culture N uptake by maize roots is independent of solution concentration down to very low solution concentrations ($21 \times 10^{-6}\ M$). This is much lower than the concentrations of mineral N found in the soil, which are normally greater than $50 \times 10^{-6}\ M$. Thus, most of the time in the field roots are probably absorbing N near their maximum rate or they are not absorbing at all. This suggests that absorption of N from a soil layer is usually limited by the length of roots in the layer, by the plant's demand for N, or by the movement of N from the bulk soil to the root surface (see Chapter 6).

Plant Demand

One of the most important factors affecting N uptake is plant demand, in this case defined as the amount of N needed to maintain near-optimal tissue concentrations during plant growth. Figure 12.2 gives the maize shoot N concentration in high-yielding treatments of a number of experiments conducted under a wide range of climatic conditions. These data suggest that when water and other stresses do not severely limited crop growth, "optimum" N concentration is a strong function of growth stage. Of course, minor variations in the optimum concentration may be found among genotypes, and stresses such as drought, salinity, and nutrient deficiencies may modify actual shoot concentrations even when soil N levels are adequate. Actual shoot N concentrations can be substantially lower than "optimum" concentrations (Fig. 12.8).

Forms of Nitrogen

NO_3^- is the predominant form of inorganic N in most well-drained soils. However, NH_4^+ may be found in relatively high concentrations (1) after recent fertilization with NH_4^+ or urea, (2) in poorly aerated soils, (3) in very acid soils in which nitrification is slowed by a paucity of nitrifying bacteria, and (4) when nitrification inhibitors (such as nitrapyrin) have been used.

Early studies indicated that C_4 grasses such as maize, grain sorghum, and coastal bermudagrass can absorb NH_4^+ and NO_3^- equally well (Burton and Jack-

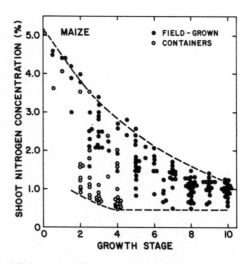

FIG. 12.8.   Relationship between maize growth stage and shoot N concentration in experiments with N fertility ranging from optimum to very deficient. From Jones (1983). Reproduced by permission of *Field Crops Research* and the author.

son, 1962b; Morris and Giddens, 1963) However, it is difficult to control the form of soil N experimentally because of microbial nitrification of NH_4^+ to NO_3^-. For dilute nutrient solutions, slightly greater maize growth is obtained when N is supplied as NO_3^- rather than as NH_4^+ (Blair et al., 1970; Warncke and Barber, 1973). However, the most rapid growth occurs when NO_3^- and NH_4^+ are supplied together (Cox and Reisenauer, 1973; Schrader et al., 1972; Wier et al., 1972).

For solution culture, maize normally takes up the two forms in proportion to their ratio in the solution (Mills and McElhannon, 1982; Schrader et al., 1972; Warncke and Barber, 1973); however, when the ratio of one form to the other exceeds about 4, the plant takes up the less concentrated form more rapidly than its concentration would suggest. After silking, uptake of NO_3^- may decrease while uptake of NH_4^+ remains high. The decrease in carbohydrate translocation to the roots during grainfill and the high respiratory cost of NO_3^- uptake may favor uptake of the NH_4^+ form at this time (Johnson, 1983; Mills and McElhannon, 1982; Veen, 1981).

Development of nitrification inhibitors has stimulated additional interest in the uptake of NH_4^+ and NO_3^-. In soils treated with different forms of N (HN_4^+ or urea with nitrification inhibitors versus NO_3^-), maize yields are not adversely affected by high soil NH_4^+ as long as adequate K^+ is supplied. However, if K^+ is inadequate, excessive NH_4^+ uptake causes soluble N compounds to increase in the tissue and NH_4^+ toxicity symptoms may appear (Barker and Bradfield, 1963; Dibb and Welch, 1976; MacLeod and Carson, 1966).

Because of the relatively low levels of NO_3^- in the soil solution and the negative electrical potential of the root cytoplasm (about -100 mV), the electrochemical gradient between the root cytoplasm and the soil solution generally favors the passive loss of NO_3^- from the root (Higinbotham, 1973). In fact, NO_3^- in the root is readily exchanged with NO_3^- in the soil solution (Morgan et al., 1973). Plants must expend metabolic energy to overcome the electrochemical gradient and accumulate NO_3^-. The respiratory cost of NO_3^- uptake is probably equivalent to 2–3 kg CO_2/kg N (Johnson, 1983; Veen, 1981). Though the mechanism of NO_3^- uptake is not well understood, its active uptake is probably mediated by protein "carriers" located in the plasma membranes of epidermal and cortical root cells. These proteins presumably associate with NO_3^- in the free space of the cell wall and, upon expenditure of metabolic energy, release it into the cytoplasm (Epstein, 1972). The active uptake of NO_3^- by maize roots is induced by NO_3^- (Jackson et al., 1973), and the induction is a function of the internal concentration of NO_3^- (Neyra and Hageman, 1975).

NH_4^+ uptake by maize roots and leaves is associated with an enzyme that requires hydrolysis of ATP and the presence of both Mg^{2+} and a monovalent cation such as K^+ or NH_4^+ for maximum activity (Leonard and Hodges, 1973; Leonard and Hotchkiss, 1976; Perlin and Spanswick, 1981). The monovalent cation requirement of this enzyme may explain the strictly competitive uptake of NH_4^+ and K^+ found in several studies (see Haynes and Goh, 1978). At high soil pH, the equilibrium between NH_3 and NH_4^+ favors NH_3 and NH_3 may be the form taken up (Bennett and Adams, 1970; Wier et al., 1972).

Nitrate Reduction

NO_3^- taken up by the plant must be reduced to NH_3 before it can be combined with carbon skeletons to form amino acids. This reduction is accomplished by two enzymes. Nitrate reductase (NR) reduces NO_3^- to NO_2^-, and nitrite reductase (NiR) reduces NO_2^- to NH_3. Because of its pivotal role in N metabolism, NR has been the subject of numerous studies and a number of reviews (Beevers and Hageman, 1969; Guerrero et al., 1981; Hewitt et al., 1979).

Both NR and NiR occur in most plant tissues, but they have been studied in greatest detail in roots and leaves. In roots NR is a soluble enzyme located in the cytosol, and NiR is located in the plastids. In photosynthetic tissues NR is soluble in the cytosol, while NiR is found in chloroplasts (Suzuki et al., 1981a). In C_4 grass seedlings NiR activity appears before NR (C. T. Tischler, 1981, personal communication). In addition, NiR activity is normally several times higher than that of NR. This assures that NO_2^-, which is toxic to plant tissues, remains at low concentrations.

When NO_3^- enters the root it may (1) be reduced in the root, (2) be stored in the vacuole, or (3) be translocated to the shoot. Since little NO_3^- can be stored in the vacuole, most is either translocated to the shoot as NO_3^- or is reduced and then translocated to the shoot. The relative proportions of NO_3^- and reduced forms of N translocated to the shoot varies with environmental conditions. Waldron (1976) found that N in the xylem exudate of sugarcane is primarily in the reduced form when soil N levels are low. However, at high soil N levels the NO_3^- content of the sap increases markedly. In addition, genotypic variation in the form of translocated N occurs at high soil NO_3^- levels.

In several studies, the relative amounts of NR activity in different tissues have been used to deduce the location of NO_3^- reduction. NR activity is typically lower in roots than shoots of sugarcane (Maretzki and Dela Cruz, 1967) and maize (Oaks, 1979). In roots, NR activity is higher in root tips than in older portions of the root (Oaks, 1979). In sugarcane stalks, NR activity is higher in the younger nodes than in the older nodes (Maretzki and Dela Cruz, 1967), and in surinamgrass (*Brachiaria decumbens*), NR activity is higher in the leaf blades than in the sheaths (Fernandes and Freire, 1976).

The energy source for maize NR can be either NADH or NADPH (Redinbaugh and Campbell, 1981). The NADH form is present in all tissues, but the NADPH form is limited to nongreen tissues such as the root and scutellum. The energy source for NiR in photosynthetic tissues is reduced ferredoxin, the generation of which depends on the light reaction of photosynthesis.

NR activity is regulated by several factors. In the photosynthetic tissues of maize and sudangrass (*Sorghum sudanense*), NR activity reaches a maximum at 35-40°C, then declines at higher temperatures (Hallmark and Huffaker, 1978). NR activity is induced by NO_3^- uptake and is correlated with the NO_3^- concentration of the tissue; however, much of the cellular NO_3^- is sequestered in the vacuole and is unavailable to NR. Thus, NR activity is more closely correlated with the concentration of NO_3^- in the cytosol and the amount of NO_3^- entering the cell than with total tissue NO_3^- concentration (Meeker et al., 1974; Shaner and Boyer, 1976). NH_3 and glutamine cause feedback inhibition of NR, though it is unclear whether the levels ordinarily found in the cytosol are adequate to inhibit activity.

Because of the dependence of NR activity on reducing power and NO_3^-, water stress (Shaner and Boyer, 1976), darkness (Duke et al., 1978), and low soil NO_3^- availability can reduce NR activity. For example, NR activity is positively related to soil N fertility in sugarcane and *Brachiaria* (Fernandes and Freire, 1976; Rosario and Sooksathan, 1977; Waldron, 1976). In addition, genotypic variation in yield, tolerance to stresses, and hybrid vigor are often correlated with NR activity (Bhatt et al., 1979; Eck et al., 1975; Pal et al., 1976), probably due to differences in plant N demand.

Ionic Balance

Nitrogen source (NO_3^- or NH_4^+) and total N uptake play pivotal roles in the plant's ionic balance. A balance between positive and negative charges must be maintained in the cytoplasm (Dijkshoorn, 1962; Ulrich, 1941). A change in the ionic form of N taken up by the plant could result in serious charge imbalance if adjustments in plant metabolism were not made to counteract the effects of the change.

Plants growing in well-aerated soil normally take up more anions than cations, primarily because NO_3^- is the predominant form of N in these soils. However, upon entering the plant, most NO_3^- is reduced and incorporated into organic compounds. When NO_3^- is reduced, its negative charge is transferred to HCO_3^-, which causes an increase in the pH of the cytoplasm. This increase in pH causes the activity of PEP carboxylase to increase (Davies, 1973) and stimulates the production of malate from phosphoenolpyruvate (PEP) and oxaloacetate (OAA) (Ben-Zioni et al., 1970).

$$
\begin{array}{ccccc}
\text{COO}- & \text{ADP} \quad \text{ATP} & \text{COO}- & & \text{COO}- \\
| & & | & & | \\
\text{C}-\text{O}-\text{P} & \longrightarrow & \text{C}=\text{O} & \longrightarrow & \text{HCOH} \\
| & & | & & | \\
\text{CH}_2 & +\text{HCO}_3^- & \text{CH}_2 & & \text{CH}_2 \\
& & | & & | \\
& & \text{COO}- & & \text{COO}- \\
\\
\text{PEP} & & \text{OAA} & & \text{Malate}
\end{array}
$$

The increase in the malate pool can provide only short-term maintenance of charge balance in the plant. When NO_3^- reduction and malate accumulation occur in the shoot, malate is translocated back to the roots via the phloem. It is accompanied by K^+ or other cations to prevent net charge separation between the roots and shoot. Upon arrival in the root, malate is decarboxylated, and the HCO_3^- produced by the decarboxylation is exchanged at the root surface for NO_3^- or other anions (Kirkby, 1974). This exchange causes an increase in rhizosphere pH when NO_3^- is the primary source of N taken up (Nye, 1981).

Though malate is the organic anion produced as a direct result of NO_3^- uptake, in the longer term its negative charges are transferred to other organic anions (Dijkshoorn, 1962; Kirkby, 1968; Tuil, 1965), and organic anions typically account for about 90% of the difference in charge due to the imbalance between inorganic anions and cations in the plant (Dijkshoorn, 1962; Smith, 1978; Tuil, 1965).

Different species accumulate different organic anions for maintenance of ionic balance. *Trans*-aconitate predominates in some cereal crops but is practically absent from legumes, in which malonate accumulates in relatively high amounts (Clark, 1969). Malate, citrate, and oxalate are the most important organic anions in ryegrass (*Lolium perenne*) (Jones and Barnes, 1967). *Trans*-aconitate, malate, and citrate are the predominant organic anions in maize (Clark, 1968); oxalate is most important in setariagrass (*Setaria anceps*) (Smith, 1972a); and aconitate and oxalate predominate in green panic (*Panicum maximum* var. *trichoglume*) (Smith, 1978).

Some C_4 forage grasses (such as setariagrass) can accumulate excessive levels of oxalate, and this can cause occasional deaths in cattle (Jones et al., 1970). Accumulation of oxalate varies among genotypes (Hacker, 1974), and high rates of nitrogen and potassium fertilization cause oxalate to increase (Jones and Ford, 1972; Roughton and Warrington, 1976; Smith, 1972).

Increased NO_3^- uptake also stimulates uptake of cations and their translocation to the shoot (Blair et al., 1970; Haynes and Goh, 1978). Thus, in field and pot experiments where NO_3^- is the predominant form of N, increased tissue N concentrations due to higher N fertilizer rates often stimulate higher cation concentrations (Terman and Allen, 1974b). For example, maize earleaf K^+ concentrations are highly correlated with N concentrations (Escano et al., 1981a), and bermudagrass (*Cynodon dactylon*) shoot N concentrations are correlated with inorganic cation (Ca + Mg + K + Na) concentrations (Thomas and Langdale, 1980) (Fig. 12.9). When soil cation supplies are inadequate, their concentrations

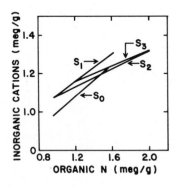

FIG. 12.9. Effect of organic N concentrations of bermudagrass on inorganic cation (K + Na + Ca + Mg) concentrations at four salinity levels. Variation in N concentration due to five fertilizer N treatments. From Thomas and Langdale (1980). Reproduced from *Agronomy Journal*, by permission of the American Society of Agronomy.

in the plant decrease and organic anion concentrations increase to provide the necessary charge balance.

In contrast to NO_3^-, uptake of NH_4^+ and its conversion to amino acids and proteins reduces cytoplasmic pH, stimulates the decarboxylation of malate to pyruvate, reduces the organic anion concentration of the cells, increases anion uptake (including HPO_4^-), and can increase levels of free NH_4^+, amides, and free basic amino acids (see Haynes and Goh, 1978).

Nutrient Interactions

Application of fertilizer N and P in a single concentrated band often enhances uptake of P, especially when N is applied as NH_4^+ (Blair et al., 1970; Engelstad and Allen, 1971; Leikam et al., 1983; Miller et al., 1970). Three explanations for the stimulatory effects of N on P uptake have been proposed: (1) increased solubility of soil P in the presence of acid-forming (NH_4^+-based) fertilizers, (2) increased root growth in the fertilized zone, and (3) increased metabolic activity of the root. Since NH_4^+ stimulates the uptake of anions (Chao, 1966), the stimulatory effect of NH_4^+ on P uptake may also be related to maintenance of charge balance (Emmert, 1982; Leonce and Miller, 1966). More recent work suggests that the general stimulatory effect of N on P uptake may also be due to an increase in the number of P carriers in the plasmalemma (Subramonia Iyer and Saxena, 1977), and the presence of NH_4^+ may increase the activity of the carrier-P complex (Barneix and Arnozis, 1980).

High N fertility in the absence of adequate soil K^+ increases in the N-K ratio of maize tissue, increases the plant's susceptibility to stalk and root rot (Martens and Arny, 1967a,b; Otto and Everett, 1956; Parker and Burrows, 1959), reduces nitrate reductase activity (Khanna-Chopra et al., 1980), and increases the reducing sugar concentration of the pith tissue (Clements, 1980; Martens and Arny, 1967a). These changes illustrate the importance of adequate K^+ nutrition for general plant vigor and disease resistance.

Both N and S are irreplaceable constituents of proteins, and the N-S ratio of particular proteins is quite constant. However, among proteins the N-S ratio varies widely. For example, the N-S ratio of chloroplast proteins and nucleoproteins is less than 10 while that of grain globulin is more than 60 (Dijkshoorn and van Wijk, 1967). Organic N-organic S and total N-S ratios have been used to diagnose sulfur deficiency in crop plants, including C_4 grasses. In wheat, a yield response to S can be expected when the N-S ratio of the grain is greater than 17 (Randall et al., 1981). In maize, S response is found when the ratio is greater than 12-16, depending on the stage of growth and plant part analyzed (Escano et al., 1981b; Rabuffetti and Kamprath, 1977; Rendig and Amparano, 1980). In sugarcane, the optimum leaf N-S ratio is 10-17 (Fox, 1976; Gosnell and Long, 1969).

Sulfur-deficient plants contain low levels of the S-containing amino acids (cysteine and methionine), and their protein synthesis is impaired. These plants are typically low in sugars and accumulate amides and NO_3^-. In contrast, N-deficient plants accumulate SO_4^{2-}. Thus, deficiency of either N or S inhibits the incorporation of the other element into protein and increases the concentration of soluble forms of the other element (Friedrich et al., 1979; Friedrich and Schrader, 1978, 1979; Rabuffetti and Kamprath, 1977).

Nitrate Accumulation

Nitrogen in the form of NO_3^- (NO_3-N) can accumulate to significant levels (1.2–1.8%) prior to silking in the roots, stem bases, and leaf sheaths of well-fertilized maize plants (Chevalier and Schrader, 1977; Hanway, 1962c; Schrader, 1978). In general, the NO_3-N concentrations of the roots, lower stems, and lower leaf sheaths are several times higher than those of the leaf blades, ears, and upper stem (Friedrich et al., 1979). Those of the lower leaves are generally greater than the upper leaves, and those of the sheaths are greater than in the blades (Chevalier and Schrader, 1977). Hanway (1962c) found that by silking the NO_3-N concentrations of sheath and stalk tissues of well-fertilized plants had dropped from more than 1% to 0.25 and 0.35%, respectively. By physiological maturity, maximum concentrations are typically less than 0.2% (Hanway, 1962c; Gonske and Keeney, 1969). Regardless of phenological stage, tissue NO_3-N concentrations begin to increase in the respective tissues when total N concentrations exceed 2.6% in the leaf blades, 0.7% in the leaf sheaths, and 0.3% in the stalks (Hanway, 1962c). This and other evidence (Sumner et al., 1965) suggests that in C_4 grasses NO_3-N accumulates to high levels only when the plant can utilize no more reduced N.

Considerable genotypic variation exists in maize NO_3-N concentrations (Gonske and Keeney, 1969; Schrader, 1978). For example, Schrader (1978) reported that 20 days after silking stem concentrations varied from 0.48% to 0.78% and leaf sheath concentrations varied from 0.04 to 0.36%. This variation suggests that genotypes could be selected on the basis of their ability to accumulate NO_3-N in their stem bases and efficiently to translocate that N to the grain during grain filling. This might reduce the probability that low soil N availability during grain filling (due to drying of superficial soil layers) would reduce grain yields.

Other C_4 grasses also accumulate high tissue NO_3-N concentrations due to high rates of N fertilizer (Harms and Tucker, 1973; Hojjati et al., 1972; Sumner et al., 1965), uptake of NO_3^- rather than NH_4^+ (Spiers and Holt, 1971), and (sometimes) high rates of K^+ uptake (Smith and Clark, 1968).

Nitrate (Nitrite) Toxicity

Nitrate is readily absorbed in the alimentary tract of mammals, is relatively nontoxic, and is excreted in the urine. However, NO_2^- is more toxic than NO_3^-, and reduction of NO_3^- to NO_2^- in the upper gastrointestinal tract can lead to acute NO_2^- toxicity. This toxicity is due to the oxidation of hemoglobin to methemoglobin, which interferes with O_2 transport in the blood and can cause death due to anoxia. In mammals, NO_3^- reduction is due almost exclusively to microflora in the alimentary tract. In humans and most other mammals, the contents of the upper gastrointestinal tract are acidic and unfavorable for bacterial colonization, and under most conditions NO_3^- is absorbed by the body before it can be reduced by bacteria of the lower tract. However, ruminants have active microflora in the upper tract where NO_3^- present in forage is readily converted to NO_2^- (Walters and Walker, 1977).

A number of studies have shown that forage NO_3-N in the range of 0.07–0.45% (0.3–2.0% NO_3^-) can be toxic to cattle (see Hojjati et al., 1972; Sumner et al., 1965). Because the NO_3^- concentrations in most tropical forage grasses are almost always low, toxicity is rare in cattle and other herbivores consuming them. However, tropical grass species vary in their tendency to accumulate NO_3^-. Sudangrass (*Sorghum sudanense*) and pearl millet (*Pennisetum americanum*) can accumulate potentially toxic concentrations of NO_3-N in their tissues under high levels of N fertilization (Harms and Tucker, 1973; Smith and Clark, 1968; Sumner et al., 1965), and in Brazil tannergrass (*Brachiaria radicans*) has been implicated in a number of cases of NO_3^- toxicity in cattle (Andrade et al., 1971). Tannergrass accumulates high levels of NO_3^- (0.3–0.9%) in its foliage under conditions in which other *Brachiaria* species accumulated much less (0.03–0.06%). In tannergrass, high levels of NO_3^- are found only when the forage is green and succulent, though drought stress has been shown to increase levels of NO_3^- in other C_4 forage grasses (Wright and Davison, 1964).

Nitrogen Fixation

Since the early 1960s numerous studies have suggested the occurrence of significant rates of N fixation by bacteria associated with the rhizosphere of C_4 grasses. Associations have been reported between C_4 grasses (*Paspalum notatum*, *Cynodon dactylon*, *Panicum maximum*, *Panicum coloratum*, *Digitaria decumbens*, *Brachiaria* spp., *Zea mays*, *Setaria italica*, *Sorghum bicolor*, *Saccharum* spp., *Spartina alterniflora*) and several genera of N-fixing bacteria (*Azotobacter*, *Campylobacter*, *Azospirillum*, *Beijerinckia*, *Enterobacter*, *Klebsiella*). These bacteria associate with grass roots without causing nodule formation; therefore, the N fixation is often described as "associative" or "nonsymbiotic." Vose and Ruschel (1981) suggest the term "diazotrophic rhizocoenosis" for associative N_2-fixing systems in, on, or close to the roots. The subject has recently been reviewed by Vose and Ruschel (1981) and Weier (1980), and the reader is referred to those reviews for additional references.

Interest in nonsymbiotic N fixation in tropical grasses was stimulated by the work of J. Dobereiner and her associates in Brazil in the early 1960s. She discovered large numbers of N-fixing bacteria in the rhizospheres of sugarcane and bahiagrass (*Paspalum notatum*) (Dobereiner, 1960). The discovery that bacterial nitrogenase reduces acetylene to ethylene and that this reduction can be measured easily with gas chromatography led to increased interest in nonsymbiotic N fixation. In early studies, washed grass roots were used to study rates of acetylene reduction by bacterial nitrogenase and to infer rates of potential N fixation. However, this methodology has been criticized on several grounds, and fixation of gaseous ^{15}N is now measured in many laboratory studies.

Azospirillum spp. are often found in the mucigel surrounding the roots, within and between the cells of the root cortex, in the root xylem, and even in the stems of C_4 grasses (De-Polli et al., 1982; Okon, 1982; Patriquin et al., 1978). *Azospirillum brasilense* adheres to and penetrates root and root hair surfaces in the early stages of infection and enters more mature roots by penetrating areas where epidermal tissues are sloughing off at the points of emergence of lateral roots. Both inter- and intracellular infections occur, and colonization of the middle lamella by hydrolysis of its pectic components is an important means of infection (Umali-Garcia et al., 1978, 1980).

Studies during the past 20 years now clearly indicate that a number of factors affect the rate of nonsymbiotic N fixation. Some plant genotypes have much more nitrogenase activity than others (Bouton and Brooks, 1982; Ela et al., 1982; Ruschel and Ruschel, 1977, 1981). Diurnal fluctuations in nitrogenase activity occur with a peak around midday. The temperature optimum of nitrogenase activity is 30°C or greater, low soil moisture reduces activity (Weier et al., 1981), low O_2 tension in the soil atmosphere is needed for maximum activity, and high light intensities stimulate bacterial populations in the rhizosphere, presumably by increasing the amount of root exudate available to the bacteria. Addition of fertilizer N sometimes reduces nitrogenase activity, presumably by end-product inhibition.

Many studies have found that inoculation of plant roots with N-fixing bacteria gives a positive yield response (see Albrecht et al., 1981; Cohen et al., 1980; Nur et al., 1980; Weaver et al., 1980). Other studies indicate little or no response to inoculation (Albrecht et al., 1977; Barber et al., 1976, 1979; Sloger and Owens, 1978), though some of these failures may have been due to incompatibility between the crop cultivar and bacterial inoculant (Baldani et al., 1981). In addition, when yield increases are reported, they cannot always be shown to be associated with N fixation (Smith et al., 1976). At least part of the increase in production of crop dry matter may be due to bacterial production of plant growth hormones (Barea and Brown, 1974; Tien et al., 1979), including IAA, cytokinins, and gibberellins, and addition of small amounts of these hormones to the rooting medium in the absence of the bacteria can produce changes in root morphology similar to those produced by the bacteria.

The genotypic specificity of the association between C_4 grass roots and N-fixing bacteria, the complex regulation of N fixation when the association exists, and the difficulty of estimating actual contribution of N fixation to the N economy of the plant make generalizations concerning the importance of N fixation difficult. However, several recent field studies suggest that inoculation of grain sorghum and maize with *Azospirillum* can cause crop N content to increase by as much as 75 kg N/ha per crop (Freitas et al., 1982; Kapulnik et al., 1981a,b; Pal and Malik, 1981), and Rennie (1981) concluded that associative N_2 fixation often results in fixation of at least 30 kg N/ha yr.

NITROGEN REMOBILIZATION

Grass tillers are determinant organs that senesce after maturation of the inflorescence. However, the process of maturation and senescence usually begins well before maturation of the seeds. This process includes the progressive senescence of the older leaves and the translocation of soluble organic compounds and minerals from the dying leaves to actively growing tissue (sinks). Depending on the reproductive status of the tiller, these sinks may be leaves, young tillers, or the inflorescence. Whatever the case, this translocation represents a means of conserving reduced carbon compounds, N, P, and some other mobile nutrients.

Nitrogen compounds are among the most important substances translocated during the maturation of the tiller. In many cases, redistribution of N from the leaves and stem is the major source of grain N. However, the extent of redistribution of N to the grain depends on several factors, including the concentration and amount of N in the plant at the beginning of grain filling, the amount of N taken up from the soil during grain filling, the size of the grain sink, the genotype, and environmental factors such as drought or frost that affect the N demand of the sink and the supply of available soil N.

Maize leaf N concentrations decrease following silking (Beauchamp and Estes, 1976; Beauchamp et al., 1976; Mite, 1980). Leaf N declines first in the older leaves and sheaths, then proceeds up the plant to successively younger tissues, and the decrease is especially large in N-deficient plants (Fig. 12.10). There are two major sources of N translocated to the grain during maturation and senescence, hydrolyzed plant proteins and vacuolar NO_3–N. The more important is hydrolyzed protein. Under normal conditions, plant proteins undergo continuous hydrolysis and resynthesis. For example, the half-life of RuBP carboxylase in the leaves of young maize plants varies from three to eight days (Simpson et al., 1981). During senescence, protein synthesis declines relative to proteolysis, and a net decrease in protein concentration is observed (Hill, 1980). NO_3–N may also serve as a significant, though less important, source of N during grain filling. Hanway (1962c) showed that in well-fertilized maize the NO_3–N

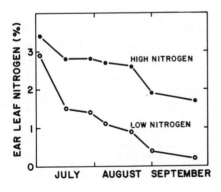

FIG. 12.10.   Effect of leaf age and N fertility on maize ear leaf N concentration. From Hanway (1962c). Reproduced from *Agronomy Journal*, by permission of the American Society of Agronomy.

concentration of the lower stem tissue can fall from 0.3% at silking to less than 0.1% at maturity.

Soluble N compounds translocated out of the older leaves and older stem internodes move to the most active sink for N. During vegetative growth, this is generally the apical meristem and young expanding leaves. This movement accounts for part of the lower nutritive value of older leaves compared with younger leaves of C_4 forage grasses. It is also responsible for the typical chlorosis and senescence of older leaves in vegetative tillers of sugarcane and other tall grasses. During reproductive growth, the leaves and stem, which previously were sinks for translocated N, become sources of N translocated to the grain (Crawford et al., 1982).

Genotype also has an important effect on the redistribution of N to the grain (Beauchamp and Estes, 1976; Beauchamp et al., 1976; Below et al., 1981). Mite (1980) found that the genotypes that are most responsive to high rates of N fertilizer are those with the lowest rate of N translocation from leaves and stalks at low fertilizer rates. This suggests that genotypes with efficient remobilization of N can achieve maximum grain yields at lower rates of fertilizer N.

NITROGEN LOSS

When plants die, their nutrients return to the soil through leaching, abrasion, and decay. These nutrients are a major factor in nutrient cycling of the biosphere. However, less is known about the loss of nutrients from living plants. Wetselaar and Farquhar (1980) recently reviewed the literature concerning N losses from plant tissues, and this section summarizes their conclusions, especially with regard to C_4 grasses.

Repeated plant harvests and chemical analyses are often used to determine the rates of dry matter and nutrient accumulation. Occasional studies have revealed that after anthesis a net loss of N occurs from maize (Firth et al., 1973; Terman and Allen, 1974a), grain sorghum (Herron et al., 1963; Norman and Wetselaar, 1960), rhodesgrass (*Chloris gayana*) (Weinmann, 1940), pearl millet (*Pennisetum americanum*) (Norman and Wetselaar, 1960), and a tropical pasture dominated by *Themeda australis*, *Sorghum plumosum*, and *Chrysopogon fallax* (Norman, 1963).

Net accumulation of N in the shoots occurs between planting (or defoliation) and anthesis. After anthesis, N may continue to accumulate in the crop or it may decrease. A net decrease often occurs in crops with high tissue N contents due to adequate N fertilization. Wetselaar and Farquhar (1980) concluded that such crops lose a mean of about 1.2 kg N/ha day. Net loss of N is much less common in N-deficient plants, where slow accumulation of N in the tops usually continues until harvest.

Losses of N can also occur from the shoots of vegetative grasses such as sugarcane. For example, Borden (1948) found that in Hawaii recovery of N from the shoots of sugarcane can decrease from about 220 kg/ha at 11 months to less than

96 kg/ha at 27 months. These dramatic losses occurred only at high rates of N application, and almost no N was lost when no fertilizer N was applied.

Several explanations have been proposed for this loss of N from plant tissues. In perennials, N is translocated to the roots when shoots become dormant and leaves senesce. However, Wetselaar and Farquhar (1980) point out that in all cases where the N content of the roots has also been measured, postanthesis decline in shoot N content is accompanied by a similar or larger decline in the roots.

One of the most obvious mechanisms of N loss is via shedding or other loss of plant parts. These losses include death and decomposition of the root system, shedding of pollen and aborted fruits, insect grazing, loss of leaf and sheath material during the normal senescence of leaves, loss of particulate material from leaf surfaces by abrasion, and loss of root mass through normal senescence and decomposition of the root system. Shedding of senescent leaves is undoubtedly an important means by which N is lost from grasses. Clements and Moriguchi (1942) found that senescent leaves still attached to the plant contain one-third to one-half the N in green leaves. This loss of N probably contributes significantly to the decrease in sugarcane N content prior to harvest (Borden, 1948). However, Wetselaar and Farquhar (1980) point out that when attempts have been made to quantify N losses due to shedding and decomposition of plant parts, they have usually failed to account for the total N loss observed (see Herron et al., 1963).

Leaching of soluble N compounds from leaves and other plant parts by rain and dew undoubtedly accounts for some loss of N, especially as membrane integrity decreases during senescence. For example, Norman and Wetselaar (1960) found greater postanthesis N losses from grain sorghum, forage sorghum, pearl millet, and sudangrass (*Sorghum sudanense*) during a high-rainfall year than during a low-rainfall year. However, in many studies losses by leaching and dew formation can be discounted either because plants were grown in greenhouses and the foliage was not sprinkled or because N losses occurred during periods in which rainfall and dew formation were minimal (Wetselaar and Farquhar, 1980).

Guttation is another means by which soluble N compounds can be transferred from the interior of the leaf to its surface. Though this mechanism has not been studied in C_4 grasses, Goatley and Lewis (1966) estimate that guttation losses by rye, wheat, and barley amount to less than 1 kg N/ha during a crop.

Gaseous N compounds such as ammonia (NH_3), nitric oxide (NO), nitrous oxide (N_2O), and nitrogen dioxide (NO_2) can be both absorbed by and lost from leaves, including those of tropical grasses such as maize and grain sorghum (Farquhar et al., 1979; Porter et al., 1972; Weiland and Stutte, 1979). Trace amounts of hydrogen cyanide (HCN) are also lost from sorghum leaves due to the decomposition of the cyanogenic glycoside dhurrin (Franzke and Hume, 1945). Considerable evidence suggests that a compensation point exists for these compounds in the leaf tissue. If ambient concentrations are above the compensation point, net N uptake occurs. If concentrations are below the compensation point, net loss occurs. Though the compensation point is probably affected by temperature, light level, plant age, and tissue N content, we do not yet understand the processes well enough to estimate accurately net long-term absorption or emission of gaseous N compounds.

GROWTH AND DEVELOPMENT

The N status of the crop affects all aspects of growth and development. The following section describes some of the effects of N on frequently measured components of the growth of grasses.

Leaf Number and Appearance Rate

Grass tillers are determinant organs in which production of leaf primordia ceases when the inflorescence is initiated. Thus, the time of inflorescence initiation and the rate of leaf primordia formation prior to panicle initiation determine the final leaf number of the tiller (Kiesselbach, 1950). In C_4 grasses, temperature is the most important environmental factor affecting the rate of leaf formation, and temperature and day length interact to affect the date of panicle initiation and final leaf number (see Chapter 5). Mineral nutrition can also affect the rate of leaf formation and final leaf number. For example, low soil fertility increases the interval between appearance of sugarcane leaves (Rosario et al., 1977; Sampaio and Beatty, 1976).

In contrast to sugarcane and most forage grasses, initiation of maize and grain sorghum leaf primordia typically occurs for only two to six weeks after planting. Thus, environmental factors such as temperature, day length, and soil fertility affect final leaf number for only a few weeks. In addition, seed and soil reserves of N and P are usually sufficient to assure adequate nutrition during the initial part of this period. Thus, under most conditions, N nutrition has little effect on the number of leaves produced by maize or grain sorghum (Cowie, 1973; Eik and Hanway, 1965). Even when severe N deficiency is imposed in solution culture, final leaf number is affected only slightly. For example, Cowie (1973) found that the rate of leaf appearance was slower in N-deficient grain sorghum plants than in plants with sufficient N. However, the appearance of the panicle was delayed in N-deficient plants, and their final leaf number was often equal to or only slightly less than those with adequate N nutrition. The author has found similar effects of severe N deficiency in maize (unpublished).

Nitrogen deficiency also reduces tillering of grasses (Escalada and Plucknett, 1975a, 1977; Krishnamurthy, 1973; Sampaio and Beaty, 1976). However, in crops such as grain sorghum, pearl millet, finger millet (*Eleusine coracana*), and sugarcane, tillers are often formed prior to the development of severe N stress. Thus, the effect of N deficiency on tillering can vary considerably depending on the timing of the stress.

Leaf Weight and Area

Leaf area is more sensitive to N stress than are leaf number and rate of appearance. For example, Cowie (1973) found that severe N stress, which had almost no effect on final leaf number, reduced the weights of the largest leaves by more than 50%. Similar results were reported by Eik and Hanway (1965), who found that nutrient stress due to lack of "starter" fertilizer reduced mean leaf numbers from 21 to 20 but reduced leaf area per plant 45 days after emergency by 50%. The effects of N stress on grain sorghum leaf weight and area are given in Figs. 12.11 and 12.12.

Leaf Senescence

Senescence of maize leaves has three primary causes. First, the sheaths of the first four to six leaves are damaged by the normal expansion of the stem and by the growth of prop roots. These roots perforate the sheaths surrounding the stalk and cause premature leaf death. The second cause of leaf senescence is the normal maturation and senescence of leaves with age. Third, normal senescence is accelerated by drought, inadequate or imbalanced nutrition, and other stresses. For example, in vegetative sugarcane 12-14 green leaves are usually found on healthy stalks; however, only 4-6 green leaves may occur on stalks suffering from nutrient or moisture stress.

Nitrogen deficiency increases the rate of leaf senescence. In maize the premature chlorosis and senescence of the lower leaves begins at the leaf tip and spreads basipetally and laterally from the midrib forming a V-shaped area of yellow, then brown tissue. On sugarcane the pattern of senescence is reversed, with tissue death in older leaves proceeding from the margin inward (Gascho and Taha, 1972). In green panic (*Panicum maximum* var. *trichoglume*), senescence begins at the leaf tip and spreads backward uniformly across the width of the leaf (Smith, 1972a,b).

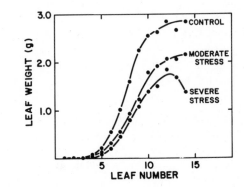

FIG. 12.11. Effect of N stress on grain sorghum leaf weights. From Cowie (1973).

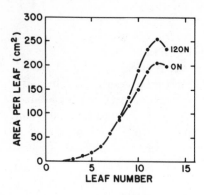

FIG. 12.12. Effect of N stress on grain sorghum leaf areas. From G. Arkin (1981), personal communication.

Maize and grain sorghum grain yields are often correlated with total plant dry weight (Shafer and Wiggans, 1941) and leaf area (Swanson, 1941). The reduction to individual leaf sizes and the premature senescence of older leaves caused by N stress cause total plant leaf area and leaf weight to decrease. This, in turn, reduces light interception and dry matter accumulation. The reduction in leaf area or leaf weight often explains most of the variation in final dry matter yield (Table 12.8).

TABLE 12.8. **Regressions of Total Plant Dry Weights on Plant Leaf Weight or Leaf Area in Experiments in Which N Limited Growth, where** Y **= Estimated Plant Dry Weight (g),** W **= Leaf Weight/Plant (g),** A **= Leaf Area/Plant (dm^2),** r **= the Correlation Coefficient, and** n **= the Number of Treatments**

EQUATION	r	n	COMMENTS
$Y = 9.45W - 70.0$	0.97	22	Hanway (1962c), nearly mature maize, 2 years data
$Y = 6.22W + 14.3$	0.99	8	Cowie (1973), nearly mature grain sorghum, greenhouse
$Y = 8.20W - 10.4$	0.99	7	El-Hattab et al. (1980), nearly mature maize, 2 years data
$Y = 4.64A - 66$	0.99	7	El-Hattab et al. (1980), maize 15 days after silking
$Y = 5.14A - 178$	0.87	9	Nunez and Kamprath (1969), mature maize. Plymouth site, 1965
$Y = 5.15A - 198$	0.82	9	Plymouth site, 1966
$Y = 7.73A - 275$	0.66	9	Clayton site, 1965
$Y = 5.78A - 210$	0.94	9	Clayton site, 1966
$Y = 7.53A - 32$	0.91	5	Gerald Arkin (personal communication); grain sorghum; leaf area at anthesis, dry matter at harvest

Under most conditions, light interception and photosynthesis depend on both the amount of leaf area produced and its duration. Swanson (1941) recognized that grain sorghum yield potential is correlated with high leaf area production and long maturity. Since N stress reduces leaf area and increases the rate of leaf senescence, leaf area duration (LAD), the integral of leaf area index (LAI) over time, is reduced by N stress. Yadov (1981) found that sugarcane leaf area duration is highly correlated with sugar yield ($r > 0.94$) in experiments with different rates of N fertilizer.

Root Growth

The amount and distribution of mineral N in the soil profile affect root growth and function. High levels of mineral N in the root zone stimulate both shoot and root growth. The increase in root growth is due to an increase in the number of root apices produced, especially those of first-order lateral roots (Drew and Saker, 1975; Maizlish et al., 1980). However, in most cases an increase in soil N fertility increases shoot growth more than root growth (Maizlish et al., 1980; Myers, 1980; Shank, 1945).

When fertilizer N is placed in a band, roots proliferate in the zones of high N concentration (Drew and Saker, 1975; Edwards and Barber, 1976; Ohlrogge, 1962). Only about one-third of the root system is needed to support the N demand of the shoot as long as adequate soil N is available (Drew and Saker, 1975; Edwards and Barber, 1976). In fact, Ohlrogge (1962) has shown that a single maize root allowed to proliferate in a zone of NO_3^--rich soil can absorb sufficient N to support luxuriant top growth.

Since so small a portion of the root system is needed to supply the demands of the top, it is no surprise that grasses can efficiently extract N from the lower parts of the soil profile and can thus support rapid top growth even though N is in low supply in the surface soil layers (Alston, 1980; Burns, 1980; Dancer and Peterson, 1969).

Photosynthesis

Nitrogen deficiency decreases rates of photosynthesis and dry matter accumulation of C_4 grasses. In both pot and field studies, high yields caused by increased N fertilization are associated with high forage N concentrations. High leaf N concentrations are required for high relative growth rate, net assimilation rate, relative leaf growth rate, and net photosynthetic rate (Brown and Wilson, 1983; Wilson, 1975; Wilson and Brown, 1983) (Fig. 12.13).

The effects of plant N concentration on net photosynthesis and growth are similar in C_3 and C_4 grasses. Leaf N concentration is highly correlated with net photosynthesis. However, the N concentration associated with maximum rates of photosynthesis and growth are generally lower in C_4 grasses than in C_4 grasses grown under similar conditions (Brown and Wilson, 1983; Schmitt and Edwards, 1981; Wilson and Brown, 1983) (Figs. 12.1 and 12.13).

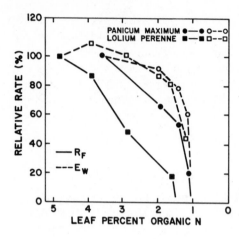

FIG. 12.13. Effect of leaf organic N concentration on leaf relative growth rate (R_F) and net assimilation rate (E_W). From Wilson (1975). Reproduced by permission of the *Netherlands Journal of Agricultural Science*.

There has been some dispute whether the effect of N deficiency on net photosynthesis is due primarily to its effect on mesophyll resistance or on stomatal resistance. For example, Brown and Wilson (1983) found that N deficiency affects guineagrass (*Pancium maximum*) mesophyll resistance more than stomatal resistance, and intercellular CO_2 concentration increases as plants become more N deficient. However, other work (Goudriaan and van Laar, 1978; Louwerse and Zweerde, 1977; Wong et al., 1979) suggests that stomatal resistance is closely coupled with mesophyll resistance such that near-constant intercellular CO_2 concentrations are maintained. For example, Goudriaan and van Keulen (1979) found the same linear relationship between net photosynthesis and water vapor conductance of maize leaves, whether N was deficient or adequate (Fig. 12.14). Similarly, Bolton (1979) found that in guineagrass the ratio of net photosynthesis to transpiration was unaffected by leaf N concentration.

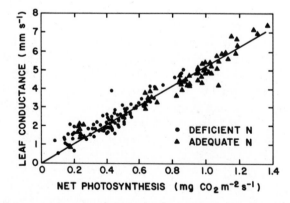

FIG. 12.14. Relationship between maize leaf net photosynthesis and stomatal conductance at two levels of N nutrition. From Goudriaan and van Keulen (1979). Reproduced by permission of the *Netherlands Journal of Agricultural Science*.

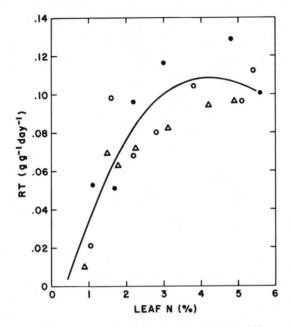

FIG. 12.15. Relationship between shoot relative growth rate and leaf N concentration of Nandi setariagrass (*Setaria anceps* cv. Nandi). From Henzell and Oxenham (1973). Reproduced by permission of *Communications in Soil Science and Plant Analysis.*

One of the most obvious effects of N deficiency in tropical grasses is a general chlorosis of the leaf tissue due to a dramatic decrease in chlorophylls a and b, carotenoids, and benzoquinones. As a result of the decrease in pigment concentration, the rate of O_2 evolution per unit leaf area decreases with N deficiency; however, the rate of O_2 evolution per unit chlorophyll increases. This increase in photosynthesis per unit chlorophyll may suggest a smaller photosynthetic unit in N-deficient leaves than in normal leaves (Baszynski et al., 1972, 1975). Hall et al. (1972) found that chloroplasts of N-deficient maize plants are only one-third to one-half the size of chloroplasts from normal plants.

Though photosynthesis is sensitive to leaf N concentration, expansion growth is even more sensitive. As a result, nonstructural carbohydrates accumulate during N stress (Gallaher and Brown, 1977; Wilson and Brown, 1983). Thus, fertilizer N is withheld from sugarcane prior to harvest. This reduces expansion growth more than photosynthesis and causes the sucrose concentration of the stalks to increase substantially (Clements, 1980).

The effects of tissue N concentration on leaf appearance and expansion rates and photosynthesis are integrated by classical growth analysis. Relative growth rate, relative leaf growth rate, and net assimilation rate per unit leaf area are curvilinear functions of leaf N content (Brown and Wilson, 1983; Henzell and Oxenham, 1973; Wilson, 1975). In several tropical forage grasses these measures of plant growth approach a maximum at 2.5–4% N in the leaf blades (Fig. 12.15).

13

PHOSPHORUS

Plant phosphorus occurs in both inorganic (orthophosphate and to a minor extent pyrophosphate) and organic forms. In the organic form it is bound to the hydroxyl groups of sugars and alcohols. These compounds include important products of intermediary carbon metabolism such as fructose-6-P, glucose-6-P, and 3-phosphoglyceric acid. Phosphorus is bound by diester linkages in DNA, RNA, and phospholipids such as phospatidyl choline and phosphatidyl ethanolamine, essential components of biological membranes. The ability of phosphorus to form pyrophosphate bonds accounts for its central role in the energy relations of the cell. The formation and cleavage of the pyrophosphate bonds of ATP and ADP control the transfer of energy in both photosynthesis and respiration. In addition, the related compounds uridine triphosphate, cytidine triphosphate, and guanosine triphosphate are involved in the synthesis of sucrose, cellulose, phospholipids, and nucleic acids. Thus, phosphorus plays a central role in almost every metabolic process in the plant.

SOIL PHOSPHORUS

A detailed discussion of soil phosphorus is beyond the scope of this book. Some aspects of P adsorption to soils, movement to roots, and uptake by the plant are discussed in Chapter 6. Recent reviews of soil phosphorus include Anderson (1980), Kamprath and Watson (1980), and Olsen and Khasawneh (1980).

Soil analysis has long been used to assess soil P fertility and make recommendations for fertilization of C_4 grasses. By the 1890s the Hawaii sugarcane industry had developed soil extractants utilizing 1% aspartic acid to make fertilizer P recommendations (Jones, 1982). As other soil extractants were developed, they were applied to high-value crops such as sugarcane and maize. Only since the 1960s have they been used routinely to make P fertilizer recommendations of other, lower-value C_4 grasses such as grain sorghum, pearl millet, and forage grasses. Critical concentrations of extractable P for several soil extractants and a variety of C_4 grasses are summarized in Table 13.1.

With some soil extractants, the critical P concentration varies among soils. For example, Woodruff and Kamprath (1965) found that for the double acid extractant ($0.05\ N\ HCl + 0.025\ N\ H_2SO_4$) the critical P concentration for pearl millet is 10 ppm on a silty clay loam and 17 ppm on a fine sandy loam.

In addition to soil type and crop species, subsoil P availability can affect the critical topsoil P concentration. Roots readily absorb P from below the plow layer (Murdock and Engelbert, 1958), and subsoil P availability is sometimes considered when making fertilizer P recommendations.

PHOSPHORUS UPTAKE

As discussed in Chapter 6, HPO_4^{2-} and $H_2PO_4^{-}$ move to the root principally by diffusion in the soil solution. They are absorbed actively against an electrochemical potential gradient, and the potential rate of P absorption is affected by the nutritional status of the plant. For example, when roots of P-deficient maize plants are exposed to high concentrations of soil solution P, they absorb that P more than 1.5 times as fast as roots of plants with adequate P. The soil solution P concentration at which uptake by 18-day-old maize is half the maximum (Km) is between 1 and $5 \times 10^{-6}\ M$ (0.03–0.15×10^{-6} mg P/liter) (Anghinoni and Barber, 1980a). This is similar to the critical P concentration in $0.01\ N\ CaCl_2$ reported in Table 13.1

Young plants usually have a high demand for P per unit root length. As a result, higher external (soil) P concentrations are required for maximum P uptake and growth in the seedling stage than at later stages (Fox and Kang, 1978; Fox et al., 1974). For example, Fox and Kang (1978) found that the external P requirement for maximum growth of maize decreases from 6 mg P/liter soil solution at 3 weeks to 0.001 mg P/liter at 9 weeks. Banding of fertilizer P near the seed (or sett) and coating the seed with fertilizer P have been used to stimulate P uptake and growth of seedlings and young sugarcane plants (Christie, 1975a;

TABLE 13.1. Critical Soil Test P Concentrations for Several C$_4$ Grasses with Five Common Extractants

CROP	LOCATION	CRITICAL P CONCENTRATION (ppm)	REFERENCES
	0.01 N CaCl$_2$		
Maize, grain sorghum	Hawaii	0.05–0.06	Fox (1979)
Pearl millet	Southeastern United States	0.07–0.20	Fox and Kamprath (1970)
Sudangrass		0.37	Soltanpur et al. (1974)
Guineagrass	Queensland, Australia	0.29	Moody and Standley (1980)
Molassesgrass		0.34	
	0.5 N NaHCO$_3$		
Maize	Goias, Brazil	20	Yost et al. (1981)
Sugarcane	South Africa	11–15	Toit et al. (1981)
Guineagrass	Queensland, Australia	20	Bruce and Bruce (1972)
	0.25 N HCl + 0.03 N NH$_4$F		
Maize	Mexico	25	Ortega (1971)
	United States	30	Thomas and Peaslee (1973)
Sugarcane	South Africa	16–25	Toit et al. (1963)
	0.05 N HCl + 0.025 N H$_2$SO$_4$		
Maize	Alabama	15	Rouse (1968)
	North Carolina	8	Kamprath (1967)
	Goias, Brazil	10–15	Yost et al. (1981)
Bermudagrass	Alabama	20	Rouse (1968)
		25	Jordan et al. (1966)
	North Carolina	25	Woodhouse (1968)
Pearl millet		10–17	Woodruff and Kamprath (1965)
	0.02 N H$_2$SO$_4$		
Sugarcane	Hawaii	18–22	Ayres (1960)
	South Africa	16–30	Toit et al. (1963)

Jones, 1982; Silcock and Smith, 1982; Welsh et al., 1966). Banding has the advantage of placing high concentrations of soil P where it can be quickly reached by the young root system. It has the additional advantages of reducing fixation of P and stimulating root growth in the band of high P concentration. However, as soil solution P concentration increases, P uptake per unit root soon becomes saturated, and P uptake per plant becomes proportional to the fraction of the root system exposed to high P concentrations. Thus, maximum P uptake requires that most of the root system be exposed to high solution P concentrations (Jungk and Barber, 1974; Robertson et al., 1966; Stryker et al., 1974). Therefore, banding is advisable only in quite P-deficient soils when low rates of P fertilizer are used. Under these conditions, the reduced P fixation and the increased root growth in the fertilizer band increase P uptake over that obtained when the same amount of P is mixed throughout the soil. However, for less P-deficient soils and when high rates of fertilizer are used, uniform incorporation of P is recommended (Anghinoni and Barber, 1980b). When P fertility is marginal, a combination of uniformly incorporated and banded P fertilizer may prove best (Welsh et al., 1966; Yost et al., 1979).

PLANT PHOSPHORUS

Plant physiologists have often contended that plant P concentration is a better indicator of P fertility than extractable soil P, and plant analysis has long been used as a substitute for (or supplement to) soil analysis. Numerous systems of plant analysis have been applied to C_4 grasses, and a brief discussion of P storage and remobilization is needed to understand the relative strengths and weaknesses of the various systems.

Growing plant meristems need a continuous supply of P for incorporation into new tissue. Soil P supply is not constant due to variation in soil water and root development; therefore, plants have developed mechanisms for taking up and storing excess P and subsequently remobilizing and using it for new growth.

In seeds, phytin, the hexaphosphate ester of inositol, is the most important reserve form of P. It accumulates during seed formation and is metabolized and converted to other P compounds during germination. In the vegetative tissues of C_4 grasses, orthophosphate is probably the most important reserve (Hanway, 1962a,b,c; Hartt, 1972; Sharpley and Reed, 1982). For example, Hanway (1962c) found that about 67% of the P in maize leaves and 87% of that in stalks is water soluble (mostly inorganic). Sharpley and Reed (1982) found that about 70% of the P in grain sorghum shoots is in the inorganic form, and this percentage decreases slightly when P is deficient. Sugarcane stem tissue accumulates inorganic P when the soil P supply permits. If P deficiency occurs subsequently, this inorganic P can be translocated to meristematic tissues (Hartt, 1972). Since much of the P in the sugarcane internodes is in the highly mobile inorganic form, the stem P concentration is more sensitive to P nutrition than is that of the leaf, where more P is in the organic form (Baver, 1960; Clements, 1955). Thus, the P concentration of the stem may decline significantly during periods of P deficiency, even though the leaf P concentration remains almost unchanged (Hartt, 1972).

Tissue Age

As grass tillers age, the ratio of cytoplasm to structural tissue decreases. This normally causes a gradual decline in shoot P concentration (Clements, 1980; Hanway, 1962c; Martin and Matocha, 1973). In sugarcane, the apical meristem and dormant auxillary buds have the highest P concentrations in the plant, normally 0.3–0.7%. Young leaf laminae, leaf sheaths, and stem internodes have higher P concentrations than their older counterparts (Clements, 1980). Similar relationships have been reported in maize, though the difference between old and young tissues tends to disappear as the crop nears maturity (Hanway, 1962c).

The decrease in shoot P concentration with increasing age is the result of a decrease in the ratio of meristematic tissues with high P concentrations to structural tissues with low P concentrations. Because of the effect of temperature on the rate of phenological development, growth stage is often a better index of phenological development than age. Maize shoot P concentration declines as the crop progresses from growth stage 1 (four fully expanded leaves) to 5 (silking) and 10 (physiological maturity) (Jones, 1983) (Figs. 13.1 and 6.10). Similar results have been found in grain sorghum (Eck and Musick, 1979b), sugarcane (Clements, 1980), and forage grasses (Table 13.2).

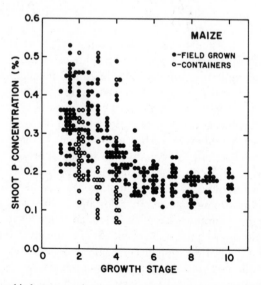

FIG. 13.1.   Relationship between maize growth stage and shoot P concentrations from experiments with a wide range of P fertility. From Jones (1983). Reprinted by permission of *Field Crops Research* and the author.

TABLE 13.2. Critical P Concentrations for Various Tissues of C_4 Grasses

SPECIES	TISSUE	STAGE OF DEVELOPMENT	CRITICAL LEVEL OR RANGE (%)	REFERENCE
Zea mays (maize)	Ear leaf	Silking or tasseling	0.10	Jones and Eck (1973)
			0.25	Melsted et al. (1969)
			0.23	Gallo et al. (1968)
			0.295	Tyner (1946)
			0.34	Hanway and Dumenil (1965)
			0.25–0.50	Neubert et al. (1969)
			0.26–0.32	Escano et al. (1969)
	Whole shoot	8-leaf	0.45	Jones (1983)
		12-leaf	0.35	
		Tip of tassel visible	0.25	
		75% plants silking	0.21	
		Early dent	0.20	
		Physiological maturity	0.18	
	Lower stems	Late summer	0.0044	Goodall and Gregory (1947)
Sorghum bicolor (grain sorghum)	Whole shoot	23–39 days	0.3–0.6	Lockman (1972c)
	Youngest leaf	37–56 days	0.2–0.6	
	Third leaf below head	Bloom stage	0.2–0.35	
		Dough stage	0.15–0.25	
	Leaves	56 days	0.16	Sharpley and Reed (1982)
Saccharum spp. hybrid (sugarcane)	First fully expanded leaf	5 months	0.20–0.30	Halais (1962)
		During rapid growth	0.22–0.30	Gascho and El Wali (1979)
	Third leaf blade	4–6 months	0.18	Gosnell and Long (1971)
		5 months	0.21	Halais (1962)
	Internodes 8–10	6–12 months	0.32–0.40	Hartt (1960)
	Apical meristem		0.45	Clements (1955)

TABLE 13.2. (*Continued*)

Species	Tissue	Stage of Development	Critical Level or Range (%)	Reference
Pennisetum americanum (pearl millet)	Whole plant	4–5 weeks	0.16–0.20	Clark et al. (1965)
Digitaria decumbens (pangolagrass)		Preflowering (45 days) 4–5 weeks	0.16 0.12–0.16	Andrew and Robins (1971) Little et al. (1959) Oakes and Skov (1962)
Melinis minutiflora (molassesgrass)		Preflowering (57 days)	0.18	Andrew and Robins (1971)
Panicum maximum var. *trichoglume* cv. Petrie (green panic)		Preflowering (57 days) 3–4 leaves 4–5 leaves 6–7 leaves	0.20 0.55 0.32 0.15	Andrew and Robins (1971) Smith (1975)
Pennisetum clandestinum (kikuyugrass)		Preflowering (45 days)	0.22	Andrew and Robins (1971)
Setaria anceps cv. Nandi (setariagrass)		4 leaves 5 leaves 6 leaves 7 leaves Preflowering (57 days)	0.46 0.36 0.24 0.14 0.21	Smith (1975) Andrew and Robins (1971)

Species	Plant part / stage	Value	Reference
Cynodon dactylon cv. Coastal (bermudagrass)	4–5 weeks	0.10–0.26	Fisher and Caldwell (1959)
C. dactylon cv. common and Midland (bermudagrass)		0.24–0.28	Adams et al. (1967a)
Cenchrus ciliaris (buffelgrass)	Preflowering (57 days)	0.25	Andrew and Robins (1971)
	(39 days)	0.26	Christie (1975a)
	(46 days)	0.30	Christie and Moorby (1975)
	Flowering	0.16	Smith (1975)
	Youngest expanding leaves	0.17	
Chloris gayana cv. Pioneer (rhodesgrass)	Whole plant Preflowering (57 days)	0.22	Andrew and Robins (1971)
Paspalum dilatatum (dallisgrass)		0.25	
Sorghum halepense (johnsongrass)	Preflowering (boot stage)	0.16–0.20	Spooner et al. (1971) cited in Martin and Matocha (1973)
Sorghum sudanense (sudangrass)	4–5 weeks	0.14–0.18	Dozenka et al. (1966) cited in Martin and Matocha (1973)

When using tissue analysis to diagnose nutrient deficiencies, the effects of age on tissue P concentration can be minimized by selecting index tissues of equivalent age. For example, leaves or stalk internodes may be chosen by counting downward from the youngest fully expanded leaf and choosing the same leaf or internode regardless of crop age or growth stage. However, even under these conditions, age affects P concentration. For example, the P concentrations of sugarcane leaf and stem index tissues decrease with increasing age. The most rapid decrease occurs from 1 to 4 months of age. This is followed by a slower decline for the remainder of the crop (Clements, 1955; Gosnell and Long, 1971; Samuels, 1969b).

Plant Analysis

Tissue P concentration has long been used to detect nutrient deficiencies and toxicities in C_4 grasses. For example, in the 1920s the P concentration of juice from sugarcane stalks was found to correlate with response to fertilizer P, and it was routinely used for fertilizer recommendations (Jones, 1982).

In order to minimize labor, most systems of tissue analysis use the same tissue sample for analysis of several nutrients. For maize and grain sorghum, an particular leaf lamina is normally the index tissue. Clements (1980) recognized that sensitivity of sugarcane internode P concentration and developed an index based on both leaf sheath and internode P concentrations. Systems developed for forage grasses normally use a sample of leaves if the plant is tufted or of leaves and stems if the plant is stoloniferous. In all systems the concentration of P (or other nutrients) should be compared with the critical concentration for plants of the same phenological stage.

Critical P concentrations of various tissues of a selection of C_4 grasses are given in Table 13.2. Almost without exception, critical concentration decreases as plants or plant parts mature. For example, Smith (1975) found that the critical P concentration of Nandi setaria (*Setaria anceps* cv. Nandi) decreases from about 0.46% at the four-leaf stage to about 0.14 at the seven-leaf stage. Jones (1983) found that the near-optimum, whole-shoot P concentration of maize decreases from 0.45% at the eight-leaf stage to 0.18% at physiological maturity.

Many systems of plant analysis have been developed to diagnose P deficiencies. Samuels (1969b) discusses no fewer than 16 different methods of sugarcane tissue diagnosis that are used worldwide. Systems developed in Jamaica, Mauritius, and Puerto Rico use various leaf blades as index tissues for P diagnosis. The Hawaii sugar industry has used several tissues for P analysis. During the 1950s a colorful controversy arose concerning the best means of assessing P fertility with tissue analysis. One school of thought (Baver, 1960; Hartt, 1955, 1958, 1960) held that the P concentration of mature internodes is the best indicator of the P status of the crop. Another school (Clements, 1955, 1958, 1962) strongly opposed the inference that stalk tissue is a better indicator of P deficiency than the leaf sheath or blade, which had previously been used (Clements, 1940). This controversy led to the recognition that sugarcane leaf sheath P concentrations are sensitive to sheath moisture and sugar levels as well as to P nutrition. When the moisture content or the sugar content of the leaf sheath is low due to drought stress or other environmental conditions, the P concentration of the sheath decreases, even on soils with high P fertility. For this reason, the "standard P index" was developed to normalize the critical sheath P concentrations of the sheath for the effects of sheath moisture and sugar contents (Clements, 1955). When it subsequently became evident that stem internodes contain a large pool of inorganic P that can be translocated on demand (Hartt, 1955, 1958, 1960), the "amplified P index" was devised (Clements, 1958). This index is essentially the product of the standard P index and the P concentration of specific mature internodes. Thus, the amplified P index can exceed the critical level when the P concentration of one or both of the index tissues is high.

Since P is utilized in the synthesis of cell membranes, DNA, and RNA, a decrease in meristem P concentration has immediate adverse effects on meristem function. However, meristems are strong sinks for plant P, and they tend to maintain their P concentration even when those of other tissues decrease. For example, Clements (1955) found that the growth of sugarcane approaches a maximum when the concentration of P in the apical meristem approaches 0.45% and the ratio of the P concentration in the apical meristem to that in the mature stem tissue is about 12. If P nutrition is inadequate, the ratio will exceed 12 due to the decrease in the P concentration of the stem tissue. If luxury P consumption occurs, the stem tissue accumulates inorganic P and the ratio declines.

Genotype Effects

Numerous studies have found genotypic differences in maize P uptake and tissue concentration (Clark and Brown, 1974b; Obreza et al., 1982; Terman et al., 1975). However, the mechanisms responsible for the differences in P uptake are unclear. Baker et al. (1970) and Phillips et al. (1971) concluded that differences among hybrids could not be explained by the P absorption characteristics of their roots. They proposed that the differences might be due to genotypic differences in depth of rooting or in the binding of P in the leaves. However, these studies were done in solution culture at P concentrations higher than those found in most soils. Nielsen and Barber (1978) found significant genotypic differences among solution-grown maize genotypes in the maximum rate of P absorption per unit root length and the solution P concentration at which uptake was half its maximum rate. When the same genotypes were grown in soils with high and low levels of available P, genotypic differences in shoot P concentration were found only at the high level of P fertility. The hybrid with the highest maximum rate of P absorption per unit root length in solution culture had the highest shoot P concentration when grown in soil. However, since this occurred only at the high fertility level, the genotypic differences in P uptake probably related to P demand of the shoot rather than the ability of the roots to absorb P.

In contrast to the marked genotypic differences in P accumulation by maize, genotypic effects on P accumulation in sugarcane are small and are probably less important than genotypic differences in N and K concentrations (Humbert, 1968).

Vesicular-Arbuscular Mycorrhizae

Numerous studies have demonstrated that vesicular-arbuscular mycorrhizae (VAM) infection of roots, including those of C_4 grasses, enhances tillering and shoot dry weight, principally by increasing plant N and P uptake (Allen, et al., 1981b; Murdock et al., 1967; Wallace, 1981) (see Chapter 6). The effect is frequently seen in soils low in plant-available N and P, and VAM infection is often inhibited in soils high in available P (Hays et al., 1982; Menge et al., 1978; Mosse, 1973). Under some circumstances, infection by VAM in soils high in P may inhibit plant growth (Hays et al., 1982; Mosse, 1973; Tinker, 1975), presumably due to fungal carbohydrate utilization (Hays et al., 1982; Stribley et al., 1980). This type of inhibition has been observed in the early stage of infection of blue grama (*Bouteloua gracilis*) grown on soils high in plant-available P (Hays et al., 1982).

EFFECTS OF PHOSPHORUS DEFICIENCY

The symptoms of P deficiency in tropical grasses are not as dramatic as those of some other nutrients. Foliar P deficiency symptoms include decreased leaf growth, purple or bronze coloration on the leaf sheaths and stems of older leaves, a blue-green coloration of young leaf blades, and drying of leaf tips and margins. In addition, elongation of secondary rootlets is reduced (Christie, 1975b; Humbert, 1968), tillering is drastically reduced (Christie and Moorby, 1975; Humbert, 1968), plant phosphatase activity increases (Besford, 1979; McLachlan, 1982), inorganic P concentration of the tissue decreases (Hartt, 1972; Hanway, 1962c; Sharpley and Reed, 1982), photosynthetic rate decreases (Hartt, 1972), translocation of carbohydrate decreases (Hartt, 1972), and phenological development slows (Peaslee et al., 1971).

POTASSIUM

Potassium is an essential element for all living organisms and is one of the three plant nutrients most likely to limit crop yields. Unlike N and P, K does not form covalent bonds with organic molecules in the plant or soil. For this reason, its behavior is quite different from theirs. The effects of soil and plant K on the growth and development of C_4 grasses are summarized in this chapter.

SOIL POTASSIUM

As in the case of phosphorus, the total K content of the soil is not a good indicator of K fertility. Three principal K fractions occur in the soil. These are soil solution K, exchangeable K adsorbed to clay minerals and humus, and nonexchangeable K within the clay mineral lattice. Solution K and exchangeable K are much more easily absorbed by the plant than is nonexchangeable K. Thus, soil extractants that remove soluble and exchangeable K are used to predict plant-available K. In general, soils with large amounts of 2:1 clay minerals, especially illite, have relatively large pools of nonexchangeable K and often "fix" large amounts of fertilizer K in the nonexchangeable form, particularly after long periods of cropping without addition of fertilizer K. Soils with small amounts of 2:1 clay minerals fix little or no K and have small pools of nonexchangeable K (Mengel and Kirkby, 1980).

The amount of extractable K in the plow layer is normally used to assess K fertility. However, several factors may complicate interpretation of soil test data. First, crops can extract K from the subsoil as well as the topsoil (Woodruff and Parks, 1980), but soils differ markedly in the amount of extractable K in the subsoil (Ayres and Fujimoto, 1944; Hanway et al., 1962). Thus, extractable K in the subsoil should be considered in making fertilizer recommendations.

Second, crop response to extractable K varies with soil texture. At the same level of extractable K, crops extract K more readily from coarse-textured soils than from fine-textured soils (Cope and Rouse, 1973; Hanway et al., 1962; Rouse, 1968). Third, small grains and most C_4 grasses have lower critical concentrations of extractable K than dicotyledonous plants such as cotton, soybeans, and forage legumes (Cope and Rouse, 1973). Table 14.1 gives critical concentrations and adequate ranges of extractable K for several C_4 grasses.

PLANT POTASSIUM

Potassium, like nitrogen and phosphorus, has many functions in the plant (Mengel and Kirkby, 1980). These include (1) its behavior as a counter ion whose uptake is coupled with the uptake of anions and other cations (see Chapter 6); (2) its role as a counter ion in the long-distance transport of organic anions from one part of the plant to another; (3) its role in the regulation of cell turgor, especially in the guard cells; (4) its important though poorly understood role in the energy metabolism of cells; and (5) its activation of numerous enzymes.

TABLE 14.1. Critical Levels of Soil Test K for Tropical Grasses

Extractant	Crop	Region	Soil	Critical K Concentration in Plow Layer	Reference
$1N$ NH$_4$OAc	Bermudagrass	Southeastern United States	Sand	27 $\mu g/g$	Woodhouse (1968)
	Four C$_4$ forage grasses	Puerto Rico	Clay (Oxisol)	140–224 kg K/ha in top 60 cm (18–28 $\mu g/g$)	Vicente-Chandler et al. (1962)
	Maize	Midwestern United States	Various	100 $\mu g/g$	Hanway et al. (1962)
0.05N HCl + 0.025N H$_2$SO$_4$	Bermudagrass	Southeastern United States	Loamy sand, sandy loam	20	Jordan et al. (1966)
	Maize	Brazil	Clay	50	Ritchey (1979)
	Maize	Southeastern United States	Coarse	30	Rouse (1968)
			Medium	45	
			Fine	60	
0.02N H$_2$SO$_4$	Sugarcane	Hawaii	Various	212–336 kg/ha in top 30 cm (54–86 $\mu g/g$ at bulk density of 1.3)	Ayres and Humbert (unpublished)

Plant Uptake

In recent years research on uptake of K by plant roots has concentrated on the roles of plasma membrane-bound carrier enzymes and ATPases (Epstein, 1976). As discussed in Chapter 6, at low external K concentrations (K < 1 mM) K influx into root cells is largely dependent on the activity of a proton-extruding ATPase at the plasmalemma (Cheesman and Hanson, 1979a,b; Glass and Dunlop, 1978; Leonard and Hotchkiss, 1976). This carrier system produces active K uptake (net K influx against an electrochemical potential gradient). As external K concentration approaches 1 mM, the K uptake activity of the ATPase declines, possibly due to substrate inhibition of the carrier function. However, the proton-extruding activity of the ATPase increases, causing an increased electopotential difference across the membrane. As the external K concentration exceeds 1 mM, K enters the root passively down an electrochemical potential gradient (Cheeseman and Hanson, 1979a,b). The effect of external K concentration on the active and passive components of K uptake in excised maize roots is shown in Fig. 14.1.

The effect of plant age or stage of development on K uptake varies among C_4 grass species. In sugarcane and other species that remain vegetative, the K content of the shoot (including dead leaves) increases steadily with time (Fig. 14.2). However, in maize a dramatic decrease in K uptake usually occurs near the beginning of grain fill (Fig. 14.3) (Hanway, 1962c; Jordan et al., 1950; Yanuka et al., 1982). In some cases (Sayre, 1955) shoot K content actually decreases during the later stages of grain fill, but in most cases this is probably due to the loss of K from senescent leaves. In grain sorghum, a similar decrease in K uptake is usually observed after flowering; however, a second period of rapid uptake and translocation to the grain can occur shortly before physiological maturity (Roy and Wright, 1974).

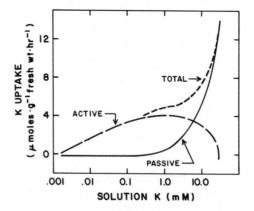

FIG. 14.1.   Effect of solution K concentration on the active and passive components of K uptake in excised maize roots. From Cheeseman and Hanson (1979b). Reproduced by permission of *Plant Physiology*.

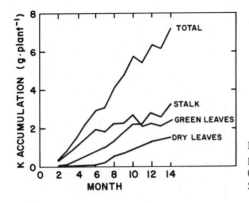

FIG. 14.2.   Accumulation of K in various parts of the sugarcane shoot. From Ayres (1937). Reproduced courtesy of the Hawaiian Sugar Planters' Association.

The decrease in K uptake during grain filling in maize and grain sorghum coincides with a general reduction in root growth (see Chapter 6). Since K uptake is strongly dependent on root metabolism, the decreases in K uptake and root growth may result from the diversion of assimilate from the root system to the rapidly growing seeds. However, this is probably not an adequate explanation since the stems normally contain significant reserves of nonstructural carbohydrates during the beginning of grain fill. Another explanation for this decrease in K uptake relates to the demand of the shoot for K. Though maize and grain sorghum plants require almost as much K as N in the vegetative parts of the plant, the grain contains only 30–40% as much K as N (Steele et al., 1981).

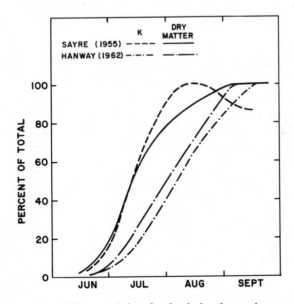

FIG. 14.3.   Dry matter and K accumulation of maize during the growing season in two studies.

Blevins (1982) suggests that in the grain K plays a role in neutralizing acidic amino acids like aspartate and glutamate in grain proteins. This suggests that the grain is a relatively weak sink for K, compared with the vegetative parts of the plant. As a result, the demand of the total plant for K may decline abruptly when vegetative growth ceases and grain growth begins.

Plant Composition

The concentration of K in the plant varies with plant part and with time. Table 14.2 shows typical values of plant K concentrations for several components of maize and sugarcane shortly prior to harvest. From Table 14.2 and more detailed analyses of maize K concentrations (Hanway, 1962c), it is clear that in maize and sugarcane leaf blades normally have higher K concentrations than stems. Younger blades have higher K concentrations than older blades, and leaf sheaths normally have higher K concentrations than leaf blades. As previously noted (and unlike N and P), K is not concentrated in the grain, which often has a lower K concentration than the leaves and leaf sheaths. Also in contrast to N and P, the concentration of K within the maize leaf lamina is greater in the midrib and near the center than near the margin (Jones, 1970; Repka and Jurekova, 1981); and unlike N and P, lamina K concentration is higher near the basal end than near the apex (Repka and Jurekova, 1981).

Critical Concentrations

As with other nutrients, many systems have been used to diagnose K deficiency of C_4 grasses. In most cases the K concentration of the leaf lamina or the whole shoot is used as an index of plant K nutrition (Table 14.3). However, since tissue

TABLE 14.2. **Typical Potassium Concentrations of Various Maize and Sugarcane Parts Prior to Harvest (Hanway, 1962c; Clements, 1980)**

SUGARCANE		MAIZE		
PLANT PART	K (%)	PLANT PART	HIGH K (%)	LOW K (%)
Meristem and expanding cane	3.55	Young blades	0.8	0.4
Young blades	1.65	Old blades	0.4	0.2
Old blades	1.31	Young sheaths	1.0	0.4
Young sheaths	2.48	Old sheaths	0.4	0.2
Old sheaths	2.05	Tassel	0.3	0.1
Internodes		Ear shoot	1.8	1.1
45–47	1.07	Husks	0.7	0.4
24–26	0.75	Cob	1.6	1.4
1–3	0.72	Grain	0.4	0.3

TABLE 14.3.   Critical K Levels and Normal Ranges for Several Tropical Grasses

Tissue and Stage of Development or Age	Critical Level or Range (%)	Source
Maize		
Ear leaf at silking	1.7-2.5	Jones (1967)
	1.9	Melsted et al. (1969)
	1.6	Gallaher et al. (1972)
	1.6-2.3	Escano et al. (1981a)
Grain Sorghum		
Whole shoot at 23-39 days	3.0-4.5	Lockman (1972c)
Third leaf at 37-56 days	2.0-3.0	
Third leaf at bloom stage	1.4-1.7	
Third leaf at dough stage	1.0-1.5	
Sugarcane		
Third leaf blade at 4-6 months	1.0	Gosnell and Long (1971)
Leaf sheaths 3-6 at 7-14 months	0.46-0.65% of sheath H_2O	Jones and Bowen (1981)
Oldest fully expanded leaf during period of rapid growth (June)	1.0-1.6	Gascho and El Wali (1979)
Leaf sheaths 3-6 at 7-14 months	2.25	Schmehl and Humbert (1964)
Internodes 8-10 at 7-14 months	1.0	

K remains largely in solution rather than immobilized in protein or structural components of the cell, Clements (1980) proposed that the K concentration be expressed as a percentage of tissue water content rather than dry weight. Clements found that this approach reduces the effects of genotype on tissue K concentration and improves the diagnosis of K deficiency in sugarcane.

Genotype Effects

As with N and P, C_4 grass genotypes sometimes differ significantly in K concentration (Farquhar and Lee, 1962; Gorsline et al., 1961; Gosnell and Long, 1971). For example, the shoot K concentrations of 12 maize genotypes 115 days after planting ranged from 1.38 to 2.01% (Bruetsch and Estes, 1976). These studies generally indicate that among genotypes accumulation of K is inversely related to accumulation of Ca + Mg. Among maize genotypes a weak, though significant, correlation exists between K uptake properties of excised roots and grain yields (Frick and Bauman, 1978, 1979). However, the weakness of the correlation ($r = 0.70$) makes its usefulness in breeding questionable.

Several studies have revealed genetic variation in the K concentrations in C_4 forage grasses (Bray and Hacker, 1981; Smith, 1972). For example, at high fertilizer N levels, the critical concentration of K in the shoots of *Setaria anceps* cv. Nandi was 1-2%. In cv. Narok it was 2-3%, and in cv. Kazungula it was 4-6% (Smith, 1972).

Tissue Age

As plants mature, the concentration of K, like those of N and P, decreases in the vegetative tissues. This is primarily due to a greater increase in structural materials (such as cellulose and lignin) than in cytoplasm. The decreases in leaf blade, leaf sheath, and stem K concentrations have been documented by Hanway (1962c) and Sayre (1955) for maize and by Lockman (1972a) for grain sorghum. For example, in the leaves of well-fertilized maize plants, K decreases from 2.3 to 2.7% at about 1 month of age to 1.1 to 1.5% at silking and 0.4 to 0.7% at maturity (Hanway, 1962c). In grain sorghum, normal leaf K concentrations decline from 2.0 to 3.0% at 37 to 56 days to 1.0 to 1.5% at bloom stage (Lockman, 1972a).

The K concentrations of leaf blade, leaf sheath, and young internode tissues of sugarcane change little after about 3 months of age (Clements, 1980; Gosnell and Long, 1971). This suggests that tissue K concentration is more sensitive to the age and maturity of the tissue sampled than to the age of the plant or the tiller on which the tissue is growing (Table 14.4).

Cation Ratios

For grasses growing under nonsaline conditions, the most abundant cations in the tissue are normally K, Ca, and Mg. The mechanism of root uptake of these nutrients is discussed in Chapter 6. Numerous studies with maize (Baker et al., 1966; Claasen and Barber, 1977; Soofi and Fuehring, 1964), grain sorghum (Berthouly and Guerrier, 1979a,b; Boawn et al., 1960a,b), sugarcane (Clements, 1980), and C_4 forage grasses (Landua et al., 1973; Smith, 1972, 1978, 1981) have demonstrated that genotype, soil aeration, soil compaction, drought stress, age of the plant, and soil N, K, Ca, and Mg concentrations affect the relative accumulation of the three cations.

Plants must maintain a balance between cations and anions in their tissues. They accomplish this by three principal mechanisms. First, uptake of mineral anions and cations are often approximately equal. Second, when uptake of positively and negatively charged nutrients are not equal, efflux of OH^- or H^+ ions may help maintain the balance of charges within the tissue. Third, when more cations than anions are taken up, organic acids accumulate in the tissues to neutralize the excess of positive charges.

All cations effectively neutralize negative charges in the tissue. However, the ratios in which different cations are absorbed by the plant varies with genotype,

TABLE 14.4.   Potassium Concentration of Sugarcane Tissues at Different Times During the Two-Year Growth of the Crop (modified from Clements, 1980)

Age (MON)	Young Blades (%)	Young Sheaths (%)	Top Internodes (%)
3	1.7	2.9	2.9
5	1.8	2.9	2.7
8	1.9	2.8	2.7
10	1.7	2.7	2.1
12	1.6	2.7	2.1
15	1.7	2.6	1.7
17	1.5	2.6	2.1
19	1.8	2.9	2.5
22	1.8	2.8	3.3
24	1.7	2.5	1.8

plant age, and environmental conditions. Munson (1970) has reviewed many of the factors affecting the concentrations of these elements in the tissue. As discussed in Chapter 6, K is normally taken up by roots against an electrochemical potential gradient (active uptake). Ca and Mg are not normally taken up actively; they move passively down an electrochemical potential gradient into the root. Because of the fundamental difference in their uptake mechanisms, environmental factors often affect K uptake differently than Ca and Mg uptake.

Poor soil aeration and soil compaction can inhibit root growth and metabolism. Since K uptake normally requires expenditure of metabolic energy, it may decrease relative to Ca + Mg uptake, and the (Ca + Mg)–K ratio may increase (Lawton, 1945).

In general, grasses have lower values of the ratio (Ca + Mg)–K than do dicotyledonous plants growing under the same environmental conditions (Bower and Pierre, 1944; Collander, 1941; Newton, 1928) (Table 14.5). This suggests that in competition with legumes, grasses have a competitive advantage for absorption of K.

Fertilization with K decreases the (Ca + Mg)–K ratio of the tissue (Clements, 1980, Larson and Pierre, 1953; Smith, 1981). In contrast, addition of Ca or Mg decreases the uptake of K and increases the (Ca + Mg)–K ratio in the tissue (Berthouly and Guerrier, 1979a; Landau et al., 1973; York et al., 1954). For example, increasing the rates of K fertilizer increases tissue K concentration while decreasing the concentration of Ca + Mg (Fig. 14.4). In addition, when soil K is adequate, high rates of N fertilizer can stimulate K uptake and decrease the (Ca + Mg)–K ratio slightly (Boawn et al., 1960a,b; Viets et al., 1954).

In maize and grain sorghum, K uptake per unit dry matter accumulation

TABLE 14.5.   Ratios of (Ca + Mg)/K in Several Plant Species

CROP	(CA + MG)/K	SOURCE
Maize	0.75	Newton (1928)
Wheat	0.67	
Barley	0.67	
Sunflowers	0.72	
Beans	0.90	
Peas	0.86	
Maize	0.43	Collander (1941)
Oats	0.36	
Buckwheat	1.56	
Spinach	0.93	

decreases as plants or tillers increase in age. Uptake of Ca and Mg is less affected by these changes; therefore, the (Ca + Mg)-K ratio increases as the plant ages (Jenne et al., 1958; Sayre, 1955) (Fig. 14.5).

EFFECTS OF POTASSIUM DEFICIENCY

Potassium deficiency has many effects on the growth and metabolism of the plant. These effects include decreases in ATP synthesis, enzyme activity, translocation of photosynthate, drought resistance, disease resistance, and winterhardiness.

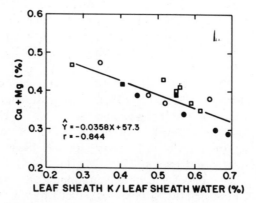

FIG. 14.4.   Relationship between sheath K (expressed as percent of sheath water content) and sheath Ca + Mg (expressed as percent of dry weight). From Clements (1980).

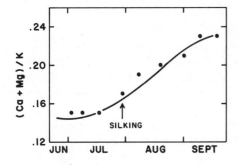

FIG. 14.5. Increase in (Ca + Mg)-K ratio during maize grain development. From Sayre (1955).

Energy Metabolism

Potassium nutrition has a profound effect on the energy metabolism of plants. In most studies, suboptimal K nutrition reduces oxidative phosphorylation, photophosphorylation, and photoreduction of NADP (Mengel and Kirkby, 1980). It is generally recognized that K plays analogous roles in photophosphylation and oxidative phosphorylation. In both systems, K moves across plastid membranes in exchange for protons. In the chloroplast, K moves out of the stroma in exchange for inward movement of protons; in the mitochondrion, K moves inward in exchange for outward movement of protons. In both cases, the pH and electropotential gradients drive ATP synthesis.

Mengel and Kirkby (1980) argue that the role of K in energy metabolism is central to its role in other plant processes. For example, the effects of K on CO_2 assimilation, protein synthesis, amino acid synthesis, and translocation can all be explained by its effects on ATP and NADPH synthesis.

Enzyme Activation

Another important function of K in plant biochemistry is enzyme activation. This subject has been reviewed by Evans and Wildes (1971), Mengel and Kirkby (1980), and Wilson and Evans (1968). More than 60 enzymes are known to require activation by monovalent cations. *In vitro*, NH_4 and Rb are toxic at the concentrations needed for activation. Thus, K is the most important monovalent cation for *in vitro* enzyme activation. Potassium activates synthetases, oxidoreductases, dehydrogenases, transferases, and kinases; and it is important in the synthesis of protein, starch, and other high-molecular-weight compounds.

Inadequate K nutrition results in the accumulation of simple sugars and amino acids (Clements, 1970; Hartt, 1934a,b, 1969, 1970; Nowakowski, 1962, 1971). This effect has been attributed both to the role of K in enzyme activation and to its role in the production of ATP through respiration and photophosphorylation. Mengel and Kirkby (1980) conclude that ATP synthesis is more sensitive to K deficiency than is enzyme activation, and the effects of K on concentrations of low-molecular-weight compounds can be explained by K-induced reduction in ATP availability.

Photosynthesis

Potassium deficiency has a pronounced effect on maize leaf photosynthesis (Estes et al., 1973; Koch and Estes, 1975; Peaslee and Moss, 1966; Smid and Peaslee, 1976). In most studies, the K concentration at which CO_2 uptake approaches a maximum is 1.5-2.0%, slightly lower than earleaf concentrations associated with maximum grain yields (Fig. 14.6).

Mesophyll chloroplasts of K-deficient maize plants have poorly developed lamellar and granal structure. Unlike P-deficient chloroplasts, they normally contain starch grains (Hall et al., 1972), which is consistent with the observation that translocation of malate and sugar phosphates is inhibited by K deficiency.

Repka and Jurekova (1981) found that in maize leaves without nutrient deficiencies the K concentration is greatest near the basal portion of the lamina. However, the photosynthetic rate and concentrations of N, P, and chlorophyll are greatest near the center of the leaf. Thus, when K nutrition is adequate, photosynthetic rate is not necessarily correlated with K concentrations along the lamina.

Translocation

It has long been known that K deficiency affects the phloem function. In sugarcane, K deficiency can cause phloem necrosis (Hartt, 1929, 1934a) and reduce

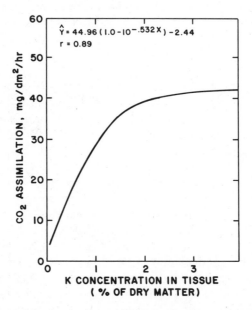

$$\hat{Y} = 44.96 (1.0 - 10^{-.532X}) - 2.44$$
$$r = 0.89$$

FIG. 14.6. Effect of maize leaf K concentration on CO_2 assimilation rate. From Smid and Peaslee (1976). Reproduced from *Agronomy Journal*, by permission of the American Society of Agronomy.

the translocation of photosynthates from the leaf (Hartt, 1969, 1970). The levels of tissue K associated with reduced translocation are higher than those associated with visual symptoms of K deficiency (Hartt, 1969). The effect of K deficiency on movement of photosynthates from the mesophyll cells to the phloem may result from a decrease in the turnover of C from malate and aspartate to sugar phosphates, thus inhibiting translocation from the mesophyll to the bundle sheath cells. Alternatively, K deficiency may reduce the release of photosynthate from the bundle sheath cells to the apoplast outside the phloem (Doman and Geiger, 1979), or it may directly affect the sugar uptake mechanism of the phloem, possibly by affecting a K-activated, plasmalemma-bound ATPase associated with sugar transport into the phloem (Giaquinta, 1977). Finally, the effect of K on phloem loading may be due to its effect on ATP synthesis in photophosphorylation (Mengel and Kirkby, 1980).

Regardless which mechanism is most important, inhibition of translocation causes decreased carbohydrate availability in other parts of the plant. For example, Clements (1980) reported that in sugarcane, K deficiency is associated with low concentrations of total sugars in both leaf sheaths and mature stalks.

Stomatal Activity

Peaslee and Moss (1966) found that leaf diffusive resistance is abnormally high in K-deficient maize. Fischer (1968) demonstrated the pivotal role of K in stomatal activity. When environmental conditions favor stomatal opening, the K concentration of the guard cells is observed to increase dramatically. This increase in positive charge in the guard cells is largely balanced by the production of malate from PEP and the movement of protons out of the guard cells (Humble and Hsiao, 1969, 1970; Humble and Raschke, 1971), though Cl uptake by the guard cells may also be involved (Smith and Raven, 1979). Photophosphorylation probably provides the ATP necessary for K uptake by the guard cells, and abscisic acid simultaneously prevents stomatal opening and K uptake by the guard cells. Fusicoccin, a toxin derived from fungi, simultaneously promotes K uptake by guard cells and stomatal opening, probably due to its selective stimulation of ATPase (Giaquinta, 1979).

Counter Ion Effects

In addition to its role in enzyme activation and the energy metabolism of the cell, an important function of K is its role as a counter ion in the synthesis, uptake, and transport of organic and inorganic anions. Potassium plays an important role in maintaining the balance of positive and negative charges within the different parts of the plant and between the plant and its soil environment. In this respect, it is important that K is extremely mobile in both the xylem and phloem. Marschner (1974) pointed out that K is the principal counter ion that accompanies the upward movement of nitrate in the xylem and the downward movement of malate in the phloem.

Potassium also plays an important role in cell expansion. Sustained growth of plant cells requires the activation of an ATPase located at the plasma membrane. Activation of this plamalemma ATPase results in inward movement of K and extrusion of protons from the cell. These counter ion movements of K and protons are correlated with cell expansion, which is probably a direct consequence of one or more of the following (Hanson and Trewavas, 1982):

1. Proton extrusion into the apoplast causes the cell wall to become more plastic.
2. Increased K influx, with accompanying anion influx or malate synthesis, maintains the cell turgor needed for cell expansion.
3. Increased electropotential difference across the plasma membrane may regulate aspects of cell wall synthesis.

Drought Resistance

Potassium plays an important role in the water economy of plants. Its role in stomatal closure has already been discussed. In addition, adequate K is needed to maintain the water content and turgor of leaf tissue (Clements, 1980; Koehler et al., 1982; Mengel and Kirkby, 1980). For example, in well-watered sugarcane plants leaf sheath water content is positively correlated with tissue K concentration (Clements, 1980), and leaf K concentration increases during osmotic adjustment due to drought stress (Koehler et al., 1982) (see Chapter 10).

Recent work also suggests that adequate K nutrition helps maintain nitrate reductase activity during water stress in maize (Khanna-Chopra et al., 1980) and is necessary for normal synthesis of proline from arginine in water-stressed finger millet (*Elusine coracana*) (Rao et al., 1981a,b).

Weimberg et al. (1982) have suggested that accumulation of proline by salt-stressed (and water-stressed?) grain sorghum plants occurs only when the concentration of monovalent cations (mostly K) in the leaves reaches a threshold of about 200 mole/Mg fresh weight. They suggest that proline accumulation in the cytoplasm osmotically balances accumulation of monovalent cations in the vacuoles.

Winterhardiness

Winterhardiness of C_4 (and C_3) grasses is strongly affected by K nutrition (Kresge, 1974). In the southeastern United States, N and K fertility have contrasting effects on winter survival of coastal bermudagrass (*Cynodon dactylon*) cv. Coastal. High N fertilizer rates often reduce winter survival, probably by stimulating top growth and reducing carbohydrate reserves in stolons and roots. However, this effect can be minimized by application of adequate levels of fertilizer K (Fig. 14.7), which promotes translocation and storage of carbohydrate reserves. In general, a fertilizer N:K ratio of 2.4:1.0 is adequate to maintain

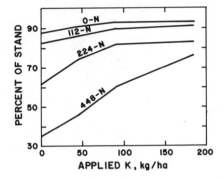

FIG. 14.7. Effect of four levels of fertilizer N and K application on winter survival of Coastal bermudagrass (*Cynodon dactylon* cv. Coastal). From Adams and Twersky (1960). Reproduced from *Agronomy Journal*, by permission of the American Society of Agronomy.

high levels of winterhardiness (Adams and Twersky, 1960; Kresge and Decker, 1966).

Disease Resistance

Many factors influence the susceptibility of plants to disease, including genetic resistance, climate, and mineral nutrition. With regard to nutrition, K often has a prophylactic effect on disease (Goss, 1968). For example, Perrenoud (1977) reviewed the effects of K nutrition on the effects of parasitic fungi in a number of crops, including tropical grasses. He reported that K reduces the incidence of dollar spot (pathogen unknown), helminthosporium leaf spot (*Helminthosporium cynodontis*), and leaf spot (pathogen unknown) in bermudagrass (*Cynodon dactylon*). In maize, adequate K nutrition reduces root rot (*Gibberella sauvinetti, G. zeae, G. roseum*), smut (*Ustilago zeae, U. maydis*), northern corn leaf blight (*Helminthosporium turcicum*), and stalk rot (*Gibberella, Diplodia, Fusarium*, and *Rhizoctonia* spp.).

The effects of K on disease severity may be numerous. It may reduce soil fungal populations, reduce the pathogenicity of the organisms, speed maturation and increase the physical resistance of the plant tissues, promote rapid healing, stimulate production of new tillers, and increase the biochemical resistance to pathogens (Leath and Ratcliffe, 1974; Munson, 1970; Perrenoud, 1977).

A typical example of the effects of K nutrition on the severity of a foliar fungal disease is found in Matocha and Smith (1980). They found that on coastal bermudagrass, helminthosporium leaf spot severity is inversely related to tissue K concentration and dry matter yield (Fig. 14.8).

One of the best-studied examples of the effects of K on disease resistance is maize stalk rot. High fertilizer N and low fertilizer K are often associated with severe stalk rot (Josephson, 1962; Liebhardt and Murdock, 1965; Otto and Everett, 1956). Its severity is also positively correlated with stalk nitrogen concentration and negatively correlated with stalk sugar concentration (Craig and Hooker, 1961; Martens and Arny, 1967a). Parker and Burrows (1959) found a correlation of 0.68 between the incidence of root and stalk rot and the N:K ratio

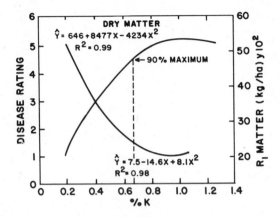

FIG. 14.8. Effect of shoot K concentration on leafspot (*Helminthosporium cynodontis*) disease rating and dry (R_1) matter production of Coastal bermudagrass (*Cynodon dactylon* cv. Coastal). From Matocha and Smith (1980). Reproduced from *Agronomy Journal*, by permission of the American Society of Agronomy.

of the leaf tissue. Younts and Musgrave (1958) and Martens and Arny (1967b) have also found that Cl has an additional prophylactic effect on stalk rot, and application of KCl often reduces the incidence of stalk rot more than application of K or Cl alone.

The stalk rot pathogen usually enters the root system, then advances to the pith of the lower stalk internodes where it destroys the tissue and weakens the stalk. Potassium and Cl appear to delay its advance from the roots to the stalk rather than prevent its growth in the stalk (Martens and Arny 1967b).

15

ACID AND HIGHLY
LEACHED SOILS

Chapters 12, 13, and 14 deal with the effects of N, P, and K nutrition on growth of C_4 grasses. Deficiencies of these "primary" nutrients can occur in most soils. In contrast, "secondary" nutrient (Ca, Mg, S) and micronutrient (Fe, Mn, Zn, B, Cu, Cl, Co, Mo) deficiencies and toxicities are most common in either acid, strongly leached soils or in alkaline or calcareous soils. For example, Ca, S, and Si deficiencies as well as Al and Mn toxicities most frequently occur on highly leached acid soils. In contrast, Fe, Zn, Mn, and Cu deficiencies more often occur on calcareous or alkaline soils, though they can also occur on acid soils that have been overlimed and on some highly leached soils such as Histosols and Quartzipsamments.

Nutrient toxicities and deficiencies commonly associated with acid and/or highly leached soils are described in this chapter. Problems normally associated with saline, calcareous, and alkaline soils are discussed in Chapter 16.

DEVELOPMENT OF SOIL ACIDITY

Developoment of strongly acid soil profiles largely depends on the biological production of H^+ by microorganisms that convert ammonium to nitrate (nitrifiers) or that oxidize reduced forms of S. Hydrogen ions are also produced when plants take up more cations than anions and secrete H^+ to prevent charge imbalance between the soil and root (Adams, 1981).

The H^+ produced in the soil may react in several ways. First, it displaces bases from cation exchange complex. These bases may then be lost by leaching where rainfall exceeds evapotranspiration. The bases lost through leaching are accompanied by soluble anions. If, over a long period of time, leaching losses exceed the rate of nutrient recycling by the vegetation, highly leached soils deficient in bases and soluble anions result.

The second major effect of low soil pH is to reduce the cation exchange capacity of soils with large amounts of pH-dependent charge, for example, Oxisols, Ultisols, Histosols, and Dystrandepts. In these soils the cation exchange capacity (CEC) declines with decreasing soil pH; therefore, soil acidity reduces their capacity to adsorb cations and prevent leaching loss of bases.

The third major effect of low soil pH is to increase the solubility of toxic cations such as Al^{3+} and Mn^{2+}. For example, kaolinite dissociates according to the reaction

$$Al_2Si_2O_5(OH)_2 + 6H^+ \leftrightarrows 2Al^{3+} + 2Si(OH)_4 + H_2O \qquad [1]$$

Soil Mn may also be solubilized:

$$MnO_2 + 4H^+ + 2e^- \leftrightarrows Mn^{2+} + 2H_2O \qquad [2]$$

Thus, the production of high concentrations of Mn in the soil solution depends on the level of readily reducible Mn oxides, the pH, and the supply of electrons. The supply of electrons is greatest when soils become anaerobic.

Soil acidity and severe leaching affect C_4 grass growth in a number of ways. The principal problems include Al and Mn toxicity and P, Ca, Si, and S deficiencies. Phosphorus deficiencies caused by strong P fixation in acid, highly leached soils were discussed in Chapter 13. The effects of Al and Mn toxicities and Ca, Si, and S deficiencies on C_4 grass growth are discussed below. Occasionally, Zn, Mn, and Cu deficiencies are found in C_4 grasses grown on highly leached soils such as Histosols, Quartzipsamments (highly leached quartz sands), and recently limed Oxisols and Ultisols. Because such deficiencies are more common on calcareous and alkaline soils, they are discussed in Chapter 16.

ALUMINUM TOXICITY

Aluminum toxicity is the most common problem affecting the growth of C_4 grasses on acid soils (Adams, 1981). When soil pH is above about 5.5, solution concentration of Al is normally very low. However, below pH 5.5 the solution concentration Al increases with decreasing soil pH, and it can become toxic to many crop species. Several indexes can be used to predict Al toxicity. These include the concentration of Al in the soil solution or its chemical activity (Adams, 1981), the amount of Al exchangeable in 1 N KCl at pH 7.0 (Fig. 15.1), soil pH (Fig. 15.2), percent base saturation (Fig. 15.3), and the percentage of the cation exchange complex occupied by Al (Adams and Pearson, 1967). The latter, called percent Al saturation, is the most commonly used index of Al toxicity.

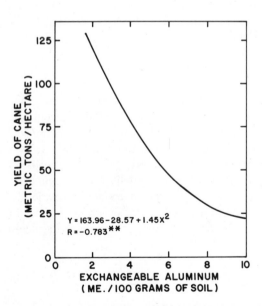

FIG. 15.1. Relationship between exchangeable Al and sugarcane yield on Cialitos clay (Oxisol). From Abruna-Rodriguez and Vicente-Chandler (1967). Reproduced from *Agronomy Journal*, by permission of the American Society of Agronomy.

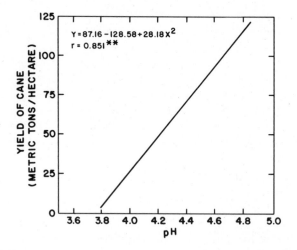

FIG. 15.2. Relationship between pH and sugarcane yield on Cialitos clay (Oxisol). From Abruna-Rodriguez and Vicente-Chandler (1967). Reproduced from *Agronomy Journal*, by permission of the American Society of Agronomy.

It is especially useful because it takes into account both the amount of Al on the exchange complex and the amount of bases available to the plant, which tend to ameliorate the toxic effects of Al (Vidal and Broyer, 1962; Rhue and Grogan, 1977a). Several studies have shown that percent Al saturation is a good index of plant response to Al toxicity in C_4 grasses (Abruna-Rodriquez and Vicente-Chandler, 1967; Abruna et al., 1974, 1982; Brenes and Pearson, 1973; Kamprath, 1970) (Fig. 15.4).

Topsoil Aluminum Toxicity

The effects of Al toxicity are most severe when high levels of Al saturation occur in the topsoil, where rooting densities and nutrient concentrations are normally high. The primary effect of Al toxicity is the inhibition of root growth and disruption of root function. This can result in inadequate water and nutrient uptake and loss of specificity in ion uptake. The loss of specificity can lead to abnormally high uptake of toxic elements such as Al, Fe, and Mn, which may have direct toxic effects on top growth. The effects of different levels of Al saturation on yields of maize and grain sorghum are shown in Figs. 15.4 and 15.5

Topsoil acidity is easily corrected by incorporation of agricultural lime, which reacts as follows:

$$CaCO_3 + 2H^+ \leftrightarrows Ca^{2+} + H_2O + CO_2 \qquad [3]$$

Application of lime increases soil pH, thus reducing the solubility of Al (reaction 1).

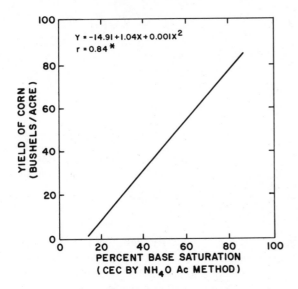

FIG. 15.3. Relationship between base saturation and maize yield on a Corozal clay subsoil (Ultisol). From Abruna et al. (1974). Courtesy of the authors and the *Journal of Agriculture*, *University of Puerto Rico.*

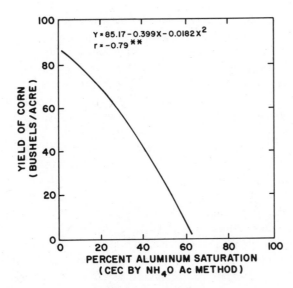

FIG. 15.4. Relationship between percent Al saturation and maize yield on a Humatas clay (Oxisol). From Abruna et al. (1974). Courtesy of the authors and the *Journal of Agriculture*, *University of Puerto Rico.*

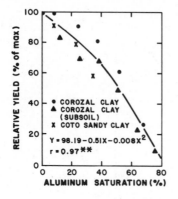

FIG. 15.5. Relationship between percent Al saturation and relative grain sorghum yield on Corozal (Ultisol) and Coto (Oxisol) soils. From Abruna et al. (1982). Courtesy of the authors and the *Journal of Agriculture, University of Puerto Rico.*

Lime-induced yield decreases have frequently been reported, particularly in highly weathered tropical soils, at pH values near 7.0. Reduced P and micronutrient availability; antagonisms between Ca, K, and Mg; and nutritional imbalances have all been proposed as causes of lime-induced yield reductions.

Subsoil Aluminum Toxicity

Many agricultural soils have low levels of Al saturation in the topsoil but higher levels in the subsoil. In these soils, Al toxicity may reduce root growth and nutrient availability in the subsoil (Fig. 15.6) (Adams, 1981; Bouldin, 1979; Brenes and Pearson, 1973).

Three strategies—deep lime incorporation, leaching of Ca below the plow layer, and use of Al-tolerant cultivars—can be used to overcome the effects of subsoil acidity. Considerable work has shown that deep incorporation of lime increases root development to the depth of lime incorporation and increases drought tolerance (Gonzalez-Erico et al., 1979; Perez-Escolar and Lugo-Lopez, 1978; Sumner, 1970). For example, incorporation of lime to 30 cm rather than 15 cm increases the amount of water extracted by maize and delays the onset of water stress (Bouldin, 1979; Gonzalez-Erico et al., 1979).

The principal problem associated with deep lime incorporation is the power required to mix the lime with a large volume of soil, and in most cases incorporation of lime below 30 cm is not feasible (Adams, 1981). However, by choosing fertilizer rates and products to encourage leaching of Ca, subsoil Al toxicity can be ameliorated. For example, Adams et al. (1967b) showed that application of calcitic limestone and high rates of NH_4NO_3 to an Ultisol increases subsoil pH. Similar effects were found with $Ca(NO_3)_2$ (Adams and Pearson, 1969). Above pH 6, HCO_3 is probably the most important anion accompanying Ca. Below pH 6, solution HCO_3 concentrations are quite low and other accompanying anions (such as NO_3 or Cl) are needed (Ritchey et al., 1980).

When Ca and HCO_3 are leached from a well-limed topsoil into a very acid

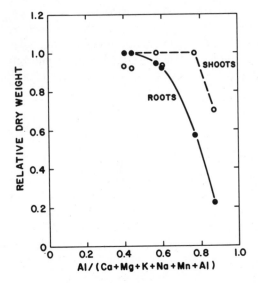

FIG. 15.6. Relationship between percent Al saturation of the subsoil and relative dry weights (at 34 days after planting) of maize roots growing into the subsoil and maize shoots. From Brenes and Pearson (1973).

lower layer, the Ca increases the extractable bases while the HCO_3 increases the pH and decreases the solubility of Al in the lower layer. When NO_3 is the accompanying anion, greater uptake of anions than cations may cause OH exudation by the root system, increase subsoil pH, and thereby decrease Al solubility. Increased maize root growth in the subsoil has been shown to result both from alleviation of Ca deficiency (Ritchey et al., 1982) and reduction in Al toxicity (Ritchey et al., 1980).

The effects of Ca leaching are not always favorable. For example, Friesen et al. (1982) found significant leaching of Ca in a Nigerian Ultisol, but it had little effect on subsoil Al saturation. In fact, over the long term, leaching of Ca from the topsoil to the subsoil reduced maize yields due to reappearance of Al toxicity in the topsoil.

Aluminum Uptake

Aluminum uptake by roots appears to be passive. Soluble Al in the soil solution first exchanges with Ca in the root free space. It then moves to and across the plasmalemma (Huett and Menary, 1979). By replacing Ca in the plasmalemma, Al may also increase permeability to toxic ions (Huett and Menary, 1980a).

Huett and Menary (1980b) studied the distribution of Al in kikuyugrass (*Pennisetum clandestinum*) roots. They found that Al is uniformly distributed along the roots. The highest concentrations are in the epidermis, with lower concentrations in the cortex and stele. This suggests that Al enters the symplast in the epidermis or cortex, thus avoiding the endodermal apoplastic barrier. Other studies (Rasmussen, 1968) suggest that the maize root epidermis is an effective barrier to Al. Aluminum did not penetrate the epidermis until it was broken by the emergence of lateral root primodia. It then entered both the cortex and the stele.

Root Growth

The first observable effect of Al toxicity is a reduction in root elongation. More severe toxicity results in morphological deformities such as brown root tips and thickened, brittle, and twisted roots with short, thick branches (Foy et al., 1978). These symptoms are recognizable both in solution culture (Rhue and Grogan, 1977a,b; Saigusa et al., 1980) and in soil (Adams and Moore, 1983). They differ dramatically from those associated with Ca deficiency, which is characterized by production of straight, small-diameter roots with brown tips.

An example of the relationship between percent Al saturation and root dry weight is shown in Fig. 15.6. In this study (Brenes and Pearson, 1973), maize plants were grown in pots filled with subsoils having variable levels of Al saturation. The subsoil was overlain with 2.5 cm of well-limed and fertilized soil, and plants were watered daily. Root growth into the acid subsoil was reduced by high levels of Al saturation even though shoot growth was hardly affected. Thus, as long as the topsoil supplies adequate water and nutrients, severe reductions in root growth may have little or no effect on shoot growth.

An example of the effect of percent Al saturation on root length is given in Fig. 15.7. In this study (Gonzalez-Erico et al., 1979), several rates of lime were incorporated to either 15 or 30 cm. As in the previous study, root growth was adversely affected by Al toxicity when percent Al saturation exceeded about 40%.

Genetic Control of Aluminum Tolerance

Four-carbon grass genotypes vary considerably in their response to Al toxicity (Duvick et al., 1981; Lafever, 1981; Lundberg et al., 1977). Though inbred lines of maize vary widely in their tolerance to Al toxicity in nutrient solutions (Rhue and Grogan, 1977a,b), segregating populations from crosses between inbreds fall into dinstinct classes with regard to Al tolerance.

Plant Al and Fe concentrations are often correlated. For example, distributions of Al and Fe among and within maize leaves are correlated (Jones, 1970). In addition, grain sorghum genotypes that are well adapted to acid soils accumulate lower concentrations of both Al and Fe than poorly adapted genotypes (Duncan, 1981b), probably because genetic mechanisms controlling Al and Fe uptake are similar (Gorsline et al., 1968).

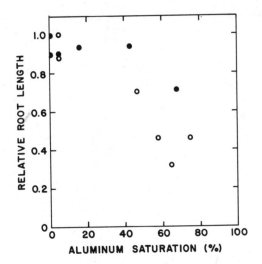

FIG. 15.7.   Effect of percent aluminum saturation on maize root growth in a Brazilian Oxisol.
○ = 0-15 cm layer; ● = 15-30 cm layer. From Gonzalez-Erico et al. (1979).

Nutrient Interactions

Several nutrients interact with Al in acid soils. Two of these interactions are described below.

Aluminum and P availability and uptake are probably interrelated in several ways. First, high concentrations of reactive Al can cause precipitation of P in the soil solution (Kittrick and Jackson, 1955) and at the root surface (Foy and Brown, 1963; McCormick and Borden, 1972; MacLeod and Jackson, 1967). Second, tolerance to toxic levels of Al is associated with tolerance of low levels of soil P (Clark and Brown, 1974b), perhaps because adequate P uptake requires normal root growth and development (Lyness, 1936; Shenk and Barber, 1980; Smith, 1934).

High concentrations of bases in nutrient solution can ameliorate the toxic effects of Al (Vidal and Broyer, 1962). Calcium and Mg are equally effective in protecting maize roots from Al toxicity (Rhue and Grogan, 1977a,b). This is consistent with field studies indicating that percent Al saturation is a better index of Al toxicity than extractable Al alone (Abruna et al., 1974; Brenes and Pearson, 1973; Kamprath, 1970).

Plant Composition

The concentration of Al in plant tissues is affected by soil solution Al, genotype, plant age, and plant part. "Normal" Al concentrations for several maize and grain sorghum parts at various stages of growth are given in Table 15.1.

TABLE 15.1.   Sufficiency Ranges of Al Concentrations for Maize and Grain Sorghum

TISSUE AND STAGE OF DEVELOPMENT OR AGE	SUFFICIENCY RANGE (g/Mg)	SOURCE
Maize		
Whole plants at 3 to 4 leaf stage	100–200	Jones and Eck (1973)
Ear leaf at silking	10–200	
Stalk above ear at silking	10–25	
Stalk below ear at silking	50–100	
Whole plant at 30–45 days	0–70	Lockman (1969) cited in Jones and Eck (1973)
Grain Sorghum		
Whole plant 23–39 days	0–375	Lockman (1972c)
Youngest fully developed leaf at 37–56 days	0–200	
Third leaf below head at bloom stage	0–220	

Jones (1970) studied the distribution of Al in normal maize plants that were not suffering from Al toxicity. He found that leaf Al concentrations are higher in lower leaves than in upper leaves, concentrations are lower in the leaf margins and midribs than in whole leaf, and concentrations are greater in leaf tips than in leaf bases.

In maize and sweet sorghum (*Sorghum bicolor*), symptoms of Al toxicity (such as leaf freckling and reductions in top growth) are associated with high Al concentrations in the leaves (Farina et al., 1980). Grain sorghum genotypes grown on acid soils can have significantly different leaf concentrations of Al. Genotypes adapted to acid soils may have leaf Al concentrations less than 500 ppm while poorly adapted genotypes have Al concentrations exceeding 5000 ppm (Duncan, 1981b).

Despite the correlation between tissue Al concentrations and toxicity symptoms, high tissue concentrations are probably only indirect effects of Al toxicity. The accumulation of Al (and probably Fe) in the shoot reflects prior root damage and a loss of selectivity in ion uptake. Tissue Al concentrations directly harmful to the tops have not been determined.

MANGANESE TOXICITY

Because soil solution manganese concentrations respond to soil pH, soil aeration, and reducible Mn content of the soil, Mn toxicity is more complex and not as well understood as Al toxicity.

Soil Manganese

Because of differences in susceptibility of species to Mn toxicity and the lack of data on soil solution Mn concentrations, no universal "critical" concentration of soil solution Mn can be given (Adams, 1981). However, available data suggest that for tolerant species, a critical solution Mn concentration is 10–20 g/Mg (Adams, 1981; Smith, 1979). Adams (1981) concludes that phytotoxicity can best be predicted by combining soil pH and easily reducible Mn. If a soil contains at least 50–100 g/Mg easily reducible Mn (Fergus, 1953) and the pH is less than 5.5, phytotoxicity is probable. Lower soil pH and/or poor aeration increase the probability of phytotoxicity (Adams, 1981).

Toxicity Symptoms

In tropical grasses, the most characteristic symptom of Mn toxicity is the appearance of small brown or reddish-brown spots 0.5–2.0 mm in diameter on the older leaves. These spots appear first near the tips of the leaves. As the leaf grows older, the spots often coalesce and form longitudinal brown streaks 2–5 mm wide and up to 5 cm in length. This is followed by leaf death progressing from the tip. In the light brown dead tissue the darker spots and streaks are still visible (Smith, 1972b, 1973, 1974; Smith and Verschoyle, 1973). Light-colored streaks similar to those caused by Fe deficiency may also occur in the upper leaves of grain sorghum plants (Clark et al., 1981).

Autoradiography and staining techniques indicate that the necrotic leaf spots contain high concentrations of Mn (Bussler, 1981).

Plant Composition

In tropical grasses, tissue Mn concentrations vary among years and among plant parts. Grain sorghum Mn concentrations may be lower in dry years than in wet ones (Jacques et al., 1975b), and sugarcane Mn concentrations fluctuate as a result of weather conditions (Bowen, 1981a; Clements, 1980). Wet, cool weather is often associated with high leaf sheath Mn concentrations, probably due to more reduced soil conditions.

Manganese concentrations are greater in leaves than in stems and inflorescences, and in leaf blades Mn is concentrated in the margins and tips (Jacques et al., 1975b). Manganese concentrations are higher in younger leaves, nodes, and internodes than in older tissues (Clark et al., 1981; Evans, 1960; Jones, 1970). This appears to be the case whether Mn is in the toxic or the deficient range (Clark et al., 1981).

In areas where low soil pH and poor soil aeration are common, tissue Mn concentrations may be negatively correlated with crop yields. This has been reported for sugarcane on Hawaii (Bowen, 1981a) and for maize in Illinois (Peck et al., 1969; Walker and Peck, 1974).

For grain sorghum and sugarcane, toxic shoot Mn concentrations are about 10 times as great as those in the deficient range. For example, Clark et al. (1981) found that Mn-deficient grain sorghum plants have shoot Mn concentrations of 24 g/Mg while those suffering from Mn toxicity have concentrations of about 220 g/Mg. Both Clements (1980) and Bowen (1983) found that sugarcane yields are severely reduced when Mn approaches 100 g/Mg in young leaf sheaths, and Bowen (1983) concluded that sheath Mn concentrations should be maintained between 5 and 50 g/Mg.

These results contrast with those of Smith (1979) who found considerably higher critical shoot Mn concentrations in several C_4 forage grasses. For example, setariagrass (*Setaria anceps*) and dallisgrass (*Paspalum dilatatum*) showed no foliar symptoms of Mn toxicity at shoot Mn concentrations over 200 g/Mg. Guineagrass (*Panicum maximum*) and columbusgrass (*Sorghum almum*) had symptoms of Mn toxicity when concentrations exceeded 1000 g/Mg and growth was reduced between 1000 and 2000 g/Mg. Rhodesgrass (*Chloris gayana*) and buffelgrass (*Cenchrus ciliaris*) were more sensitive to high levels of tissue Mn and regrowth was severely inhibited when tissue Mn exceeded 100 g/Mg.

Malavolta et al. (1979) report critical Mn concentrations in sweet sorghum (*Sorghum bicolor*) of 445 g/Mg in upper leaves and 1440 in lower leaves. Edwards and Asher (1982) report critical maize leaf Mn concentrations of 200 g/Mg, while Tanner (1977) reports much higher values of 2500–6500 g/Mg. It is unclear whether genotype or growing conditions were responsible for these large differences.

Numerous studies have shown that cultivars of tropical grasses differ in their uptake of Mn and their sensitivity to Mn toxicity (Foy et al., 1980; Jacques et al., 1975b; Naismith et al., 1974). Brown et al. (1972) and Smith (1979) found that tolerance is associated with high tissue Mn concentrations.

Because of their dependence on pH, soil solution concentrations of Al and Mn are often correlated. However, among grain sorghum and wheat genotypes, uptake of and tolerance to Al and Mn are not correlated. Foy et al. (1973) found opposite Al and Mn tolerances for two wheat genotypes, and Duncan (1981b) found no relationship between tissue concentrations of Mn and Al among several grain sorghum cultivars grown on acid soils.

CALCIUM DEFICIENCY

Calcium is the fifth most abundant element (by weight) in the earth's crust. However, because of differences in parent materials and weathering, soils differ widely in their Ca contents. Poorly weathered soils that have developed from parent materials high in Ca may have high Ca contents, contain free gypsum ($CaSO_4 \cdot 2H_2O$) or calcite ($CaCO_3$), and have high pH. On the other hand, in highly leached soils Ca may move beyond the root zone and Ca deficiency can occur. Calcium deficiencies are not uncommon in Oxisols, Ultisols, Andepts, and Histosols.

The primary role of Ca in plants is membrane stabilization. Adequate Ca helps maintain membrane integrity, especially during periods of cold temperatures, O_2 deficiency, and senescence (Garrard and Humphrey, 1967; Christiansen et al., 1970; Poovaiah, 1979). Ultrastructural symptoms of Ca deficiency include disintegration of the plasmalemma, tonoplast, and endoplasmic reticulum (Wyn Jones and Lunt, 1967; Hecht-Buchholz, 1979).

Uptake and Translocation

When soil solution concentrations of Ca are very low (0.005-0.05 mM Ca), well below the range of solution Ca found in most soils (1-20 mM Ca), Ca uptake by excised maize roots is dependent on metabolism. However, at higher solution concentrations (5-50 mM Ca), passive uptake is probably more important than active uptake (Higinbotham, 1973; Maas, 1969). When mass flow of Ca in the transpiration stream is greater than plant uptake, Ca may precipitate as $CaSO_4$ at the root surface (Malzer and Barber, 1975).

In contrast to P and K, which are taken up along the length of the root, most Ca is taken up and transferred to the stele near the root apex where the endodermal cell walls are not yet suberized. For example, in barley and maize there is an inverse relationship between suberization of the endodermis and Ca movement into the stele (Ferguson and Clarkson, 1975; Robards et al., 1973). In maize roots with well-developed endodermis, Ca is concentrated at the inner tangential wall of the epidermis and in the endodermis, suggesting that the epidermis (or the outermost layer of the cortex) as well as the endodermis act as barriers to Ca movement (Chino, 1979; Gielink et al., 1966; Lauchli, 1967). The movement of Ca into the stele is also facilitated by rupture of the endodermal cylinder and epidermis by young branch roots (Karas and McCulley, 1973).

Once Ca has reached the stele, its movement is unidirectional. It moves upward in the xylem by a series of exchange reactions along negatively charged sites on the walls of the xylem vessels (Iserman, 1970; Geijn and Smeulders, 1981). In these exchange reactions, Ca competes with other cations, and high concentrations of other divalent cations cause more rapid Ca movement. Calcium moves to meristematic and young tissue where it is deposited in membranes and cell walls. It is adsorbed to insoluble negative charges such as carboxyl, phosphoryl, and phenolic hydroxyl groups. In the cell well it forms an integral part of the pectate cross linkages. Growing tissues are important sinks for Ca because they are sites of rapid production of insoluble negative charges.

Most evidence suggests that Ca, like all divalent alkali earth cations except Mg, are immobile in the phloem (Stencek and Koontz, 1970; Ziegler, 1975). The Ca in older leaves is not remobilized and translocated to growing points (Loneragan and Snowball, 1969), and maize seedlings cannot transport Ca against the transpiration stream from the point of entry in one root tip to other tips located farther down the root system (Marschner and Richter, 1974).

Since Ca moves in the xylem, when the supply of Ca is more than adequate to satisfy the insoluble negative charges in the tissue, it accumulates near the end of the transpiration path. This Ca can be lost from the plant in droplets of gutta-tion fluid or by leaching (Christiansen and Foy, 1979).

Plant Composition

Factors affecting tissue Ca concentrations include plant species, availability of soil Ca, availability of other cations, plant age, plant part, and drought stress. Critical levels and normal ranges of tissue Ca for several C_4 grasses are given in Table 15.2.

The total cation content of the plant is determined by the amount of anions absorbed and amount of anion equivalents (carboxyl, phosphoryl, phenolic hydroxyl groups, and others) produced by the plant. When the supply of one cation is interrupted, the uptake of others will normally increase (Forster and Mengel, 1969). For example, a decrease in the supply of K to the roots often causes uptake of Ca, Mg, and/or Na to increase.

In C_4 grasses the effect of plant age on tissue Ca concentration has long been recognized (Ayres, 1936). Gosnell and Long (1971) found that for sugarcane

TABLE 15.2. Critical Ca Levels and Normal Ranges for Several Tropical Grasses

TISSUE AND STAGE OF DEVELOPMENT OR AGE	CRITICAL LEVEL OR RANGE (%)	SOURCE
Maize		
Ear leaf at silking	0.21–1.0	Jones (1967)
Ear leaf at silking	0.21–1.0	Neubert et al. (1969)
Ear leaf at silking	0.4	Melsted et al. (1969)
Sugarcane		
Third leaf blade at 4–6 months	0.18	Gosnell and Long (1971)
Top visible dewlap leaf during period of rapid growth (June)	0.20–0.45	Gascho and El Wali (1979)
Leaf Sheaths during period of rapid growth	0.20 in soils with pH < 5.8; 0.10 in other soils	Clements (1980) Bowen (1983)
Grain Sorghum		
Whole shoot at 23–39 days	0.9–1.3	Lockman (1972c)
Third leaf at 37–56 days	0.15–0.90	
Third leaf at bloom stage	0.30–0.60	
Third leaf at dough stage	0.20–0.60	

grown in Uganda the Ca content of the youngest fully expanded leaf decreased from 0.37% at 1.0 month to 0.25% at 5.0 months. It then remained almost constant through 10.0 months. Bowen (1975) showed that for sugarcane grown on the acid soils of the island of Hawaii, the Ca content of the immature leaf sheaths normally decreases from above 0.30% in the first 4 months to between 0.15 and 0.20% after 18 months. Lockman (1972a,b,c) found that the normal range of Ca in grain sorghum tops is 0.9–1.3% at 23–39 days, but it decreases to 0.20–0.60% at the dough stage. Jacques et al. (1975a) found that the Ca concentration of the tops declines from 0.36 to 0.55% about one month after planting to 0.15–0.21% at maturity.

In grain sorghum, tissue Ca concentrations usually decline in the order: leaf blades > leaf sheaths > culms > heads > grain (Jacques et al., 1975a). Calcium concentrations of maize leaves are higher in the tips then in the bases, are higher near the margins than in the midribs, and are higher in old leaves than young leaves (Jones, 1970). This is due to the accumulation of Ca near the end of the translocation path and in older leaves, which have transpired more than younger leaves.

Drought stress often causes shoot Ca concentration to increase (Gosnell and Long, 1971; Jacques et al., 1975a; Lockman, 1972a,b). For example, Gosnell and Long (1971) reported that drought stress increased the Ca concentration of the youngest fully expanded sugarcane leaf from 0.22 to 0.27%. Calcium was the only element whose concentration consistently increased with drought stress.

Like other elements, tissue Ca concentrations are under partial genetic control. For example, grain sorghum cultivars grown on acid soils differ significantly in leaf Ca concentration (Duncan, 1981b). Maize leaf Ca concentrations appear to be affected by three major genes located on chromosome 9 (Gorsline et al., 1968; Naismith et al., 1974). Other genes on chromosome 9 may control leaf P and Mn concentrations (Naismith et al., 1974).

Deficiency Symptoms

Applications of high rates of N, P, and K fertilizers to acid soils without addition of Ca can produce Ca deficiency in maize (Melsted, 1953). Deficiency symptoms referred to as "bull-whip" develop in susceptible cultivars. This condition results from the failure of the young expanding leaves to unfold normally as they emerge from the whorl. The tips of the younger leaves remain caught inside the rolled tips of the older leaves, and fully expanded leaves remain attached to the older leaves by their tips. Another characteristic symptom in both maize and grain sorghum is curling and serration of the leaf margins. These symptoms are unaffected by high levels of micronutrients and Si, and in nutrient solution culture they are accentuated by high levels of other cations (Mg, K, NH_4, and Na) in the solution (Kawasaki and Moritsugu, 1979). The symptoms are rapidly corrected in subsequent leaves by addition of Ca to the nutrient solution.

In field-grown sugarcane, Ca deficiency also causes leaf freckling, reduced tillering, thin woody stalks, and rapid leaf senescence (Clements, 1980).

SILICON DEFICIENCY

The beneficial effects of silicates on grass growth in acid soils have long been known (Fisher, 1929; McGeorge, 1924; Toth, 1939). Most of the research on the effects of silicates has been done on sugarcane and rice, principally because these crops require large amounts of Si, and their growth is adversely affected by Si deficiency. Since the work on rice deals with submerged soils, it will not be discussed here.

Silicon is the second most abundant element by weight (28%) in the earth's crust. It occurs as quartz (SiO_2), as a structural component of layer silicate clay minerals, as amorphous SiO_2, and as monomeric H_4SiO_4. Soluble H_4SiO_4 is in equilibrium with SiO_2 and is the form of Si taken up by plants.

A wide variety of soils can be deficient in Si (Ayres, 1966; Fox et al., 1967; Gascho and Andreis, 1974; Tamimi and Matsuyama, 1973). Response to calcium silicate or other silicate-containing amendments is often found on highly weathered and leached tropical soils with low pH, low extractable Si, low cation exchange capacity and low base saturation (Oxisols and Ultisols). Volcanic ash soils formed under high rainfall (Hydrandepts and Dystrandepts) and some organic soils (Histosols) may also respond to Si amendments.

Plant response to added Si is not well correlated with total soil Si. However, Si-deficient soils are consistently acid (pH less than about 6.5) with low levels (less than 16 g/Mg) of H_2O-extractable Si (Fox et al., 1967; Gascho and Andreis, 1974).

The early studies of McGeorge (1924), Fisher (1929), and Toth (1939) and much subsequent work in Hawaii (Suehisa et al., 1963; Monteith and Sherman, 1962; Roy et al., 1971) has shown that addition of Si to soils that are low in available P increases extractable P and P uptake by the plant. In these soils the effect of Si may be primarily one of improving P availability by occupying exchange sites that otherwise fix P. However, in Florida peat soils Si application increases sugarcane yields even though the P content of the tissue is high (Gascho and Andreis, 1974). In these and other Si-deficient soils, Si may have several roles. First, it may decrease the toxicity of Mn and Al in both the soil and the plant. Thus, Clements (1980) suggests that the ratio of Mn to Si in sugarcane tissue is a better indicator of Mn toxicity than is the concentration of Mn alone. Second, Si strengthens a number of plant tissues by forming insoluble deposits in the walls and lumens of certain cells. This increases the mechanical resistance of the tissue and may reduce the movement of ions and water through the apoplast of the endodermis.

Uptake and Translocation

The mechanism of Si uptake is not well understood. Jones and Handreck (1965) concluded that uptake of Si by oat plants can be explained by passive uptake in the transpiration stream. In contrast, Barber and Shone (1966) found that mass flow of Si into the root system is inadequate to explain its uptake in barley. In addition, Si uptake is reduced by both low temperatures and metabolic inhibitors, suggesting that metabolic energy is required for uptake. Vorm (1980) varied the concentration of Si in nutrient solution and thereby varied the possible supply of Si due to mass flow. He found that absorption of Si varied among species, increasing in the order soybean < sunflower < wheat < sugarcane < rice. Silicon uptake by soybean could be explained by passive uptake in the transpiration stream. However, Si uptakes by sugarcane and rice was greater than that supplied by transpiration at all solution concentrations. At the lowest solution Si concentrations (0.75 mg/liter), sugarcane and rice absorbed 31 and 52 times the amount supplied through transpiration. Thus, it appears clear that Si uptake by grasses depends on metabolic activity. This is consistent with the results of Takahashi and Miyake (1977) (cited in Vorm, 1980). They reviewed information on Si uptake by 175 plant species and distinguished Si accumulators from Si nonaccumulators. The accumulators were found in the monocotyledonous families Gramineae, Cyperaceae, and Musaceae (which includes bananas and plantains).

Most mechanisms used to explain active nutrient uptake involve movement of ionic solutes. However, at pH below about 9, the principal form of soluble Si is H_4SiO_4, which is not an ion (Lindsay, 1979). Thus the mechanism of active Si uptake is still unclear (Barber and Shone, 1966).

Plant Composition

Silicon is found in all parts of the grass plant, but the greatest concentrations occur in inflorescence bracts, specialized cells of the leaf and stem epidermis (trichomes, stomata, silica cells), xylem vessels and sclerenchyma cells of the leaf, and the endodermis of the root. In higher plants, Si occurs primarily in the form of opal ($SiO_2 \cdot nH_2O$) (Harper et al., 1982; Jones and Milne, 1963; Jones et al., 1963; Lanning et al., 1958).

In roots, Si is concentrated in the endodermis. However, the morphology of endodermal Si deposits varies among species. In grain sorghum and sugarcane, Si forms aggregates on the inner tangential wall (Sangster and Wynn Parry, 1976b,c). In sugarcane the aggregates are also produced on the radial walls and Si is deposited along the middle lamella of the radial walls (Wynn Parry and Kelso, 1977). In pangolagrass (*Digitaria decumbens*), rice, wheat, barley, and oats, aggregates do not form, but Si is deposited in a diffuse manner in the endodermal cell walls. The inner tangential and radial walls are silicified in pangolagrass and rice, and the outer tangential wall is also silicified in the winter cereals (Bennett, 1982; Wynn Parry and Soni, 1972; Sangster, 1977, 1978).

In sugarcane and grain sorghum, endodermal silicification occurs at about the same time as suberization and thickening of the inner tangential wall (Sangster and Wynn Parry, 1976a). Deposition of both suberin and Si are apparently coordinated with the collapse of the cortical cells and the development of large lysigenous lacunae in the cortex (Wynn Parry and Kelso, 1975). Endodermal silicification may strengthen the root so that it can resist soil shrinkage, movement of the shoot, and insect and pathogen attack. In addition, Si (along with suberin, lignin, and polyphenols) may make the endodermis a more effective barrier between the cortex and stele (Ponnaiya, 1960).

In grass leaves Si accumulates in several types of specialized cells, including marginal sclerenchyma cells, some of the long epidermal cells, xylem vessels, sclerenchyma, and special silica cells. In the basal part of the leaf sheath, Si accumulates in trichomes, most long epidermal cells, special silica cells, and silica cells of cork-silica cell pairs (Da Silva and Labouriau, 1970; Lanning and Garabedian, 1963; Soni et al., 1970).

The greatest concentrations of Si are normally found in inflorescence bracts, where it protects the caryopsis from insect and disease attack (Wynn Parry and Smithson, 1966; Takeoka et al., 1979). In contrast, caryopses of most species are free of silica. One exception is foxtail millet (*Setaria italica*), which has granular silica deposits in the outer layers of the caryopsis.

In sugarcane, the highest density of silica-filled cells is found in the stem epidermis, where Si fills the lumen of the short rectangular silica cells paired with cork cells.

Sugarcane leaves with normal Si concentrations contain Si in the outer tangential walls of long epidermal cells, in epidermal trichomes, and in internal dumbbell-shaped cells. However, Si-deficient plants have little Si in these cells, and that Si which is present concentrates first in the dumbbell-shaped cells and last in the stomates. This may account for the insensitivity of the stomata of Si-deficient plants to drought stress (Wong You Cheong et al., 1971a).

Two types of silicification occur in leaves. In cells such as trichomes, the secondary cell wall becomes impregnated with Si between the cellulose strands. This may occur before the secondary wall has completely developed. In opal-filled silica cells, the lumen of the cell fills with Si, even though the secondary cell wall may not be silicified. Silica granules first appear around the periphery of the lumen and eventually fill the entire lumen. They then polymerize to form a solid mass. Silica may also accumulate in intercellular spaces between the walls of silicifying cells (Bennett, 1982).

Silicification of the lumen of silica cells occurs only when these cells senesce. It does not normally occur in tissues that remain potentially meristematic (root tips, nodal tissue, leaf bases) (Dayanander et al., 1976; Kaufman et al., 1969). Both gibberellic acid and sucrose inhibit the development of the silica cells (Soni et al., 1972), and sugarcane cultivars with low Si concentrations have greater gibberellin-stimulated stem elongation than those with high Si concentrations (Moore and Buren, 1978).

Grass cultivars differ in the Si concentration of their leaves and stems (Artschwager, 1930; Lanning and Garabedian, 1963). The Si concentration of stem internodes varies among sugarcane cultivars and is correlated with the number of silica cells in the stem epidermis (Artschwager, 1930). The Si concentrations of leaf blades and sheaths differ among grass genotypes, and cultivars with the lowest leaf Si concentrations are most susceptible to chinch bug damage (Lanning and Linko, 1961).

As the plant ages, total Si concentrations of blades and sheaths increase (Lanning and Linko, 1961). This is the result of the continued uptake and translocation of soluble Si by the root system and its concentration in the shoot where, after deposition, retranslocation does not occur. Even though soluble Si concentrations may be higher in young than in old leaves, the total Si concentration of older sugarcane and maize leaves is always higher than that of younger leaves (Clements, 1980; Jones, 1970). The total Si concentration in leaf sheaths is normally higher than in leaf blades (Clements, 1980), and the highest concentrations in the blade are found in the tip and margins where trichomes and marginal sclerenchyma are heavily silicified (Jones, 1970).

Because of its structural role in strengthening cell walls, the Si concentration of forages has been used as an index of forage quality. Like other components of forage quality, Si concentration is affected by temperature, undoubtedly because high temperatures accelerate both transpiration and plant development. Henderson and Robinson (1982a,b) studied the effects of temperature, light, and soil moisture on the Si content of bermudagrass (*Cynodon dactylon* cv. common and cv. Coastal), bahiagrass (*Paspalum notatum*), and dallisgrass (*Paspalum dilatatum*). In all species high Si concentrations were associated with high temperatures (35°C). When soil moisture and temperature were high, increasing light intensities caused Si concentrations to decrease, probably due to dilution of Si in the tissue due to increased growth.

Deficiency Symptoms

A number of researchers have concluded that sugarcane has a basic requirement for Si, and satisfactory growth is not possible until that requirement is satisfied (Ayres, 1966; Wong You Cheong and Halais, 1970; Fox et al., 1969). Application of $CaSiO_3$ and other soluble siliceous materials have produced large yield increases on acid, low-Si soils for maize, grain sorghum, sudangrass (*Sorghum sudanense*), and kikuyugrass (*Pennisetum clandestinum*) (Clements et al., 1967; Khalid et al., 1978; Monteith and Sherman, 1962; Tamimi and Matsuyama, 1973). However, most of the research on the effects of Si has been conducted on sugarcane, where significant responses have been reported in Australia (Haysom and Chapman, 1975; Hurney, 1973), Florida (Elawad et al., 1982a,b; Gascho and Andreis, 1974), Hawaii (Ayres, 1966; Clements, 1965; Fox et al., 1969), Mauritius (Ross et al., 1974; Wong You Cheong and Halais, 1970), and Puerto Rico (Samuels, 1969a). In some areas of Florida and Hawaii, sugar yields are highly correlated with Si concentration of index tissues (Bowen, 1981a; Elawad et al., 1982a,b; Gascho and Andreis, 1974) (Fig. 15.8). Khalid et al. (1978) found similar growth responses to Si application in sugarcane, maize, and kikuyugrass in a study of residual Si effects.

Silicon deficiency reduces sugarcane stalk number, height, diameter, and strength. This results in decreased cane and sugar yields (Elawad et al., 1982; Sherman, 1969). However, the most striking visual symptom of Si deficiency is "leaf freckle" (Clements et al., 1974; Elawad et al., 1982b). In Hawaii, freckling

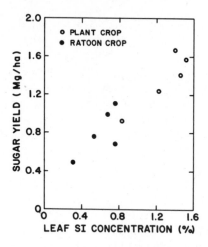

FIG. 15.8. Relationship between leaf Si concentration and sugarcane sugar yield in Florida. From Elawad et al. (1982a,b).

is most severe on poorly drained, highly weathered acid soils during cool, wet weather. During warm periods, sugarcane grows vigorously in these areas. However, as temperatures decline and rainfall increases in winter, small elongated chorotic spots develop on the leaves. As the leaf ages, these spots become reddish, then brown, then necrotic. The necrotic areas spread, especially on leaves exposed to direct sunlight, and older leaves die prematurely. Leaf chlorophyll content decreases as freckling spreads, and photosynthesis undoubtedly decreases. Root growth is severely restricted, and shoot growth and stalk strength decrease (Sherman, 1969).

Lesion formation (freckling) due to Si deficiency does not occur under glass or plastic roofs, and only the side of the leaf exposed to sunlight develops freckles (Wong You Cheong et al., 1971a,b). This suggests that lesion formation may require a specific radiation regime, perhaps including normal levels of ultraviolet radiation. The lesions on sugarcane leaves are the result of degeneration of trichomes, bulliform cells, and stomata. The reddish brown mass in the lesions is soluble in polar solvents, and it contains little if any Fe or Mn. Visible degeneration of the cells follows a decrease in the rate of sugar formation in the leaf and an increase in leaf peroxidase activity. This suggests that lesion formation is not the immediate cause of reduced growth rates due to Si deficiency (Wong You Cheong et al., 1971a,b).

Leaves of freckled plants have high levels of Al, Fe, Mn, and B, while concentrations of Si and Ca are low (Clements et al., 1974; Elawad et al., 1982b). Roots have similar nutrient imbalances. Yields in affected areas are inversely related to leaf concentrations of Al, Fe, Mn, and B, and they are positively correlated with Ca and Si concentrations (Bowen, 1981a; Clements et al., 1974). Uptake of toxic levels of Al, Fe, Mn, and B is probably related to both high concentrations of these elements in the soil solution of acid, poorly drained soils and their toxic effects on the selectivity of the roots. Presence of adequate Si and Ca in the soil solution increases root selectivity and thus reduces uptake of toxic elements.

SULFUR DEFICIENCY

Total S ranges from 50 to 400 g/Mg in most soils. This S occurs in both organic and inorganic forms, and in both highly weathered sandy soils and in Histosols the organic pool is by far the larger pool (Mitchell and Blue, 1981a). Because the ratio of organic N to organic S is quite stable in most proteins, organic N and organic S contents of soils are usually highly correlated, and soil organic S content is usually 10–15% that of organic N. For example, in Florida Ultisols, Spodosols, and Quartzipsamments, total S was highly correlated ($r = 0.95$) with total N (Mitchell and Blue, 1981a). Since organic S constitutes most of the total soil S, mineralization and immobilization of S play important roles in soil S fertility (Saggar et al., 1981).

In well-drained soils, inorganic S occurs primarily as SO_4^{2-}. In neutral soils almost all this SO_4 is in solution and can readily be lost by leaching, but in acid soils some SO_4 is adsorbed on Fe- and Al-oxides. Adsorption is greatest in clayey horizons, especially those with 1:1 clays. Thus, SO_4 leached from sandy surface horizons often accumulates in deeper clayey horizons (Chao et al., 1962); Kamprath et al., 1956; Mitchell and Blue, 1981a,b), where it is available to plants whose root systems reach those horizons.

Highly weathered acid soils are often S deficient, at least in sandy surface horizons. Such deficiencies have been frequently reported in Africa (Oke, 1970), Australia (Jones et al., 1975; Sanchez and Isbell, 1979), South America (Sanchez and Isbell, 1979), and the southern United States (Matocha, 1971; Mitchell and Blue, 1981a,b; Mitchell and Gallaher, 1980).

Plants take up S as SO_4. The most important losses of SO_4 from the system are by leaching, removal by crops, and immobilization by decomposing crop residues. In addition, SO_4 can be reduced in flooded soils and lost by volatilization of H_2S. In most soils these losses of SO_4 are offset by additions of S in rainfall and by direct absorption of SO_2 from the atmosphere. However, these additions decrease with distance from industrial centers where fossil fuels (primarily coal) are burned in large quantities.

Sulfur is an "incidental" constituent of many common fertilizers, including single superphosphate, ammonium sulfate, iron sulfate, and zinc sulfate. Thus, S is often added to soils as a component of N, P, and micronutrient fertilizers. It is also commonly added as finely divided elemental S or as gypsum. However, in recent years increasing crop yields and more intensive agricultural production have increased the demand for S while the use of S-containing fertilizers has decreased in favor of fertilizers with higher N and P concentrations. This has resulted in more frequent reports of S deficiency.

Plant Composition

Sulfur is absorbed primarily as SO_4^{2-}. Uptake is probably by active transport mediated by an enzyme carrier. After SO_4 is absorbed, it replaces the pyrophosphoryl group of ATP to form adenosine phosphosulfate (APS). The sulfuryl

($-SO_3$) group of APS is transferred to a carrier, reduced to SH by ferredoxin, and is then used to synthesize cysteine (see Mengel and Kirkby, 1982).

In addition to cysteine, S is a component of the important amino acid methionine and a number of compounds involved in redox reactions, such as lipoic acid, glutathione, and ferredoxin. It is also a constituent of coenzyme A and the vitamins biotin and thiamine (Mengel and Kirkby, 1982).

Plants absorb approximately equal amounts of S and P. However, S and N are irreplaceable constituents of proteins, whose amino acid composition is strictly determined. Under normal conditions the organic N–organic S ratio of plants is quite constant, but when S is deficient, the N–S ratio increases due to accumulation of NO_3 and soluble amino acids (Friederich et al., 1979; Friedrich

TABLE 15.3.   Critical Concentrations, Sufficiency Ranges of S and Critical N–S Ratios in Selected C$_4$ Grasses

TISSUE AND STAGE OF DEVELOPMENT	CRITICAL CONCENTRATION OR SUFFICIENCY RANGE (%)	CRITICAL N–S RATIO	REFERENCE
Maize			
Whole plants, 3 to 4 leaves	0.2–0.3	—	Jones and Eck (1973)
Ear leaf at silking	0.1–0.3	—	
	0.21–0.50	—	Neubert et al. (1969)
	0.24	12	Daigger and Fox (1971)
	—	12–15	Stewart and Porter (1969)
Sugarcane			
Blades 4, 5 at 3 months	0.10	—	Schmehl and Humbert (1964)
Blades at boom stage	0.15	—	Halais and Figon (1969)
Blade with top visible ligule	0.17	17	Gosnell and Long (1969)
Leaves 3–6 at 70 days	0.08	—	Fox (1976)
Whole plant at 70 days	0.10	—	
Sheaths 3, 4, 5, 6 at 7 months	0.5–0.8	—	Bonnett (1967)
Sheaths 3, 4, 5, 6	0.2	—	Clements (1980)
Sudangrass			
Tops at 4 weeks	0.12	16	Mitchell and Blue (1981b)
Bermudagrass			
Tops	0.2	—	Woodhouse (1969a,b)

and Schrader, 1978, 1979; Rabuffetti and Kamprath, 1977) (see Chapter 12). Thus, the N–S ratio of the tissue as well as its S concentration can be used to diagnose S deficiency (Table 15.3).

Since N and S are both constituents of proteins, their distributions within the plant and the effects of age and phenological development on their concentrations are similar (see Chapter 12). Due to the normal turnover of proteins, S (like N) is continually remobilized and can be translocated to sites of S demand. For example, Friedrich and Schrader (1979) found that most of the N and S in mature maize grain is taken up by the plant prior to silking and is remobilized and translocated to the grain during grain filling.

Deficiency Symptoms

The visual symptoms of S deficiency in C_4 grasses are similar to N deficiency and also have some characteristics in common with Fe and Zn deficiencies. In both N and S deficiency, uniform chlorosis develops over the entire leaf surface; however, in N-deficient plants the chlorosis develops first on older leaves. In S-deficient plants younger leaves develop the symptoms first, reflecting the somewhat lower mobility of S than N. In maize and sugarcane, S deficiency also produces interveinal chlorosis similar to that found with Fe and Zn deficiency. However, S deficiency usually occurs on highly weathered sandy soils, while Fe and Zn deficiencies are more common on alkaline and calcareous soils. In species such as sugarcane, buffelgrass (*Cenchrus ciliaris*), and dallisgrass (*Paspalum dilatatum*), a fine line of anthocyanin pigmentation may develop on leaf margins. This is not diagnostic, however, because the same pigmentation may occur with N deficiency (Clements, 1980; Gascho and Taha, 1972; Smith, 1974; Smith and Verschoyle, 1973). Because of the similarities among S, N, Zn, and Fe deficiencies, plant analysis is usually needed to verify S deficiency.

Hidden symptoms of maize S deficiency include an increase in NO_3-N content of the tissue and a decrease in nitrate reductase activity, soluble leaf protein, and chlorophyll contents (Friedrich and Schrader, 1978). Sulfur deficiency also produces low grain S and N concentrations (Rabuffetti and Kamprath, 1977). Low grain concentrations can be rapidly diagnosed with a simple staining technique in cool-season cereals (Moss et al., 1982).

16

SALINE AND ALKALINE SOILS

The acid, highly leached soils discussed in Chapter 15 normally develop where the supply of bases by weathering of primary minerals or from other sources is less than the losses suffered by leaching. In contrast, alkaline soils normally develop where the addition of bases is greater than their loss by leaching.

Strictly speaking, any soil or soil layer with a pH greater than 7.0 is alkaline. Practically, the term is normally applied to soils or layers with pH greater than about 7.3. Important subgroups of alkaline soils are saline, sodic, saline-sodic, and calcareous soils. Saline soils contain concentrations of neutral soluble salts that seriously interfere with the growth of most plants. However, less than 15% of the cation exchange capacity (CEC) is occupied by Na and the pH is usually below 8.5. The high concentrations of soluble salts reduce crop growth primarily by reducing osmotic potential of the soil solution, thus inhibiting water absorption.

Sodic soils do not contain large amounts of neutral soluble salts, but more than 15% of the CEC is occupied by Na. Their pH is usually between 8.5 and 10. The high pH reduces the availability of micronutrients such as Fe, Mn, and Zn. In addition, the high levels of exchangeable Na result in deflocculation of clays and the breakdown of soil structure. This results in poor soil aeration, the effects of which were discussed in Chapter 11.

Saline-sodic soils are characterized both by high concentrations of soluble salts and by more than 15% of the CEC occupied by Na. The high concentration of soluble salt prevents pH from exceeding 8.5, and their clays are normally flocculated. However if the soluble salts are leached out of the profile without addition of sufficient Ca, their pH may increase above 8.5 and deflocculation may occur.

Calcareous soils contain sufficient free calcium carbonate ($CaCO_3$) to effervesce visibly when treated with 0.1 N hydrochloric acid. Their pH is normally buffered between 7.2 and 8.3 by the hydrolysis of $CaCO_3$. Crops growing on calcareous soils often suffer from micronutrient (Fe, Zn, Mn, Cu) deficiencies caused by nutrient insolubility at high pH and P deficiency caused by the formation of insoluble calcium phosphate complexes. Overliming can cause similar problems by creating an artificially calcareous topsoil.

The effects of salinity and micronutrient deficiencies induced by high soil pH are discussed in this chapter.

SALINITY

All soils contain soluble salts, some of which are essential for crop growth. However, under certain conditions the concentration of these salts becomes so high that it suppresses crop growth. Salinity is defined as the presence of excessive concentrations of soluble salts (U. S. Salinity Laboratory Staff, 1954). Ions contributing appreciably to salinity include Cl^-, SO_4^{2-}, HCO_3^-, Na^+ Ca^{2+}, Mg^{2+}, and occasionally NO_3^- and K^+. Salinity normally occurs in arid and semiarid regions where rainfall is insufficient to leach salts from the root zone. Sources of salts include rainfall and ocean spray, irrigation water, weathering of primary minerals, and upward movement of salts from below the root zone due to rising water tables. A wide variety of agricultural and nonagricultural soils are affected by salinity, about one-third of the developed agricultural land in arid and semi-arid regions (Allison, 1964).

The most common method of quantifying soil salinity is to measure the electrical conductivity (EC, mmho/cm) of saturation extracts of the soil in the root zone. The osmotic potential (ψ_o, MPa), total dissolved solids (TDS, mg/liter), and total cations or anions (TI, meq/liter) of the soil solution are also used as indexes of salinity. For many soils and waters the following relationships can be used (U.S. Salinity Laboratory Staff, 1954).

$$\psi_o = -0.036 \text{ EC} \tag{1}$$

$$\text{TDS} = 640 \text{ EC} \tag{2}$$

$$\text{TI} = 10 \text{ EC} \tag{3}$$

Numerous aspects of soil salinity and its management are reviewed by Bernstein (1974, 1975), Bresler et al. (1982), Maas and Hoffman (1977), and Maas and Nieman (1978).

Glycophytes are plants that can tolerate only low concentrations of salts; halophytes tolerate relatively higher concentrations. Two general types of halophytes have been recognized. Salt includers absorb large amounts of salt and store it in their stems and leaves. They are often succulents. Salt excluders efficiently take up K and exclude or actively secrete Na before it reaches the shoot. Halophytic grasses are salt excluders and are often sclerophyllous plants with small, highly lignified leaves.

Plants affected by salinity are generally stunted and have smaller leaves than normal plants. They are usually dark green or their leaves may have a bluish-green cast due to wax accumulation. Maas and Nieman (1978) cite these general rules for the effects of sublethal salinity on plant growth.

1. Growing cells are not visibly damaged.
2. Growth is retarded, not prevented or distorted.
3. Ontogeny is normal, but it is delayed.
4. Concentrations of specific enzymes, total protein, RNA, and DNA are not strongly affected.
5. Roots are inhibited less than shoots.
6. Within limits, inhibition of growth by salinity is reversible.

Ion Uptake and Excretion

At low solution salt concentrations (0–0.5 mM), active ion uptake predominates, but above about 1 mM passive uptake becomes relatively more important (see Chapter 6). In saline soils, cation and anion concentrations are well into the range at which passive uptake predominates. For example, when the electrical conductivity of the soil solution is 1.0 mmho/cm (too low to reduce growth of most species), the total cation concentration is about 10 meq/liter (from Eq. 3). Thus, salt-excluding halophytes have greater ion selectivity than glycophytes in the range of passive ion uptake.

Salt exclusion in grasses may be aided by two specialized cell types. In the halophytic C$_3$ grass *Puccinellia peisonis*, the inner cells of the root cortex have abnormally large amounts of cytoplasm and numerous mitochondria. In addition, many plasmodesmata connect them to the endodermal cells. This suggests that the symplasm of these cells regulates movement of ions into the stele (Stelzer and Lauchli, 1977). In addition, maize plants grown in saline conditions develop specialized xylem parenchyma cells rich in cytoplasm with abundant mitochondria and rough endoplasmic reticulum. This specialized xylem parenchyma develops in the older parts of the root and apparently mediates the exchange of K for Na in the ascending xylem sap (John and Lauchli, 1980; Yeo et al., 1977). This Na may be translocated to the root surface and excreted (Lessani and Marschner, 1978).

In contrast to excretion from the roots, salt excretion from glands on the leaves and stems is undoubtedly an important means of ridding halophytes of large amounts of salt. For example, Sandu et al. (1981) studied salt accumulation and excretion in the halophytic C$_4$ grass kallargrass (*Diplachne fusca*). When the EC of the rooting medium was 3 mmho/cm, the total Na inside of the leaf was only 12% of that on its surface.

Osmotic Adjustment

Soil salinity has both osmotic and specific-ion effects on plant growth. The osmotic effects are the result of lowering the soil water potential due to increasing solute concentration in the root zone. This interferes with soil water extraction and turgor maintenance. However, plants adjust osmotically to increasing salinity in order to maintain the gradient of water potential between the soil and plant. Three principal mechanisms are used to reduce plant water potential. First, tissue water content decreases. Second, tissue inorganic ion concentrations increase. Third, nontoxic organic osmotica such as proline and glycinebetaine increase, primarily in the cytoplasm. For example, in the study with kallargrass cited above, increasing the EC of the rooting medium from 3 to 40 mmho/cm caused tissue water content to decrease from 3.1 to 1.9 g water/g dry matter. Glycinebetaine and proline increased from 21 to 57 and from 2 to 9 mol/ Mg dry matter, respectively. Similar results have been obtained with another halophytic C$_4$ grass, *Spartina townsendii* (Storey and Wyn Jones, 1978).

Genotypic Variation

Genotypic variation in salinity tolerance is well documented. Bresler et al. (1982) have reported the relative tolerance of numerous species, including many C_3 and C_4 grasses (Table 16.1). These grasses vary widely in their tolerance to salinity. Among agronomically important C_4 grasses, bermudagrass (*Cynodon dactylon*) and grain sorghum are quite tolerant, while maize is relatively susceptible (Fig. 16.1). However, barley (*Hordeum vulgare*), the crop most commonly grown on saline soils, is even more tolerant than bermudagrass.

Variation in salinity tolerance among genotypes of the same species has been found in grain sorghum (Heilman, 1973; Ratandilok, 1978), maize (Siegel et al., 1980), sugarcane (Bernstein et al., 1966; Syed and El-Swaify, 1972) (Fig. 16.2), and bermudagrass (Youngner and Lunt, 1967). In C_3 grasses, genotypic variation occurs in rice (Fageria et al., 1981), barley (Chauhan et al., 1980), and wheat (Abdul-Kadir and Paulsen, 1982).

Relative or absolute yield decrease due to salinity is normally used to evaluate tolerance. Studies with both C_3 and C_4 grasses suggest that tolerance is associated with the genotype's ability to maximize K accumulation, minimize Na and/or Cl accumulation, and/or increase free proline concentration (Ahmad, 1978; Chauhan et al., 1980; Storey and Wyn Jones, 1978).

Other Factors

Other factors affecting plant response to salinity include temperature, humidity, light intensity, soil fertility, and stage of development. Factors that increase

TABLE 16.1.   Salinity Tolerance of Several C_3 and C_4 Grasses (modified from Bresler et al., 1982)

COMMON NAME	SPECIES	ELECTRICAL CONDUCTIVITY AT 90% RELATIVE YIELD (mmho/cm)
	Moderately Sensitive	
Bentgrass	*Agrostis palustris* (C_3)	—
Canarygrass	*Phalaris arundinaceae* (C_3)	—
Lovegrass	*Eragrostis* spp. (C_4)	2.0
Maize (forage, grain)	*Zea mays* (C_4)	1.8, 1.7
Millet	*Setaria italica* (C_4)	—
Rice (paddy)	*Oryza sativa* (C_3)	3.0
Sugarcane	*Saccharum* spp. hybrid (C_4)	1.7
Timothy	*Phleum pratense* (C_3)	—

TABLE 16.1. *(Continued)*

Common Name	Species	Electrical Conductivity at 90% Relative Yield (mmho/cm)
	Moderately Tolerant	
Alkali sacaton	*Sporobolus airoides* (C_4)	—
Barley (forage)	*Hordeum vulgare* (C_3)	6.0
Bromegrass	*Bromus inermis* (C_3)	—
Dallisgrass	*Paspalum dilatatum* (C_4)	—
Fescue	*Festuca elatior* (C_3)	3.9
Hardinggrass	*Phalaris tuberosa* (C_3)	4.6
Orchardgrass	*Dactylis glomerata* (C_3)	1.5
Oats	*Avena sativa* (C_3)	—
Rhodesgrass	*Chloris gayana* (C_4)	—
Rye (hay)	*Secale cereale* (C_3)	—
Ryegrasss (perennial)	*Lolium perenne* (C_3)	5.6
Sorghum (grain)	*Sorghum bicolor* (C_4)	4.8
Sudangrass	*Sorghum sudanense* (C_4)	2.8
Wheat	*Triticum aestivum* (C_3)	6.0
Wheatgrass, slendor	*Agropyron trachycaulum* (C_3)	—
Wheatgrass, western	*Agropyron smithi* (C_3)	—
Wildrye, beardless	*Elymus triticoides* (C_3)	2.7
Wildrye, Canada	*Elymus canadensis* (C_3)	—
	Tolerant	
Barley (grain)	*Hordeum vulgare* (C_3)	8.0
Bermudagrass	*Cynodon dactylon* (C_4)	6.9
Nuttall alkaligrass	*Puccinellia airoides* (C_4)	—
Rescuegrass	*Bromus catharticus* (C_3)	—
Saltgrass	*Distichlis stricta* (C_4)	—
Wheatgrass, crested	*Agropyron desertorum* (C_3)	3.5
Wheatgrass, fairway	*Agropyron cristatum* (C_3)	7.5
Wheatgrass, tall	*Agropyron elongatum* (C_3)	7.5
Wildrye, altai	*Elymus angustus* (C_3)	—
Wildrye, Russian	*Elymus junceus* (C_3)	—

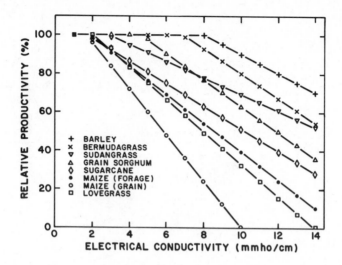

FIG. 16.1.   Effect of salinity on the relative productivity of several C₄ grasses. From Bresler et al. (1982).

evaporative demand (low humidity, high temperature, and high light intensity) tend to aggravate the effects of salinity (Hoffman and Jobes, 1978; Meiri et al., 1982). Salinity causes relatively greater yield reductions under fertile condtions than under infertile conditions. Finally, salinity tolerance varies with development stage. For example, maize and rice (but not grain sorghum) are more tolerant of salinity at germination and emergence than later in development. In addi-

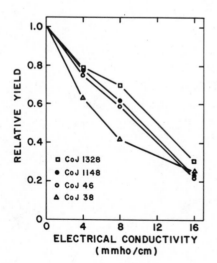

FIG. 16.2.   Genotypic variation in salt tolerance of sugarcane cultivars. From Dev and Bajwa (1972).

tion, maize grain production is more sensitive than forage production to salinity (Bresler et al., 1982).

IRON DEFICIENCY

Iron is the fourth most abundant mineral in the earth's crust. Most temperate soils contain 1.5-4.0% Fe, and some tropical soils contain much more. In contrast, most crops accumulate less than 10 kg Fe/ha; therefore, it is ironic that Fe should ever limit crop growth. However, several factors often limit Fe availability to crops. First, the solubility of Fe in the soil solution is very low, especially in well-aerated soils of high pH. Second, plants can absorb Fe only in the ferrous (Fe^{2+}) form, not the ferric (Fe^{3+}) form.

Lindsay (1979) and Lindsay and Schwab (1982) have recently reviewed the role of chemical equilibria in determining Fe availability to plants. Soil pH determines the concentration of Fe^{3+} and its hydrolysis products in the soil solution. Redox reactions determine the equilibrium between Fe^{3+} and Fe^{2+}. The activity of Fe^{2+} is much lower in well-aerated layers (pe + pH = 16) than in highly reduced layers (pe + pH = 8). In fact, for each unit decrease in pH, the activity of Fe^{2+} decreases 100-fold. As a result, Fe deficiency often occurs on well-aerated calcareous and alkaline soils and on soils that have been heavily limed (lime chlorosis).

Effects of Iron Deficiency

Iron deficiency on calcareous and alkaline soils causes dramatic, but usually localized, decreases in crop growth and yield. Iron-deficient areas of grain sorghum, sugarcane, and maize fields are visible from a distance because of the dramatic decreases in growth and the marked chlorosis of the leaves. These small areas, often referred to as "hot spots", are irregular in shape, and severely deficient plants may be quite close to normal plants. On most C_4 grasses, visible symptoms of Fe deficiency include development of pronounced chlorotic stripes on the leaves. Striping begins at the tips of the young leaves and extends to their bases. The symptoms are usually most severe on the most recently expanded leaves; however, if deficiencies are severe, all leaves are affected. If severe deficiencies are not corrected, the leaves may become entirely yellow, turn white, then die.

When plants become Fe deficient, the roots excrete H^+, which reduces the pH of the rhizoplane and solubilizes soil Fe (Kannan, 1980, 1981). At the same time, the roots excrete organic compounds that reduce Fe^{3+} to Fe^{2+} (Uren, 1982). One important Fe-reducing agent is caffeic acid, a derivative of *p*-coumaric acid and phenlalanine metabolism (Bennett et al., 1982a; Olsen et al., 1982). Excretion of H^+ and reducing agents are very sensitive to the Fe status of the plant. Excretion begins as soon as Fe deficiency occurs and ceases when the plant's immediate need for Fe is satisfied.

Plant citric and malic acid contents are also sensitive of Fe. Iron-deficient plants accumulate higher concentrations of these compounds than nondeficient plants (Hofner and Grieb, 1979; Landsburg, 1981). These organic acids may be the source of H^+ excreted by the roots, and they play an important role in Fe translocation within the plant.

Other root responses to Fe deficiency include increased root hair production, decreased root tip diameter, and (in some species) production of specialized transfer cells in the root epidermis (Romheld et al., 1982).

Iron Uptake and Utilization

Iron absorption by the root is dependent on plant metabolism (Moore, 1972). Young Fe-deficient maize plants growing in flowing nutrient solution absorb and translocate Fe to the shoot as a linear function of solution Fe^{2+} concentration up to about 0.5 mg/liter. Above this concentration, absorption increases curvilinearly with solution concentration to a maximum rate at about 8 mg Fe/liter. Iron-deficient, three-week-old maize plants absorbed enough Fe to completely recover from Fe deficiency within 72 hr, even when the solution concentration was only 0.1 mg/liter.

Within the root system Fe^{2+} is reoxidized to Fe^{3+}. It is then transported to the shoot as chelated Fe^{3+}-citrate. When it reaches the Fe sink in the shoot, it is again reduced to Fe^{2+}, either chemically or photochemically by ultraviolet or blue light (Bennett et al., 1982b; Krizek et al., 1982). Thus, Fe absorbed by the roots (or absorbed from foliar sprays of $FeSO_4$ or Fe-chelates) is relatively mobile within the plant. It is rapidly translocated through the xylem and phloem to meristematic tissues (Hsu et al., 1982; Kannan and Pandely, 1982). However, once Fe is incorporated into enzymes and storage proteins it is quite immobile. Retranslocation can occur (Kannan and Pandley, 1982), but its magnitude is small. Thus, newly emerging leaves of Fe-deficient grasses are more chlorotic than older leaves. Iron applied to chlorotic leaves as foliar sprays is rapidly incorporated into proteins, but translocation to younger leaves is usually insufficient to prevent their chlorosis (Hsu et al., 1982). Thus, multiple foliar applications of Fe are usually needed. For example, Withee and Carlson (1959) reported that on some Fe-deficient calcareous soils three foliar applications of $FeSO_4$ are needed for maximum grain sorghum yields.

Iron is an important component of many enzymes. It is necessary for the function of haem enzymes such as catalase, peroxidase, and cytochrome oxidase. It also occurs in nonhaem proteins such as ferredoxin, which is involved in photosynthesis, nitrite reduction, and sulfate reduction. Within the plant Fe is stored in the form of the ferric phosphoprotein phytoferritin (Mengel and Kirkby, 1982).

Most of the iron in leaves is found in the chloroplasts. The typical chlorosis of Fe-deficient leaves is due to the inability of these leaves to form precursors of chlorophyll (Miller et al., 1982), and it prevents thylakoid development (Terry and Low, 1982). In fact, in severely Fe-deficient leaves chloroplasts may break down completely leaving colorless cells without chloroplasts (Bussler, 1981).

Genotype Effects

Genotypic differences in Fe uptake and utilization have been found in many crops, including C_4 grasses. Genotypes that exhibit few symptoms of Fe deficiency when grown in Fe-deficient soils or nutrient solutions are described as "efficient." Those that have symptoms of Fe deficiency are termed "inefficient." In general, monocots are somewhat more susceptible to Fe deficiency than dicots (Brown, 1978a). For example, maize and grain sorghum are often more sensitive to Fe deficiency than are soybeans, tomatoes, and sunflowers (Brown and Bell, 1969; Brown and Jones, 1975a; Brown et al., 1972; Clark and Brown, 1974; Marschner et al., 1974). However, wide variation in Fe efficiency exists among cultivars of the same crop. Among C_4 grasses, these differences have been described in grain sorghum (Brown and Jones, 1975a; Kannan, 1982; Williams et al., 1982), maize (Brown and Ambler, 1970; Clark and Brown, 1974), and forage grasses (Foy et al., 1977; Voigt, et al., 1982).

For grain sorghum and maize grown in Fe-deficient nutrient culture, genotypes that accumulate high concentrations of Fe in the leaves are more efficient (have less chlorosis) than those that accumulate lower concentrations. This is probably due to the inability of inefficient genotypes to excrete H^+ and reducing compounds from the roots. In addition, a negative relationship is often found between Fe and P concentrations in these genotypes, possibly because P inhibits the reduction of Fe^{3+} at the root surface or because high concentrations of P in the plant cause precipitation of Fe (Brown and Jones, 1975b; Esty et al., 1980).

Plant roots absorb Fe both as the intact Fe chelate molecule and as ionic Fe after chelate splitting (Romheld and Marschner, 1981). In addition, the roots of efficient species such as sunflower adapt to Fe deficiency by increasing chelate splitting and taking up more Fe in the ionic form. In contrast, the relative uptake of Fe chelate and ionic Fe by maize plants is the same whether the plant is Fe deficient or not. Romhold and Marschner (1981) speculate that increased Fe uptake by Fe-efficient plants is due to (1) sorption of chelate molecules at the plasmalemma of the root cells, (2) weakening of chelate–Fe bonds, (3) reduction of Fe, (4) dissociation of Fe from the chelate, and (5) uptake of the Fe. In these Fe-efficient plants, accumulation of phenols in the swollen tips of Fe-deficient roots may be involved in chelate sorption.

In contrast to maize and grain sorghum, lovegrass (*Eragrostis curvula*) genotypes that develop severe Fe-chlorosis do not necessarily have lower foliar Fe concentrations than genotypes with less severe symptoms (Foy et al., 1977; Voigt et al., 1982). However, inefficient genotypes have higher tissue Mn levels than efficient genotypes, suggesting that the balance between Fe and Mn is responsible for the Fe deficiency. These results are like those of studies that suggest that a "biochemically active fraction" of leaf Fe (e.g., water- or acid-soluble fractions) are better correlated with chlorophyll content than is total leaf Fe (DeKock, 1955; DeKock et al., 1974).

Nutrient Interactions

Iron uptake and metabolism are affected by other mineral nutrients (see Mengel and Kirkby, 1982; Murphy et al., 1981). Iron uptake is often negatively correlated with uptake and/or soil availability of P, Mn, Cu, Ca, Mg, K, and Zn. However, multiple interactions among these nutrients are possible, both in the soil and in the plant. Some interactions that have been documented in C_4 grasses are discussed below.

Iron phosphates can precipitate either in the soil or in the plant. For example, Pandley and Kannan (1979) found that high levels of KH_2PO_4 in the nutrient solution cause dramatic decreases in maize shoot Fe concentrations, while causing slight increase in root Fe concentrations. Esty et al. (1980) found a strong negative relationship between Fe and P concentrations for grain sorghum grown in P-deficient nutrient solutions. Brown et al. (1977) found that P-efficient grain sorghum cultivars are more susceptible to Fe chlorosis than P-inefficient cultivars. Iron chlorosis is probably caused not only by low Fe concentrations but also by high P-Fe ratios in the tissue (Kashirad and Marschner, 1974). For example, Walker and Sarkar (1979) found that maize P-Fe ratios above 17 are associated with Fe chlorosis while ratios below 4 are associated with Fe toxicity. In contrast to the effects found in solution culture, on P- and Fe-deficient calcareous soils, addition of fertilizer P can stimulate Fe uptake. In fact, P fertilization may be necessary before grain sorghum will respond to added Fe (Datin and Westerman, 1982).

Multiple interactions are also found between Fe and Cu. Copper and Zn can replace Fe on organic acid chelating agents, and Cu and Fe are antagonistic in some cases (Olsen, 1972). Copper deficiency reduces translocation of Fe from the roots to the shoots of grain sorghum (Clark et al., 1981), and causes Fe to accumulate in the nodes of maize stems (Brown and Jones,1975b). This can result in chlorosis due to Fe deficiency in the leaves. In contrast, toxic concentrations of Cu can increase Fe accumulation in maize roots and shoots (Dragun et al., 1976).

In acid soils containing large amounts of Mn, Fe-Mn interactions are sometimes found (Olsen, 1972; Somer and Shive, 1942). Iron deficiency symptoms can be caused by high Mn concentrations in the soil or in nutrient solutions. In these cases leaf Fe concentrations may be high even when Fe-deficiency symptoms are present, but foliar application of Fe^{2+} eliminates the symptoms (Evans, 1960; Humbert and Martin, 1955; Olsen, 1972).

Iron and Zn uptake and metabolism are interrelated (Lucas and Knezek, 1972; Olsen, 1972). High levels of Zn can induce Fe chlorosis in maize and millet even though Fe uptake and tissue concentrations are not affected (Brown and Tiffen, 1962; Watanabe et al., 1965). In contrast, plant Fe concentrations are often very high in Zn-deficient maize (Jackson et al., 1967, Warnock, 1970).

Correction of Iron Deficiency

Serious problems have been encountered in the application of inorganic Fe salts to calcareous and alkaline soils. In general, the rates of Fe needed to obtain significant plant response are prohibitive (Anderson, 1982; Mathers and Pennington, 1982). In addition, when the compounds are applied as bands, subsequent tillage disrupts the bands and decreases their residual effectiveness.

Iron supplied in fluid polyphosphate or as granular acidic urea phosphates can increase grain sorghum growth in pots. However, these results need field confirmation and economic evaluation (Mortvedt, 1982; Mortvedt and Giordano, 1971).

Direct soil application of Fe chelates such as FeEDTA, FeDTPA, and FeEDDHA can increase Fe availability if the chelating agent retains Fe in the soil. However, only EDDHA appears to have long-term stability above pH 8 (Lindsay and Schwab, 1982).

A promising method of applying $FeSO_4$ to the soil has recently been reported. Matocha and Pennington (1982) found that $FeSO_4$ applied to grain sorghum stubble or to weeds in grain sorghum fields is chelated by the plant material. When the residues are incorporated into calcareous soils, the Fe is available to the subsequent grain sorghum crop. These results are consistent with those of Patil and Patil (1981), who found that incorporating organic matter (without added Fe) into Fe-deficient soils causes increased Fe uptake and yields in grain sorghum.

Foliar applications of $FeSO_4$ and Fe chelates have long been used to correct Fe deficiencies in grain sorghum, sugarcane, and other crops. Solutions containing 6–12 g Fe (from $FeSO_4$)/100 liter H_2O are normally applied two or three times to grain sorghum. Numerous Fe chelates have also been used (Murphy and Walsh, 1972). Though these methods of alleviating Fe deficiency are undoubtedly effective, the Fe affects only those leaves that are wetted by the solution. Leaves that emerge subsequently are chlorotic, and reapplication is necessary. Limited work suggests that Fe chelates can also be applied with sprinkler irrigation water (Murphy and Walsh, 1972).

The best long-term solution to the problem of Fe deficiency in most soils is the development of Fe-efficient genotypes with high yield potential and good agronomic characteristics. This is particularly important in grain sorghum, one of the most susceptible major crops and one that is often grown on calcareous soils with low available Fe.

ZINC DEFICIENCY

Total soil Zn concentrations are normally between 10 and 300 g/Mg. Soils derived from basic igneous rocks are usually high in Zn while those derived from siliceous materials often have low Zn concentrations (Mengel and Kirkby, 1982). Even soils low in total Zn have much more total Zn than the critical concentration of DTPA-extractable Zn considered adequate for crop growth (0.8 g/Mg) (Boawn, 1971). Thus, (except in highly leached sandy soils and some Histosols) Zn deficiencies are not normally due to inadequate soil Zn. Rather, they are due to low Zn solubility resulting from high pH (Mortvedt, 1976).

Extractable Zn is negatively correlated with soil pH (Saeed and Fox, 1977; Stewart and Berger, 1965), and Wear (1956) found that 92% of the variation in plant Zn uptake is explained by soil pH. Lindsay (1979) points out that soluble soil Zn decreases 100-fold for each unit increase in pH, from almost 10^{-2} M at pH 4 to less than 10^{-13} M at pH 9. Thus, Zn deficiency is likely to occur on calcareous and alkaline soils, especially those in which land leveling or erosion has exposed high-pH subsoils low in organic matter (Grunes et al., 1961).

Some of the worst cases of Zn deficiency occur when soils low in soluble Zn are fertilized with large amounts of P and/or lime (Friesen et al., 1980; Takkar et al., 1976). Overliming highly weathered acid soils increases soil pH, thereby reducing Zn concentration in the soil solution. Fertilizer P usually has little effect on Zn absorption by the root system, but it inhibits Zn translocation to the leaves (Sharma et al., 1968; Stukenholtz et al., 1966). In maize, P fertilization causes Zn immobilization in root cell walls (Youngdahl et al., 1977) and in stem nodes (Dwivedi et al., 1975).

Despite the effects of fertilizer P on Zn translocation, Zn deficiency is readily overcome by low rates of fertilizer Zn. In addition, fertilizer Zn usually has good residual activity for several years (Viets and Lindsay, 1973).

Zinc Uptake and Translocation

Zinc absorption by plant tissues and its translocation from the roots to the shoots are active processes that are reduced by metabolic inhibitors and cold temperatures (Bowen, 1969; Moore, 1972). Cool soil temperatures reduce the utilization of soil Zn by maize (Ellis et al., 1964; Martin et al., 1965), primarily by inhibiting translocation to the shoot (Edwards and Kamprath, 1975). However, once translocation to the shoot has occurred, Zn can readily be translocated from older to younger tissues (Isarangkura et al., 1978; Singh and Steenberg, 1974).

In grasses with normal Zn concentrations, the highest concentrations are normally found in the apical meristem with decreasing concentrations in the elongating stems, leaf blades, and leaf sheaths. Younger leaves normally have higher Zn concentrations than older leaves, and leaf tips have higher concentrations than other parts (Clements, 1980; Jones, 1970; Leece, 1978b).

Critical tissue Zn concentrations and sufficiency ranges for maize, grain sorghum, and sugarcane are given in Table 16.2. In general, critical concentrations of index tissues decline with increasing age. Zinc deficiency is likely when whole shoot concentrations fall below 25–30 g/Mg for young plants and 15 g/Mg in plants at heading or tasseling.

Zinc deficiency is often associated with high shoot concentrations of P, Fe, Mn, Cu, Na, and B (Clark, 1978; Shukla and Mukhi, 1979, 1980). Part of this effect is due to concentration of these elements in the tissue as a result of slow

TABLE 16.2.   Critical Concentrations and Sufficiency Ranges for Zn in Several C₄ Grasses and Index Tissues

Tissue and Stage of Development	Critical Concentration or Range (g/Mg)	Reference
Maize		
Ear leaf at tasseling	15	Melsted et al. (1969)
	17.2	Beyers and Coetzer (1969)
Second leaf below ear at tasseling	15	Viets et al. (1953)
Second leaf below ear at silking	14.9	Grunes et al. (1961)
Ear leaf at silking	20–70	Jones (1967)
	50–150	Neubert et al. (1969)
Whole plant, 30–45 days after emergence	20–50	Lockman (1969) cited in Jones and Eck (1973)
Whole plant, 6 leaf stage	25	Isarangkura et al. (1978)
Grain Sorghum		
Whole plant, 23–39 days after planting	30–60	Lockman (1972c)
Youngest fully developed leaf, 37–56 days after planting	20–40	
Third leaf below head, 66–70 days after planting	15–30	
Third leaf below head, dough stage	7–16	Lockman (1972c)
Last fully extended leaves	15–40	Whitney (1970) cited in Jones and Eck (1973)
Sugarcane		
Sheaths 3, 4, 5, 6	9	Clements (1980)
	10	Marzola (1978)
Youngest collared leaf	15	
Apical meristem	150	

growth of Zn-deficient plants (Christiansen and Jackson, 1981; Murphy et al., 1981). However, in some cases total uptake and/or translocation of these nutrients, especially P, increases as a result of Zn deficiency (Khan and Zende, 1977; Safaya and Gupta, 1979). Because of the effects of Zn on uptake and translocation of other elements, conventional methods of plant analysis are sometimes inadequate for diagnosis of Zn deficiency. Ratios of index tissue nutrient concentrations such as P–Zn and Fe–Zn have been used as indicators of Zn deficiency. Takkar et al. (1976) concluded that maize will likely respond to fertilizer

Zn when the P–Zn ratio exceeds 150 in the grain, 90 in the stover, or 100 in the leaves.

Maize grown on alkaline soils of northwestern New South Wales, Australia, may have visual symptoms of Zn deficiency and respond to Zn fertilization even though leaf Zn concentrations, P–Zn ratios, and Fe–Zn ratios are normal (Leece, 1976, 1978a,b). Much of the Zn in such plants is apparently inactive due to chelation by organic ligands. Under such conditions, the "activity" of Zn can be estimated from the activity of Zn-requiring enzymes such as carbonic anhydrase (Bar-Akiva and Lavon, 1969; Gibson and Leece; 1981).

Like Fe deficency, Zn deficiency causes an adaptive response in plant roots. Kannan and Ramani (1982) found that the roots of Zn-deficient grain sorghum plants excrete H^+ and thereby reduce the pH of the nutrient solution. This excretion ceases when Zn is added to the nutrient solution.

Zinc Deficiency Symptoms

Zinc deficiency symptoms vary somewhat among C_4 grasses (Clements, 1980; Gascho and Taha, 1972; Smith, 1972b, 1973, 1974; Smith and Verschoyle, 1973). In general, mild Zn deficiency produces interveinal chlorosis similar to that caused by Mn and Fe deficiencies; however, the chlorosis occurs primarily on the lower half of the leaf. This contrasts with the symptoms of Mn and Fe deficiencies exhibited on the upper half and throughout the leaf, respectively.

In maize, more severe Zn deficiency produces a broad band of chlorotic tissue on each side of the midrib and on the lower half of the leaf. The midrib and the margin remain green. Internode elongation is also reduced. The somewhat condensed whorl of basally chorotic leaves was described by early workers as "whitebud." Zinc deficiency symptoms of grain sorghum are similar, except that less interveinal striping occurs, and more definite white bands are formed.

In sugarcane, the stripes may be wide and light green in color (Clements, 1980), or they may be light green or whitish lines above the major veins (Evans, 1960). Severe deficiency may produce severely chlorotic young tillers, leaf death from the top down, deformed spindle leaves with ragged edges and distinct piping of internode pith tissue (Clements, 1980; Evans, 1960).

Somewhat similar symptoms are found in Zn-deficient green panic (*Panicum maximum* var. *trichoglume*), where prominent interveinal chlorosis develops on the central portion of the leaf blade. Light brown spots 2–5 mm in diameter may occur in the strips. It is accompanied by generally pale color, necrosis of the leaf tips and margins, and nocturnal production of a light brown exudate on the leaf margins. Rhodesgrass (*Chloris gayana*) produces a similar nocturnal exudate. However, the most characteristic symptom of Zn deficiency in rhodesgrass, dallisgrass (*Paspalum dilatatum*), and bufflegrass (*Cenchrus ciliaris*) is the production of small brown necrotic spots (0.5–2.0 mm wide and 2–20 mm long) on the leaves. These are usually most numerous around the leaf margins and tips. The necrotic spots often coalesce to form necrotic patches surrounded by a characteristic yellow "halo" in buffelgrass.

Zinc deficiency also causes abnormally shaped leaf blades due to unequal lateral expansion along the length of the blade, leading to alternating wide and narrow sections.

Even when moderate, Zn deficiency causes dramatic reductions in growth and yield of annual crops like maize and grain sorghum. It also can severely delay their phenological development. However, as in the case of P deficiency, perennial crops such as sugarcane often recover from Zn deficiency as their root systems expand.

Genotype Effects

All grasses that have been carefully studied exhibit genotypic variability for tolerance to Zn deficiency. This variability has been found in maize (Clark, 1978; Kannan, 1983; Safaya and Gupta, 1979), grain sorghum (Kannan and Ramani, 1982; Williams et al., 1982), and wheat (Shukla and Raj, 1974).

Both maize and grain sorghum exhibit heterosis for tolerance of Zn deficiency, that is, hybrids tend to be less sensitive to low Zn availability than inbreds (Kannan, 1983; Kannan and Ramani, 1982).

Halim et al. (1968) found that the degree and pattern of Zn deficiency vary among maize genotypes. For example, some inbreds have early resistance but become susceptible later. Others are susceptible early but gain resistance later. Inbreds with early resistance are often obtained from seeds with high Zn concentrations, suggesting that they were able to utilize sufficient Zn from the seed to avoid Zn deficiency symptoms. Resistance is also correlated with root development. Resistant genotypes have healthy roots, while susceptible genotypes have reduced root development and decayed root tips.

Some studies suggest that genotypic differences in resistance to Zn deficiency are correlated with the ability of the genotype to absorb Zn from the soil (Shukla and Raj, 1974, 1976). Others suggest that efficient utilization of Zn that is absorbed is more important than the total amount of Zn absorbed (Halim et al., 1968). Clark (1978) found that a "Zn-efficient" maize inbred absorbed more Zn, translocated more Zn to the shoot, and produced more dry matter per unit Zn absorbed than an inefficient inbred. Thus, both uptake and utilization may be important factors in genotypic tolerance of Zn deficiency.

MANGANESE DEFICIENCY

Soils generally contain 20–3000 g Mn/Mg soil with an average of about 600 g/Mg. If it were all dissolved, the solution concentration would be about 0.1 M at 10% soil water content (Lindsay, 1979). Mn^{2+} is the most important form of Mn in the soil solution and is the form absorbed by plants. It is in equilibrium with Mn oxides in which Mn is in trivalent and tetravalent forms. Redox reactions govern the concentrations of Mn^{2+}; therefore, as with Fe, redox potential and pH are important determinants of Mn availability to plants. As discussed in

Chapter 14, Mn toxicity can occur in very acid and in poorly drained soils. However, Mn deficiency can occur in well-drained, high-pH soils (including calcareous soils), in organic soils with low native Mn contents, and in leached acid soils that have been overlimed. Viets and Lindsay (1973) report that DTPA-extractable Mn is a good indicator of Mn availability in calcareous and alkaline soils, with a critical soil concentration of about 1 g/Mg. In high-pH soils, banded $MnSO_4$ is a good source of Mn, but it has little residual activity. Foliar application of Mn chelates can also be used to correct Mn deficiency.

Manganese Uptake and Translocation

Uptake of Mn^{2+} by both grass roots and leaf tissue is active (Bowen, 1969, 1981b). As with Cu^{2+} and Zn^{2+}, two concentration-dependent phases of uptake occur in barley roots and sugarcane leaf tissue. The transition from one phase to the other occurs at 9.6×10^{-15} M in barley roots and 7.5×10^{-15} M in sugarcane leaf tissue (Bowen, 1981b).

As in the case for Fe, plant roots are able to excrete reducing agents and reduce the pH of the rhizoplane to increase the concentration of Mn^{2+} at the root surface (Uren, 1981); however, the importance of this mechanism has not been studied in detail.

The normal Mn concentration of the whole grain sorghum plant decreases by more than 50% between the seedling stage and maturity, primarily due to the increase in the structural material in the stem (Jacques et al., 1975b). When Mn is deficient, sugarcane plant parts differ little in Mn concentration, ranging from 2 g/Mg in stalk tissue to 4–6 g/Mg in leaf blades. This indicates that Mn is readily translocated from older to younger tissues. However, when Mn is at normal or high levels, it tends to accumulate at the end of the transpiration path. It is higher in old leaves than in young leaves, and it is higher in leaf tips than in leaf bases (Clements, 1980).

Table 16.3 gives critical concentrations and sufficiency ranges of Mn concentrations for several C_4 grasses. Though critical leaf concentrations range from about 6–20 g/Mg, severely Mn-deficient sugarcane leaves have concentrations as low as 2–3 g/Mg (Clements, 1980; Evans, 1960).

Manganese Deficiency Symptoms

In most C_4 grasses the symptoms of Mn deficiency are somewhat indistinct. Interveinal chlorosis is found in maize, grain sorghum, and sugarcane. It is more severe on the apical half of the blade, a characteristic differentiating if from the whole-leaf interveinal chlorosis caused by Fe deficiency (Evans, 1960; Krantz and Melsted, 1964).

In sugarcane and grain sorghum, severe Mn deficiency causes red freckles or long, uniform necrotic lesions surrounded by dark red areas. Old leaves are more severely affected than young leaves, indicating that Mn is relatively mobile.

TABLE 16.3.   Critical Concentrations and Sufficiency Ranges for Mn in Selected C$_4$ Grasses

Tissue and Stage of Development	Critical Concentration or Sufficiency Range (g/Mg)	Reference
	Maize	
Ear leaf at tasseling	15	Melsted et al. (1969)
Whole plant, 3 to 4 leaves	50–160	Jones and Eck (1973)
Ear leaf at silking	20–250	
Stalk above ear at silking	20–70	
	50–100	
Grain at maturity	5–15	
Ear leaf at silking	20–150	Jones (1967)
	34–200	Neubert et al. (1969)
	Grain Sorghum	
Whole plant 23–39 days after planting	40–150	Lockman (1972c)
Youngest fully developed leaf, 37–56 days after planting	6–100	
Third leaf below head at bloom stage	8–190	
Third leaf below head at dough stage	8–40	
	Sugarcane	
Blades 3, 4, 5, 6	20–400	Schmehl and Humbert (1964)

Roots of Mn-deficient plants are smaller, darker in color, and slightly brittle compared to normal roots (Clark et al., 1981; Clements, 1980).

In dallisgrass (*Paspalum dilatatum*), rhodesgrass (*Chloris gayana*), buffelgrass (*Cenchrus ciliaris*), and green panic (*Panicum maximum* var. *trichoglume*) the leaves become pale green and the younger leaves and tillers are weak. Young leaves may remain folded or rolled. The apical meristem may die, and, as a result, tillering may increase.

COPPER DEFICIENCY

Copper deficiency is quite rare in C$_4$ grasses, occurring infrequently on calcareous or overlimed soils, in highly leached sandy soils, and in organic soils. The Cu content of soils generally ranges from 2 to 100 g/Mg, slightly less than that of Zn. However, like Zn^{2+} and Mn^{2+}, soil Cu^{2+} is quite insoluble at high soil pH. Lindsay (1979) shows that soluble Cu^{2+} decreases 100-fold for each unit increase in soil pH, from about 10^{-6} M in a typical soil at pH 4.4 to about 10^{-14} M at pH 8.4. However, unlike Zn and Mn, the low solution Cu^{2+} concentration at high soil pH is due to Cu adsorption on soil particles rather than formation of insoluble salts (Lindsay, 1979; Mengel and Kirkby, 1982).

Copper deficiency also occurs in acid, highly leached sandy soils and in organic soils. In acid, sandy soils Cu^{2+} is relatively soluble and can be leached below the root zone. Since Cu^{2+} is also strongly adsorbed by organic matter, it can also be leached from soils in soluble organic matter complexes (Hodgson et al., 1966). This leaching and the strong adsorption of Cu to organic matter may explain the low availability of Cu to sugarcane and other C_4 grasses grown on Florida Histosols and other soils rich in organic matter (Allison et al., 1927; Anderson, 1956).

Copper Uptake and Translocation

Copper uptake appears to be an active process in grasses and other plants (Bowen, 1969, 1981b; Bowen and Nissen, 1976). Bowen (1981b) reports that, like Zn^{2+} and Mn^{2+}, Cu^{2+} uptake is characterized by two concentration-dependent phases operating in different concentration ranges. Cu^{2+} and Zn^{2+} compete for the same uptake mechanism, which may explain the frequently cited negative correlation between Cu and Zn uptake by plants (Clark, 1978; Leece, 1978a; Safaya and Gupta, 1979). In fact, Clements (1980) reports that application of Zn to Zn-deficient soil can induce Cu and Mn deficiencies in sugarcane. Similarly, P–Cu interactions, which are similar to P–Zn interactions, have been described (Murphy et al., 1981), suggesting that high P concentrations can inhibit Cu uptake and/or translocation to the shoot.

Like Zn and Mn, Cu concentrations of leaf blades and sheaths decrease slightly as the plant ages (Jacques et al., 1975b); however, the relative difference in Cu concentration between young and old leaves and between leaf tips and bases is less than the difference in Zn, Mn, or B concentrations (Jones, 1970). Thus, Cu may be relatively less mobile in the plant than other micronutrients, perhaps due to its tendency to form strong complexes with organic compounds in the plant.

Table 16.4 gives critical concentrations and sufficiency ranges for several C_4 grasses.

Copper Deficiency Symptoms

In Cu-deficient grasses the most prominent symptoms are drooping rolled leaves, described as "droopy top" in sugarcane. In maize this is accompanied by

TABLE 16.4.   Critical Concentrations and Sufficiency Ranges for Cu in Selected C$_4$ Grasses

TISSUE AND STAGE OF DEVELOPMENT	CRITICAL CONCENTRATION OR SUFFICIENCY RANGE (g/Mg)	REFERENCE
Maize		
Ear leaf at tasseling	5	Melsted et al. (1969)
Whole plant, 3 to 4 leaves	7–20	Jones and Eck (1973)
Ear leaf at silking	3–15	
Stalk above ear at silking	3–15	
	3–10	
Grain at maturity	1–5	
Ear leaf at silking	6–20	Jones (1967)
	8–20	Neubert et al. (1969)
Grain Sorghum		
Whole plant 23–39 days after planting	8–15	Lockman (1972c)
Youngest fully developed leaf, 37–56 days after planting	2–15	
Third leaf below head at bloom stage	2–7	
Third leaf below head at dough stage	1–3	
Sugarcane		
Blades 3, 4, 5, 6	5	Schmehl and Humbert (1964)
Sheaths 3, 4, 5, 6	2	Clements (1980)

chlorosis of young leaves, light green streaks on the leaves similar to those in Fe-deficient plants, and (in the case of severe deficiency) senescent leaf margins similar to symptoms of K deficiency (Mohr and Dickinson, 1979). In sugarcane, "droopy top" is accompanied by slow unfurling of the spindle leaves, unusually wide and soft leaves, chlorotic stripes between the major vascular bundles, and reduced growth (Anderson, 1956; Evans, 1960). In both maize and sugarcane the stalks are soft and rubbery. In narrow-leafed C$_4$ forage grasses, young leaves are chlorotic, never unroll, roll inward, or bend backward forming charcteristic "hooks." In all cases the apparent symptoms of water stress are associated with adequate soil and plant water contents.

OTHER MICRONUTRIENT DEFICIENCIES

Other micronutrient deficiencies such as B, Mo, and Cl can be produced by growing C_4 grasses in nutrient solutions deficient in these elements. These micronutrients and their effects on C_4 grass growth are discussed by Clements (1980), Evans (1960), Gupta (1979), Jones and Eck (1973), Martin (1934), and others. However, these deficiencies occur so rarely in C_4 grasses that they are primarily of academic interest. Boron and Mo deficiencies have much more practical importance in N-fixing legumes and vegetables, many of which have high requirements for these elements.

REFERENCES

Abbe, E. C., L. F. Randolf, and J. Einset. 1941. The developmental relationship between shoot apex and growth pattern of leaf blade in diploid maize. *Am. J. Bot.* **28**:778-784.

Abbott, L. K. and A. D. Robson. 1982. The role of vesicular arbuscular mycorrhizal fungi in agriculture and the selection of fungi for inoculation. *Aust. J. Agric. Res.* **33**:389-408.

Abdul-Kadir, S. M. and G. M. Paulsen. 1982. Effect of salinity on nitrogen metabolism in wheat. *J. Plant Nutr.* **5**:1141-1151.

Abdullahi, A. and R. L. Vanderlip. 1972. Relationships of vigor tests and seed source and size to sorghum seedling establishment. *Agron. J.* **64**:143-144.

Abruna, F., R. Perez-Escolar, J. Vicente-Chandler, R. W. Pearson, and S. Silva. 1974. Response of corn to acidity factors in eight tropical soils. *J. Agric. Univ. P. R.* **58**:59-77.

Abruna, F., J. Rodriguez, and S. Silva. 1982. Crop response to soil acidity factors in Ultisols and Oxisols in Puerto Rico. VI. grain sorghum. *J. Agric. Univ. of Puerto Rico.* **61**:28-38.

Abruna-Rodriguez, F. and J. Vicente-Chandler. 1967. Sugarcane yields as related to acidity of a humid tropic Ultisol. *Agron. J.* **59**:330-331.

Acevedo, E., E. Fereres, T. C. Hsiao, and D. W. Henderson. 1979. Diurnal growth trends, water potential, and osmotic adjustment of maize and sorghum leaves in the field. *Plant Physiol.* **64**:476-480.

Acevedo, E., T. C. Hsiao, and D. W. Henderson. 1971. Immediate and subsequent growth responses of maize leaves to changes in water status. *Plant Physiol.* **48**:631-636.

Ackerson, R. C. and D. R. Krieg. 1977. Stomatal and nonstomatal regulation of water use in cotton, corn, and sorghum. *Plant Physiol.* **60**:850-853.

Ackerson, R. C., D. R. Krieg, and F. J. M. Sung. 1980. Leaf conductance and osmoregulation of field-grown sorghum genotypes. *Crop Sci.* **20**:10-14.

Adams, F. 1981. Alleviating chemical toxicities: liming acid soils. In G. F. Arkin and H. M. Taylor (eds.), *Modifying the Root Environment to Reduce Crop Stress*, pp. 269-301. American Society of Agricultural Engineers, St. Joseph, MI.

Adams, F. and B. L. Moore. 1983. Chemical factors affecting root growth in subsoil horizons of Coastal Plains soils. *Soil Sci. Soc. Am. J.* **47**:99-102.

Adams, F. and R. W. Pearson. 1969. Neutralizing soil acidity under bermudagrass sod. *Soil Sci. Soc. Am. Proc.* **33**:737-742.

Adams, F. and R. W. Pearson. 1967. Crop response to lime in the Southern United States and Puerto Rico. In R. W. Pearson and F. Adams (eds.), *Soil Acidity and Liming*, pp. 161-202. Agronomy 12. American Society of Agronomy, Madison, WI.

Adams, W. E., M. Stelly, H. D. Morris, and C. B. Elkins. 1967a. A comparison of coastal and common bermudagrasses [*Cynodon dactylon* (L.) Pers.] in the Piedmont Region. II. Effect of fertilization and crimson clover (*Trifolium incarnatum*) on nitrogen, phosphorus, and potassium contents of the forage. *Agron. J.* **59**:281–284.

Adams, W. E. and M. Twersky. 1960. Effect of soil fertility on winter killing of coastal bermudagrass. *Agron. J.* **52**:325–326.

Adams, W. E., A. W. White, Jr., and R. N. Dawson. 1967b. Influence of lime sources and rates on "Coastal" bermudagrass production, soil profile reaction, exchangeable Ca and Mg. *Agron. J.* **59**:147–149.

Adegbola, A. 1966. Preliminary observations on the reserve carbohydrate and regrowth potential of tropical grasses. Proceedings of the 10th International Grassland Congress, Helsinki, Finland, pp. 933–936.

Addicott, F. T. 1969. Ageing, senescence and abscision in plants, phytogerontology. *Hort. Sci.* **4**:14–16.

Afuakwa, J. J. and R. K. Crookston. 1984. Using the kernel milk line to visually monitor grain maturity in maize. *Crop Sci.* **24**:687–691.

Afuakwa, J. J., R. K. Crookston, and R. J. Jones. 1984. Effect of temperature and sucrose availability on kernel black layer development in maize. *Crop Sci.* **24**:285–288.

Ahmad, I. 1978. Some aspects of salt tolerance in *Agrostis stolonifera*. Ph.D. dissertation, University of Wales, Swansea.

Aho, N. 1980. Effet de la secheresse atmospherique sur la photosynthese nette de deux especes de type C₄: le sorgho et le mais. *C. R. Acad. Sc. Paris, Serie D.* **290**:543–546.

Ahring, R. M. and G. W. Todd. 1977. The bur enclosure of the caryopses of buffalograss as a factor affecting germination. *Agron. J.* **69**:15–17.

Aitken, Y. 1962. Shoot apex accessibility and pasture management. *J. Aust. Inst. Agric. Sci.* **28**:50–52.

Akin, D. E., J. R. Wilson, and W. R. Windham. 1983. Site and rate of tissue digestion in leaves of C_3, C_4, and C_3/C_4 intermediate *Panicum* species. *Crop Sci.* **23**:147–155.

Akita, S. and D. N. Moss. 1973. Photosynthetic responses to CO_2 and light by maize and wheat leaves adjusted for constant stomatal apertures. *Crop Sci.* **13**:234–237.

Albrecht, S. L., Y. Okon, and R. H. Burris. 1977. Effects of light and temperature on the association between *Zea mays* and *Spirillum lipoferum*. *Plant Physiol.* **60**:528–531.

Albrecht, S. L., Y. Okon, J. Lonnquist, and R. H. Burris. 1981. Nitrogen fixation by corn—*Azospirillum* associations in a temperate climate. *Crop Sci.* **21**:301–306.

Aldrich, S. R. 1943. Maturity measurements in corn and an indication that grain development continues after premature cutting. *J. Am. Soc. Agron.* **35**:667–680.

Alexander, D. E. and R. G. Creech. 1977. Breeding special industrial and nutritional types. In G. F. Sprague (ed.), *Corn and Corn Improvement*, pp. 363–390. American Society of Agronomy, Madison, WI.

Allen, L. H., Jr. 1971. Variations in carbon dioxide concentration over an agricultural field. *Agric. Meteorol.* **8**:5–24.

Allen, M. F., T. S. Moore, Jr., and M. Christensen. 1980. Phytohormone changes in *Bouteloua gracilis* infected by vesicular-arbuscular mycorrhizae: 1. Cytokinin increases in the host plant. *Can. J. Bot.* **58**:371–374.

Allen, M. F., J. C. Sexton, T. S. Moore, Jr., and M. Christensen. 1981a. Influence of phosphate source on vesicular arbuscular mycorrhizae of *Bouteloua gracilis*. *New Phytol.* **87**:687–694.

Allen, M. F., J. C. Sexton, T. S. Moore, Jr., and M. Christensen. 1981b. Comparative water relations and photosynthesis of mycorrhizal and non-mycorrhizal *Bouteloua gracilis* H.B.K. Lag ex Steud. *New Phytol.* **88**:683–693.

Allen, R. 1980. The impact of CO_2 on world climate. *Environment* **22**:6–13, 37–38.

Allison, J. C. S. and D. J. Watson. 1966. The production and distribution of dry matter in maize after flowering. *Ann. Bot. N. S.* **30**:365–381.

Allison, J. C. S., J. H. Wilson, and J. H. Williams. 1975a. Effect of defoliation after flowering on changes in stem and grain mass of closely and widely spaced maize. *Rhod. J. Agric. Res.* **13**:145–147.

Allison, J. C. S., J. H. Wilson, and J. H. Williams. 1975b. Effect of partial defoliation during the vegetative phase on subsequent growth and grain yield of maize. *Ann. Appl. Biol.* **81**:367–375.

Allison, L. E. 1964. Salinity in relation to irrigation. *Adv. Agron.* **16**:139–180.

Allison, R. W., O. C. Bryan, and J. H. Hunter. 1927. The stimulation of plant response on the raw peat soils of the Florida Everglades through the use of copper sulfate and other chemicals. *Fla Agr. Exp. Stn. Bull.* **190**.

Allmaras, R. R., W. W. Nelson, and W. B. Voorhees. 1975. Soybean and corn rooting in southwestern Minnesota: II. Root distributions and related water inflow. *Soil Sci. Soc. Am. Proc.* **39**:771–777.

Allred, K. W. 1982. Describing the grass inflorescence. *J. Range Manage.* **35**:672–675.

Alofe, C. O. and L. E. Schrader. 1975. Photosynthate translocation in tillered *Zea mays* following $^{14}CO_2$ assimilation. *Can. J. Plant Sci.* **55**:407–414.

Alston, A. M. 1980. Response of wheat to deep placement of nitrogen and phosphorus fertilizers on a soil high in phosphorus in the surface layer. *Aust. J. Agric. Res.* **31**:13–24.

Anderson, E. R. 1972a. Flooding tolerance of *Panicum coloratum*. *Queensl. J. Agric. Anim. Sci.* **29**:173–179.

Anderson, E. R. 1974. The reaction of seven *Cenchrus ciliaris* L. cultivars to flooding. *Trop. Grassl.* **8**:33–40.

Anderson, G. 1980. Assessing organic phosphorus in soils. In F. E. Khasawach, E. C. Sample, and E. J. Kamprath (eds.), pp. 411–432. *The Role of Phosphorus in Agriculture.* Am. Soc. Agron., Crop Sci. Soc. Am., Soil Sci. Soc. Am., Madison.

Anderson, J. 1956. Droopy top in cane. *Cane Grow. Q. Bull.* **19**:146–148.

Anderson, W. B. 1982. Diagnosis and correlation of iron deficiency in field crops—an overview. *J. Plant Nutr.* **5**:785–795.

Anderson, W. P. 1972b. Ion transport in the cells of higher plant tissues. *Ann. Rev. Plant Physiol.* **23**:51–72.

Anderson, W. P. 1976. Transport through roots. In U. Luettge and M. J. Pitman (eds.), *Encyclopedia of Plant Physiology, Vol. 2, Transport in Plants II. Part B. Tissues and Organs,* pp. 129–156. Springer-Verlag, New York.

Anderson, W. P. and C. R. House, 1967. A correlation between structure and function in the root of *Zea mays. J. Exp. Bot.* **18**:544–555.

Andrade, S. O., L. Retz, and O. Marmo. 1971. Estudos sobre *Brachiaria* sp. (Tanner grass) III. Ocorrencias de intoxicacoes de bovinos durante um ano (1970–1971) e niveis de nitrato em amostras de graminea. *Arq. Inst. Biol., Sao Paulo* **38**:239–252.

Andre, M., J. Massimino, A. Daguenet, D. Massimino, and J. Thiery. 1982. The effect of a day at low irradiance of a maize crop. II. Photosynthesis, transpiration, and respiration. *Physiol. Plant.* **54**:283–288.

Andreis, H., Jr. 1976. A water table study on an Everglades peat soil: Effects on sugarcane and on soil subsidence. *Sugar J.* **39**:(6):8–11.

Andrew, C. S. and M. F. Robins. 1971. The effect of phosphorus on the growth, chemical composition, and critical phosphorus percentages of some tropical pasture grasses. *Aust. J. Agric. Res.* **22**:693–703.

Anghinoni, I. and S. A. Barber. 1980a. Phosphorus influx and growth characteristics of corn roots as influenced by phosphorus supply. *Agron. J.* **72**:685–688.

Anghinoni, I. and S. A. Barber. 1980b. Predicting the most efficient phosphorus placement for corn. *Soil Sci. Soc. Am. J.* **44**:1016–1020.

Angus, J. F., R. B. Cunningham, M. W. Moncur, and D. H. Mackenzie. 1981. Phasic development in field crops. I. Thermal response in the seedling phase. *Field Crops Res.* **3**::365–378.

Angus, J. F. and M. W. Moncur. 1977. Water stress and phenology in wheat. *Aust. J. Agric. Res.* **28**:117–181.

Aparacio-Tejo, P. M. and J. S. Boyer. 1983. Significance of accelerated leaf senescence at low water potentials for water loss and grain yield in maize. *Crop Sci.* **23**:1198–1202.

Arad, S. and A. E. Richmond. 1976. Leaf cell water and enzyme activity. *Plant Physiol.* **57**:656–658.

Arisz, W. H. 1956. Significance of the symplasm theory for transport across the root. *Protoplasma* **46**:5–62.

Ariyanayagam, R. P., C. L. Moore, and V. R. Carangal. 1974. Selection for leaf angle in maize and its effects on grain yield and other characters. *Crop Sci.* **14**:551–556.

Arkel, H. van. 1980. The forage and grain yield of cold-tolerant sorghum and maize as affected by time of planting in the highlands of Kenya. *Neth. J. Agric. Sci.* **28**:63–77.

Arnon, I. 1975. Mineral nutrition of maize. International Potash Institute. Berne, Switzerland.

Artschwager, E. 1925. Anatomy of the vegetative organs of sugarcane. *J. Agric. Res.* **30**:197–242.

Artschwager, E. 1930. A comparative study of the stem epidermis of certain sugarcane varieties. *J. Agric. Res.* **41**:853–865.

Artschwager, E. 1940. Morphology of the vegetative organs of the sugarcane. *J. Agric. Res.* **60**:503–549.

Ashton, F. M. 1956. Effects of a series of cycles of alternating low and high soil water contents on the rate of apparent photosynthesis in sugar cane. *Plant Physiol.* **31**:266–274.

Avdulov, N. P. 1931. Karyo-systematische Untersuchungen der Familie Gramineen. (Russian with German summary). *Bull. Appl. Bot. Suppl.*, **44.**

Ayres, A. 1936. Factors influencing the mineral composition of sugar cane. *Rep. Assoc. Hawaii. Sugar Technol.* **15**:29–41.

Ayres, A. 1937. Absorption of mineral nutrients by sugar cane at successive stages of growth. *Hawaii. Plant. Rec.* **41**:335–351.

Ayres, A. S. 1960. The growth of sugar cane as influenced by phosphorus. The critical range of soil phosphorus. *Int. Soc. Sugar Cane Technol. Proc.* **10**:462–467.

Ayres, A. S. 1966. Calcium silicate slag as a growth stimulant for sugarcane on low silicon soils. *Soil Sci.* **101**:216–227.

Ayres, A. S. and C. K. Fujimoto. 1944. The vertical distribution of available (exchangeable) potassium in Oahu soils. *Hawaii. Plant. Rec.* **48**:249–269.

Azcon, R. and J. A. Ocampo. 1981. Factors affecting the vesicular-arbuscular infection and mycorrhizal dependency of thirteen wheat cultivars. *New Phytol.* **87**:677–685.

Babalola, O. and R. Lal. 1977a. Subsoil gravel horizon and maize root growth: 1. Gravel concentrations and bulk density effects. *Plant Soil* **46**:337–346.

Babalola, O. and R. Lal. 1977b. Subsoil gravel horizon and maize root growth: 2. Effects of gravel size, inter-gravel texture and natural gravel horizon. *Plant Soil* **46**:347–357.

Badu-Apraku, B., R. B. Hunter, and M. Tollenaar. 1983. Effect of temperature during grain filling on whole plant and grain yield in maize (*Zea mays* L.). *Can. J. Plant Sci.* **63**:357–363.

Bagnall, D. 1979. Low temperature responses of three *Sorgum* species. In J. M. Lyons, D. Graham, and J. K. Raison (eds.), *Low Temperature Stress in Crop Plants. The Role of the Membrane*, pp. 67–80. Academic Press. New York.

Baker, D. E., B. R. Bradford, and W. I. Thomas. 1966. Leaf analysis for corn—tool for predicting soil fertility needs. *Better Crops with Plant Food* 50:36-40.

Baker, D. E., A. E. Jarrell, L. E. Marshall, and W. I. Thomas. 1970. Phosphorus uptake from soils by corn hybrids selected for high and low phosphorus accumulation. *Agron. J.* 62:103-106.

Balasko, J. A. and D. Smith. 1971. Influence of temperature and nitrogen fertilization on the growth and composition of switchgrass (*Panicum virgatum* L.) and timothy (*Phleum pratense* L.) at anthesis. *Agron. J.* 63:853-857.

Baldani, J. I., P. A. A. Pereira, R. E. M. da Rocha, and J. Dobereiner. 1981. Host plant specificity in the infection of C_3 and C_4 plants by *Azospirillum* spp. *Pesq. Agropec. Bras.* 16:325-330.

Balko, L. G. and W. A. Russell. 1980a. Response of maize inbred lines to nitrogen fertilizer. *Agron. J.* 72:723-728.

Balko, L. G. and W. A. Russell. 1980b. Effects of rates of nitrogen fertilizer on maize inbred lines and hybrid progeny. I. Prediction of yield response. *Maydica* 25:67-79.

Banath, C. L. and N. H. Monteith. 1966. Soil oxygen deficiency and sugar cane root growth. *Plant Soil* 25:143-149.

Bandurski, R. S., A. Schulze, P. Dayanandan, and P. B. Kaufman. 1984. Response to gravity by *Zea mays* seedlings. I. Time course of the response. *Plant Physiol.* 74:284-288.

Banks, P. A. 1981. Weed control. In R. R. Duncan (ed.), Ratoon cropping of sorghum for grain in the southeastern United States, pp. 21-25. University of Georgia, Coll. Agric. Exp. Stn., Res. Bull. 269, Nov. 1981.

Bar-Akiva, A. and R. Lavon. 1969. Carbonic anhydrase activity as an indicator of zinc deficiency in citrus leaves. *J. Hort. Sci.* 44:359-362.

Baran, R., D. Bassereau, and N. Gillet. 1974. Measurement of available water and root development on an irrigated sugar-cane crop in the Ivory Coast. *Proc. ISSCT* 15:726-735.

Barber, D. A. and M. G. T. Shone. 1966. The absorption of silica from aqueous solutions by plants. *J. Exp. Bot.* 17:569-578.

Barber, L. E., S. A. Russell, and H. J. Evans. 1979. Inoculation of millet with *Azospirillum. Plant Soil* 52:49-57.

Barber, L. E., J. D. Tjepkema, S. A. Russel, and H. J. Evans. 1976. Acetylene reduction (nitrogen fixation) associated with corn inoculated with *Spirillum. Appl. Environ. Microbiol.* 32:108-113.

Barber, S. A. 1962. Diffusion and mass-flow concept of soil nutrient availability. *Soil Sci.* 93:39-49.

Barber, S. A. 1971. Effect of tillage practice on corn (*Zea mays* L.) root distribution and morphology. *Agron. J.* 63:724-726.

Barber, S. A., J. M. Walker, and E. H. Vasey. 1963. Mechanisms for the movement of plant nutrients from the soil and fertilizer to the plant root. *J. Agric. Food Chem.* 11:204-207.

Barea, J. M. and M. E. Brown. 1974. Effects on plant growth produced by *Azotobacter paspali* related to synthesis of plant growth regulating substances. *J. Appl. Bacteriol.* 40:583-593.

Barger, G. L. 1969. Total growing degree days. *Weekly Weather Crop Bull.* 56:(18, May 5):10.

Barker, A. V. and R. Bradfield. 1963. Effect of potassium and nitrogen on the free amino acid content of corn plants. *Agron. J.* 55:465-470.

Barley, K. P. 1962. The effects of mechanical stress on the growth of roots. *J. Exp. Bot.* 13:95-110.

Barley, K. P. 1963. Influence of soil strength on growth of roots. *Soil Sci.* 96:175-180.

Barley, K. P. 1970. The configuration of the root system in relation to nutrient uptake. *Adv. Agron.* 22:159-201.

Barlow, E. W. R. and L. Boersma. 1976. Interaction between leaf elongation, photosynthesis, and carbohydrate levels of water-stressed corn seedlings. *Agron. J.* 68:923-926.

Barlow, E. W. R., L. Boersma, and J. L. Young. 1976a. Root temperature and soil water potential effects on growth and soluble carbohydrate concentration of corn seedlings. *Crop. Sci.* **16**:59-62.

Barlow, E. W. R., T. M. Ching, and L. Boersma. 1976b. Leaf growth in relation to ATP levels in water stressed corn plants. *Crop. Sci.* **16**:405-407.

Barlow, E. W. R., L. Boersma, and J. L. Young. 1977. Photosynthesis, transpiration, and leaf elongation in corn seedlings at suboptimal temperatures. *Agron. J.* **69**:95-100.

Barnard, C. 1964. Form and structure. In C. Barnard (ed.), *Grasses and Grasslands*, pp. 47-72. MacMillan, London.

Barneix, A. J. and P. A. Arnozis. 1980. Effect of nitrate and ammonium on phosphate absorption by wheat seedlings. *Oyton* **39**:7-13.

Barnes, D. L. and D. G. Woolley. 1969. Effect of moisture stress at different stages of growth. I. Comparison of a single-eared and a two-eared corn hybrid. *Agron. J.* **61**:788-790.

Barnett, K. H. and R. B. Pearce. 1983. Source-sink ratio alteration and its effect on physiological parameters in maize. *Crop. Sci.* **23**:294-299.

Barnett, N. M. and A. W. Naylor. 1966. Amino acid and protein metabolism in Bermuda grass during water stress. *Plant Physiol.* **41**:1222-1230.

Bartholomew, P. E. and P. de V. Booysen. 1969. The influence of clipping frequency on reserve carbohydrates and regrowth of *Eragrostis curvula*. *Proc. Grassl. Soc. S. Afr.* **4**:35-43.

Barton, H., W. G. McCully, H. M. Taylor, and J. E. Box, Jr. 1966. Influence of soil compaction on emergence and first-year growth of seeded grasses. *J. Range Manage.* **19**:118-121.

Baser, D. K., I. K. Jaggi, and S. B. Sinha. 1981. Root and shoot growth and transpiration from maize plant as influenced by various depths of water table. *Z. Acker Pflanzenb.* **150**:390-399.

Bashaw, E. C. 1975. Problems and possibilities of apomixis in the improvement of tropical forage grasses. In E. C. Doll and G. O. Mott (eds.), *Tropical Forages in Livestock Production Systems*, pp. 23-30. ASA Special Pub. No. 24, American Society of Agronomy, Crop Sci. Soc. Am., Soil Sci. Soc. Am., Madison, WI.

Bashaw, E. C. 1980. Apomixis and its application to crop improvement. In W. R. Fehr and H. H. Hadley (eds.), *Hybridization of Crop Plants*, pp. 45-63. American Society of Agronomy and Crop Science Society of America, Madison, WI.

Bassham, J. A. 1964. Kinetic studies of the photosynthetic carbon reduction cycle. *Annu. Rev. Plant Physiol.* **15**:101-120.

Bassham, J. A. 1977. Increasing crop production through more controlled photosynthesis. *Science* **197**:630-638.

Baszynski, T., J. Brand, R. Barr, D. W. Krogmann, and F. L. Crane. 1972. Some biochemical characteristics of chloroplasts from mineral-deficient maize. *Plant Physiol.* **50**:410-411.

Baszynski, T., B. Pancyzyk, M. Krol, and Z. Krupa. 1975. The effect of nitrogen deficiency on some aspects of photosynthesis in maize leaves. *Z. Pflanzenphysiol.* **74S**:200-207.

Baver, L. D. 1960. Plant and soil composition relationships as applied to cane fertilization. *Hawaii. Plant. Rec.* **56**:1-86.

Beadle, C. L., K. R. Stevenson, H. H. Neumann, C. W. Thurtell, and K. M. King. 1973. Diffusive resistance, transpiration and photosynthesis in single leaves of corn and sorghum in relation to leaf water potential. *Can. J. Plant Sci.* **53**:537-544.

Beard, J. B., S. M. Batten, and G. Pittman. 1981. The comparative low temperature hardiness of 19 bermudagrasses. *Tex. Agric. Exp. Stn. Rep.*, No. 3831. Jan. 1981.

Beardsell, M. F. and D. Cohen. 1974. Endogenous abscisic acid-plant water stress relationships under controlled environmental conditions. In R. L. Bieleski, A. R. Ferguson, and M. M. Cresswell (eds.), *Mechanisms of Regulation of Plant Growth*, Bulletin 12, pp. 411-415. Roy. Soc. N. Z., Wellington.

Beardsell, M. F. and D. Cohen. 1975. Relationships between leaf water status, abscisic acid levels, and stomatal resistance in maize and sorghum. *Plant Physiol.* **56**:2207-2212.

Beater, B. E. 1964. Symposium on soil compaction. Proceedings of the South African Sugar Technology Association. April, 1964. pp. 144-153.

Beaty, E. R., A. E. Smith, and E. E. Worley. 1973. Growth and survival of perennial tropical grasses in north Georgia. *J. Range Manage.* **26**:204-206.

Beauchamp, T. F. and G. O. Estes. 1976. Genotype variation in nutrient uptake efficiency in corn. *Agron. J.* **68**:521-523.

Beauchamp, E. G., L. W. Kannenberg, and R. B. Hunter. 1976. Nitrogen accumulation and translocation in corn genotypes following silking. *Agron. J.* **68**:418-422.

Beauchamp, E. G. and D. J. Lathwell. 1966. Effect of root zone temperatures on corn leaf morphology. *Can. J. Plant Sci.* **46**:593-601.

Beckel, D. K. B. 1956. Cortical disintegration in the roots of *Bouteloua gracilis* (H.B.K.) Lag. *New Phytol.* **55**:183-190.

Beetle, A. A. 1955. The four subfamilies of the Gramineae. *Bull. Torrey Bot. Club.* **82**:196-197.

Beevers, L. and R. H. Hageman. 1969. Nitrate reduction in higher plants. *Ann. Rev. Plant Physiol.* **20**:495-522.

Beffagna, N., E. Marre, and S. M. Cocucci. 1979. Cation-activated ATPase activity of plasmalemma-enriched membrane preparations from maize coleoptiles. *Planta* **146**:387-391.

Begg, J. E., J. F. Bierhuizen, E. R. Lemon, D. K. Misra, R. O. Slatyer, and W. R. Stern. 1964. Diurnal energy and water exchanges in bulrush millet in an area of high solar radiation. *Agric. Meteorol.* **1**:294-312.

Begg, J. E. and G. W. Burton. 1973. Comparative study of five genotypes of bulrush millet (*Pennisetum typhoides* Burm. S. & H.) under a range of photoperiods and temperatures. In R. O. Slatyer (ed.), *Plant Response to Climatic Factors*, pp. 141-144. Proceedings of the Uppsala Symposium, 1970. UNESCO, Paris.

Below, F. E., L. E. Christiansen, A. J. Reed, and R. H. Hageman. 1981. Availability of reduced N and carbohydrates for ear development of maize. *Plant Physiol.* **68**:1186-1190.

Benedict, C. R. 1978. The fractionation of stable carbon isotopes in photosynthesis. *What's New in Plant Phys.* **9**:13-16.

Bennett, A. C. and F. Adams. 1970. Concentration of NH_3 (aq) required for incipient NH_3 toxicity to seedlings. *Soil Sci. Soc. Am. Proc.* **34**:259-263.

Bennett, D. M. 1982. Silicon deposition in the roots of *Hordeum sativum* Jess, *Avena sativa* L. and *Triticum aestivum* L. *Ann. Bot.* **50**:239-245.

Bennett, J. H., R. A. Olsen, and R. B. Clark. 1982a. Modification of soil fertility by plant roots: Iron stress-response mechanism. *What's New in Plant Phys.* **13**:1-4.

Bennett, J. H., E. H. Lee, D. L. Krizek, R. A. Olsen, and J. C. Brown. 1982b. Photochemical reduction of iron. II. Plant related factors. *J. Plant Nutr.* **5**:335-344.

Bennett, J. M. and C. Y. Sullivan. 1978a. Effects of water stress during panicle development and at bloom on grain sorghum. *Project Completion Report: Physiological Aspects of Water Use Efficiency*, pp. 1-16. Nebraska Water Resources Center, University of Nebraska, Lincoln.

Bennett, J. M. and C. Y. Sullivan. 1978b. The effect of water stress on grain sorghum during vegetative growth, panicle development, and bloom. *Project Completion Report: Physiological Aspects of Water Use Efficiency*, pp. 17-31. Nebraska Water Resources Center, University of Nebraska, Lincoln.

Bennett, J. M. and C. Y. Sullivan. 1978c. Conditioning of grain sorghum photosynthesis to water deficits. *Project Completion Report: Physiological Aspects of Water Use Efficiency*, pp. 32-43. Nebraska Water Resources Center, University of Nebraska, Lincoln.

Bennett, J. M. and C. Y. Sullivan. 1981. Effects of water stress preconditioning on net photosynthetic rate of grain sorghum. *Photosynthetica* **15**:330-337.

Bennett, K. J., H. G. McPherson, and I. J. Warrington. 1982c. Effect of pretreatment temperature on response of photosynthesis rate in maize to current temperature. *Aust. J. Plant Physiol.* **9**:773-781.

Bennett, O. L., B. D. Doss, D. A. Ashley, V. J. Kilmer, and E. C. Richardson. 1964. Effects of soil moisture regime on yield, nutrient content, and evapotranspiration for three annual forage species. *Agron. J.* **56**:195-198.

Bentham, G. 1881. Notes on Gramineae. *J. Linn. Soc. Bot.* **19**:14-134.

Ben-Zioni, A., Y. Vaadia, and S. H. Lips. 1970. Correlations between nitrate reduction, protein synthesis, and malate accumulation. *Physiol. Plant.* **23**:1039-1047.

Bernal, E. J. 1972. El pasto para (*Brachiaria mutica* (Forsk) Stapf). III. Efecto de la profundidad del nivel freatico y de la aplicacion de nitrogeno bajo condiciones controladas. *Revista ICA* **7**:191-204.

Bernardon, A. E., D. L. Huss, and W. G. McCully. 1967. Effects of herbage removal on seedling development in cane bluestem. *J. Range Manage.* **20**:69-72.

Bernstein, L. 1974. Crop growth and salinity. In J. van Schilfgaard (ed.), *Drainage for Agriculture*, pp. 39-54. American Society of Agronomy, Madison, WI.

Bernstein, L. 1975. Effects of salinity and sodicity on plant growth. *Annu. Rev. Phytopathol.* **13**:295-312.

Bernstein, L., R. A. Clark, L. E. Francois, and M. D. Derderian. 1966. Salt tolerance of N.Co. varieties of sugar-cane. II. Effects of soil salinity and sprinkling on chemical composition. *Agron. J.* **58**:503-507.

Berthouly, M. and G. Guerrier. 1979a. Interactions calcium-potassium-magnesium chez le sorghograin. I. Influence d'une carence en calcium sur la distribution de ces elements. *Commun. Soil Sci. Plant Anal.* **10**:1523-1539.

Berthouly, M. and G. Guerrier. 1979b. Interactions calcium-potassium-magnesium chez le sorghograin. 3. Influence d'une carence en potassium sur la distribution de ces elements. *Commun. Soil. Sci. Plant Anal.* **10**:1557-1571.

Besford, R. T. 1979. Phosphorus nutrition and acid phosphatase activity in the leaves of seven plant species. *J. Sci. Food Agric.* **30**:281-285.

Bewley, J. D. and K. M. Larsen. 1982. Differences in the responses to water stress of growing and non-growing regions of maize mesocotyls: protein synthesis on total, free and membrane-bound polyribosome fractions. *J. Exp. Bot.* **33**:406-415.

Bews, J. G. 1929. *The World's Grasses: Their Differentiation, Distribution, Economics, and Ecology.* Longman's Green, London.

Beyers, C. P. de L. and F. J. Coetzer. 1969. The influence of zinc fertilizing on the yield of maize and the mineral composition of the leaves. *Agr. Sci. S. Afr. Agro.* **1**:(2):41-46.

Bhatt, K. C., P. P. Vaishnav, Y. D. Singh, and J. J. Chinoy. 1979. Nitrate reductase activity: a biochemical criterion of hybrid vigor in *Sorghum bicolor* (L.) Moench. *Ann. Bot.* **44**:495-502.

Bhatt, K. C., P. P. Vaishnav, Y. D. Singh, and J. J. Chinoy. 1981. Biochemical basis of heterosis in sorghum: changes in chlorophylls and ascorbic acid turnover during seedling growth. *Ann. Bot.* **47**:321-328.

Bjorkman, O., M. R. Badger, and P. A. Armond. 1980. Response and adaptation of photosynthesis to high temperatures. In N. C. Turner and P. J. Kramer (eds.), *Adaptation of Plants to Water and High Temperature Stress*, pp. 233-249. Wiley, New York.

Bjorkman, O., J. Boynton, and J. Berry. 1976. Comparison of the heat stability of photosynthesis, chloroplast membrane reactions, photosynthetic enzymes, and soluble protein in leaves of heat-adapted and cold-adapted C_4 species. *Carnegie Inst. Wash. Yearbook* **75**:400-407.

Blacklow, W. M. 1972. Influence of temperature on germination and elongation of the radicle and shoot of corn (*Zea mays* L.). *Crop Sci.* **12**:647-650.

Blair, G. J., M. H. Miller, and W. A. Mitchell. 1970. Nitrate and ammonium as sources of nitrogen for corn and their influence on the uptake of other ions. *Agron. J.* **62**:530-532.

Blenkinsop, P. G. and J. E. Dale. 1974. The effects of nitrate supply and grain reserves on Fraction I protein level in the first leaf of barley. *J. Exp. Bot.* **25**:913-926.

Blevins, D. G. 1982. Protein yield of crops is highly correlated to potassium removed in the crops. *Agron. Abst.* p. 93.

Blum, A. 1972. Effect of planting date on water use and its efficiency in dryland grain sorghum. *Agron. J.* **64**:775-778.

Blum, A. 1974. Genotypic responses in sorghum to drought stress. I. Response to soil moisture stress. *Crop. Sci.* **14**:361-364.

Blum, A. 1975a. Effect of the Bm gene on epicuticular wax and water relations of *Sorghum bicolor*. *Israel J. Bot.* **24**:50.

Blum, A. 1975b. Effect of the Bm gene on epicuticular wax deposition and the spectral characteristics of sorghum leaves. *SABRAO J.* **7**:45-52.

Blum, A. 1979. Genetic improvement of drought resistance in crop plants: A case for sorghum. In H. Mussell and R. C. Staples (eds.), *Stress Physiology in Crop Plants*, pp. 430-445. Wiley-Interscience, New York.

Blum, A. 1983. Genetic and physiological relationships in plant breeding for drought resistance. *Agric. Water Manage.* **7**:195-205.

Blum, A., G. F. Arkin, and W. R. Jordan. 1977a. Sorghum root morphogenesis and growth. I. Effect of maturity genes. *Crop. Sci.* **17**:149-153.

Blum, A. and A. Ebercon. 1976. Genotypic responses in sorghum to drought stress. III. Free proline accumulation and drought resistance. *Crop Sci.* **16**:428-431.

Blum, A., W. R. Jordan, and G. F. Arkin. 1977b. Sorghum root morphogenesis and growth. II. Manifestation of heterosis. *Crop. Sci.* **17**:153-157.

Blunt, C. G. and R. J. Jones. 1980. The use of leaf development rate to determine time to irrigate panola grass. *Aust. J. Exp. Agric. Anim. Husb.* **20**:556-560.

Boawn, L. C. 1971. Evaluation of DTPA-extractable zinc as a zinc soil test for Washington soils. Proceedings of the 22nd Annual Pacific Northwest Fertilizer Conference, pp. 143-146. Pacific Northwest Plant Food Association, Portland, OR.

Boawn, L. C., C. E. Nelson, F. G. Viets, Jr., and C. L. Crawford. 1960a. Nitrogen carrier and nitrogen rate influence on soil properties and nutrient uptake by crops. *Wash. Agr. Exp. Stn. Bull.* 614.

Boawn, L. C., C. E. Nelson, F. G. Viets, Jr., and C. L. Crawford. 1960b. Effect of nitrogen carrier, nitrogen rate, zinc rate and soil pH on zinc uptake by sorghum, potatoes and sugar beets. *Soil Sci.* **90**:329-337.

Bogdan, A. V. 1977. *Tropical Pasture and Fodder Plants*. Longman, London.

Bolton, J. K. 1979. The effects of nitrogen nutrition on photosynthesis and associated characteristics in C_3, C_4, and intermediate grass species. M.S. thesis, University of Georgia.

Bonhomme, R., F. Ruget, M. Derieux, and P. Vincourt. 1982. Relations entre production de matiere seche aerienne et energie interceptee chez differents genotypes de mais. *C. R. Acad. Sc. Paris* **294**:393-398.

Bonnett, J. A. 1967. Sulphur deficiency in the sheath related to sugarcane yield decline in a Puerto Rico soil. *Int. Soc. Sugar Cane Technol. Proc.* **12**:244.

Bonnett, O. T. 1966. Inflorescences of maize, wheat, rye, barley and oats: Their initiation and development. Univ. Ill., Coll. Agric. Exp. Stn. Bulletin 721.

Booysen, P. de V., N. M. Tainton, and J. D. Scott. 1963. Shoot-apex development in grasses and its importance in grassland management. *Herb. Abst.* **33**:209-213.

Borden, R. J. 1948. Nitrogen effects upon the yields and composition of sugar cane. *Hawaii. Plant Rec.* **52**:1-51.

Botha, F. C. and P. J. Botha. 1979. The effect of water stress on the nitrogen metabolism of two maize lines. II. Effects on the rate of protein synthesis and chlorophyll content. *Z. Pflanzenphysiol.* **94**:179-183.

Bouldin, D. R. 1979. The influence of subsoil acidity on crop yield potential. Cornell International Agriculture Bulletin 34, Cornell University, Ithaca, New York.

Bouton, J. H. and C. O. Brooks. 1982. Screening pearl millet for variability in supporting bacterial acetylene reduction activity. *Crop Sci.* **22**:680-683.

Boutton, T. W., A. T. Harrison, and B. N. Smith. 1980. Distribution of biomass of species differing in photosynthetic pathway along an altitudinal transect in southeastern Wyoming grassland. *Oecologia (Berl.)* **45**:287-298.

Bowden, B. N. 1965. Modern grass taxonomy. *Outlook on Agriculture* **4**:243-253.

Bowen, C. E. 1975. Micronutrient composition of sugar-cane sheaths as affected by age. *Trop. Agric.* **52**:131-137.

Bowen, G. D. and S. E. Smith. 1981. The effects of mycorrhizas on nitrogen uptake by plants. In F. E. Clark and T. Rosswall (eds.), *Terrestrial Nitrogen Cycles*, pp. 237-247. *Ecol. Bull.* **33**:237-247.

Bowen, H. D. 1981. Alleviating mechanical impedance. In G. F. Arkin and H. M. Taylor (eds.), *Modifying the Root Environment to Reduce Crop Stress*, pp. 21-57. American Society of Agricultural Engineers, St. Joseph, MI.

Bowen, J. E. 1969. Absorption of copper, zinc, and manganese by sugarcane leaf tissue. *Plant Physiol.* **44**:255-261.

Bowen, J. E. 1981a. Micro-element nutrition of sugarcane. I. Effect of micro-elements on growth and yield. *Trop. Agric. (Trinidad)* **58**:13-21.

Bowen, J. E. 1981b. Kinetics of active uptake of boron, zinc, copper, and manganese in barley and sugarcane. *J. Plant Nutr.* **3**:215-223.

Bowen, J. E. 1983. Micro-element nutrition of sugar-cane 3. Critical micro-element levels in immature leaf sheaths. *Trop. Agric.* **60**:133-138.

Bower, C. A. and W. H. Pierre. 1944. Potassium response of various crops on a high-lime soil in relation to their contents of potassium, calcium, magnesium and sodium. *Agron. J.* **36**:608-614.

Bowling, D. J. F. 1981. Release of ions to the xylem in roots. *Physiol. Plant.* **53**:392-397.

Bowling, D. J. F. and A. Q. Ansari. 1971. Evidence for a sodium influx pump in sunflower roots. *Planta* **98**:323-329.

Bowling, D. J. F. and A. Q. Ansari. 1972. Control of sodium transport in sunflower roots. *J. Exp. Bot.* **23**:241-246.

Boyd, L. and G. S. Avery, Jr. 1936. Grass seedling anatomy: The first internode of *Avena* and *Triticum*. *Bot. Gaz.* **97**:765-779.

Boyer, J. S. 1968. Relationship of water potential to growth of leaves. *Plant Physiol.* **43**:1056-1062.

Boyer, J. S. 1970. Leaf enlargement and metabolic rates in corn, soybean, and sunflower at various leaf water potentials. *Plant Physiol.* **46**:233-235.

Boyer, J. S. 1971. Resistances to water transport in soybean, bean, and sunflower. *Crop Sci.* **11**:403-407.

Bradford, J. M. and R. W. Blanchar. 1977. Profile modification of a Fragiudalf to increase crop production. *Soil Sci. Soc. Am. J.* **41**:127-131.

Branson, F. A. 1953. Two new factors affecting resistance of grasses to grazing. *J. Range Manage.* **6**:165–171.

Branson, F. A. 1956. Quantitative effects of clipping treatments on five range grasses. *J. Range Manage.* **9**:86–88.

Bray, R. A. and J. B. Hacker. 1981. Genetic analysis in the pasture grass *Setaria sphacelata*. II. Chemical composition, digestibility and correlations with yield. *Aust. J. Agric. Res.* **32**:311–323.

Breaux, R. D. 1984. Breeding to enhance sucrose content of sugarcane varieties in Louisiana. *Field Crops Res.* **9**:59–67.

Brecke, B. J. and W. B. Duke. 1980. Dormancy, germination and emergence characteristics of fall panicum (*Panicum dichotomiflorum*) seed. *Weed Sci.* **28**:683–685.

Brenes, E. and R. W. Pearson. 1973. Root responses of three Gramineae species to soil acidity in an Oxisol and an Ultisol. *Soil Sci.* **116**:295–302.

Bresler, E., B. L. McNeal, and D. L. Carter. 1982. *Saline and Sodic Soils.* Springer-Verlag, Berlin.

Brevedan, E. R. and H. F. Hodges. 1973. Effects of moisture deficits on ^{14}C translocation in corn (*Zea mays* L.). *Plant Physiol.* **52**:436.

Briske, D. D. and A. M. Wilson. 1978. Moisture and temperature requirements for adventitious root development in blue grama seedlings. *J. Range Manage.* **31**:174–178.

Brodie, H. W., R. Yoshida, and L. G. Nickell. 1969. Effect of air and root temperatures on growth of four sugarcane clones. *Hawaiian Planters' Rec.* **58**:21–52.

Brougham, R. W. 1960. The relationship between the critical leaf area, total chlorophyll content, and maximum growth-rate of some pasture and crop plants. *Ann. Bot.* **24**:463–474.

Brouwer, R. 1954. The regulating influence of transpiration and suction tension on the water and salt uptake by the roots of intact *Vicia faba* plants. *Acta Botan. Neerl.* **3**:264–312.

Brouwer, R. 1956a. Water uptake from water and salt solutions. *Acta Botan. Neerl.* **5**:268–276.

Brouwer, R. 1956b. Investigations into the occurrence of active and passive components in the ion uptake by *Vicia faba*. *Acta Botan. Neerl.* **5**:287–314.

Brouwer, R. 1965. Ion absorption and transport in plants. *Ann. Rev. Plant Physiol.* **16**:241–266.

Brown, J. C. 1978a. Mechanism of iron uptake by plants. *Plant Cell Environ.* **1**:249–257.

Brown, J. C. and J. E. Ambler. 1970. Further characterization of iron uptake in two genotypes of corn. *Soil Sci. Soc. Am. Proc.* **34**:249–254.

Brown, J. C. and W. D. Bell. 1969. Iron uptake dependent upon genotype of corn. *Soil Sci. Soc. Am. Proc.* **33**:99–101.

Brown, J. C., J. E. Ambler, R. L. Chaney, and C. D. Foy. 1972. Differential responses of plant genotypes to micronutrients. In J. J. Mortvedt, P. M. Giordana, and W. L. Lindsay (eds.), *Micronutrients in Agriculture*, pp. 389–418. Soil Science Society of America, Madison, WI.

Brown, J. C., R. B. Clark, and W. E. Jones. 1977. Efficient and inefficient use of phosphorus by sorghum. *Soil Sci. Soc. Am. J.* **41**:747–750.

Brown, J. C. and W. E. Jones. 1975a. Phosphorus efficiency as related to iron inefficiency in sorghum. *Agron. J.* **67**:468–472.

Brown, J. C. and W. E. Jones. 1975b. Heavy metal toxicity in plants. 1. A crisis in embryo. *Comm. Soil Sci. Plant Anal.* **6**:421–438.

Brown, J. C. and L. O. Tiffin. 1962. Zinc deficiency and iron chlorosis dependent on the plant species and nutrient-element balance in Tulare clay. *Agron. J.* **54**:356–358.

Brown, R. H. 1978b. A difference in the N use efficiency of C_3 and C_4 plants and its implication in adaptation and evolution. *Crop Sci.* **18**:93–98.

Brown, R. H. and R. E. Blaser. 1970. Soil moisture and temperature effects on growth and soluble carbohydrates of orchardgrass (*Dactylis glomerata*). *Crop Sci.* **10**:213–216.

Brown, R. H. and W. V. Brown. 1975. Photosynthetic characteristics of *Panicum milioides*, a species with reduced photorespiration. *Crop Sci.* **15**:681–685.

Brown, R. H., R. E. Blaser, and H. L. Dunton. 1966. Leaf area index and apparent photosynthesis under various microclimates for different pastures species. Proceedings of the 10th International Grassland Congress, Helsinki, 1966, pp. 108–113.

Brown, R. H. and J. R. Wilson. 1983. Nitrogen response of *Panicum* species differing in CO_2 fixation pathways. II. CO_2 exchange characteristics. *Crop Sci.* **23**:1154–1159.

Brown, W. V. 1974. Another cytological difference among the Kranz subfamilies of the Gramineae. *Bull. Torrey Bot. Club* **101**:120–124.

Brown, W. V. 1975. Variations in anatomy, associations, and origins of Kranz tissue. *Am. J. Bot.* **62**:395–402.

Brown, W. V. 1977. The Kranz syndrome and its subtypes in grass systematics. *Mem. Torrey Bot. Club* **23**:1–97.

Brown, W. V. and B. N. Smith. 1972. Grass evolution, the kranz syndrome, $^{13}C/^{11}C$ ratios, and continental drift. *Nature* **239**:345–346.

Bruce, R. C. and I. J. Bruce. 1972. The correlation of soil test phosphorus analysis with response of tropical pastures to superphosphates on some north Queensland soils. *Aust. J. Exp. Agric. Anim. Husb.* **12**:188–194.

Bruetsch, T. F. and G. O. Estes. 1976. Genotype variation in nutrient uptake efficiency in corn. *Agron. J.* **68**:521–523.

Bull, T. A. 1964. The effects of temperature, variety, and age on the response of *Saccharum* spp. to applied gibberellic acid. *Aust. J. Agric. Res.* **15**:77–84.

Bull, T. A. 1969. Photosynthetic efficiencies and photorespiration in Calvin cycle and C_4-dicarboxylic acid plants. *Crop Sci.* **9**:726–729.

Bull, T. A. 1971. The C_4 pathway related to growth rates in sugarcane. In M. D. Hatch, C. B. Osmond, and R. O. Slatyer (eds.), *Photosynthesis and Photorespiration*, pp. 68–75. Wiley-Interscience, New York.

Bunting, E. S. 1973. Plant density and yield of grain maize in England. *J. Agric. Sci.* **81**:455–463.

Burch, G. J., R. C. G. Smith, and W. K. Mason. 1978. Agronomic and physiological responses of soybean and sorghum crops to water deficits. II. Crop evaporation, soil water depletion and root distribution. *Aust. J. Plant Physiol.* **5**:169–177.

Buren, L. L., P. H. Moore, and Y. Yamasaki. 1979. Gibberellin studies with sugarcane. II. Hand-sampled field trials. *Crop Sci.* **19**:425–428.

Burnett, E. and J. L. Tackett. 1968. Effect of soil profile modification on plant root development. *Int. Congr. Soil Sci. Trans.*, 9th, Adelaide, **III**:329–337.

Burns, I. G. 1980. Influence of the spatial distribution of nitrate on the uptake of N by plants: A review and a model for rooting depth. *J. Soil Sci.* **31**:155–173.

Burns, R. E. 1972. Environmental factors affecting root development and reserve carbohydrates of bermudagrass cuttings. *Agron. J.* **64**:44–45.

Burr, G. O., C. E. Hartt, H. W. Brodie, T. Tanimoto, H. P. Kortschak, D. Takahashi, F. M. Ashton, and R. E. Coleman. 1957. The sugar cane plant. *Ann. Rev. Plant Physiol.* **8**:275–308.

Burris, J. S., R. H. Brown, and R. E. Blaser. 1967. Evaluation of reserve carbohydrates in midland bermudagrass (*Cynodon dactylon* L.). *Crop Sci.* **7**:22–24.

Burris, J. S. and R. J. Navritil. 1979. Relationship between laboratory cold-test methods and field emergence in maize inbreds. *Agron. J.* **71**:985–988.

Burson, B. 1980. Warm season grasses. In W. R. Fehr and H. H. Hadley (eds.), *Hybridization of Crop Plants*, pp. 695–708. American Society of Agronomy and Crop Science, Society of America, Madison, WI.

Burson, B. L., J. Correa, and H. C. Potts. 1978. Anatomical study of seed shattering in bahiagrass and dallisgrass. *Crop. Sci.* 18:122-125.

Burton, G. W. 1965. Photoperiodism in pearl millet, *Pennisetum typhoides*. *Crop Sci.* 5:333-335.

Burton, G. W. 1980. Pearl millet. In W. R. Fehr and H. H. Hadley (eds.), *Hybridization of Crop Plants*, pp. 457-470. Am. Soc. Agron. and Crop Sci. Soc. Am., Madison, WI

Burton, G. W. and J. E. Jackson. 1962a. A method for measuring sod reserves. *Agron. J.* 54:53-55.

Burton, G. W. and J. E. Jackson. 1962b. Effect of rate and frequency of applying six nitrogen sources on Coastal bermudagrass. *Agron. J.* 54:40-43.

Burton, G. W., J. E. Jackson, and F. E. Knox. 1959. Influence of light reduction upon the production, persistence, and chemical composition of coastal bermuda grass (*Cynodon dactylon*). *Agron. J.* 52:537-542.

Bussler, W. 1981. Microscopical possibilities for the diagnosis of trace element stress in plants. *J. Plant Nutr.* 3:115-128.

Burzynski, W. and Z. Lechowski. 1983. The effect of temperature and light intensity on the photosynthesis of *Panicum* species of the C_3, C_3-C_4, and C_4 type. *Acta Physiol. Plant.* 5:93-104.

Bykov, O. D., V. A. Koshkin, and J. Catsky. 1981. Carbon dioxide compensation concentration of C_3 and C_4 plants: Dependence on temperature. *Photosynthetica* 15:114-121.

Cable, D. R. 1982. Partial defoliation stimulates growth of Arizona cottontop. *J. Range Manage.* 35:591-593.

Caddel, J. L. and D. E. Weibel. 1972. Photoperiodism in sorghum. *Agron. J.* 64:473-476.

Calder, F. W., L. B. MacLeod, and R. I. Hayden. 1966. Electrical resistance in alfalfa roots as affected by temperature and light. *Can. J. Plant Sci.* 46:185-194.

Calvin, M. and A. A. Benson. 1948. The path of carbon in photosynthesis. *Science* 107:476-480.

Camery, M. P. and C. R. Weber. 1953. Effects of certain components of simulated hail injury on soybeans and corn. Iowa Agric. and Home Econ. Exp. Stn. Res. Bull. 400.

Campbell, C. S. 1982. Cleistogamy in *Andropogon* (Gramineae). *Am. J. Bot.* 1625-1635.

Campbell, R. B. and C. J. Phene. 1977. Tillage, matric potential, oxygen and millet yield relations in a layered soil. *Trans. ASAE* 20:271-275.

Canvin, D. T. 1979. Photorespiration: Comparison between C_3 and C_4 plants. In M. Gibbs and E. Latzko (eds.), *Encyclopedia of Plant Physiology*. New Series. Vol. 6, *Photosynthesis. II. Photosynthetic Carbon Metabolism and Related Processes*, pp. 368-396. Springer-Verlag, Berlin.

Carceller, M. and A. Fraschina. 1980a. Accumulation of free proline in young corn seedlings and its relation to drought resistance. *Turrialba.* 30:231-233.

Carceller, M. and A. Fraschina. 1980b. The free proline content of water stressed maize roots. *Z. Pflanzenphysiol.* 100:43-49.

Carlson, R. W. and F. A. Bazzaz. 1980. The effects of elevated CO_2 concentration on growth, photosynthesis, transpiration, and water use efficiency of plants. In J. J. Sigh and A. Deepak (eds.), *Environmental and Climatic Impact of Coal Utilization*, pp. 609-623. Academic Press, New York.

Carman, J. G. and D. D. Briske. 1982. Root initiation and root and leaf elongation of dependent little bluestem tillers following defoliation. *Agron. J.* 74:432-435.

Carter, C. E. 1976. Drainage parameters for sugarcane in Louisiana. Proc. 3rd National Drainage Symposium. Am. Soc. Agric. Eng. Pub. 77-1. pp. 135-138.

Carter, D. R. and K. M. Peterson. 1983. Effects of a CO_2-enriched atmosphere on the growth and competitive interaction of a C_3 and a C_4 grass. *Oecologia* 58:188-193.

Carter, J. L., L. A. Garrard, and S. H. West. 1972. Starch degrading enzymes of temperate and tropical species. *Phytochemistry* 11:2423-2428.

Carter, M. W. and C. G. Poneleit. 1973. Black layer maturity and filling period variation among inbred lines of corn (*Zea mays* L.). *Crop Sci.* 13:436-439.

Castleberry, R. M., C. W. Crum, and C. F. Krull. 1984. Genetic yield improvement of U.S. maize cultivars under varying fertility and climatic environments. *Crop Sci.* **24**:33–36.

Chakravarty, D. N. and R. M. Karmakar. 1980. Root development of different maize varieties in Diphu area of Assam. *Indian J. Agric. Sci.* **50**:527–531.

Chalmers, D. R. and R. E. Schmidt. 1979. Bermudagrass survival as influenced by deacclimation, low temperatures, and dormancy. *Agron. J.* **71**:947–949.

Chamberlin, R. J. and G. L. Wilson. 1982. Development of yield in two grain sorghum hybrids. I. Dry weight and carbon-14 studies. *Aust. J. Agric. Res.* **33**:1009–1018.

Chanson, A. and P. E. Pilet. 1981. Effect of abscisic acid on maize root georeaction. *Plant Sci. Lett.* **22**:1–5.

Chao, T. T. 1966. Effect of nitrogen forms on the absorption of bromide by sorghum. *Agron. J.* **58**:595–596.

Chao, T. T., M. E. Harward, and S. C. Fang. 1962. Adsorption and desorption phenomena of sulfate ions in soils. *Soil Sci. Soc. Am. Proc.* **26**:234–237.

Chatterton, N. J., C. E. Carlson, W. E. Hungerford, and D. R. Lee. 1972. Effect of tillering and cool nights on photosynthesis and chloroplast starch in Pangola. *Crop Sci.* **12**:206–208.

Chatterton, N. J., W. W. Hanna, J. B. Powell, and D. R. Lee. 1975. Photosynthesis and transpiration of bloom and bloomless sorghum. *Can. J. Plant Sci.* **55**:641–643.

Chaudhary, T. N., V. K. Bhatnagar, and S. S. Prihar. 1975. Corn yield and nutrient uptake as affected by water table depth and soil submergence. *Agron. J.* **67**:745–749.

Chauhan, R. P. S., C. P. S. Chauhan, and D. Kumar. 1980. Free proline accumulation in cereals in relation to salt tolerance. *Plant Soil* **57**:167–175.

Chazdon, R. L. 1978. Ecological aspects of the distribution of C_4 grasses in selected habitats of Costa Rica. *Biotropica* **10**:265–269.

Cheeseman, J. M. and J. B. Hanson. 1979a. Mathematical analysis of the dependence of cell potential on external potassium in corn roots. *Plant Physiol.* **63**:1–4.

Cheeseman, J. M. and J. B. Hanson. 1979b. Energy-linked potassium influx as related to cell potential in corn roots. *Plant Physiol.* **64**:842–845.

Chevalier, P. and L. E. Schrader. 1977. Genotypic differences in nitrate absorption and partitioning of N among plant parts of maize. *Crop Sci.* **17**:897–901.

Chino, M. 1979. Calcium localization with plant roots by electroprobe X-ray microanalysis. *Commun. Soil Sci. Plant Anal.* **10**:443–458.

Chollet, R. and W. L. Ogren. 1975. Regulation of photorespiration in C_3 and C_4 species. *Bot. Rev.* **41**:137–179.

Christensen, N. W. and T. L. Jackson. 1981. Potential for phosphorus toxicity in zinc-stressed corn and potato. *Soil Sci. Soc. Am. J.* **45**:904–909.

Christiansen, M. N. and C. D. Foy. 1979. Fate and function of calcium in tissue. *Commun. Soil Sci. Plant Anal.* **10**:427–442.

Christiansen, M. N., H. R. Carns, and D. J. Slyter. 1970. Stimulation of solute loss from radicles of *Gossypium hirsutum* L. by chilling, anacrobiosis, and low pH. *Plant Physiol.* **46**:53–56.

Christie, E. K. 1975a. Physiological responses of semi-arid grasses. II. The pattern of root growth in relation to external phosphorus concentration. *Aust. J. Agric. Res.* **26**:437–446.

Christie, E. K. 1975b. A study of phosphorus nutrition and water supply on the early growth and survival of buffelgrass grown on a sandy red earth from Southwest Queensland. *Aust. J. Exp. Agric. Anim. Husb.* **15**:239–249.

Christie, E. K. and J. Moorby. 1975. Physiological responses of semiarid grasses. I. The influence of phosphorus supply on growth and phosphorus absorption. *Aust. J. Agric. Res.* **26**:423–426.

Chudleigh, P. D., J. G. Boonman, and P. J. Cooper. 1977. Environmental factors affecting herbage yield of Rhodes grass (*Chloris gayana*) at Kitale, Kenya. *Trop. Agric. (Trin.)* **54**:193–204.

Churchward, E. H. 1967. Mechanical harvesting in the Bundaberg District, Queensland. In J. Bague (ed.), Proceedings of the International Society Sugar Cane Technology, 12th Congress, San Juan, 1965. pp. 347-353. Elsevier Pub. Co., Amsterdam, London, New York.

Claasen, N. and S. A. Barber. 1977. Potassium influx characteristics of corn roots and interaction with N, P, Ca, and Mg influx. *Agron. J.* **69**:860-864.

Clark, N. A., R. W. Hemken, and J. H. Vandersall. 1965. A comparison of pearl millet, sudangrass, and sorghum-sudangrass hybrids as pasture for lactating dairy cows. *Agron. J.* **57**:266-268.

Clark, R. B. 1968. Organic acids of maize (*Zea mays* L.) as influenced by mineral deficiencies. *Crop Sci.* **8**:165-167.

Clark, R. B. 1969. Organic acids from leaves of several crop plants by gas chromatography. *Crop Sci.* **9**:341-343.

Clark, R. B. 1978. Differential response of maize inbreds to Zn. *Agron. J.* **70**:1057-1060.

Clark, R. B. and J. C. Brown. 1974a. Internal root control of iron uptake and utilization in maize genotypes. *Plant Soil* **40**:669-677.

Clark, R. B. and J. C. Brown. 1974b. Differential phosphorus uptake by phosphorus-stressed corn inbreds. *Crop Sci.* **14**:505-508.

Clark, R. B., P. A. Pier, D. Knudsen, and J. W. Maranville. 1981. Effect of trace element deficiencies and excesses on mineral nutrients in sorghum. *J. Plant Nutr.* **3**:357-374.

Clarke, C. and B. Mosse. 1981. Plant growth responses to vesicular-arbuscular mycorrhiza. XII. Field inoculation responses of barley at two soil P levels. *New Phytol.* **87**:695-703.

Clarkson, D. T. and A. W. Robards. 1975. The endodermis, its structural development and physiological role. In J. G. Torrey and D. T. Clarkson (eds.), *The Development and Function of Roots*, pp. 415-436. Academic Press, London.

Cleland, R. E., H. B. Prins, J. R. Harper, and N. Higinbotham. 1977. Rapid hormone-induced hyperpolarization of the oat coleoptile transmembrane potential. *Plant Physiol.* **59**:393-397.

Clements, H. F. 1940. Integration of climatic and physiologic factors with reference to the production of sugar cane. *Hawaii. Plant. Rec.* **44**:201-233.

Clements, H. F. 1955. The absorption and distribution of phosphorus in the sugar cane plant. *Hawaii. Plant. Rec.* **55**:17-32.

Clements, H. F. 1958. Recent developments in the crop-logging of sugar cane—phosphorus and calcium. Hawaii Agric. Exp. Stn. Prog. Notes No. 114.

Clements, H. F. 1962. Crop logging vs. 8-10 and soil analysis. *Int. Sugar J.* **64**:129-130 and 159-162.

Clements, H. F. 1965. Effects of silicate on the growth and leaf freckle of sugarcane in Hawaii. *Int. Soc. Sugar Cane Technol. Proc.* **12**:197-215.

Clements, H. F. 1968. Lengthening versus shortening dark periods and blossoming in sugar-cane as affected by temperature. *Plant Physiol.* **43**:57-60.

Clements, H. F. 1970. Crop logging of sugarcane: nitrogen and potassium requirements and interactions using two varieties. Hawaii Agr. Exp. Stn. Tech. Bull. 81.

Clements, H. F. 1980. *Sugarcane Crop Logging and Crop Control: Principles and Practice.* University of Hawaii Press, Honolulu.

Clements, H. F. and M. Awada. 1967. Experiments on the artificial induction of flowering in sugarcane. *Int. Soc. Sugar Cane Technol.* **12**:795-812.

Clements, H. F. and T. Kubota. 1943. The primary index—its meaning and application to crop management with special reference to sugarcane. *Hawaii. Plant Rec.* **47**:257-297.

Clements, H. F. and S. Moriguchi. 1942. The nitrogen index and certain quantitative field aspects. *Hawaii. Plant. Rec.* **46**:163-190.

Clements, H. F. and S. Nakata. 1967. Minimum temperatures for sugar cane germination. *Int. Soc. Sugar Cane Technol. Proc.* **12**:554-560.

Clements, H. F., E. W. Putman, R. H. Suehisa, G. L. N. Yee, and M. L. Wehling. 1974. Soil toxicities as causes of sugarcane leaf freckle, macadamia leaf chlorosis (Keaau), and Maui sugarcane growth failure. Univ. Hawaii, Hawaii Agric. Exp. Stn. Tech. Bull. 88.

Clements, H. F., E. W. Putman, and J. R. Wilson. 1967. Eliminating soil toxicities with calcium metasilicate. *Hawaii. Sugar Technol. Rep.*, pp. 43–54.

Clifford, H. T. 1961. Floral evolution in the family Gramineae. *Evolution* **15**:455–460.

Cockburn, W. 1983. Stomatal mechanism as the basis of the evolution of CAM and C_4 photosynthesis. *Plant Cell Environ.* **6**:275–279.

Coelho, D. T. and R. F. Dale. 1980. An energy-crop growth variable and temperature function for predicting corn growth and development: Planting to silking. *Agron. J.* **72**:503–510.

Cohen, E., Y. Okon, J. Kigel, I. Nur, and Y. Henis. 1980. Increases in dry weight and total nitrogen content in *Zea mays* and *Setaria italica* associated with nitrogen fixing *Azospirillum* spp. *Plant Physiol.* **66**:246–249.

Cole, N. H. A. 1977. Effect of light, temperature, and flooding on seed germination of the neotropical *Panicum laxum* SW. *Biotropica* **9**:191–194.

Coleman, R. E. 1969. Physiology of flowering in sugar cane. *Int. Soc. Sugar Cane Technol. Proc.* **13**:992–999.

Collander, R. 1941. Selective absorption of cations by higher plants. *Plant Physiol.* **16**:691–720.

Collier, J. W. 1963. Caryopsis development in several grain sorghum varieties and hybrids. *Crop Sci.* **3**:419–422.

Colman, R. L. and A. Lazenby. 1970. Factors affecting the response of tropical and temperate grasses to fertilizer nitrogen. In Proceedings of the XI International Grassland Congress, Surfers Paradise, Australia, pp. 393–397.

Colman, R. L. and G. P. M. Wilson. 1960. The effect of floods on pasture plants. *Agric. Gaz. New South Wales.* **71**:337–347.

Connor, H. E. 1979. Breeding systems in the grasses: A survey. *N. Z. J. Bot.* **17**:547–574.

Cook, R. E. 1979. Asexual reproduction: A further consideration. *Am. Nat.* **113**:769–772.

Cooke, J. R. and R. H. Rand. 1980. Diffusion resistance models. In J. D. Hesketh and J. W. Jones (eds.), *Predicting Photosynthesis for Ecosystem Models*, Vol. I, pp. 93–121. CRC Press, Boca Raton, FL.

Cooper, C. S. and P. W. MacDonald. 1970. Energetics of early seedling growth in corn (*Zea mays* L.). *Crop Sci.* **10**:136–139.

Cooper, J. P. and N. M. Tainton. 1968. Light and temperature requirements for the growth of tropical and temperate grasses. *Herb. Abstr.* **38**:167–176.

Cooper, P. and T. D. Ho. 1983. Heat shock proteins in maize. *Plant Physiol.* **71**:215–222.

Cope, J. T., Jr., and R. D. Rouse. 1973. Interpretation of soil test results. In L. M. Walsh and J. D. Beaton (eds.), *Soil Testing and Plant Analysis*, pp. 35–54. Soil Science Society of America, Madison, WI.

Cowan, I. R., J. A. Raven, W. Hartung, and G. D. Farquhar. 1982. A possible role for abscisic acid in coupling stomatal conductance and photosynthetic carbon metabolism in leaves. *Aust. J. Plant Physiol.* **9**:489–498.

Cowie, A. M. 1973. Effects of nitrogen supply on grain yield and protein content in hybrid grain sorghum. Ph.D. thesis, University of Queensland, Australia.

Cox, W. J. and H. M. Reisenauer. 1973. Growth and ion uptake of wheat supplied nitrogen as nitrate, or ammonium, or both. *Plant Soil* **38**:363–380.

Coyne, P. I., J. A. Bradford, and C. L. Dewald, 1982. Leaf water relations and gas exchange in relation to forage production in four Asiatic bluestems. *Crop. Sci.* **22**:1036–1040.

Crafts, A. S. and T. C. Broyer. 1938. Migration of salts and water into xylem of the roots of higher plants. *Am. J. Bot.* **25**:529–535.

Craig, H. 1953. The geochemistry of the stable carbon isotopes. *Geochim. Cosmochim. Acta* **3**: 53–92.

Craig, J. and A. L. Hooker. 1961. Relation of sugar trends and pith density to Diplodia stalk rot in dent corn. *Phytopathology* **51**:376–382.

Cram, W. J. and M. G. Pitman. 1972. The action of abscisic acid on ion uptake and water flow in plant roots. *Aust. J. Biol. Sci.* **25**:1125–1132.

Crawford, R. M. M. 1978. Metabolic indicators in the prediction of soil anaerobiosis. In D. R. Nielsen and J. G. MacDonald. *Nitrogen in the Environment.* Vol. 1. *Nitrogen Behavior in Field Soil*, pp. 427–447. Academic Press, New York.

Crawford, T. W., Jr., V. V. Rendig, and F. E. Broadbent. 1982. Sources, fluxes, and sinks of nitrogen during early reproductive growth of maize (*Zea mays* L.). *Plant Physiol.* **70**:1654–1660.

Creech, R. G. and D. E. Alexander. 1978. Breeding for industrial and nutritional quality in maize. In D. B. Walden (ed.), *Maize Breeding and Genetics*, pp. 249–264. Wiley, New York.

Crider, F. J. 1955. Root growth stoppage resulting from defoliation of grasses. USDA Tech. Bull. 1102.

Crockett, R. P. and R. K. Crookston. 1980. Tillering of sweet corn reduced by clipping of early leaves. *J. Am. Soc. Hort. Sci.* **105**:565–567.

Crockett, R. P. and R. K. Crookston. 1981. Accounting for increases in the harvest index of sweet corn following early leaf clipping. *J. Am. Soc. Hort. Sci.* **106**:117–120.

Crookston, R. K. 1980. The structure and function of C_4 vascular tissue—some unanswered questions. *Ber. Dtsch. Bot. Ges* **93**:71–78.

Crookston, R. K. and D. R. Hicks. 1978. Early defoliation affects corn grain yields. *Crop Sci.* **18**:485–489.

Crosbie, T. M., J. J. Mock, and R. B. Pearce. 1977. Variability and selection advance for photosynthesis in Iowa Stiff Stalk Synthetic maize population. *Crop Sci.* **17**:511–514.

Crosbie, T. M., J. J. Mock, and R. B. Pearce. 1978. Inheritance of photosynthesis in a diallel among eight maize inbred lines from Iowa Stiff Stalk Dynthetic. *Euphytica.* **27**:657–664.

Crosbie, T. M., R. B. Pearce, and J. J. Mock. 1981a. Selection for high CO_2-exchange rates among inbred lines of maize. *Crop Sci.* **21**:629–631.

Crosbie, T. M., R. B. Pearce, and J. J. Mock. 1981b. Recurrent phenotypic selection for high and low photosynthesis in two maize populations. *Crop Sci.* **21**:736–740.

Dagg, M. 1969. Hydrological implications of grass root studies at a site in East Africa. *J. Hydrol.* **9**:438–444.

Daghlian, C. P. 1981. A review of the fossil record of monocotyledons. *Bot. Rev.* **47**:517–555.

Daigger, L. A. and R. L. Fox. 1971. Nitrogen and sulfur nutrition of sweet corn in relation to fertilization and water composition. *Agron. J.* **63**:729–730.

Damptey, H. B. and D. Aspinall. 1976. Water deficit and inflorescence development in *Zea mays* L. *Ann. Bot.* **40**:23–35.

Dancer, W. S. and L. A. Peterson. 1969. Recovery of differentially placed NO_3-N in a silt loam soil by five crops. *Agron. J.* **61**:893–895.

Das, U. K. 1933a. Measuring production in terms of temperature. *Hawaii. Plant. Rec.* **37**:32–53.

Das, U. K. 1933b. How to measure effective temperature in terms of day-degrees. *Hawaii. Plant. Rec.* **37**:174–178.

Das, U. K. 1936. The day-degree in Mauritius. *Hawaii. Plant. Rec.* **40**:103–104.

Da Silva, S. T. and L. G. Labouriau. 1970. Corpas silicosos de Gramineas das cerrados. III. *Pesq. Agropec. Bras.* **5**:167–182.

Datin, C. L. and R. L. Westerman. 1982. Effect of phosphorus and iron on grain sorghum. *J. Plant Nutr.* **5**:703–714.

Daubenmire, R. 1972. Ecology of *Hyparrhenia rufa* (Nees) in derived savanna in north-western Costa Rica. *J. Appl. Ecol.* **9**:11-23.

Davidson, J. L. and F. L. Milthorpe. 1966. Leaf growth in *Dactylis glomerata* following defoliation. *Ann. Bot. (London) (N.S.)* **30**:173-184.

Davies, D. D. 1973. Control of and by pH. *Symp. Soc. Exp. Biol.* **27**:513-529.

Davis, D. L. and W. B. Gilbert. 1970. Winter hardiness and changes in soluble protein fractions of bermudagrass. *Crop Sci.* **10**:7-9.

Davis, R. E., C. M. Harrison, and A. E. Erickson. 1959. Growth responses of alfalfa and sudan grass in relation to cutting practices and soil moisture. *Agron. J.* **51**:617-621.

Davis, R. F. and N. Higinbotham. 1976. Electrochemical gradients and K^+ and Cl^- fluxes in excised corn roots. *Plant Physiol.* **57**:129-136.

Dayan, E., H. Van Keulen, and A. Dovrat. 1981. Tiller dynamics and growth of Rhodes grass after defoliation: A model named TILDYN. *AgroEcosystems* **7**:101-112.

Dayanandan, P., F. Y. Hebard, and P. B. Kaufman. 1976. Cell elongation in the grass pulvinus in response to geotropic stimulation and auxin application. *Planta* **131**:245-252.

Daynard, T. B. 1972. Relationships among black layer formation, grain moisture percentage, and heat unit accumulation. *Agron. J.* **64**:716-719.

Daynard, T. B. and W. G. Duncan. 1969. The black layer and grain maturity in corn. *Crop Sci.* **9**:473-476.

Daynard, T. B. and L. W. Kannenberg. 1976. Relationships between length of the actual and effective grain filling periods and the grain yield of corn. *Can. J. Plant Sci.* **56**:237-242.

Daynard, T. B., J. W. Tanner, and D. J. Hume. 1969. Contribution of stalk soluble carbohydrates to grain yield in corn (*Zea mays* L.). *Crop Sci.* **9**:831-834.

De Datta, S. K. 1981. *Principles and Practices of Rice Production.* Wiley, New York.

DeKock, P. C. 1955. Iron nutrition of plants at high pH. *Soil Sci.* **79**:167-175.

DeKock, P. C., A. Hall, R. H. E. Inkson, R. C. Little, and R. R. Charlesworth. 1974. A study of iron chlorosis in pear leaves. *Ann. Edafol. Agrobiol.* **32**:101-108.

Denmead, O. T. and R. H. Shaw. 1960. The effects of soil moisture stress at different stages of growth on development and yield of corn. *Agron. J.* **52**:272-274.

De-Polli, H., C. D. Boyer, and C. A. Neyra. 1982. Nitrogenase activity associated with roots and stems of field-grown corn (*Zea mays* L.) plants. *Plant Physiol.* **70**:1609-1613.

DeWit, C. T. 1965. Photosynthesis of leaf canopies. *Versl. Landbouwkd. Onderz.* **663**:1-47.

Dexter, S. T. 1956. The evaluation of crop plants for winter hardiness. *Adv. Agron.* **8**:203-240.

Dhindsa, R. S. and R. E. Cleland. 1975. Water stress and protein synthesis. I. Differential inhibition of protein synthesis. *Plant Physiol.* **55**:778-781.

Dibb, D. W. and L. F. Welch. 1976. Corn growth as affected by ammonium vs. nitrate absorbed from soil. *Agron. J.* **68**:89-94.

Dijkshoorn, W. 1962. Metabolic regulation of the alkaline effect of nitrate utilization in plants. *Nature* **194**:165-167.

Dijkshoorn, W. and A. L. van Wijk. 1967. The sulphur requirements of plants as evidenced by the sulphur-nitrogen ratio in the organic matter—a review of published data. *Plant Soil* **26**:129-157.

Dillewijn, C. van. 1952. *Botany of Sugarcane.* Chronica Botanica, Waltham, MA.

Dirven, J. G. P. and B. Deinum. 1977. The effect of temperature on the digestibility of grasses. An analysis. *Forage Res.* **3**:1-17.

Diseker, E. G. and J. van Schilfgaarde. 1958. Field experiments with tile and ditch drainage. N. C. Agr. Exp. Stn. Tech. Bull. 133.

Dobereiner, J. 1960. Nitrogen-fixing bacteria of the genus *Beijerinckia* Derx in the rhizosphere of sugar cane. *Plant Soil* **15**:211-216.

Doering, E. J., G. A. Reichman, L. C. Benz, and R. F. Follett. 1976. Drainage requirement for corn grown on sandy soil. Proceedings of the 3rd National Drainage Symposium. ASAE Pub. 77-1. pp. 144-146.

Dolinar, L. H. and P. A. Jolliffe. 1979. Ecological evidence concerning the adaptive significance of the C_4 dicarboxylic acid pathway of photosynthesis. *Oecologia (Berl.)* **38**:23-34.

Doman, D. C. and D. R. Geiger. 1979. Effect of exogenously supplied foliar potassium on phloem loading in *Beta vulgaris* L. Plant Physiol. **64**:528-535.

Done, A. A., R. J. K. Myers, and M. A. Foale. 1984. Responses of grain sorghum to varying irrigation frequency in the Ord irrigation area. I. Growth, development, and yield. *Aust. J. Agric. Res.* **35**:17-29.

Doss, D. B., D. A. Ashley, O. L. Bennett, R. M. Patterson, and L. E. Ensminger. 1964. Yield, nitrogen content, and water use of sart sorghum. *Agron. J.* **56**:589-592.

Dovrat, A., E. Dayan, and H. van Keulen. 1980. Regrowth potential of shoot and of roots of Rhodes grass (*Chloris gayana* Kunth) after defoliation. *Neth. J. Agric. Sci.* **28**:185-199.

Downes, R. W. 1968. The effect of temperature on tillering of grain sorghum seedlings. *Aust. J. Agric. Res.* **19**:59-64.

Downes, R. W. 1972. Effect of temperature on the phenology and grain yield of *Sorghum bicolor. Aust. J. Agric. Res.* **23**:585-594.

Downey, L. A. 1971. Plant density-yield relations in maize. *J. Aust. Inst. Agric. Sci.* June:138-146.

Downton, W. J. S. 1970. Preferential C_4 dicarboxylic acid synthesis, the post-illumination CO_2 burst, carboxyl transfer step, and grana configuration in plants with C_4 photosynthesis. *Can. J. Bot.* **48**:1795-1800.

Downton, W. J. S. 1971. Adaptive and evolutionary aspects of C_4 photosynthesis. In M. D. Hatch, C. B. Osmond, and R. O. Slatyer (eds.), *Photosynthesis and Photorespiration*, pp. 3-17. Wiley-Interscience, New York.

Downton, W. J. S. and J. S. Hawker. 1975. Response of starch synthesis to temperature in chilling-sensitive plants. In R. Marcelle (ed.), *Environmental and Biological Control of Photosynthesis*, pp. 81-88. Junk, The Hague.

Dragun, J., D. E. Baker, and M. L. Risius. 1976. Growth and element accumulation by two single-cross corn hybrids as affected by copper in solution. *Agron. J.* **68**:466-470.

Drew, M. C., M. B. Jackson, and S. Giffard. 1979. Ethylene-promoted adventitious rooting and development of cortical air spaces (aerenchyma) in roots may be adaptive responses to flooding in *Zea mays* L. *Planta* **147**:83-88.

Drew, M. C., M. B. Jackson, S. C. Giffard, and R. Campbell. 1981. Inhibition by silver ions of gas space (aerenchyma) formation in adventitious roots of *Zea mays* L. subjected to exogenous ethylene or to oxygen deficiency. *Planta* **153**:217-224.

Drew, M. C. and L. R. Saker. 1975. Nutrient supply and the growth of the seminal root system in barley. II. Localized, compensating increases in lateral root growth and rates of nitrate uptake when nitrate supply is restricted to only part of the root system. *J. Exp. Bot.* **26**:79-90.

Duke, S. H., J. W. Friedrich, L. E. Schrader, and W. L. Koukkari. 1978. Oscillations in the activities of enzymes of nitrate reduction and ammonia assimilation in *Glycine max* and *Zea mays. Physiol. Plant.* **42**:269-276.

Dumbroff, E. B. and D. R. Peirson. 1971. Probable sites of passive movement of ions across the endodermis. *Can. J. Bot.* **49**:35-38.

Duncan, R. A. 1960. The mechanical harvesting of long crop sugar cane. In Proceedings of the International Society of Sugar Cane Technology, 10th Congress, Hawaii, 1959, pp. 613-620. Elsevier, New York.

Duncan, R. R. 1981a. Production principles. In R. R. Duncan (ed.), *Ratoon Cropping of Sorghum for Grain in the Southeastern United States*, pp. 1-6. University of Georgia, Coll. Agric. Exp. Stn., Res. Bull. 269, Nov. 1981.

Duncan, R. R. 1981b. Variability among sorghum genotypes for uptake of elements under acid soil field conditions. *J. Plant Nutr.* **4**:21-32.

Duncan, W. G. 1971. Leaf angles, leaf area, and canopy photosynthesis. *Crop Sci.* **11**:482-485.

Duncan, W. G. and J. D. Hesketh. 1968. Net photosynthetic rates, relative leaf growth rates, and leaf numbers of 22 races of maize grown at eight temperatures. *Crop Sci.* **8**:670-674.

Dungey, N. O. and D. D. Davies. 1982. Protein turnover in isolated barley leaf segments and the effects of stress. *J. Exp. Bot.* **33**:12-20.

Dunlop, J. 1974. A model of ion translocation by roots which allows for regulation by physical influences. In R. L. Bieleski, A. R. Ferguson, and M. M. Cresswell (eds.), *Mechanisms of Regulation of Plant Growth*, Bulletin 12, pp. 133-138. Roy. Soc. N. Z.

Dunlop, J. and D. J. F. Bowling. 1971a. The movement of ions to the xylem exudate of maize roots. I. Profiles of membrane potential and vacuolar potassium activity across the root. *J. Exp. Bot.* **22**:434-444.

Dunlop, J. and D. J. F. Bowling. 1971b. The movement of ions to the xylem exudate of maize roots. II. A comparison of the electrical potential and electrochemical potentials of ions in the exudate and in the root cells. *J. Exp. Bot.* **22**:445-452.

Dunlop, J. and D. J. F. Bowling. 1971c. The movement of ions to the xylem exudate of maize roots. III. The location of the electrical and electrochemical potential differences between the exudate and the medium. *J. Exp. Bot.* **22**:453-464.

Dunn, G. A. 1921. Note on the histology of grain roots. *Am. J. Bot.* **8**:207-211.

Dunn, J. H. and C. J. Nelson. 1974. Chemical changes occurring in three bermudagrass turf cultivars in relation to cold hardiness. *Agron. J.* **66**:28-31.

Dure, L. S. 1960. Gross nutritional contributions of maize endosperm and scutellum to germination growth of maize axis. *Plant Physiol.* **35**:919-925.

Durley, R. C., T. Kannangara, N. Seetharama, and G. M. Simpson. 1983. Drought resistance of *Sorghum bicolor*. 5. Genotypic differences in the concentrations of free and conjugated abscisic, phaseic, and indole-3-acetic acids in leaves of field-grown drought-stressed plants. *Can. J. Plant Sci.* **63**:131-145.

Duval-Jouve, J. 1875. Histotaxie des feuilles de Graminees. *Ann. Sci. Nat. Bot. Ser. 6* **1**:227-346.

Duvick, D. N., R. A. Klesse, and N. M. Frey. 1981. Breeding for tolerance of nutrient imbalances and constraints to growth in acid, alkaline and saline soils. *J. Plant Nutr.* **4**:111-129.

Dwivedi, R. S., N. S. Randhawa, and R. L. Bansal. 1975. Phosphorus-zinc interation. I. Sites of immobilization of zinc in maize at a high level of phosphorus. *Plant Soil* **43**:639-648.

Dyer, M. I. 1980. Mammalian epidermal growth factor promotes plant growth. *Proc. Nat. Acad. Sci. USA* **77**:4836-4837.

Dyer, M. I. and U. G. Bokhari. 1976. Plant-animal interactions: Studies of the effects of grasshopper grazing on blue grama grass. *Ecology* **57**:762-772.

Eagles, H. A. and A. K. Hardacre. 1979. Genetic variation in maize (*Zea mays* L.) for germination and emergence at 10°C. *Euphytica* **28**:287-295.

Earley, E. B., J. C. Lyons, E. Inselberg, R. H. Maier, and E. R. Leng. 1974. Earshoot development of Midwest dent corn (*Zea mays* L.). Ill. Exp. Stn. Bull. 747.

Earley, E. B., R. J. Miller, G. L. Reichert, R. H. Hageman, and R. D. Seif. 1966. Effects of shade on maize production under field conditions. *Crop Sci.* **6**:1-7.

Earley, E. B., W. O. McIlrath, R. D. Seif, and R. H. Hageman. 1967. Effects of shade applied at different stages in corn (*Zea mays* L.) production. *Crop Sci.* **7**:151-156.

Eastin, J. A. 1969. Leaf position and leaf function in corn—carbon-14 labeled photosynthate distribution in corn in relation to leaf position and leaf function. Proceedings of the 24th Annual Corn and Sorghum Research Conference, American Seed Trade Association, Publ. No. 24, p. 81-89.

Eastin, J. A. 1970. C-14 labeled photosynthate export from fully expanded corn (*Zea mays* L.) leaf blades. *Crop Sci.* **10**:415–418.

Eastin, J. D., J. E. Hultquist, and C. Y. Sullivan. 1973. Physiologic maturity in grain sorghum. *Crop Sci.* **13**:175–178.

Eavis, B. W. and D. Payne. 1968. Soil physical conditions and root growth. In W. J. Whittington (ed.), *Root Growth*, pp. 247–255. Butterworths, London.

Ebercon, A., A. Blum, and W. R. Jordan. 1977. A rapid colorimetric method for epicuticular wax content of sorghum leaves. *Crop Sci.* **17**:179–180.

Eck, H. V. and J. T. Musick. 1979a. Plant water stress effects on irrigated grain sorghum. I. Effects on yield. *Crop Sci.* **19**:589–592.

Eck, H. V. and J. T. Musick. 1979b. Plant water stress effects on irrigated grain sorghum. II. Effects on nutrients in plant tissues. *Crop. Sci.* **19**:592–598.

Eck, H. V., G. C. Wilson, and T. Martinez. 1975. Nitrate reductase activity of grain sorghum leaves as related to yields of grain, dry matter, and nitrogen. *Crop Sci.* **15**:557–561.

Edmeades, G. O. and T. B. Daynard. 1979. Relationship between final yield and photosynthesis at flowering of individual maize plants. *Can. J. Plant Sci.* **59**:585–601.

Edwards, D. G. and C. J. Asher. 1982. Tolerance of crop and pasture species to manganese toxicity. In A. Scaife (ed.), Proceedings of the Ninth International Mineral Nutrition Colloquium, Vol. 1, pp. 145–150. Commonwealth Agricultural Bureaux, UK.

Edwards, G. and D. A. Walker. 1983. *C_3, C_4: Mechanisms, and Cellular and Environmental Regulation of Photosynthesis*. University of California Press, Los Angeles.

Edwards, J. H. and E. J. Kamprath. 1975. Zinc accumulation by corn as influenced by phosphorus, temperature, and light intensity. *Agron. J.* **66**:479–482.

Edwards, J. H. and S. A. Barber. 1976. Nitrogen uptake characteristics of corn roots at low N concentration as influenced by plant age. *Agron. J.* **68**:17–19.

Egharevba, P. N., R. D. Harrocks, and M. S. Zuber. 1976. Dry matter accumulation in maize in response to defoliation. *Agron. J.* **68**:40–43.

Ehara, K., Y. Yamada, and N. Maeno. 1967. Physiological and ecological studies on the regrowth of herbage plants. IV. The evidence of utilization of food reserves during the early stage of regrowth in bahiagrass. *J. Jpn. Soc. Grassl. Sci.* **12**:1–4.

Ehleringer, J. 1980. Leaf morphology and reflectance in relation to water and temperature stress. In N. C. Turner and P. J. Kramer (eds.), *Adaptation of Plants to Water and High Temperature Stress*, pp. 295–308. Wiley, New York.

Ehleringer J. and R. W. Pearcy. 1983. Variation in quantum yield for CO_2 uptake among C_3 and C_4 plants. *Plant Physiol.* **73**:555–559.

Eik, K. and J. J. Hanway. 1965. Some factors affecting development and longevity of leaves of corn. *Agron. J.* **57**:7–12.

Ela, S. W., M. A. Anderson, and W. J. Brill. 1982. Screening and selection of maize to enhance associative bacterial nitrogen fixation. *Plant Physiol.* **70**:1564–1567.

Elawad, S. H., G. J. Gascho, and J. J. Street. 1982a. Response of sugarcane to silicate source and rate. I. Growth and yield. *Agron. J.* **74**:481–484.

Elawad, S. H., J. J. Street, and G. J. Gascho. 1982b. Response of sugarcane to silicate source and rate. II. Leaf freckling, and nutrient content. *Agron. J.* **74**:484–487.

Eldredge, J. C. 1935. The effect of injury in imitation of hail damage on the development of the corn plant. Iowa Agric. and Home Econ. Exp. Stn. Res. Bull. 185.

El-Hattab, H. S., M. A. Hussein, A. H. El-Hattab, M. S. A. Raouf, and A. A. El-Nomany. 1980. Growth analysis of maize plant in relation to grain yield as affected by nitrogen levels. *Z. Acker Pflanzenb.* **149**:46–57.

Ellis, Jr., R., J. F. Davis, and D. L. Thurlow. 1964. Zinc availability in calcareous Michigan soils as influenced by phosphorus level and temperature. *Soil Sci. Soc. Am. Proc.* **28**:83-87.

Ellis, R. P., J. C. Vogel, and A. Fuls. 1980. Photosynthetic pathways and geographical distribution of grasses in South-West Africa/Nambia. *S. Afr. J. Sci.* **76**:307-314.

Elmore, C. D. 1980. The paradox of no correlation between leaf photosynthetic rates and crop yields. In J. D. Hesketh and J. W. Jones (eds.), *Predicting Photosynthesis in Ecosystem Models*, Vol. II, pp. 155-167. CRC Press, Boca Raton, FL.

El-Sharkawy, M. A., J. H. Cock, and A. A. Held K. 1984. Water use efficiency of cassava (*Manihot esculenta* Crantz) II. Differing sensitivity of stomata to air humidity in cassava and other warm-climate species. *Crop Sci.* **24**:503-507.

Emmert, F. H. 1982. The stoichiometry of interactions between phosphorus and nutrient cations. *J. Plant Nutr.* **5**:1171-1176.

Engelstad, D. P. and S. E. Allen. 1971. Effect of form and proximity of added N on crop uptake of P. *Soil Sci.* **112**:330-337.

Epstein, E. 1966. Dual pattern of ion absorption by plant cells and by plants. *Nature* **212**:1324-1327.

Epstein, E. 1972. *Mineral Nutrition of Plants.* Wiley, New York.

Epstein, E. 1976. Kinetics of ion transport and the carrier concept. In U. Luttge and M. G. Pitman (eds.), *Transport in Plants IIB, Encycl. Plant Physiol.* Vol. 2, pp. 70-94. Springer-Verlag, Berlin.

Epstein, E. and C. E. Hagen. 1972. A kinetic study of the absorption of alkali cations by barley roots. *Plant Physiol.* **27**:457-474.

Eriksen, F. I. and A. S. Whitney. 1981. Effects of light intensity on growth of some tropical forage species. I. Interaction of light intensity and nitrogen fertilization on six forage grasses. *Agron. J.* **73**:427-433.

Esau, K. 1953. *Plant Anatomy.* Wiley, New York.

Escalada, R. G. and D. L. Plucknett. 1975a. Ratoon cropping of sorghum. I. Origin, time of appearance, and fate of tillers. *Agron. J.* **67**:473-478.

Escalada, R. G. and D. L. Plucknett. 1975b. Ratoon cropping of sorghum. II. Effect of daylength and temperature on tillering and plant development. *Agron. J.* **67**:479-484.

Escalada, R. G. and D. L. Plucknett. 1977. Ratoon cropping of sorghum. III. Effect of nitrogen and cutting height on ratoon performance. *Agron. J.* **69**:341-346.

Escano, C. R., C. A. Jones, and G. Uehara. 1981a. Nutrient diagnosis in corn grown on Hydric Dystrandepts: I. Optimum tissue nutrient concentrations. *Soil Sci. Soc. Am. J.* **45**:1135-1139.

Escano, C. R., C. A. Jones, and G. Uehara. 1981b. Nutrient diagnosis Dystrandepts: II. Comparison of two systems of tissue diagnosis. *Soil Sci. Soc. Am. J.* **45**:1140-1144.

Escolar, R. P. and F. W. Allison. 1976. Effect of water table depth on the yield of seven sugar cane varieties in Puerto Rico. *J. Agric. Univ. P. R.* **10**:228-237.

Eshel, A. and Y. Waisel. 1972. Variations in sodium uptake along primary roots of corn seedlings. *Plant Physiol.* **49**:585-589.

Eshel, A. and Y. Waisel. 1973. Variations in uptake of sodium and rubidium along barley roots. *Physiol. Plant.* **28**:557-560.

Estes, G. O., D. W. Koch, and T. F. Bruetsch. 1973. Influence of potassium nutrition on net CO_2 uptake and growth in maize (*Zea mays* L.). *Agron. J.* **65**:972-975.

Esty, J. C., A. B. Onken, L. R. Hossner, and R. Matheson. 1980. Iron use efficiency in grain sorghum hybrids and parental lines. *Agron. J.* **72**:589-592.

Etherton, B. and B. Rubinstein. 1978. Evidence for amino acid-H^+ co-transport in oat coleoptiles. *Plant Physiol.* **61**:933-937.

Evans, H. 1935. The root system of the sugar cane. *Emp. J. Exp. Agric.* **3**:351-362.

Evans, H. 1936. The root system of the sugar cane. II. Some typical root systems. *Emp. J. Exp. Agric.* **4**:208-221.

Evans, H. 1937. Further investigations on the root system of sugarcane. Mauritius Sugar Cane Res. Stn. Bull. 12.

Evans, H. 1960. Elements other than nitrogen, potassium and phosphorus in the mineral nutrition of sugar cane. *Int. Soc. Sugar Cane Technol. Proc.* **10**:473-508.

Evans, H. 1964a. The root system of sugarcane—An evaluation of its salient features. *Indian J. Sugarcane Res. Dev.* **8**:160-171.

Evans, H. F. and R. A. Wildes. 1971. Potassium and its role in enzyme activation. In *Potassium in Biochemistry and Physiology*, pp. 13-39. Proceedings of the 8th International Colloquium of the Potash Institute, Berne, Switzerland.

Evans, L. T. 1964b. Reproduction. In C. Barnard (ed.), *Grasses and Grasslands*, pp. 126-153. MacMillan, London.

Evans, L. T. 1975a. The physiological basis of crop yield. In L. T. Evans (ed.), *Crop Physiology: Some Case Histories*, pp. 327-355. Cambridge University Press, Cambridge.

Evans, L. T. 1975b. Beyond photosynthesis—the role of respiration, translocation and growth potential in determining productivity. In J. P. Cooper (ed.), International Biological Programme 3. *Photosynthesis and Productivity in Different Environments*, pp. 501-507. Cambridge University Press, Cambridge.

Evans, L. T. and R. B. Knox. 1969. Environmental control of reproduction in *Themeda australis*. *Aust. J. Bot.* **17**:375-389.

Evans, W. F. and F. C. Stickler. 1961. Grain sorghum seed germination under moisture and temperature stresses. *Agron. J.* **53**:369-372.

Evers, G. W. and E. C. Holt. 1972. Effects of defoliation treatments on morphological characteristics and carbohydrate reserves in kleingrass (*Panicum coloratum* L.). *Agron. J.* **64**:17-20.

Exner, B. B. 1971. Anatomy of the branch roots of sugarcane. *Int. Soc. Sugar Cane Tech. Proc.* **14**:739-745.

Fageria, N. K., M. P. B. Filho, and H. R. Gheyi. 1981. Avaliacao de cultivares de arroz para tolerancia a salinidade. *Pesq. Agropec. Bras.* **16**:677-681.

Fakorede, M. A. B. and J. J. Mock. 1978. Changes in morphological and physiological traits associated with recurrent selection for grain yield in maize. *Euphytica* **27**:397-409.

Falvey, J. L. 1981. *Imperata cylindrica* and animal production in south-east Asia: A review. *Tropical Grasslands* **15**:52-56.

Farina, M. P. W., M. E. Sumner, C. O. Plank, and W. S. Letzsch. 1980. Aluminum toxicity at near neutral soil pH levels. *J. Plant Nutr.* **2**:683-697.

Faris, J. A. 1927. Cold chlorosis of sugar cane. *Phytopath.* **16**:885-891.

Farquhar, R. H. and J. B. Lee. 1962. Variability associated with sugar cane leaf sampling for foliar diagnosis. *Int. Soc. Sugar Cane Technol. Proc.* **11**:203-214.

Farquhar, G. D. and T. D. Sharkley. 1982. Stomatal conductance and photosynthesis. *Annu. Rev. Plant Physiol.* **33**:163-203.

Farquhar, G. D., R. Wetselaar, and P. M. Firth. 1979. Ammonia volatilization from senescing leaves of maize. *Science* **203**:1257-1258.

Fehrenbacher, J. B., J. P. Vavra, and A. L. Lang. 1958. Deep tillage and deep fertilization experiments on a claypan soil. *Soil Sci. Soc. Am. Proc.* **22**:553-557.

Feldman, L. J. 1975. Cytokinins and quiescent center activity in roots of *Zea mays*. In J. G. Torry (ed.), *The Development and Function of Roots*, pp. 55-72. Academic Press, New York.

Fenton, R., W. J. Davies, T. A. Mansfield. 1977. The role of farnesol as a regulator of stomatal opening in *Sorghum*. *J. Exp. Bot.* **28**:1043-1053.

Fenton, R., T. A. Mansfield, and A. R. Wellburn. 1976. Effects of isoprenoid alcohols on oxygen exchange of isolated chloroplasts in relation to their possible physiological effects on stomata. *J. Exp. Bot.* **27**:1206-1214.

Fereres, E., E. Acevedo, D. W. Henderson, and T. C. Hsaio. 1978. Seasonal changes in water potential and turgor maintenance in sorghum and maize under water stress. *Physiol. Plant.* **44**:261-267.

Fergus, I. F. 1953. Manganese toxicity in an acid soil. *Queensl. J. Agric. Sci.* **10**:15-27.

Ferguson, I. B. 1979. The movement of calcium in non-vascular tissue of plants. *Commun. Soil Sci. Plant Anal.* **10**:217-224.

Ferguson, I. B. and D. T. Clarkson. 1975. Ion transport and endodermal suberization in the roots of *Zea mays*. *New Phytol.* **75**:69-79.

Ferguson, I. B. and D. T. Clarkson. 1976a. Ion uptake in relation to the development of a root hypodermis. *New Phytol.* **77**:11-14.

Ferguson, I. B. and D. T. Clarkson. 1976b. Simultaneous uptake and translocation of magnesium and calcium in barley (*Hordeum vulgare* L.) roots. *Planta* **128**:267-269.

Fernandes, M. S. and L. R. Freire. 1976. Efeitos de nitrogenio nitrico aplicado ao solo na actividade de nitrato-reductase e na acumulacao de N-soluvel em *Brachiaria* sp. *Turrialba* **26**:268-273.

Ferraris, R. and D. F. Sinclair. 1980. Factors affecting the growth of *Pennisetum purpureum* in the wet tropics. I. Short-term growth and regrowth. *Aust. J. Agric. Res.* **31**:899-913.

Firth, P., H. Thitipoca, S. Suthipradit, R. Wetselaar, and D. F. Beech. 1973. Nitrogen balance studies in the central plain of Thailand. *Soil Biol. Biochem.* **5**:41-46.

Fischer, K. S. and G. L. Wilson. 1971a. Studies of grain production in *Sorghum vulgare*. I. The contribution of pre-flowering photosynthesis to grain yield. *Aust. J. Agric. Res.* **22**:33-37.

Fischer, K. S. and G. L. Wilson. 1971b. Studies of grain production in *Sorghum vulgare*. II. Sites responsible for grain dry matter production during the post-anthesis period. *Aust. J. Agric. Res.* **22**:39-47.

Fischer, K. S. and G. L. Wilson. 1975. Studies of grain production in *Sorghum bicolor* (L. Moench). III. The relative importance of assimilate supply, grain growth capacity and transport system. *Aust. J. Agric. Res.* **26**:11-23.

Fischer, K. S., G. L. Wilson, and I. Duthie. 1976. Studies of grain production in *Sorghum bicolor* (L. Moench). VII. Contribution of plant parts to canopy photosynthesis and grain yield in field stations. *Aust. J. Agric. Res.* **27**:235-242.

Fischer, R. A. 1968. Stomatal opening: Role of potassium uptake by guard cells. *Science* **160**:784-785.

Fisher, F. L. and A. G. Caldwell. 1959. The effect of continued use of heavy rates of fertilizers on forage production and quality of Coastal bermudagrass. *Agron. J.* **51**:99-102.

Fisher, J. D., D. Hansen, and T. K. Hodges. 1970. Correlation between ion fluxes and ion-stimulated adenosine triphosphatase activity of plant roots. *Plant Physiol.* **46**:812-814.

Fisher, R. A. 1929. A preliminary note on the effect of sodium silicate on increasing the yield of barley. *J. Agr. Sci.* **19**:132-139.

Fleming, A. A. and J. H. Palmer. 1975. Variation in chlorophyll content of maize lines and hybrids. *Crop Sci.* **15**:617-620.

Flowers, T. J., P. F. Troke, and A. R. Yeo. 1977. The mechanism of salt tolerance in halophytes. *Annu. Rev. Plant Physiol.* **28**:89-121.

Follett, R. F., R. R. Allmaras, and G. A. Reichman. 1974. Distribution of corn roots in sandy soil with a declining water table. *Agron. J.* **66**:288-292.

Ford, C. W. and J. R. Wilson. 1981. Changes in levels in solutes during osmotic adjustment to water stress in leaves of four tropical pasture species. *Aust. J. Plant Physiol.* **8**:77-91.

Forde, B. J., H. C. M. Whitehead, and J. A. Rowley. 1975. Effect of light intensity and temperature on photosynthetic rate, leaf starch content, and ultrastructure *Paspalum dilatatum*. *Aust. J. Plant Physiol.* **2**:185-195.

Forster, H. and K. Mengel. 1969. The effect of a short term interruption in the K supply during the early stage on yield formation, mineral content, and soluble amino acid content. *Z. Acker Pflanzenb.* **130**:203-213.

Foth, H. D. 1962. Root and top growth of corn. *Agron. J.* **54**:49-52.

Fox, R. L. 1976. Sulfur and nitrogen requirements of sugarcane. *Agron. J.* **68**:891-896.

Fox, R. L. 1979. Comparative responses of field grown crops to phosphate concentrations in soil solutions. In H. Mussell and R. C. Staples (eds.), *Stress Physiology in Crop Plants*, pp. 81-106. Wiley, New York.

Fox, R. L. and E. J. Kamprath. 1970. Phosphate sorption isotherms for evaluating the phosphorus requirement of soils. *Soil Sci. Soc. Am. Proc.* **34**:902-907.

Fox, R. L. and B. T. Kang. 1978. Influence of phosphorus fertilizer placement and fertilization rate on maize nutrition. *Soil Sci.* **125**:34-40.

Fox, R. L., R. K. Nishimoto, J. R. Thompson, and R. S. de la Pena. 1974. Comparative external phosphorus requirements of plant growing in tropical soils. *Trans. Int. Cong. Soil Sci. (Moscow)* **4**:232-239.

Fox, R. L., J. A. Silva, D. L. Plucknett, and D. Y. Teranishi. 1969. Soluble and total silicon in sugar cane. *Plant Soil* **30**:81-92.

Fox, R. L., J. A. Silva, O. R. Younge, D. L. Plucknett, and G. D. Sherman. 1967. Soil and plant silicon and silicate response by sugarcane. *Soil Sci. Soc. Am. Proc.* **31**:775-779.

Foy, C. D. and J. C. Brown. 1963. Toxic factors in acid soils: I. Characterization of aluminum toxicity in cotton. *Soil Sci. Soc. Am. Proc.* **28**:403-407.

Foy, C. D., R. L. Chaney, and M. C. White. 1978. The physiology of metal toxicity in plants. *Ann. Rev. Plant Physiol.* **29**:511-566.

Foy, C. D., A. L. Fleming, and J. W. Schwartz. 1973. Opposite aluminum and manganese tolerances of two wheat varieties. *Agron. J.* **65**:123-126.

Foy, C. D., P. W. Voigt, and J. W. Schwartz. 1977. Differential susceptibilities of weeping lovegrass strains to an iron-related chlorosis on calcareous soils. *Agron. J.* **69**:491-496.

Foy, C. D., P. W. Voigt, and J. W. Schwartz. 1980. Differential tolerance of weeping lovegrass genotypes to acid coal mine spoils. *Agron. J.* **72**:859-862.

Francis, C. A. 1969. Identification of photoperiod insensitive strains of maize (*Zea mays* L.). *Crop Sci.* **9**:675-677.

Francis, C. A. 1973. The effects of photoperiod on growth and morphogenesis in maize (*Zea mays* L.) field trials in Colombia. In *Plant Response to Climatic Factors*, pp. 57-60. Proc. Uppsala Symp. 1970. UNESCO.

Frankland, B. 1981. Germination in shade. In H. Smith (ed.), *Plants and the Daylight Spectrum*, pp. 187-204. Academic Press, New York.

Fransolet, S., R. Deltour, R. Bronchart, and C. van de Walle. 1979. Changes in ultrastructure and transcription induced by elevated temperature in *Zea mays* embryonic root cells. *Planta* **146**:7-18.

Franzke, C. J. and A. M. Hume. 1945. Liberation of HCN in sorghum. *J. Am. Soc. Agron.* **37**:848-851.

Freitas, J. L. M. de, R. E. M. da Rocha, P. A. A. Pereira, and J. Dobereiner. 1982. Organic matter and inoculation with *Azospirillum* on nitrogen incorporation by field grown maize. *Pesq. Agropec. Bras.* **17**:1423-1432.

Frey, N. M. 1981. Dry matter accumulation in kernals of maize. *Crop Sci.* **21**:118-122.

Frick, H. and L. F. Bauman. 1978. Heterosis in maize as measured by K uptake characteristics of seedling roots. *Crop Sci.* **18**:99–103.

Frick, H. and L. F. Bauman. 1979. Heterosis in maize as measured by K uptake properties of seedling roots: Pedigree analyses of inbreds with high or low augmentation potential. *Crop Sci.* **19**:707–710.

Friedrich, J. W. and L. E. Schrader. 1978. Sulphur deprivation and nitrogen metabolism in maize seedlings. *Plant Physiol.* **61**:900–903.

Friedrich, J. W. and L. E. Schrader. 1979. N deprivation in maize during grain filling. II. Remobilization of ^{15}N and ^{35}S and the relationship between N and S accumulation. *Agron. J.* **71**:466–472.

Friedrich, J. W., L. E. Schrader, and E. V. Nordheim. 1979. N deprivation in maize during grain filling. I. Accumulation of dry matter, nitrate-N, and sulfate-S. *Agron. J.* **71**:461–465.

Friesen, D. K., A. E. R. Juo, and M. H. Miller. 1982. Residual value of lime and leaching of calcium in a kaolinitic Ultisol in the high rainfall tropics. *Soil Sci. Soc. Am. J.* **46**:1184–1189.

Friesen, D. K., M. H. Miller, and A. S. R. Juo. 1980. Liming and lime-phosphorus-zinc interactions in two Nigerian Ultisols: II. Effects on maize root and shoot growth. *Soil Sci. Sco. Am. J.* **44**:1227–1232.

Frolich, W. G., W. G. Pollmer, and D. Klein. 1980. Dry matter and protein accumulation in maize hybrids diverse for protein content under different Western European environments. In W. G. Pollmer and Phipps (eds.), *Improvement of Quality Traits of Maize for Grain and Silage Use*, pp. 199–220. Martinus Nijhoff, The Hague.

Fryrear, D. W. and W. G. McCully. 1972. Development of grass root systems as influenced by soil compaction. *J. Range Manage.* **25**:254–257.

Fuchs, A. 1968. Beziehungen swischen der Organogenese und der Ertragsbildung bei *Zea mays*. *Z. Pflanzenzuch.* **60**:260–283.

Fujii, T. and S. Isikawa. 1962. Effects or after-ripening on photoperiodic control of seed germination in *Eragrostis ferruginea* Beauv. *Bot. Mag. Tokyo* **75**:296–301.

Furbank, R. T. and M. R. Badger. 1982. Photosynthetic oxygen exchange in attached leaves of C_4 monocotyledons. *Aust. J. Plant Physiol.* **9**:553–558.

Fussell, L. K., C. J. Pearson, and M. J. T. Norman. 1980. Effect of temperature during various growth stages on grain development and yield of *Pennisetum americanum*. *J. Exp. Bot.* **31**:621–633.

Gaastra, P. 1959. Photosynthesis of crop plants are influenced by light, carbon dioxide, temperature, and stomatal diffusion resistance. *Meded. Landbouwhogesch. Wageningen* **59**:1–68.

Gallaher, R. N. and R. H. Brown. 1977. Starch storage in C_4 vs. C_3 grass leaf cells as related to nitrogen deficiency. *Crop Sci.* **17**:85–88.

Gallaher, R. N., W. L. Parks, and L. M. Josephson. 1972. Effect of levels of soil potassium, fertilizer potassium, and season on yield and ear leaf potassium content of corn inbreds and hybrids. *Agron. J.* **64**:645–647.

Gallais, A., M. Kellerhals, and F. Philippe. 1982. Etude des interactions entre epis chez le mais. *Agronomie* **2**:995–1004.

Gallo, J. R., R. Hiroce, and L. T. DeMiranda. 1968. Leaf analysis in corn plant nutrition. Part I. Correlations of leaf analysis and yield. *Bragantia* **27**:(15):177–186.

Gallopin, I. G. and P. A. Jolliffe. 1973. Effects of low nonfreezing temperatures on chlorophyll accumulation in corn and other grasses. *Crop Sci.* **13**:466–468.

Garber, P. G. 1977. Effect of light and chilling temperature on chilling-sensitive and chilling-resistant plants. *Plant Physiol.* **59**:981–985.

Gardner, J. C. 1980. The effect of seed size and density on field emergence and yield of pearl millet [*Pennisetum americanum* (L.) K. Schum.]. Master thesis, Kansas State University, Manhattan, KS.

Gardner, W. A. 1981. Insect control. In R. R. Duncan (ed.), *Ratoon Cropping of Sorghum for Grain in the Southeastern Untied States*, pp. 26–37. University of Georgia, Coll. Agric. Exp. Stn. Res. Bull. 269, Nov.

Garg, B. K., S. Kathju, A. N. Lahiri, and S. P. Vyas. 1981. Drought resistance in pearl millet. *Biol. Plant.* **23**:182–185.

Garner, W. W. and H. A. Allard. 1920. Effect of the relative length of day and night and other factors of the environment on growth and reproduction in plants. *J. Agric. Res.* **18**:553–606.

Garner, W. W. and H. A. Allard. 1923. Further studies on photoperiodism, the response of the plant to relative length of day and night. *J. Agric. Res.* **23**:871–920.

Garrard, L. A. and T. E. Humphrey. 1967. Effect of divalent cations on the leakage of sucrose from corn scutellum slices. *Phytochemistry* **6**:1085–1095.

Garrard, L. A. and S. H. West. 1972. Suboptimal temperature and assimilate accumulation in leaves of 'Pangola' digitgrass (*Digitaria decumbens* Stent.). *Crop Sci.* **12**:621–623.

Garrity, D. P., C. Y. Sullivan, and D. G. Watts. 1984. Changes in grain sorghum stomatal and photosynthetic response to moisture stress across growth stages. *Crop Sci.* **24**:441–446.

Garrity, D. P., D. G. Watts, C. Y. Sullivan, and J. R. Gilley. 1982a. Moisture deficits and grain sorghum performance: Effect of genotype and limited irrigation strategy. *Agron. J.* **74**:808–814.

Garrity, D. P., D. G. Watts, C. Y. Sullivan, and J. R. Gilley. 1982b. Moisture deficits and grain sorghum performance: Evapotranspiration—yield relationships. *Agron. J.* **74**:815–820.

Gascho, G. J. and H. J. Andreis. 1974. Sugarcane response to calcium silicate slag applied to organic and sand soils. *Int. Soc. Sugar Cane Technol. Proc.* **15**:543–551.

Gascho, G. J. and A. M. O. El Wali. 1979. Tissue testing in Florida sugarcane. *Sugar J.* **42**:(3):15–16.

Gascho, G. J. and S. F. Shih. 1979. Varietal response of sugarcane to water table depth. 1. Lysimeter performance and plant response. *Soil Crop Sci. Soc. Fla. Proc.* **38**:23–27.

Gascho, G. J. and F. A. Taha. 1972. Nutritional deficiency symptoms of sugarcane. Agricultural Experiment Stations, Institute of Food and Agricultural Sciences, Univ. Fla. Circ. 5-221.

Gaskel, M. L. and R. B. Pearce. 1980. Photosynthetic acclimation of maize to solar radiation level. *Maydica* **25**:55–64.

Gaskel, M. L. and R. B. Pearce. 1981. Growth analysis of maize hybrids differing in photosynthetic capability. *Agron. J.* **73**:817–821.

Gates, C. T. 1968. Water deficits and growth of herbaceos plants. In T. T. Kozlowski (ed.), *Water Deficits and Plant Growth*, pp. 135–190. Academic Press, New York.

Gaultney, L., G. W. Krutz, G. C. Steinhardt, and J. B. Liljedahl. 1982. Effects of subsoil compaction on corn yields. *Trans. ASAE* **25**:563–569.

Geijn, S. C. van de and F. Smeulders. 1981. Diurnal changes in the flux of calcium toward meristems and transpiring leaves in tomato and maize plants. *Planta* **151**:265–271.

Gerik, T. J. and F. R. Miller, 1984. Photoperiod and temperature effects on tropically and temperately adapted sorghum. *Field Crops Res.* **9**:29–40.

Giaquinta, R. T. 1977. Possible role of pH gradient and membrane ATPase in the loading of sucrose into the sieve tubes. *Nature (London)* **267**:369–370.

Giaquinta, R. T. 1979. Phloem loading of sucrose. Involvement of membrane ATPase and proton transport. *Plant Physiol.* **63**:744–748.

Gibson, T. S. and D. R. Leece. 1981. Estimation of physiologically active zinc in maize by biochemical assay. *Plant Soil* **63**:395–406.

Gielink, A. J., G. Sauer, and A. Ringoet. 1966. Histoautoradiographic localization of calcium in oat plant tissues. *Stain Technol.* **41**:281–286.

Gifford, R. M. 1974. A comparison of potential photosynthesis, productivity and yield of plant species with differing photosynthetic metabolism. *Aust. J. Plant Physiol.* **1**:107–117.

Giles, K. L., C. M. Bassett, and J. D. Eastin. 1975. The structure and ontogeny of the hilum region in *Sorghum bicolor. Aust. J. Bot.* **23**:795-802.

Gill, W. R. and R. D. Miller. 1956. A method for study of the influence of mechanical impedance and aeration on the growth of seedling roots. *Soil Sci. Soc. Am. Proc.* **20**:154-157.

Gilmore, E. C. and J. S. Rogers. 1958. Heat units as a method of measuring maturity in corn. *Agron. J.* **50**:611-615.

Gingrich, J. R. and M. B. Russell. 1956. Effect of soil moisture tension and oxygen concentration on the growth of corn roots. *Agron. J.* **48**:517-520.

Ginsburg, H. and B. Z. Ginzburg. 1974. Radial water and solute flows in roots of *Zea mays.* IV. Electrical potential profiles across the root. *J. Exp. Bot.* **25**:28-35.

Glass, A. D. M. and J. Dunlop. 1978. The influence of potassium content on the kinetics of K^+ influx into excised ryegrass and barley roots. *Planta* **141**:117-119.

Glasziou, K. T. and T. A. Bull. 1971. Feedback control of photosynthesis in sugarcane. In M. D. Hatch, C. B. Osmond, and R. O. Slatyer (eds.), *Photosynthesis and Photorespiration*, pp. 82-88. Wiley-Interscience, New York.

Glasziou, K. T., T. A. Bull, M. D. Hatch, and P. C. Whiteman. 1965. Physiology of sugarcane. VII. Effects of temperature, photoperiod duration, and diurnal and seasonal temperature changes on growth and ripening. *Aust. J. Biol. Sci.* **18**:53-66.

Glover, J. 1967. The simultaneous growth of sugarcane roots and tops in relation to soil and climate. *Proc. S. Afr. Sugar Cane Technol. Assoc.* **41**:143-159.

Glover, J. 1968. Further results from the Mount Edgecombe Root Laboratory. Proceedings of the South African Sugar Technology Association, April 1968, pp. 123-132.

Goatley, J. L. and R. W. Lewis. 1966. Composition of guttation fluid from rye, wheat and barley seedlings. *Plant Physiol.* **41**:373-375.

Goeschl, J. D., L. Rappaport, and H. K. Pratt. 1966. Ethylene as a factor regulating the growth of pea epicotyls subjected to physical stress. *Plant Physiol.* **41**:877-884.

Goins, T., J. Lunin, and H. L. Worley. 1966. Water-table effects on growth of tomatoes, snap beans, and sweet corn. *Trans. ASAE* **9**:530-533.

Goldsworthy, A. 1970. Photorespiration. *Bot. Rev.* **36**:321-340.

Goldsworthy, A. 1975. Photorespiration in relation to crop yield. In U. S. Gupta (ed.), *Physiological Aspects of Dryland Farming*, pp. 329-348. Oxford and IBH Publishing, New Delhi.

Goldsworthy, P. R. and M. Colegrove. 1974. Growth and yield of highland maize in Mexico. *J. Agric. Sci., Camb.* **83**:213-221.

Goldsworthy, P. R., A. F. E. Palmer, and D. W. Sparling. 1974. Growth and yield of lowland tropical maize in Mexico. *J. Agric. Sci., Camb.* **83**:223-230.

Gonske, R. G. and D. R. Keeney. 1969. Effect of fertilizer nitrogen, variety, and maturity on the dry matter yield and nitrogen fractions of corn grown for silage. *Agron. J.* **61**:72-76.

Gonzalez-Erico, E., E. J. Kamprath, G. C. Naderman, and W. V. Soares. 1979. Effect of depth of lime incorporation on the growth of corn on an Oxisol of Central Brazil. *Soil Sci. Soc. Am. J.* **43**:1155-1158.

Goodall, D. W. and F. G. Gregory. 1947. Chemical composition of plants as an index to their nutritional status. *Imp. Bur. Hort. Plantation Crops Tech. Comm.* **17**.

Goodman, M. M. 1976. Maize. In N. W. Simmonds (ed.), *Evolution of Crop Plants*, pp. 128-136. Longman, London.

Gorski, T. 1980. Annual cycle of the red and far red radiation. *Int. J. Biometeorol.* **24**:361-365.

Gorsline, G. W., J. L. Ragland, and W. I. Thomas. 1961. Evidence for inheritance of differential accumulation of calcium, magnesium and potassium by maize. *Crop Sci.* **1**:155-156.

Gorsline, G. W., W. I. Thomas, and D. E. Baker. 1968. Major gene inheritance of Sr-Ca, Mg, K, P, Zn, Cu, B, Al-Fe, and Mn concentration in corn (*Zea mays* L.). Pa. Agr. Exp. Stn. Bull. 746.

Gosnell, J. M. 1971. Some effects of a water-table level on the growth of sugarcane. *Int. Soc. Sugar Cane Technol. Proc.* **14**:841–849.

Gosnell, J. M. and A. C. Long. 1969. A sulphur deficiency in sugar cane. *S. Afr. Sugar Cane Technol. Assoc. Proc.* **43**:26–29.

Gosnell, J. M. and A. C. Long. 1971. Some factors affecting foliar analysis of sugarcane. *Proc. S. Afr. Sugar Technol. Assoc.* **45**:1–16.

Goss, R. L. 1968. The effects of potassium on disease resistance. In *Role of Potassium in Agriculture*, pp. 221–241. American Society of Agronomy, Madison, WI.

Goudriaan, J. and H. van Keulen. 1979. The direct and indirect effects of nitrogen shortage on photosynthesis and transpiration in maize and sunflower. *Neth. J. Agric. Sci.* **27**:227–234.

Goudriaan, J. and H. H. van Laar. 1978. Relations between leaf resistance, CO_2-concentration and CO_2-assimilation in maize, beans, lalang grass, and sunflower. *Photosynthetica* **12**:241–249.

Gould, F. W. 1968. *Grass Systematics.* McGraw-Hill, New York.

Gould, F. W. 1978. *Common Texas Grasses: An Illustrated Guide.* Texas A&M University Press, College Station.

Gould, F. W. and T. W. Box. 1965. *Grasses of the Texas Coastal Bend.* Texas A&M University Press, College Station.

Graber, L. F., N. T. Nelson, W. A. Lenke, and W. B. Albert. 1927. Organic food reserves in relation to the growth of alfalfa and other perennial herbaceous plants. *Wis. Agr. Exp. Stn. Bull.* **80**:1–128.

Grable, A. R. and E. G. Seimer. 1968. Effects of bulk density, aggregate size and soil water suction on oxygen diffusion, redox potential and elongation of corn roots. *Soil Sci. Soc. Am. Proc.* **32**:180–186.

Graham, E. R., P. L. Lopez, and T. M. Dean. 1972. Artificial light as a factor influencing yields of high-population corn. *Trans. Am. Soc. Agric. Eng.* **15**:576–579.

Graham, J. H., R. T. Leonard, and J. A. Menge. 1981. Membrane-mediated decrease in root exudation responsible for phosphorus inhibition of vesicular-arbuscular mycorrhiza formation. *Plant Physiol.* **68**:548–552.

Green, J. M. 1949. Effect of flaming on the growth of inbred lines of corn. *Agron. J.* **41**:144–146.

Grimes, D. W., R. J. Miller, and P. L. Wiley. 1975. Cotton and corn root development in two field soils of different strength characteristics. *Agron. J.* **67**:519–523.

Grobellaar, W. P. 1963. Responses of young maize plants to root temperature. *Meded. Landboushogesch. Wageningen* **63**:1–71.

Groves, R. H., M. W. Hagon, and P. S. Ramakrishnan. 1982. Dormancy and germination of seed of eight populations of *Themeda australis. Aust. J. Bot.* **30**:373–386.

Grunes, D. L., L. C. Boawn, C. W. Carlson, and F. G. Viets, Jr. 1961. Zinc deficiency of corn and potatoes as related to soil and plant analyses. *Agron. J.* **53**:68–71.

Guerrero, M. G., J. M. Vega, and M. Losada. 1981. The assimilatory nitrate-reducing system and its regulation. *Ann. Rev. Plant Physiol.* **32**:169–204.

Gumbs, F. A. and L. A. Simpson. 1981. Influence of flooding and soil moisture content on elongation of sugar cane in Trinidad. *Expl. Agric.* **17**:403–406.

Gupta, D. and I. Kovacs. 1974. Cold wave tolerance of maize seedlings. *Z. Acker Pflanzenb.* **140**:306–311.

Gupta, U. C. 1979. Boron nutrition of crops. *Adv. Agron.* **31**:273–307.

Gutierrez, M., V. E. Gracen, and G. E. Edwards. 1974a. Biochemical and cytological relationships in C_4 plants. *Planta* **119**:297–300.

Gutierrez, M., R. Kanai, C. B. Huber, S. B. Ku, and G. E. Edwards. 1974b. Photosynthesis in mesophyll protoplasts and bundle sheath cells of various types of C_4 plants. I. Carboxylases and CO_2 fixation studies. *Z. Pflanzenphysiol.* **72**:305–319.

Haberlandt, D. G. 1882. Vergleichende Anatomie des assimilatorischen Gewebesystems der Pflanzen.

Haberlandt, D. G. 1900. Ueber die Perception des geotropischen Reizes. *Ber. Deut. Bot. Ges.* **18**:261–272.

Hackel, E. 1887. Gramineae. In A. Engler and R. Prantl (eds.), *Die Naturlichen Pflanzenfamilien.* Vol. II, pp. 1–97.

Hackel, E. 1889. Andropogoneae. In A. DeCandolle and C. DeCandolle (eds.), Monogr. Phan., Vol. VI, pp. 1–716.

Hacker, J. B. 1974. Variation in oxalate, major cations, and dry matter digestibility of 47 introductions of the tropical grass *Setaria. Trop. Grassl.* **8**:145–154.

Hacker, J. B., B. J. Forde, and J. M. Gow. 1974. Simulated frosting of tropical grasses. *Aust. J. Agric. Sci.* **25**:45–57.

Halais, P. 1935. Un nouvel indice de climatologie agricole. *Rev. Agric. Maurice* **80**:44–49.

Halais, P. 1962. The detection of NPK deficiency trends in sugar cane crops by means of foliar diagnosis run from year to year on a follow-up basis. *Proc. ISSCT* **11**:214–221.

Halais, P. and C. Figon. 1969. Nitrogen and sulphur status of cane leaves as influenced by sulfate of ammonia applications. Mauritius Sugar Ind. Res. Inst. Annu. Rep. pp. 90–92.

Halim, A. H., C. E. Wassom, and R. Ellis, Jr. 1968. Zinc deficiency symptoms and zinc and phosphorus interactions in several strains of corn (*Zea mays* L.). *Agron. J.* **60**:267–271.

Hall, A. J., H. D. Ginzo, J. H. Lemcoff, and A. Soriano. 1980. Influence of drought during pollen-shedding on flowering, growth and yield of maize. *Z. Acker Pflanzenb.* **149**:287–298.

Hall, A. J., J. H. Lemcoff, and N. Trapani. 1981. Water stress before and during flowering in maize and its effect on yield, its components, and their determinants. *Maydica* **26**:19–38.

Hall, A. J., F. Vilella, N. Trapani, and C. Chimenti. 1982. The effect of water stress and genotype on the dynamics of pollen-shedding and silking in maize. *Field Crops Res.* **5**:349–363.

Hall, J. D., R. Barr, A. H. Al-Abbas, and F. L. Crane. 1972. The ultrastructure of chloroplasts in mineral-deficient maize leaves. *Plant Physiol.* **50**:404–409.

Hallmark, W. B. and R. C. Huffaker. 1978. The influence of ambient nitrate, temperature and light on nitrate assimilation in sudangrass seedlings. *Physiol. Plant.* **44**:147–152.

Hallock, D. L., R. H. Brown, and R. E. Blaser. 1965. Relative yield and composition of Ky. 31 fescue and coastal bermudagrass at four nitrogen levels. *Agron. J.* **57**:539–542.

Hansen, G. K. 1974a. Resistance to water transport in soil and young wheat plants. *Acta Agric. Scand.* **24**:37–48.

Hansen, G. K. 1974b. Resistance to soil water flow in soil and plants, plant water status, stomatal resistance and transpiration of Italian ryegrass, as influenced by transpiration demand and soil water depletion. *Acta Agric. Scand.* **24**:83–92.

Hansen, G. K. and C. R. Jensen. 1977. Growth and maintenance respiration in whole plants, tops, and roots of *Lolium multiflorum. Physiol. Plant.* **39**:155–164.

Hansen, J., D. Johnson, A. Lacis, S. Lebedeff, P. Lee, D. Rind, and G. Russell. 1981. Climate impact of increasing atmospheric carbon dioxide. *Science* **213**:957–966.

Hanson, W. D. 1971. Selection for differential productivity among juvenile maize plants; associated net photosynthetic rate and leaf area changes. *Crop Sci.* **11**:334–339.

Hanson, W. D. 1973. Changes in efficiencies and numbers of chloroplasts associated with divergent selections for juvenile productivity in *Zea mays* (L.). *Crop Sci.* **13**:386–387.

Hanson, A. D. and W. D. Hitz. 1983. Whole-plant response to water deficits: Water deficits and the nitrogen economy. In H. M. Taylor, W. R. Jordan, and T. R. Sinclair (eds.), *Limitations to Efficient Water Use in Crop Production*, pp. 331–343. American Society of Agronomy, Crop Science Society of Agronomy, Soil Science Society of America, Madison, WI.

Hanson, A. D., C. E. Nelson, A. R. Pedersen, and E. H. Everson. 1979. Capacity for proline accumulation during water stress in barley and its implications for breeding for drought resistance. *Crop Sci.* **19**:489–493.

Hanson, J. B. and A. J. Trewavas. 1982. Regulation of plant cell growth: The changing perspective. *New Phytol.* **90**:1–18.

Hanway, J. J. 1962a. Corn growth and composition in relation to soil fertility: I. Growth of different plant parts and relation between leaf weight and grain yield. *Agron. J.* **54**:145–217.

Hanway, J. J. 1962b. Corn growth and composition in relation to soil fertility: II. Uptake of N, P, and K and their distribution in different plant parts during the growing season. *Agron. J.* **54**:217–222.

Hanway, J. J. 1962c. Corn growth and composition in relation to soil fertility: III. Percentages of N, P, and K in different plant parts in relation to stage of growth. *Agron. J.* **54**:222–229.

Hanway, J. J. 1969. Defoliation effects on different corn (*Zea mays* L.) hybrids as influenced by plant population and stage of development. *Agron. J.* **61**:534–538.

Hanway, J. J., S. A. Barber, R. H. Bray, A. C. Caldwell, M. Fried, L. T. Kurtz, K. Lawton, J. T. Pesek, K. Pretty, M. Reed, and F. W. Smith. 1962. North Central regional potassium studies. III. Field studies with corn. Iowa Agr. and Home Econ. Exp. Stn. Bull. 503.

Hanway, J. J. and L. Dumenil. 1965. Corn leaf analysis—the key is correct interpretation. *Plant Food Rev.* **11**:(12):5–8.

Hardacre, A. K. and H. A. Eagles. 1980. Comparisons among populations of maize at 13°C. *Crop Sci.* **20**:780–784.

Harder, H. J., R. E. Carlson, and R. H. Shaw. 1982. Yield, yield components, and nutrient content of corn grain as influenced by post-silking moisture stress. *Agron. J.* **74**:275–278.

Hardjoamidjojo, S., R. W. Skaggs, and G. O. Schwab. 1982. Corn yield response to excessive soil water conditions. *Trans. ASAE* **25**:922–927, 934.

Harlan, J. R. 1976. Tropical and sub-tropical grasses. In N. W. Simmonds (ed.), *Evolution of Crop Plants*, pp. 142–144. Longman.

Harlan, J. R. and J. M. J. de Wet. 1972. A simplified classification of cultivated sorghum. *Crop Sci.* **12**:172–175.

Harms, C. L. and B. B. Tucker. 1973. Influence of nitrogen fertilization and other factors on yield, prussic acid, nitrate, and total nitrogen concentrations of sudangrass cultivars. *Agron. J.* **65**:21–26.

Harper, J. L. 1955. Studies in seed and seedling mortality. V. Direct and indirect influences of low temperatures on the mortality of maize. *New Phytol.* **55**:35–44.

Harper, R. J., R. J. Gilkes, and A. D. Robson. 1982. Biocrystalization of quartz and calcium phosphates in plants—a reexamination of the evidence. *Aust. J. Agric. Res.* **33**:565–571.

Harradine, A. R. 1980. The biology of African feather grass (*Pennisetum macrourum* Trin.) in Tasmania, II. Rhizome biology. *Weed Res.* **20**:171–175.

Harradine, A. R. and R. D. B. Whalley. 1981. A comparison of the root growth, root morphology, and root response to defoliation of *Aristida ramosa* R. Br. and *Danthonia linkii* Kunth. *Aust. J. Agric. Res.* **32**:565–574.

Harrison, M. A. and P. B. Kaufman. 1980. Hormonal regulation of lateral bud (tiller) release in oats (*Avena sativa* L.). *Plant Physiol.* **66**:1123–1127.

Harrison-Murray, R. S. and D. T. Clarkson. 1973. Relationships between structural development and the absorption of ions by the root system of *Cucurbita pepo*. *Planta* **114**:1–16.

Harrison-Murray, R. S. and R. Lal. 1979. High soil temperature and the response of maize to mulching in the lowland humid tropics. In R. Lal and D. J. Greenland (eds.), *Soil Physical Properties and Crop Production in Tropics*, pp. 285–304. Wiley, New York.

Hartley, W. 1950. The global distribution of tribes of the Gramineae in relation to historical and environmental factors. *Aust. J. Agric. Res.* **1**:355–373.

Hartley, W. 1958a. Studies on the origin, evolution, and distribution of the Gramineae. I. The tribe Andropogoneae. *Aust. J. Bot.* **6**:115–128.

Hartley, W. 1958b. Studies on the origin, evolution, and distribution of the Gramineae. II. The tribe Paniceae. *Aust. J. Bot.* **6**:343–357.

Hartley, W. and C. Slater. 1960. Studies on the origin, evolution, and distribution of the Gramineae. III. The tribes of the Subfamily Eragrostoideae. *Aust. J. Bot.* **8**:256–276.

Hartt, C. E. 1929. Potassium deficiency in sugar cane. *Bot. Gaz.* **88**:229–261.

Hartt, C. E. 1934a. Some affects of potassium upon the growth of sugar cane and upon the absorption and migration of ash constituents. *Plant Physiol.* **9**:399–452.

Hartt, C. E. 1934b. Some effects of potassium upon the amounts of protein and amino forms of nitrogen, sugars, and enzyme activity of sugar cane. *Plant Physiol.* **9**:453–490.

Hartt, C. E. 1955. The phosphorus nitrition of sugar cane. *Hawaii. Plant. Rec.* **55**:33–46.

Hartt, C. E. 1958. Total phosphorus in internodes 8–10 as a guide to the phosphorus fertilization of sugar cane. *Hawaii. Plant. Rec.* **55**:243–270.

Hartt, C. E. 1960. The growth of sugar cane as influenced by phosphorus. The critical range of plant phosphorus. *Int. Soc. Sugar Cane Technol. Proc.* **10**:467–473.

Hartt, C. E. 1963. Translocation as a factor in photosynthesis. *Naturwissenschaften* **50**:666–667.

Hartt, C. E. 1967. Effect of moisture supply upon translocation and storage of ^{14}C in sugarcane. *Plant Physiol.* **42**:338–346.

Hartt, C. E. 1969. Effect of potassium deficiency upon translocation of ^{14}C in attached blades and entire plants of sugarcane. *Plant Physiol.* **44**:1461–1469.

Hartt, C. E. 1970. Effect of potassium deficiency upon translocation of ^{14}C in detached blades of sugarcane. *Plant Physiol.* **45**:183–187.

Hartt, C. E. 1972. Translocation of carbon-14 in sugarcane plants supplied with or deprived of phosphorus. *Plant Physiol.* **49**:569–571.

Hartt, C. E. and H. P. Kortschak. 1967. Radioactive isotopes in sugarcane physiology. *Int. Soc. Sugar Cane Technol. Proc.* **12**:647–662.

Harty, R. L. and J. E. Butler. 1975. Temperature requirements for germination of green panic, *Panicum maximum* var. *trichoglume*, during the after-ripening period. *Seed Sci. Technol.* **3**:529–536.

Hatch, M. D. 1976. The C_4 pathway of photosynthesis: Mechanism and function. In R. H. Burris and C. C. Black (eds.), *CO₂ Metabolism and Plant Productivity*, pp. 59–81. University Park Press, Baltimore.

Hatch, M. D., T. Kagawa, and S. Craig. 1975. Subdivision of C_4-pathway species based on differing C_4 acid decarboxylating systems and ultrastructural features. *Aust. J. Plant Physiol.* **2**:111–128.

Hatch, M. D. and C. R. Slack. 1966. Photosynthesis by sugar-cane leaves. A new carboxylation reaction and the pathway of sugar formation. *Biochem. J.* **101**:103–111.

Hatch, M. D. and C. R. Slack. 1968. New enzyme for the interconversion of pyruvate and phosphopyruvate and its role in the carbon (4) dicarboxylic acid pathway of photosynthesis. *Biochem. J.* **106**:141–146.

Hatch, M. D. and C. R. Slack. 1970. Photosynthetic CO_2 fixation pathways. *Ann. Rev. Plant Physiol.* **21**:141–162.

Hatfield, J. L. 1977. Light response in maize: A review. In *Agrometeorology of the Maize (Corn) Crop*, pp. 199–206. World Meteorological Organization, Geneva, Switzerland.

Hattersley, P. W. 1982. $\delta^{13}C$ values of C_4 types in grasses. *Aust. J. Plant Physiol.* **9**:139–154.

Hattersley, P. W. and L. Watson. 1975. Anatomical parameters for predicting photosynthetic pathways of green leaves: The "maximum lateral cell count" and the "maximum cells distant count." *Phytomorphology* **25**:325–333.

Hattersley, P. W. and L. Watson. 1976. C_4 grasses: An anatomical criterion for distinguishing between NADP-malic enzyme species and PCK or NAD-malic enzyme species. *Aust. J. Bot.* **24**:297–308.

Hattersley, P. W. and Z. Roksandic. 1983. $\delta^{13}C$ values of C_3 and C_4 species of Australian *Neurachne* and its allies (Poaceae). *Aust. J. Bot.* **31**:317–321.

Hauser, V. L. and H. M. Taylor. 1964. Evaluation of deep tillage treatments on a slowly permeable soil. *Trans. Am. Soc. Agr. Eng.* **7**:134–136. 141.

Hawkins, R. C. and P. J. M. Cooper. 1981. Growth, development and grain yield of maize. *Expl. Agric.* **17**:203–207.

Haynes, R. J. and K. M. Goh. 1978. Ammonium and nitrate nutrition of plants. *Biol. Rev.* **53**:465–510.

Hays, R., C. P. P. Reid, T. V. St. John, and D. C. Coleman. 1982. Effects of nitrogen and phosphorus on blue grama growth and mycorrhizal infection. *Oecologia* **54**:260–265.

Haysom, M. B. C. and L. S. Chapman. 1975. Some aspects of the calcium silicate trials at Mackay. *Queensl. Soc. Sugar Cane Technol. Proc.* **42**:117–122.

Hecht-Buchholz, C. 1979. Calcium deficiency and plant ultrastructure. *Commun. Soil Sci. Plant Anal.* **10**:67–82.

Heichel, G. H. and R. B. Musgrave. 1969. Varietal differences in net photosynthesis of *Zea mays* L. *Crop Sci.* **9**:483–486.

Heilman, M. D. 1973. Salinity and iron effects on nutrient uptake by sorghum (*Sorghum bicolor*, L. Moench). Ph.D. dissertation, Texas A&M University, College Station.

Heinricher, E. 1884. Ueber isolaterolen Blattbaru mit besonderer Berucksichtigung der europaischen, speciell der deutschen Flora. *Jahrb. Wiss. Bot.* **15**:502–567.

Heirman, A. L. and H. A. Wright. 1973. Fire in the medium fuels of West Texas. *J. Range Manage.* **26**:331–335.

Hellebust, J. A. 1976. Osmoregulation. *Annu. Rev. Plant Physiol.* **27**:485–505.

Hellmers, H. and G. W. Burton. 1972. Photoperiod and temperature manipulation induces early anthesis in pearl millet. *Crop Sci.* **12**:198–200.

Henderson, M. S. and D. L. Robinson. 1982a. Environmental influences cn fiber component concentrations of warm-season perennial grasses. *Agron. J.* **74**:573–579.

Henderson, M. S. and D. L. Robinson. 1982b. Environmental influences on yield and *in vitro* true digestibility of warm-season perennial grasses and the relationships to fiber components. *Agron. J.* **74**:943–946.

Henry, J. E. and J. S. McKibben. 1966. The effect of soil strength on corn root penetration. ASAE paper no. 66–620.

Henson, I. E. 1981a. Changes in abscisic acid content during stomatal closure in pearl millet (*Pennisetum americanum* (L.) Leeke). *Plant Sci. Lett.* **21**:121–127.

Henson, I. E. 1981b. Abscisic acid and after-effects of water stress in pearl millet (*Pennisetum americanum* (L.) Leeke). *Plant Sci. Lett.* **21**:129–135.

Henson, I. E., G. Alagarswamy, V. Mahalakshmi, and F. R. Bidinger. 1982a. Diurnal changes in abscisic acid in leaves of pearl millet (*Pennisetum americanum* (L.) Leeke) under field conditions. *J. Exp. Bot.* **33**:416–425.

Henson, I. E., V. Mahalakshmi, G. Alagarswamy, and F. R. Bidinger. 1983. An association between flowering and reduced stomatal sensitivity to water stress in pearl millet (*Pennisetum americanum* (L.) Leeke). *Ann. Bot.* **52**:641–648.

Henson, I. E., V. Mahalakshmi, F. R. Bidinger, and G. Alagarswamy. 1982b. Osmotic adjustment to water stress in pearl millet (*Pennisetum americanum* (L.) Leeke) under field conditions. *Plant Cell Environ.* **5**:147–154.

Henson, I. E. and S. A. Quarrie. 1981. Abscisic acid accumulation in detached cereal leaves in response to water stress. I. Effects of incubation time and severity of stress. *Z. Pflanzenphysiol.* **101**:431–438.

Henzell, E. F. and D. J. Oxenham. 1973. The relation between growth and leaf nitrogen concentration in Nandi setaria and some other tropical grasses. *Commun. Soil Sci. Plant Anal.* **4**:147–154.

Henzell, R. G., K. J. McCree, C. H. M. van Bavel, and K. F. Schertz. 1975. Method for screening sorghum genotypes for stomatal sensitivity to water deficits. *Crop Sci.* **15**:516–518.

Henzell, R. G., K. J. McCree, C. H. M. van Bavel, and K. F. Schertz. 1976. Sorghum genotype variation in stomatal sensitivity to leaf water deficit. *Crop Sci.* **16**:660–662.

Herbel, C. H. and R. E. Sosebee. 1969. Moisture and temperature effects on emergence and initial growth of two range grasses. *Agron J.* **61**:628–631.

Herold, A. 1980. Regulation of photosynthesis by sink activity—the missing link. *New Phytol.* **86**:131–144.

Herrero, M. P. and R. R. Johnson. 1980. High temperature stress and pollen viability in maize. *Crop Sci.* **20**:796–800.

Herron, G. M., D. W. Grimes, and J. T. Musick. 1963. Effects of soil moisture and nitrogen fertilization of irrigated grain sorghum on dry matter production and nitrogen uptake at selected stages of plant development. *Agron. J.* **55**:393–396.

Hesketh, J. D. and D. Baker. 1967. Light and carbon assimilation by plant communities. *Crop Sci.* **7**:285–293.

Hesketh, J. D. and D. N. Moss. 1963. Variation in the response of photosynthesis to light. *Crop Sci.* **3**:107–110.

Heuer, B., Z. Plaut, and E. Federman. 1979. Nitrate and nitrite reduction in wheat leaves as affected by different types of water stress. *Physiol. Plant.* **46**:318–323.

Hewitt, E. J., D. P. Hucklesby, A. F. Mann, B. A. Notton, and G. J. Ricklidge. 1979. Regulation of nitrate assimilation in plants. In E. J. Hewitt and C. V. Cutting (eds.), *Nitrogen Assimilation of Plants*, pp. 255–287. Academic Press, London.

Hicks, D. R. and R. K. Crookston. 1976. Defoliation boosts corn yield. *Crops Soils* **29**:(3):12–13.

Hicks, D. R., W. W. Nelson, and J. H. Ford. 1977. Defoliation effects on corn hybrids adapted to the northern corn belt. *Agron. J.* **69**:387–390.

Hicks, D. R. and R. E. Stucker. 1972. Plant density effect on grain yield of corn hybrids diverse in leaf orientation. *Agron. J.* **64**:484–487.

Higinbotham, N. 1973. The mineral absorption process in plants. *Bot. Rev.* **99**:15–69.

Higinbotham, N. 1974. Conceptual developments in membrane transport, 1924–1974. *Plant Physiol.* **54**:454–462.

Higinbotham, N. and W. P. Anderson. 1974. Electrogenic pumps in higher plant cells. *Can. J. Bot.* **52**:1011–1021.

Higinbotham, N., R. F. Davis, S. M. Mertz, Jr., and L. K. Shumway. 1973. Some evidence that radial transport in corn roots is into living vessels. In W. P. Anderson (ed.), *Ion Transport in Plants*. Academic Press, New York.

Higinbotham, N., B. Etherton, and R. J. Foster. 1964. Effect of external K, NH_4, Na, Ca, Mg, and H ions on the cell transmembrane electropotential of *Avena* coleoptile. *Plant Physiol.* **39**:196–203.

Higinbotham, N., B. Etherton, and R. J. Foster. 1967. Mineral ion contents and cell transmembrane electropotentials of pea and oat seedling tissue. *Plant Physiol.* **42**:37–46.

Hill, J. 1980. The remobilization of nutrients from leaves. *J. Plant Nutr.* **2**:407–444.

Hilliard, J. H. 1975. *Eragrostis curvula*: Influence of low temperatures on starch accumulation, amylolytic activity and growth. *Crop Sci.* **15**:293–294.

Hilliard, J. H. and S. H. West. 1970. Starch accumulation associated with growth reduction at low temperatures in a tropical plant. *Science* **168**:494-496.

Hiron, R. W. P. and S. T. C. Wright. 1973. The role of endogenous abscisic acid in the response of plants to stress. *J. Exp. Bot.* **24**:769-781.

Hirose, M. 1973. Comparison of physiological and ecological characteristics between tropical and temperate grass species. Extension Bulletin No. 26, Food and Fertilizer Technology Center, Taipei City, Taiwan.

Hiroto, H. and S. Watanabe. 1980. Endogenous factors affecting the varietal differences in the curvature of seminal roots of *Zea mays* L. seedlings in water culture. *Jpn. J. Crop Sci.* **50**:148-152.

Hitchcock, A. S. 1935. Manual of the Grasses of the United States. U.S. Dept. Agr. Misc. Publ. 200.

Hitchcock, A. S. 1951. *Manual of the Grasses of the United States*, 2nd ed. (Revised by Agnes Chase). U.S. Dept. Agr. Misc. Publ. 200.

Hodges, T. K. 1973. Ion absorption by plant roots. *Adv. Agron.* **25**:163-207.

Hodges, T. K. 1976. ATPase associated with membranes of plant cells. In Lüttge, U. and M. G. Pitman (eds.), *Encyclopedia of Plant Physiology, New Series*, Vol. 2A, pp. 260-283. Springer-Verlag, Berlin.

Hodges, T. K. and Y. Vaadia. 1964. Uptake and transport of radiochloride and tritiated water by various zones of onion roots of different chloride status. *Plant Physiol.* **39**:104-108.

Hodgkinson, K. C. and H. G. Baas Becking. 1977. Effect of defoliation on root growth of some arid zone perennial plants. *Aust. J. Agric. Res.* **29**:31-42.

Hoffman, G. J. and J. A. Jobes. 1978. Growth and water relations of cereal crops as affected by salinity and relative humidity. *Agron. J.* **70**:765-769.

Hofmann, W. C., M. K. O'Neill, and A. K. Dobrenz. 1984. Physiological responses of sorghum hybrids and parental lines to soil moisture stress. *Agron. J.* **76**:223-228.

Hofner, W. and R. Grieb. 1979. Effect of Fe- and Mo-deficiency on the iron content of monocotyledons and dicotyledons with different susceptibility to chlorosis. *Z. Pflanzen. Bodenkd.* **142**:626-638.

Hofstra, G. and C. D. Nelson. 1969. The translocation of photosynthetically assimilated ^{14}C in corn. *Can. J. Bot.* **47**:1435-1442.

Hojjati, S. M., T. H. Taylor, and W. C. Templeton, Jr. 1972. Nitrate accumulation in rye, tall fescue, and bermudagrass as affected by nitrogen fertilization. *Agron. J.* **62**:624-627.

Holm, L. G., D. L. Plucknett, J. V. Pancho, and J. P. Herberger. 1977. *The World's Worst Weeds*. The University Press of Hawaii, Honolulu.

Holobrada, M., I. Mistrik, and J. Kolek. 1981. The effect of temperature on the uptake and loss of anions by seedling roots of *Zea mays* L. *Biol. Plant.* **23**:241-248.

Holt, E. C. and E. C. Bashaw. 1963. Factors affecting seed production of dallisgrass. Texas Agric. Exp. Stn. Misc. Publ. 662.

Horowitz, M. 1972. Seasonal development of established johnsongrass. *Weed Sci.* **20**:392-395.

Hosegood, P. H. 1963. The root distribution Kikuyu grass and wattle trees. *E. Afr. Agric. For. J.* **29**:60-61.

Hoshikawa, K. 1969. Underground organs of the seedlings and the systematics of the Gramineae. *Bot. Gaz.* **130**:192-203.

Hoshino, T. and R. R. Duncan. 1982. Sorghum tannin content during maturity under different environmental conditions. *Jpn. J. Crop Sci.* **51**:178-184.

Howe, O. W. and H. F. Rhodes. 1955. Irrigation practice for corn in relation to stage of plant development. *Soil Sci. Soc. Am. Proc.* **19**:94-98.

Hsiao, T. C. 1970. Rapid changes in levels of polyribosomes in *Zea mays* in response to water stress. *Plant Physiol.* **46**:281-285.

Hsiao, T. C. 1973. Plant response to water stress. *Annu. Rev. Plant Physiol.* **24**:519–570.

Hsiao, T. C., E. Acevedo, E. Fereres, and O. W. Henderson. 1976. Water stress, growth, and osmotic adjustment. *Phil. Trans. R. Soc. Lond. B* **273**:479–500.

Hsu, H. H., H. D. Ashmeed, and D. J. Graff. 1982. Absorption and distribution of foliar iron by plants. *J. Plant Nutr.* **5**:969–974.

Huett, D. O. and R. C. Menary. 1979. Aluminum uptake by excised roots of cabbage, lettuce, and kikuyu grass. *Aust. J. Plant Physiol.* **6**:643–653.

Huett, D. O. and R. C. Menary. 1980a. Effect of aluminum on growth and nutrient uptake of cabbage, lettuce, and kikuyu grass in nutrient solution. *Aust. J. Agric. Res.* **31**:749–761.

Huett, D. O. and R. C. Menary. 1980b. Aluminum distribution in freeze-dried roots of cabbage, lettuce, and kikuyu grass by energy-dispensive X-ray analysis. *Aust. J. Plant Physiol.* **7**:101–111.

Huffaker, R. C., T. Radin, G. E. Kleinkopf, and E. L. Cox. 1970. Effects of mild water stress on enzymes of nitrate assimilation and of the carboxylative phase of photosynthesis in barley. *Crop Sci.* **10**:471–474.

Hull, H. M., L. N. Wright, and C. A. Bleckman. 1978. Epicuticular wax ultrastructure among lines of *Eragrostis lehmanniana* Nees. developed for seedling drought tolerance. *Crop Sci.* **18**:699–704.

Humbert, R. P. 1968. *The Growing of Sugar Cane.* Elsevier, New York.

Humbert, R. P. and J. P. Martin. 1955. Nutritional deficiency symptoms in sugarcane: molybdenum, calcium, zinc, boron, iron, manganese, magnesium. *Hawaiian Planters' Record* **60**:95.

Humble, G. D. and T. C. Hsiao. 1969. Specific requirement of potassium for light-activated opening of stomata in epidermal strips. *Plant Physiol.* **44**:230–234.

Humble, G. D. and T. C. Hsiao. 1970. Light-dependent influx and efflux of potassium of guard cells during stomatal opening and closing. *Plant Physiol.* **46**:483–487.

Humble, G. D. and K. Raschke. 1971. Stomatal opening quantitatively related to potassium transport. *Plant Physiol.* **48**:447–453.

Hume, D. J. and D. K. Campbell. 1972. Accumulation and translocation of soluble solids in corn stalks. *Can. J. Plant Sci.* **52**:363–368.

Humphreys, L. R. 1975. *Tropical Pasture and Seed Production*, 116 pp. F.A.O. Publication, Rome.

Humphreys, L. R. and A. R. Robinson. 1966. Interrelations of leaf area and non-structural carbohydrate status as determinants of the growth of sub-tropical grasses. Proceedings of the 10th International Grassland Congress (Helsinki) pp. 113–116.

Hunter, R. B., M. Tollenaar, and C. M. Breuer. 1977. Effect of photoperiod and temperature on vegetative and reproductive growth of a maize (*Zea mays*) hybrid. *Can. J. Plant Sci.* **57**:1127–1133.

Hurney, A. P. 1973. A progress report on calcium silicate investigations. *Queensl. Soc. Sugar Cane Technol. Proc.* **40**:109–113.

Hyder, D. N., A. C. Everson, and R. E. Bement. 1971. Seedling morphology and seedling failures with blue grama. *J. Range Manage.* **24**:287–292.

Hylmo, B. 1953. Transpiration and ion absorption. *Physiol. Plant.* **6**:333–405.

Hylmo, B. 1955. Passive components in the ion absorption of the plant. I. The zonal ion and water absorption in Brouwer's experiments. *Physiol. Plant.* **8**:433–449.

Hylmo, B. 1958. Passive components in the ion absorption of the plant. II. The zonal water flow, ion passage and pore size in roots of *Vicia faba*. *Physiol. Plant.* **11**:382–400.

Idso, S. B. 1980. Carbon dioxide and climate. *Science* **210**:7–8.

Idso, S. B. 1981. Carbon dioxide—an alternative view. *New Sci.* **92**:444–446.

Idso, S. B. 1983a. Carbon dioxide and global temperature: What the data show. *J. Env. Qual.* **12**:159–163.

Idso, S. B. 1983b. Stomatal regulation of evapotranspiration from well-watered plant canopies: A new synthesis. *Agric. Meteorol.* **29**:213–217.

Ilahi, I. and K. Dorffling. 1982. Changes in abscisic acid and proline levels in maize varieties of different drought resistance. *Physiol. Plant.* **55**:129–135.

Imai, H., M. Fukuyama, Y. Yamada, and T. Harada. 1973. Comparative studies on the photosynthesis of higher plants. III. Differences in response to various factors affecting the photosynthetic rate between C_4 and C_3 plants. *Soil Sci. Plant Nutr.* **19**:61–71.

Ingle, J., L. Beevers, and R. H. Hageman. 1964. Metabolic changes associated with germination of corn. I. Changes in weight and metabolites and their redistribution in the embryo axis, scutellum and endosperm. *Plant Physiol.* **39**:735–740.

Inosaka, M., K. Ito, H. Numaguchi, and M. Misumi. 1977. Studies on the productivity of some tropical grasses. 4. Effect of shading on heading habit of some tropical grasses. *Jpn. J. Trop. Agric.* **20**:236–239.

Inuyama, S., J. T. Musick, and D. A. Dusek. 1976. Effect of plant water deficits at various growth stages on growth, grain yield, and leaf water potential of irrigated grain sorghum. *Proc. Crop Sci. Soc. Jpn.* **45**:298–307.

Iremiren, G. D. and G. M. Milbourn. 1980. Effects of plant density on ear barrenness in maize. *Expl. Agric.* **16**:321–326.

Irvine, J. E. 1967. Photosynthesis in sugarcane varieties under field conditions. *Crop Sci.* **7**:297–300.

Isarangkura, R., D. Peaslee, and R. Lockhard. 1978. Utilization and redistribution of Zn during vegetative growth of corn. *Agron. J.* **70**:243–246.

Isbell, V. R. and P. W. Morgan. 1982. Manipulation of apical dominance in sorghum with growth regulators. *Crop Sci.* **22**:30–35.

Iserman, K. 1970. The effect of adsorption processes in the xylem on the calcium distribution in higher plants. *Z. Pflanzenernaehr. Bodenk.* **126**:191–203.

Isley, D. 1950. The cold-test for corn. *Proc. Int. Seed Test Assoc.* **16**:299–311.

Ivory, D. A., T. H. Stobbs, M. N. McLeod, and P. C. Whiteman. 1974. Effect of day and night temperatures on estimated dry matter digestibility of *Cenchrus ciliaris* and *Pennisetum clandestinum*. *J. Aust. Inst. Agric. Sci.* **40**:156–158.

Ivory, D. A. and P. C. Whiteman. 1978a. Effects of environmental and plant factors on foliar freezing resistance in tropical grasses. I. Precondition factors and conditions during freezing. *Aust. J. Agric. Res.* **29**:243–259.

Ivory, D. A. and P. C. Whiteman. 1978b. Effects of environmental and plant factors on foliar freezing resistance in tropical grasses. II. Comparison of frost resistance between cultivars of *Cenchrus ciliaris, Chloris gayana* and *Setaria anceps*. *Aust. J. Agric. Res.* **29**:261–266.

Ivory, D. A. and P. C. Whiteman. 1978c. Effect of temperature on growth of five subtropical grasses. I. Effect of day and night temperature on growth and morphological development. *Aust. J. Plant Physiol.* **5**:131–148.

Ivory, D. A. and P. C. Whiteman. 1978d. Effect of temperature on growth of five subtropical grasses. II. Effect of low night temperature. *Aust. J. Plant Physiol.* **5**:149–157.

Iwata, F. 1975. Ear barrenness of corn as affected by plant population. *JARQ* **9**:13–17.

Jackson, M. B. and P. W. Barlow. 1981. Root geotropism and the role of growth regulators from the cap: A reexamination. *Plant Cell Environ.* **4**:107–123.

Jackson, M. B., M. C. Drew, and S. C. Gifford. 1981. Effects of applying ethylene to the root system of *Zea mays* on growth and nutrient concentration in relation to flooding tolerance. *Physiol. Plant.* **52**:23–28.

Jackson, M. B., B. Herman, and A. Goodenough. 1982. An examination of the importance of ethanol in causing injury to flooded plants. *Plant Cell Environ.* **5**:163–172.

Jackson, T. L., J. Hay, and D. P. Moore. 1967. The effect of Zn on yield and chemical composition of sweet corn in the Willamette Valley. *Am. Soc. Hort. Sci.* **91**:462–471.

Jackson, W. A., D. Flesher, and R. H. Hageman. 1973. Nitrate uptake by dark grown corn seedlings. *Plant Physiol.* 51:120-127.

Jacques, G. L., R. L. Vanderlip, and D. A. Whitney. 1975a. Growth and nutrient accumulation and distribution in grain sorghum. I. Dry matter production and Ca and Mg uptake and distribution. *Agron. J.* 67:607-611.

Jacques, G. L., R. L. Vanderlip, and R. Ellis, Jr. 1975b. Growth and nutrient accumulation and distribution in grain sorghum. II. Zn, Cu, Fe, and Mn uptake and distribution. *Agron. J.* 67:611-616.

Jacques-Felix, H. 1955. Notes sur les Graminees d'Afrique Tropicale. VIII. Les tribus de la serie oryzoide. *J. Agr. Trop. Bot. Appl.* 2:600-619.

Jain, S. K. 1976. The evolution of inbreeding in plants. *Annu. Rev. Ecol. Syst.* 7:469-495.

Jain, T. C. 1971. Contribution of stem, laminae and ears to the dry matter production of maize (*Zea mays* L.) after ear emergence. *Indian J. Agric. Sci.* 41:579-583.

Jambunathan, R. 1980. Improvement of nutritional quality of sorghum and pearl millet. In W. Santos, N. Lopres, J. J. Barbosa, and D. Chaves (eds.), *Nutrition and Food Science*, Vol. 2, pp. 39-53. Plenum, New York.

James, N. I. 1980. Sugarcane. In W. R. Fehr and H. H. Hadley (eds.), *Hybridization of Crop Plants*. American Society of Agronomy and Crop Science Society of America, Madison, WI.

Jarvis, P. and C. R. House. 1970. Evidence for symplasmic ion transport in maize roots. *J. Exp. Bot.* 21:83-90.

Jat, R. L., M. S. Dravid, D. K. Das, and N. N. Goswami. 1975. Effect of flooding and high soil water condition on root porosity and growth of maize. *J. Indian Soc. Soil Sci.* 23:291-297.

Jenison, J. R., D. B. Shank, and L. H. Penny. 1981. Root characteristics of 44 maize inbreds evaluated in four environments. *Crop Sci.* 21:233-237.

Jenne, E. A., H. F. Rhoades, C. H. Yien, and O. W. Howe. 1958. Change in nutrient element accumulation by corn with depletion of soil moisture. *Agron. J.* 50:71-74.

Jenny, H. 1966. Pathway of ions from soil into root according to diffusion models. *Plant Soil* 25:265-289.

Jensen, C. R., J. Letey, and L. H. Stolzy. 1964. Labelled oxygen: transport through growing corn roots. *Science* 144:550-552.

Jensen, S. D. and A. J. Cavalieri. 1983. Drought tolerance in U.S. maize. *Agric. Water Manage.* 7:223-236.

Johann, H. 1935. The histology of the caryopsis of yellow dent corn with reference to resistance and susceptibility to kernel rots. *J. Agr. Res.* 51:855-883.

John, C. D. and A. Lauchli. 1980. Metabolic adaptation in mature roots of salt-stressed *Zea mays*. *Ann. Bot.* 46:395-400.

John, C. D. and H. Greenway. 1976. Alcoholic fermentation and activity of some enzymes in rice roots under anaerobiosis. *Aust. J. Plant Physiol.* 3:325-336.

Johnson, I. R. 1983. Nitrate uptake and respiration in roots and shoots: A model. *Physiol. Plant.* 58:145-147.

Johnson, R. R. 1978. Growth and yield of maize as affected by early-season defoliation. *Agron. J.* 70:995-998.

Johnson, R. R. and D. N. Moss. 1976. Effect of water stress on $^{14}CO_2$ photosynthesis and translocation in wheat during grain filling. *Crop Sci.* 16:697-701.

Johnson, W. J. and R. Dickens. 1976. Centipede cold tolerance as affected by environmental factors. *Agron. J.* 68:83-85.

Johnston, M. A. and R. A. Wood. 1971. Soil compaction studies at Pangola. Proceedings of the South African Sugar Technology Association, June, 1971.

Jones, C. A. 1979. The potential of *Andropogon gayanus* Kunth in the Oxisol and Ultisol savannas of Tropical America. *Herb. Abstr.* 49:1-8.

Jones, C. A. 1980. A review of evapotranspiration studies in irrigated sugarcane in Hawaii. *Hawaii. Plant. Rec.* **59**:195-214.

Jones, C. A. 1982. The role of phosphorus in the Hawaii sugar industry. *Hawaii. Plant. Rec.* **59**:229-264.

Jones, C. A. 1983. A survey of the variability in tissue nitrogen and phosphorus concentrations in maize and grain sorghum. *Field Crops Res.* **6**:133-147.

Jones, C. A. and J. E. Bowen. 1981. Comparative DRIS and Crop Log diagnosis of sugarcane tissue analyses. *Agron. J.* **73**:941-944.

Jones, C. A. and A. Carabaly. 1981. Some characteristics of the regrowth of 12 tropical grasses. *Trop. Agric.* **58**:37-44.

Jones, E. C. and R. J. Barnes. 1967. Non-volatile organic acids of grasses. *J. Sci. Food Agric.* **18**:321-324.

Jones, J. B., Jr. 1970. Distribution of fifteen elements in corn leaves. *Commun. Soil Sci. Plant Anal.* **1**:27-33.

Jones, J. B., Jr. 1967. Interpretation of plant analysis for several crops. In L. M. Walsh and J. B. Beaton (eds.), *Soil Testing and Plant Analysis, Part 2*, pp. 40-58. SSSA Special Publication, Series No. 2, Soil Science Society of America, Madison, WI.

Jones, J. B. and H. V. Eck. 1973. Plant analysis as an aid in fertilizing corn and grain sorghum. In L. M. Walsh and J. D. Beaton (eds.), *Soil Testing and Analysis*. Rev. ed., pp. 349-363. Soil Science Society of America, Madison, WI.

Jones, L. H. P. and K. A. Handreck. 1965. Studies of silica in the oat plant. III. Uptake of silica from soils by the plant. *Plant Soil* **23**:79-92.

Jones, L. H. P. and A. A. Milne. 1963. Studies of silica in the oat plant. I. Chemical and physical properties of the silica. *Plant Soil* **18**:207-220.

Jones, L. H. P., A. A. Milne, and S. M. Wadham. 1963. Studies of silica in the oat plant. II. Distribution of the silica in the plant. *Plant Soil* **18**:358-371.

Jones, M. M., C. B. Osmond, and N. C. Turner. 1980. Accumulation of solutes in leaves of sorghum and sunflower in response to water deficits. *Aust. J. Plant Physiol.* **7**:193-205.

Jones, M. M. and H. M. Rawson. 1979. Influence of rate of development of leaf water deficits upon photosynthesis, leaf conductance, water use efficiency, and osmotic potential in sorghum. *Physiol. Plant.* **45**:103-111.

Jones, M. M. and N. C. Turner. 1978. Osmotic adjustment in leaves of sorghum in response to water deficits. *Plant Physiol.* **61**:122-126.

Jones, R. K., M. E. Probert, and B. J. Crack. 1975. The occurrence of sulphur deficiency in the Australian tropics. In K. K. McLachlan (ed.), *Sulphur in Australasian Agriculture*. Sydney University Press, Sydney.

Jones, R. J. and C. W. Ford. 1972. Some factors affecting the oxalate content of the tropical grass *Setaria sphacelata*. *Aust. J. Exp. Agric. Anim. Husb.* **12**:400-406.

Jones, R. J., B. G. Gengenbach, and V. B. Cardwell. 1981. Temperature effects on *in vitro* kernal development of maize. *Crop Sci.* **21**:761-766.

Jones, R. J., A. A. Seawright, and D. A. Little. 1970. Oxalate poisoning in animals grazing the tropical grass *Setaria sphacelata*. *J. Aust. Inst. Agric. Sci.* **36**:41.

Jones, R. J. and S. R. Simmons. 1983. Effect of altered source-sink ratio on growth of maize kernals. *Crop Sci.* **23**:129-134.

Jordan, C. W., C. E. Evans, and R. D. Rouse. 1966. Coastal bermudagrass response to applications of P and K as related to P and K levels in the soil. *Soil Sci. Soc. Am. Proc.* **30**:477-480.

Jordan, H. V., K. D. Laird, and D. D. Ferguson. 1950. Growth rates and nutrient uptake by corn in a fertilizer-spacing experiment. *Agron. J.* **42**:261-268.

Jordan, W. R., W. A. Dugas, Jr., and P. J. Shouse. 1983a. Strategies for crop improvement for drought-prone regions. *Agric. Water Manage.* **7**:281-299.

Jordan, W. R., R. L. Monk, F. R. Miller, D. T. Rosenow, L. E. Clark, and P. J. Shouse. 1983b. Environmental physiology of sorghum. I. Environmental and genetic control of epicuticular wax load. *Crop Sci.* **23**:552-558.

Josephson, L. M. 1962. Effects of potash on premature stalk dying and lodging in corn. *Agron. J.* **54**:179-180.

Juang, T. C. and G. Uehara. 1971. Effect of ground-water table and soil compaction on nutrient element uptake and growth of sugarcane. *Int. Soc. Sugar Cane Technol. Proc.* **14**:679-687.

Julander, O. 1945. Drought resistance in range and pasture grasses. *Plant Physiol.* **20**:573-599.

Julien, M. H. R. and G. C. Soopramanien. 1975. Effects of night breaks on floral initiation and development in *Saccharum*. *Crop Sci.* **15**:625-629.

Jungk, A. and S. A. Barber. 1974. Phosphate uptake rate of corn roots as related to the proportion of the roots exposed to phosphate. *Agron. J.* **66**:554-557.

Juniper, B. E. and G. Pask. 1973. Directional secretion by the Golgi bodies in maize root cells. *Planta* **109**:225-231.

Jurgens, S. K., R. R. Johnson, and J. S. Boyer. 1978. Dry matter production and translocation in maize subjected to drought during grain fill. *Agron. J.* **70**:678-682.

Kadman-Zahavi, A. 1977. Dependence of the effects of end-of-day red of far red irradiations on the duration of the following dark periods. *Plant Physiol. Suppl.* **59**:49.

Kaigama, B. K., I. D. Teare, L. R. Stone, and W. L. Powers. 1977. Root and top growth of irrigated and nonirrigated grain sorghum. *Crop Sci.* **17**:555-559.

Kamprath, E. J. 1967. Residual effect of large applications of phosphorus on high phosphorus fixing soils. *Agron. J.* **59**:25-27.

Kamprath, E. J. 1970. Exchangeable aluminum as a criterion for liming leached mineral soils. *Soil Sci. Soc. Am. Proc.* **34**:252-254.

Kamprath, E. G., R. H. Moll, and N. Rodriguez. 1982. Effects of nitrogen fertilization and recurrent selection on performance of hybrid populations of corn. *Agron. J.* **74**:955-958.

Kamprath, E. J., W. L. Nelson, and J. W. Fitts. 1956. The effect of pH, sulfate and phosphate concentrations on the adsorption of sulfate by soil. *Soil Sci. Soc. Am. Proc.* **20**:463-466.

Kamprath, E. J. and M. E. Watson. 1980. Conventional soil and tissue tests for assessing the phosphorus status of soils. In F. E. Khasowneh, E. C. Sample, and E. J. Kamprath (eds.), *The Role of Phosphorus in Agriculture*, pp. 433-469. Am. Soc. Agron. and Crop Sci. Soc. Am., Soil Sci. Soc. Am., Madison, WI.

Kanai, R. and M. Kashiwagi. 1975. *Panicum milioides*, a Gramineae plant having Kranz leaf anatomy without C_4 photosynthesis. *Plant Cell Physiol.* **16**:669-679.

Kanemasu, E. T. and C. K. Hiebsch. 1975. Net carbon exchange of wheat, sorghum and soybean. *Can. J. Bot.* **53**:382-389.

Kannan, S. 1980. Correlative influence of pH reduction on recovery from iron chlorosis in sorghum varieties. *J. Plant Nutr.* **2**:507-516.

Kannan, S. 1981. The reduction of pH and recovery from chlorosis in Fe-stressed sorghum seedlings: The principal role of adventitious roots. *J. Plant Nutr.* **4**:73-78.

Kannan, S. 1982. Genotypic differences in iron uptake and utilization in some crop cultivars. *J. Plant Nutr.* **5**:531-542.

Kannan, S. 1983. Cultivar differences for tolerance to Fe and Zn deficiency: A comparison of two maize hybrids and their parents. *J. Plant Nutr.* **6**:323-337.

Kannan, S. and D. P. Pandley. 1982. Absorption and transport of iron in some crop cultivars. *J. Plant Nutr.* **5**:395-403.

Kannan, S. and S. Ramani. 1982. Zinc-stress response in some sorghum hybrids and parent cultivars: Significance of pH reduction and recovery from chlorosis. *J. Plant Nutr.* **5**:219-227.

Kannangara, T., R. C. Durley, G. M. Simpson, and D. G. Stout. 1982. Drought resistance of *Sorghum bicolor*. 4. Hormonal changes in relation to drought stress in field-grown plants. *Can. J. Plant Sci.* **62**:317-330.

Kannangara, T., N. Seetharama, R. C. Durley, and G. M. Simpson. 1983. Drought resistance of *Sorghum bicolor*. 6. Changes in endogenous growth regulators of plants grown across an irrigation gradient. *Can. J. Plant Sci.* **63**:147-155.

Kapulnik, Y., J. Kigel, Y. Okon, I. Nur, and Y. Henis. 1981a. Effect of *Azospirillum* inoculation on some growth parameters and N-content of wheat, sorghum, and panicum. *Plant Soil* **61**:65-70.

Kapulnik, Y., S. Sarig, I. Nur, Y. Okon, J. Kigel, and Y. Henis. 1981b. Yield increases in summer cereal crops in Israeli fields inoculated with *Azospirillum*. *Exp. Agric.* **17**:179-187.

Karamanos, A. J. 1979. Water stress: A challenge for the future of agriculture. In T. K. Scott (ed.), *Plant Regulation and World Agriculture*, pp. 415-450. Plenum Press, New York.

Karas, I. and M. E. McCulley. 1973. Further studies of the histology of lateral root development in *Zea mays*. *Protoplasma* **77**:243-269.

Karbassi, P., L. A. Garrard, and S. H. West. 1971a. Reversal of low temperature effects on a tropical plant by gibberellic acid. *Crop Sci.* **11**:755-757.

Karbassi, P., S. H. West, and L. A. Garrard. 1971b. Amylolytic activity in leaves of a tropical and a temperate grass. *Crop Sci.* **12**:58-60.

Karnok, K. J. and J. B. Beard. 1983. Effects of gibberellic acid on the CO_2 exchange rates of bermudagrass and St. Augustinegrass when exposed to chilling temperatures. *Crop Sci.* **23**:514-517.

Kashirad, A. and H. Marschner. 1974. Iron nutrition of sunflower and corn plants in mono and mixed culture. *Plant Soil* **41**:91-101.

Kathju, S., A. N. Lahiri, and K. A. Shankarnarayan. 1978. Influence of seed size and composition on the dry matter yield of *Cenchrus ciliaris*. *Experientia* **34**:848-849.

Kaufman, P. B., W. C. Bigelow, L. B. Petering, and F. B. Drogosz. 1969. Silica in developing epidermal cells of *Avena* internodes: Electron microprobe analysis. *Science* **166**:1015-1017.

Kawanabe, S. 1968. Temperature responses and systematics of the Gramineae. *Proc. Jap. Soc. Plant Taxon.* **2**:17-20.

Kawasaki, T. and M. Moritsugu. 1979. A characteristic symptom of calcium deficiency in maize and sorghum. *Commun. Soil Sci. Plant Anal.* **10**:41-56.

Kays, S. J., C. W. Nicklow, and D. H. Simons. 1974. Ethylene in relation to the response of roots to physical impedance. *Plant Soil* **40**:565-571.

Kemp, D. R. and W. M. Blacklow. 1982. The responsiveness to temperature of the extension rates of leaves of wheat growing in the field under different levels of nitrogen fertilizer. *J. Exp. Bot.* **33**:29-36.

Kemp, P. R., G. L. Cunningham, and H. P. Adams. 1983. Specialization of mesophyll morphology in relation to C_4 photosynthesis in the Poaceae. *Am. J. Bot.* **70**:349-354.

Kersting, J. F., A. W. Pauli, and R. C. Stickler. 1961b. Grain sorghum caryopsis development. II. Changes in chemical composition. *Agron. J.* **53**:74-77.

Kersting, J. F., F. C. Stickler, and A. W. Pauli. 1961a. Grain sorghum caryopsis development. I. Changes in dry weight, moisture percentage and viability. *Agron. J.* **53**:36-38.

Khalid, R. A., J. A. Silva, and R. L. Fox. 1978. Residual effects of calcium silicate in tropical soils: I. Fate of applied silicon during five years cropping. *Soil Sci. Soc. Am. Proc.* **42**:89-94.

Khan, A. A. and G. K. Zende. 1977. The site for Zn-P interactions in plants. *Plant Soil* **46**:259-262.

Khan, A. G. 1972. The effect of vesicular-arbuscular mycorrhizal associations on growth of cereals. I. Effects on maize growth. *New Phytol.* **71**:613-619.

Khan, A. G. 1975. The effect of VA mycorrhizal associations on growth of cereals. II. Effects on wheat growth. *Ann. Appl. Biol.* **80**:27-36.

Khanna-Chopra, R., G. S. Chaturverdi, P. K. Aggarwal, and S. K. Sinha. 1980. Effect of potassium on growth and nitrate reductase during water stress and recovery in maize. *Physiol. Plant.* **49**:495–500.

Kiesselbach, T. A. 1948. Endosperm type as a physiologic factor in corn yield. *J. Am. Soc. Agron.* **40**:216–236.

Kiesselbach, T. A. 1950. Progressive development and seasonal variation of the corn crop. *Neb. Agr. Exp. Stn. Res. Bull.* **166**:43–45.

Kiesselbach, T. A. and E. R. Walker. 1952. Structure of certain specialized tissues in the kernel of corn. *Am. J. Bot.* **39**:561–569.

Kigel, J. and A. Dotan. 1982. Effect of different durations of water withholding on regrowth potential and non-structural carbohydrate content in Rhodes grass. *Aust. J. Plant Physiol.* **9**:113–120.

Kimball, B. A. 1983. Carbon dioxide and agricultural yield: An assemblage and analysis of 430 prior observations. *Agron. J.* **75**:779–787.

Kiniry, J. R., J. T. Ritchie, and R. L. Musser. 1983a. Dynamic nature of the photoperiod response in maize. *Agron. J.* **75**:700–703.

Kiniry, J. R., J. T. Ritchie, R. L. Musser, E. P. Flint, and W. C. Iwig. 1983b. The photoperiod sensitive interval in maize. *Agron. J.* **75**:637–690.

Kirkby, E. A. 1968. Ion uptake and ionic balance in plants in relation to the form of nitrogen nutrition. In I. H. Rorison (ed.), *Ecological Aspects of the Mineral Nutrition of Plants*, pp. 214–222. Br. Ecol. Soc. Symp. No. 9. Blackwell, Oxford.

Kirkby, E. A. 1974. Recycling of potassium in plants considered in relation to ion uptake and organic acid accumulation. In J. Wehrmann (ed.), *Plant Analysis and Fertilizer Problems*, Vol. 2, pp. 557–558. Proceedings of the 7th International Plant Analysis and Fertilizer Problems Colloquium, Hannover.

Kittrick, J. A. and M. L. Jackson. 1955. Common ion effect on phosphate solubility. *Soil Sci.* **79**:415–421.

Klar, A. E., J. A. Usberti, Jr., and D. W. Henderson. 1978. Differential responses of guinea grass populations to drought stress. *Crop Sci.* **18**:853–857.

Kleinendorst, A. 1975. An explosion of leaf growth after stress conditions. *Neth. J. Agric. Sci.* **23**:139–144.

Kleinendorst, A. and R. Brouwer. 1970. The effect of temperature on the root medium and of the growing point of the shoot on growth, water content, and sugar content of maize leaves. *Neth. J. Agric. Sci.* **18**:140–148.

Kleinendorst, A. and R. Brouwer. 1972. The effect of local cooling on growth and water content of plants. *Neth. J. Agric. Sci.* **20**:203–217.

Klett, W. E., D. Hollingsworth, and J. L. Schuster. 1971. Increasing utilization of weeping lovegrass by burning. *J. Range Manage.* **24**:22–24.

Kneebone, W. R. and C. L. Cremer. 1955. The relationship of seed size to seedling vigor in some native grass species. *Agron. J.* **47**:472–477.

Knox, R. B. and J. Heslop-Harrison. 1963. Experimental control of aposporous apomixis in a grass of the Andropogoneae. *Bot. Not.* **116**:139–141.

Koch, D. W. and G. O. Estes. 1975. Influence of potassium stress on growth, stomatal behavior, and CO_2 assimilation in corn. *Crop Sci.* **15**:697–699.

Koehler, P. H., P. H. Moore, C. A. Jones, A. Dela Cruz, and A. Maretzki. 1982. Response of drip-irriated sugarcane to drought stress. *Agron. J.* **74**:906–911.

Konings, H. and G. Verschuren. 1980. Formation of aerenchyma in roots of *Zea mays* in aerated solutions, and its relation to nutrient supply. *Physiol. Plant.* **49**:265–270.

Kortschak, H. P., C. E. Hartt, and G. O. Burr. 1954–1959. Ann. Rep. Hawaii. Sugar Plant Assoc.

Kortschak, H. P., C. E. Hartt, and G. O. Burr. 1965. Carbon dioxide fixation in sugar cane leaves. *Plant Physiol.* **40**:209-213.

Kramer, P. J. 1981. Carbon dioxide concentration, photosynthesis, and dry matter production. *BioSci.* **31**:29-33.

Kramer, P. J. 1938. Root resistance as a cause of the absorption lag. *Am. J. Bot.* **25**:110-113.

Krantz, B. A. and S. W. Melsted. 1964. Nutrient deficiencies of corn, sorghums, and small grains. In H. B. Sprague (ed.), *Hunger Signs in Crops*, 3rd ed., pp. 25-57. David McKay, New York.

Kresge, C. B. 1974. Effect of fertilization on winterhardiness of forages. In D. A. Mays (ed.), *Forage Fertilization*, pp. 437-453. American Society of Agronomy, Crop Science Society of America, Soil Science Society of America, Madison, WI.

Kresge, C. B. and A. M. Decker. 1966. Nutrient balance in 'Midland' bermudagrass as affected by differential nitrogen and potassium fertilization: I. Forage yields and persistence. Proceedings of the 9th International Grassland Congress (Sao Paulo, Brazil). pp. 671-674.

Kriedemann, P. E., B. R. Loveys, G. L. Fuller, and A. C. Leopold. 1972. Abscisic acid and stomatal regulation. *Plant Physiol.* **49**:842-847.

Krieg, D. R. 1983. Photosynthetic activity during stress. *Agric. Water Manage.* **7**:249-263.

Krieg, D. R. and J. R. Rice. 1975. Seed development of four sorghum cultivars. Prog. Rep. 3312. Tex. Agric. Exp. Stn.

Krishna, K. R. and D. J. Bagyaraj. 1981. Note on the effect of VA mycorrhiza and soluble phosphate fertilizer on sorghum. *Indian J. Agric. Sci.* **51**:688-690.

Krishna, K. R., H. M. Suresh, J. Syamsunder, and D. J. Bagyaraj. 1981. Changes in the leaves of finger millet due to VA mycorrhizal infection. *New Phytol.* **87**:717-722.

Krishnamurthy, K. 1973. Partial analysis of grain yield from different shoots as influenced by nitrogen in finger millet. *Agron. J.* **65**:856-858.

Krishnamurthy, K., B. G. Rajashekara, M. K. Jagamath, A. B. Gowda, G. Raghunatha, and N. Venugopal. 1973. Photosynthetic efficiency of sorghum genotypes after head emergence. *Agron. J.* **65**:858-860.

Krizek, D. T., J. H. Bennett, J. C. Brown, T. Zaharieva, and K. H. Norris. 1982. Photochemical reduction of iron. I. Light reactions. *J. Plant Nutr.* **5**:323-333.

Ku, S. B., G. E. Edwards, and D. Smith. 1978. Photosynthesis and non-structural carbohydrates concentration in leaf blades of *Panicum virgatum* as affected by night temperature. *Can. J. Bot.* **56**:63-68.

Kuiper, P. J. C. 1964. Water uptake of higher plants as affected by root temperatures. *Meded. Landbouwhogesch. Wageningen* **64**:1-11.

Kumar, K. A., S. C. Gupta, and D. J. Andrews. 1983. Relationship between nutritional quality characters and grain yield in pearl millet. *Crop Sci.* **23**:232-235.

Kunth, C. S. 1833. Enumeration plantarum. 1. Agrostographic enumeratio Graminearum.

Kuznetsova, G. A., M. G. Kuznetsova, and G. M. Grineva. 1981. Characteristics of water exchange and anatomical-morphological structure in corn plants under conditions of flooding. *Soviet Plant Physiol.* **28**:241-248.

Laetsch, W. M. 1974. The C_4 syndrome: A structural analysis. *Ann. Rev. Plant Physiol.* **25**:27-52.

Lafever, H. N. 1981. Genetic differences in plant response to soil nutrient stress. *J. Plant Nutr.* **4**:89-109.

Lahiri, A. N. and B. C. Kharabanda. 1961. Dimorphic seeds in some arid zone grasses and their significance of growth differences in their seedlings. *Sci. Cult. (Calcutta)* **27**:448-450.

Lahiri, A. N. and B. C. Kharabanda. 1963. Germination studies on arid zone plants. II. Germination inhibitors in the spikelet glumes of *Lasiurus sindicus, Cenchrus ciliaris,* and *Cenchrus setigerus. Ann. Arid Zone* **1**:114-126.

Lahiri, A. N. and B. C. Kharabanda. 1964. Germination studies on arid zone plants. III. Some factors influencing the germination of grass seeds. *Nat. Inst. Sci. India Proc.* **30B**:186-196.

Lahiri, A. N. and S. Singh. 1968. Studies on plant-water relationships. IV. Impact of water deprivation on the nitrogen metabolism of *Pennisetum typhoides*. *Proc. Nat. Inst. Sci. India.* **34B**:313-322.

Lahiri, A. N. and S. Singh. 1969. Effect of hyperthermia on the nitrogen metabolism of *Pennisetum typhoides*. *Proc. Nat. Inst. of Sci. India.* **35B**:131-138.

Lahiri, A. N. and S. Singh. 1970. Studies on plant-water relationships. V. Influence of soil moisture on plant performance and nitrogen status of the shoot tissue. *Proc. Indian Nat. Sci. Acad.* **46B**:112-125.

Lai, T. M. and M. M. Mortland. 1961. Diffusion of ions in bentonite and vermiculite. *Soil Sci. Soc. Am. Proc.* **25**:353-356.

Lal, R. and P. R. Maurya. 1982. Root growth of some tropical crops in uniform columns. *Plant Soil* **68**:193-206.

Lal, R. and G. S. Taylor. 1969. Drainage and nutrient effects in a field lysimeter study: I. Corn yield and soil conditions. *Soil Sci. Soc. Am. Proc.* **33**:937-941.

Lal, R. and G. S. Taylor. 1970. Drainage and nutrient effects in a field lysimeter study: II. Mineral uptake by corn. *Soil Sci. Soc. Am. Proc.* **34**:245-248.

Lambert, R. J. and R. R. Johnson. 1978. Leaf angle, tassel morphology, and the performance of maize hybrids. *Crop Sci.* **18**:499-502.

Landsburg, E.-Ch. 1981. Organic acid synthesis and release of hydrogen ions in response to Fe deficiency stress of mono- and dicotyledonous plant species. *J. Plant Nutr.* **3**:579-581.

Landua, D. P., A. R. Swoboda, and G. W. Thomas. 1973. Response of coastal bermudagrass to soil applied sulfur, magnesium, and potassium. *Agron. J.* **65**:541-544.

Lanning, F. C. and T. Garabedian. 1963. Distribution of ash, calcium, iron, and silica in the tissues of young sorghum plants. *Trans. Kans. Acad. Sci.* **66**:443-448.

Lanning, F. C. and Y. Y. Linko. 1961. Absorption and deposition of silica by four varieties of sorghum. *J. Agr. Food Chem.* **9**:463-465.

Lanning, F. C., B. W. X. Ponnaiya, and F. C. Crumpton. 1958. The chemical nature of silica in plants. *Plant Physiol.* **33**:339-343.

Larson, W. E. and W. H. Pierre. 1953. Interaction of sodium and potassium on yield and cation composition of selected crops. *Soil Sci.* **76**:51-64.

Lauchli, A. 1967. Investigations on the distribution and transport of ions in plant tissues with the X-ray Microanalyzer. I. Experiments on vegetative organs of *Zea mays. Planta* **75**:185-207.

Lauchli, A., A. R. Spurr, and E. Epstein. 1971. Lateral transport of ions into the xylem of corn roots. II. Evaluation of a stelar pump. *Plant Physiol.* **48**:118-124.

Lavy, T. L. and S. A. Barber. 1964. Movement of molybdenum in soil and its effects on availability to the plant. *Soil Sci. Soc. Am. Proc.* **28**:93-97.

Lawanson, A. O. 1976. Effect of prior heat stress on photophosphorylation in seedlings of *Zea mays. Phyton* **34**:51-54.

Lawton, K. 1945. The influence of soil aeration on the growth and absorption of nutrients by corn plants. *Soil Sci. Soc. Am. Proc.* **10**:263-268.

Leath, K. T. and R. H. Ratcliffe. 1974. The effect of fertilization on disease and insect resistance. In D. A. Mays (ed.), *Forage Fertilization*, pp. 481-503. American Society of Agronomy, Crop Science Society, Soil Science Society of American, Madison, WI.

LeCroy, W. C. and J. R. Orsenigo. 1964. Sugarcane culture in the Florida Everglades. *Soil Crop Sci. Soc. Fla., Proc.* **24**:436-440.

Lee, H. A. 1926a. Progress report on the distribution of cane roots in the soil under plantation conditions. *Hawaii. Plant Rec.* **30**:511-519.

Lee, H. A. 1926b. A comparison of the root weights and distribution of H109 and D1135 cane varieties. *Hawaii. Plant Rec.* **30**:520-523.

Lee, K. K., R. W. Holst, I. Watanabe, and A. App. 1981. Gas transport through rice. *Soil Sci. Plant Nutr.* **27**:151-158.

Lee, K., R. C. Lommasson, and J. D. Eastin. 1974. Developmental studies on the panicle initiation in sorghum. *Crop Sci.* **14**:80-84.

Lee, S. S. and G. O. Estes. 1982. Corn physiology in short season and low temperature environments. *Agron. J.* **74**:325-331.

Leece, D. R. 1976. Occurrence of physiologically inactive zinc in maize on a black earth soil. *Plant Soil* **44**:481-486.

Leece, D. R. 1978a. Effects of boron on the physiological activity of zinc in maize. *Aust. J. Agric. Res.* **29**:739-747.

Leece, D. R. 1978b. Distribution of physiologically inactive zinc in maize growing on a black earth soil. *Aust. J. Agric. Res.* **29**:749-758.

Leikam, D. F., L. S. Murphy, D. E. Kissel, D. A. Whitney, and H. C. Moser. 1983. Effects of nitrogen and phosphorus application method and nitrogen source on winter wheat grain yield and leaf tissue phosphorus. *Soil Sci. Soc. Am. J.* **47**:530-535.

Lemeur, R. and B. L. Blad. 1974. A critical review of major types of light models for estimating the shortwave regime of plant canopies. *Meded. Fac. Landbouwwet. Rijksuniv. Gent.* **39**:1535-1585.

Leonard, R. T. and T. K. Hodges. 1973. Characterization of plasma membrane-associated adenosine triphosphatase activity of oat roots. *Plant Physiol.* **52**:6-12.

Leonard, R. T. and C. W. Hotchkiss. 1976. Cation-stimulated adenosine triphosphatase activity and cation transport in corn roots. *Plant Physiol.* **58**:331-335.

Leonce, F. S. and M. H. Miller. 1966. A physiological effect of nitrogen on phosphorus absorption by corn. *Agron. J.* **58**:245-249.

Leopold, A. C. 1949. The control of tillering in grasses by auxin. *Am. J. Bot.* **36**:437-440.

Lessani, H. and H. Marschner. 1978. Relation between salt tolerance and long-distance transport of sodium and chloride in various crop species. *Aust. J. Plant Physiol.* **5**:27-37.

Letey, J., L. H. Stolzy, and G. B. Blanck. 1962. Effect of duration and timing of low soil oxygen content on shoot and root growth. *Agron. J.* **54**:34-37.

Levitt, J. 1980. *Responses of Plants to Environmental Stresses*, Vol. II: *Water, Radiation, Salt, and Other Stresses*, 2nd edition. Academic Press, New York.

Lewis, D. G. and J. P. Quirk. 1967. Phosphate diffusion in soil and uptake by plants. III. P^{31} movement and uptake by plants as indicated by P^{32}-autoradiography. *Plant Soil* **26**:445-453.

Lewis, R. B., E. A. Hiler, and W. R. Jordan. 1974. Susceptibility of grain sorghum to water deficit at three growth stages. *Agron. J.* **66**:589-591.

Liebhardt, W. C. and J. T. Murdock. 1965. Effect of potassium on morphology and lodging of corn. *Agron. J.* **57**:325-328.

Lindsay, W. L. 1974. Role of chelation in micronutrient availability. In E. W. Carson (ed.), *The Plant Root and Its Environment*, pp. 507-514. University Press of Virginia, Charlottesville.

Lindsay, W. L. 1979. *Chemical Equilibria in Soils*. Wiley, New York.

Lindsay, W. L. and A. P. Schwab. 1982. The chemistry of iron in soils and its availability to plants. *J. Plant Nutr.* **5**:821-840.

Linn, D. M. and J. W. Doran. 1984. Effect of water-filled pore space on carbon dioxide and nitrous oxide production in tilled and nontilled soils. *Soil Sci. Soc. Am. J.* **48**:1267-1272.

Lockman, R. B. 1972a. Mineral composition of grain sorghum plant samples. Part I: Comparative analysis with corn at various stages of growth and under different environments. *Commun. Soil Sci. Plant Anal.* **3**:271-281.

Lockman, R. B. 1972b. Mineral composition of grain sorghum plant samples. Part II: As affected by soil acidity, soil fertility, stage of growth, variety and climatic factors. *Commun. Soil Sci. Plant Anal.* **3**:283–293.

Lockman, R. B. 1972c. Mineral composition of grain sorghum plant samples. Part III: Suggested nutrient sufficiency limits at various stages of growth. *Commun. Soil Sci. Plant Anal.* **3**:295–303.

Loneragan, J. F. and K. Snowball. 1969. Calcium requirements of plants. *Aust. J. Agric. Res.* **20**:465–478.

Lonnquist, J. H. and R. W. Jugenheimer. 1943. Factors affecting the success of pollination in corn. *J. Am. Soc. Agron.* **35**:923–933.

Loomis, W. E. 1934–1935. The translocation of carbohydrates in maize. *Iowa State Coll. J. Sci.* **9**:509–520.

Loomis, W. E. 1945. Translocation of carbohydrates in maize. *Science* **101**:398–400.

Louwerse, W. and W. vd. Zweerde. 1977. Photosynthesis, transpiration, and leaf morphology of *Phaseolus vulgaris* and *Zea mays* grown at different sunlight. *Photosynthetica* **11**:11–21.

Lucas, R. E. and B. D. Knezek. 1972. Climatic and soil conditions promoting micronutrient deficiencies in plants. In J. J. Mortvedt, P. M. Giordano, and W. L. Lindsay (eds.), *Micronutrients in Agriculture*, pp. 265–288. Soil Science Society of America, Madison, WI.

Ludlow, M. M. 1975. Effect of water stress on the decline of leaf net photosynthesis with age. In R. Marcelle (ed.), *Environmental and Biological Control of Photosynthesis*, pp. 123–134. Junk, The Hague.

Ludlow, M. M. and A. O. Taylor. 1974. Effect of sub-zero tempratures on the gas exchange of buffel grass. In R. L. Bieleski, A. R. Ferguson, and M. M. Cresswell (eds.), *Mechanisms of Regulation of Plant Growth. Royal Soc. New Zealand Bull.* **12**:513–518.

Ludlow, M. M. and D. A. Charles-Edwards. 1980. Analysis of the regrowth of a tropical grass/legume sward subjected to different frequencies and intensities of defoliation. *Aust. J. Agric. Res.* **31**:673–692.

Ludlow, M. M. and T. T. Ng. 1974. Water stress suspends leaf ageing. *Plant Sci. Lett.* **3**:235–240.

Ludlow, M. M. and T. T. Ng. 1976. Effect of water deficit on carbon dioxide exchange and leaf elongation rate of *Panicum maximum* var. *trichoglume. Aust. J. Plant Physiol.* **3**:401–413.

Ludlow, M. M. and T. T. Ng. 1977. Leaf elongation rate in *Panicum maximum* var. *trichoglume* following removal of water stress. *Aust. J. Plant Physiol.* **4**:263–272.

Ludlow, M. M., T. T. Ng, and C. W. Ford. 1980. Recovery after water stress of leaf gas exchange in *Panicum maximum* var. *trichoglume. Aust. J. Plant Physiol.* **7**:299–313.

Ludlow, M. M., T. H. Stobbs, R. Davis, and D. A. Charles-Edwards. 1982. Effect of sward structure of two tropical grasses with contrasting canopies on light distribution, net photosynthesis and size of bite harvested by grazing cattle. *Aust. J. Agric. Res.* **33**:187–201.

Ludlow, M. M. and G. L. Wilson. 1971a. Photosynthesis of tropical pasture plants. I. Illuminance, carbon dioxide concentration, leaf temperature, and leaf-air vapor pressure difference. *Aust. J. Biol. Sci.* **24**:449–470.

Ludlow, M. M. and G. L. Wilson. 1971b. Photosynthesis of tropical pasture plants. II. Temperature and illuminance history. *Aust. J. Biol. Sci.* **24**:1065–1075.

Ludlow, M. M. and G. L. Wilson. 1971c. Photosynthesis of tropical pasture plants. III. Leaf age. *Aust. J. Biol. Sci.* **24**:1077–1087.

Ludlow, M. M., G. L. Wilson, and M. R. Heslehurst. 1974. Studies on the productivity of tropical pasture plants. Effect of shading on growth, photosynthesis, and respiration in two grasses and two legumes. *Aust. J. Agric. Res.* **25**:425–433.

Lundberg, P. E., O. L. Bennett, and E. L. Mathias. 1977. Tolerance of bermudagrass selections to acidity. I. Effects of lime on plant growth and mine spoil material. *Agron. J.* **69**:913–916.

Lüttge, U. and N. Higinbotham. 1979. *Transport in Plants*. Springer-Verlag, Berlin.

Lüttge, U. and M. G. Pitman. 1976. *Transport in Plants. II. Tissues and Organs*. Part 3. Springer-Verlag, Berlin.

Lüttge, U. and J. Weigl. 1962. Mikroautoradiographische Untersuchungen der Aufnahme und des Transportes von $^{35}SO_4^{2-}$ und $^{45}Ca^{2+}$ in Keimwurzeln von *Zea mays* L. und *Pisum sativum* L. *Planta* **58**:113–126.

Luxmoore, R. J., L. H. Stolzy, and J. Letey. 1970. Oxygen diffusion in the soil-plant system. III. Oxygen concentration profiles, respiration rates, and the significance of plant aeration prediced for maize roots. *Agron. J.* **62**:325–329.

Lyness, A. S. 1936. Varietal differences in the phosphorus feeding capacity of plants. *Plant Physiol.* **11**:665–688.

Lyons, J. M., J. K. Raison, and P. L. Steponkus. 1979. The plant membrane in response to low temperature: An overview. In J. M. Lyons, D. Graham, and J. K. Raison (eds.), *Low Temperature Stress in Crop Plants. The Role of the Membrane*, pp. 1–24. Academic Press, New York.

Maas, E. V. 1969. Calcium uptake by excised maize roots and interactions with alkali cations. *Plant Physiol.* **44**:985–989.

Maas, E. V. and G. J. Hoffman. 1977. Crop salt tolerance—current assessment. *J. Irrig. Drain. Div. Proc. Am. Soc. Civil Eng.* **103**:115–134.

Maas, E. V., D. P. Moore, and B. J. Mason. 1968. Manganese absorption by excised barley roots. *Plant Physiol.* **43**:527–530.

Maas, E. V. and R. H. Nieman. 1978. Physiology of plant tolerance to salinity. In G. A. Jung (ed.), *Crop Tolerance to Suboptimal Land Conditions*, pp. 277–299. ASA Spec. Pub. No. 32. American Society of Agronomy Crop Science Society of America, Soil Science Society of America, Madison, WI.

MacLeod, L. B. and L. P. Jackson. 1967. Aluminum tolerance of two barley varieties in nutrient solution, peat, and soil culture. *Agron. J.* **59**:359–363.

Mack, R. N. and J. N. Thompson. 1982. Evolution in steppe with few large, hoofed mammals. *Am. Nat.* **119**:757–773.

MacLeod, L. B. and R. B. Carson. 1966. Influence of K on the yield and chemical composition of grasses grown in hydroponic culture with 12, 50, and 75% of the N supplied as NH_4^+. *Agron. J.* **58**:52–57.

Maizlish, N. A., D. D. Fritton, and W. A. Kendall. 1980. Root morphology and early development of maize at varying levels of nitrogen. *Agron. J.* **72**:25–31.

Major, D. J. 1980. Photoperiod response characteristics controlling flowering of nine crop species. *Can. J. Plant Sci.* **60**:777–784.

Major, D. J. and T. B. Daynard. 1972. Hyperbolic relation between leaf area index and plant population in corn (*Zea mays*). *Can. J. Plant Sci.* **52**:112–115.

Major, D. J., W. M. Hamman, and S. B. Rood. 1982. Effects of short-duration chilling temperature exposure on growth and development of sorghum. *Field Crops Res.* **5**:129–136.

Major, R. L. and L. N. Wright. 1974. Seed dormancy characteristics of sideoats gramagrass *Bouteloua curtipendula* (Michx.) Torr. *Crop Sci.* **14**:37–40.

Malavolta, E., E. L. M. Coutinho, G. C. Vitti, N. V. Alejo, N. J. Novaes, and V. L. Furlani Netto. 1979. Studies on the mineral nutrition of sweet sorghum. 1. Deficiency of macro and micronutrients and toxicity of aluminum, chlorine, and manganese. *Anais da E. S. A. "Luis de Quiroz." Univ. Sao Paulo, Brazil* **36**:173–202.

Malzer, G. L. and S. A. Barber. 1975. Precipitation of calcium and strontium sulfates around plant roots and its evaluation. *Soil Sci. Soc. Am. Proc.* **39**:492–495.

Mangelsdorf, P. C. 1958. Ancestor of corn. *Science* **128**:1313–1320.

Mansfield, T. A., A. R. Wellburn, and T. S. Moreira. 1978. The role of abscisic acid and farnesol in the alleviation of water stress. *Phil. Trans. Roy. Soc. London, Ser. B.* **284**:471-483.

Maranville, J. W. and G. M. Paulsen. 1972. Alteration of protein composition of corn (*Zea mays* L.) seedlings during moisture stress. *Crop Sci.* **12**:660-663.

Maretzki, A. and A. Dela Cruz. 1967. Nitrate reductase in sugarcane tissues. *Plant Cell Physiol.* **8**:605-611.

Markhart, A. H., E. L. Fiscus, A. W. Naylor, and P. J. Kramer. 1979. Effect of temperature on water and ion transport in soybean and broccoli root systems. *Plant Physiol.* **64**:83-87.

Marschner, H. 1974. Mechanisms of regulation of mineral nutrition in higher plants. In R. L. Bieleski, A. R. Ferguson, and M. M. Cresswell (eds.), *Mechanisms of Regulation of Plant Growth, Roy. Soc. N. Z. Bull.* **12**:99-109.

Marschner, H., R. Handley, and R. Overstreet. 1966. Potassium loss and changes in the fine structure of corn root tips induced by H-ion. *Plant Physiol.* **41**:1725-1735.

Marschner, H., A. Kalisch, and V. Romheld. 1974. Mechanism of iron uptake in different plant species. In S. Wehrman (ed.), *Plant Analysis and Fertilizer Problems*, Vol. 2, pp. 273-381. German Soc. Plant Nutr., Hanover, Germany.

Marschner, H. and C. Richter. 1974. Calcium translocation in roots of maize and bean seedlings. *Plant Soil* **40**:193-210.

Marshall, B. and R. W. Willey. 1983. Radiation interception and growth in an intercrop of pearl millet/groundnut. *Field Crops Res.* **7**:141-160.

Marshall, D. R., P. Broue, and A. J. Pryor. 1973. Adaptive significance of alcohol dehydrogenase isozymes in maize. *Nature (New Biology)* **244**:16-17.

Martens, J. W. and D. C. Arny. 1967a. Nitrogen and sugar levels of pith tissue in corn as influenced by plant age and by potassium and chloride ion fertilization. *Agron. J.* **59**:332-334.

Martens, J. W. and D. C. Arny. 1967b. Effects of potassium and chloride ion on root necrosis, stalk rot, and pith condition in corn (*Zea mays* L.). *Agron. J.* **59**:499-502.

Martin, J. P. 1938. Sugar cane diseases in Hawaii. Exp. Stn., Hawaii Sugar Planters' Assoc.

Martin, J. P. 1934. Boron deficiency symptoms in sugar cane. *Hawaii. Plant. Rec.* **38**:95-107.

Martin, W. E. and J. E. Matocha. 1973. Plant analysis as an aid in the fertilization of forage crops. In L. M. Walsh and J. D. Beaton (eds.), *Soil Testing and Plant Analysis*, pp. 393-426. Soil Science Society of America, Madison, WI.

Martin, W. E., J. G. McClean, and J. Quick. 1965. Effect of temperatures on the occurrence of phosphorus induced zinc deficiency. *Soil Sci. Soc. Am. Proc.* **29**:411-413.

Marzola, D. L. 1978. Effect of soil pH, phosphorus, and zinc fertilization on corn and sugarcane and an evaluation of extractants for available soil zinc. MS thesis, University of Hawaii.

Mathers, J. E. and D. Pennington. 1982. Effect of ferrous sulfate and sulfuric acid on grain sorghum yields. *Agron. J.* **62**:555-556.

Matocha, J. E. 1971. Influence of sulfur sources and magnesium on forage yields of coastal bermudagrass [*Cynodon dactylon* (L.) Pers.]. *Agron. J.* **63**:493-496.

Matocha, J. E. and D. Pennington. 1982. Effects of plant iron recycling on iron chlorosis of grain sorghum grown on calcareous soils. *J. Plant Nutr.* **5**:869-882.

Matocha, J. E. and L. Smith. 1980. Influence of potassium on *Helminthosporium cyndontis* and dry matter yields of "coastal bermudagrass." *Agron. J.* **72**:565-567.

Mattas, R. E. and A. W. Pauli. 1965. Trends in nitrate reduction and nitrogen fractions in young corn (*Zea mays* L.) plants during heat and moisture stress. *Crop Sci.* **5**:181-184.

May, L. H. 1960. The utilization of carbohydrate reserves in pasture plants after defoliation. *Herb. Abstr.* **30**:239-245.

McBee, G. G., R. M. Waskom III, F. R. Miller, and R. A. Creelman. 1983. Effect of senescence and nonsenescence on carbohydrates in sorghum during late kernel maturity states. *Crop Sci.* **23**:372-376.

McCormick, L. H. and F. Y. Borden. 1972. Phosphate fixation by aluminum in plant roots. *Soil Sci. Soc. Am. Proc.* **36**:799-802.

McCosker, T. H. and J. K. Teitzel, 1975. A review of guinea grass (*Panicum maximum*) for the wet tropics of Australia. *Trop. Grassl.* **9**:177-190.

McCree, K. J. 1972. The action spectrum, absorbance and quantum yield of photosynthesis in crop plants. *Agric. Meteorol.* **9**:191-216.

McCree, K. J. 1974a. Changes in the stomatal response characteristics of grain sorghum produced by water stress during growth. *Crop Sci.* **14**:273-278.

McCree, K. J. and S. D. Davis. 1974. Effect of water stress and temperature on leaf size and on size and number of epidermal cells in grain sorghum. *Crop Sci.* **14**:751-755.

McCree, K. J. and S. Kresovich. 1978. Growth and maintenance requirements of white clover as a function of daylength. *Crop Sci.* **18**:11-25.

McDermitt, D. K. and R. S. Loomis. 1981. Elemental composition of biomass and its relation to energy content, growth efficiency, and growth yield. *Ann. Bot.* **48**:275-290.

McDermitt, D. K. and R. S. Loomis. 1981. Elemental composition of biomass and its relation toi energy content, growth efficiency, and growth yield. *Ann. Bot.* **48**:275-290.

McGarrahan, J. P. and R. F. Dale. 1984. A trend toward a longer grain-filling period for corn: A case study in Indiana. *Agron. J.* **76**:518-522.

McGeorge, W. T. 1924. The influence of silica, lime and soil reaction upon the availability of phosphates on highly ferruginous soils. *Soil Sci.* **17**:463-468.

McIlroy, R. J. 1967. Carbohydrates of grassland herbage. *Herb. Abstr.* **37**:79-87.

McKell, C. M., V. A. Younger, F. J. Nudge, and N. J. Chatterton. 1969. Carbohydrate accumulation of coastal bermudagrass and Kentucky bluegrass in relation to temperature regimes. *Crop Sci.* **9**:534-537.

McLachlan, K. D. 1982. Leaf acid phosphatase activity and the phosphorus status of field-grown wheat. *Aust. J. Agric. Res.* **33**:453-464.

McManmon, M. and R. M. M. Crawford. 1971. A metabolic theory of flooding tolerance: The significance of enzyme distribution and behavior. *New Phytol.* **70**:299-306.

McNaughton, S. J. 1979. Grazing as an optimization process: grass-ungulate relationships in the Serengeti. *Am. Natur.* **113**:691-703.

McPherson, D. C. 1939. Cortical air spaces in the roots of *Zea mays* L. *New Phytol.* **38**:190-202.

McPherson, H. G. and J. S. Boyer. 1977. Regulation of grain yield by photosynthesis in maize subjected to a water deficiency. *Agron. J.* **69**:714-718.

McPherson, H. G. and R. O. Slatyer. 1973. Mechanisms regulating photosynthesis in *Pennisetum typhoides*. *Aust. J. Biol. Sci.* **26**:329-379.

McWilliam, J. R. 1978. Response of pasture plants to temperature. In J. R. Wilson (ed.), *Plant Relations in Pastures*, pp. 17-33. CSIRO, Melbourne.

McWilliam, J. R. and P. J. Ferrar. 1974. Photosynthetic adaptation of higher plants to thermal stress. In R. L. Bieleski, A. R. Ferguson, and M. M. Cresswell (eds.), *Mechanisms of Regulation of Plant Growth. Roy. Soc. N. Z. Bull.* **12**:467-476.

McWilliam, J. R. and B. Griffing. 1965. Temperature dependent heterosis in maize. *Aust. J. Biol. Sci.* **18**:569-583.

McWilliam, J. R., P. J. Kramer, and R. L. Musser. 1982. Temperature-induced water stress in chilling-sensitive plants. *Aust. J. Plant Physiol.* **9**:343-352.

McWilliam, J. R., W. Manokaran, and T. Kipnis. 1979. Adaptation to chilling stress in sorghum. In J. M. Lyons, D. Graham, and J. K. Raison (eds.), *Low Temperature Stress in Crop Plants. The Role of the Membrane*, pp. 491-505. Academic Press, New York.

McWilliam, J. R. and K. Mison. 1974. Significance of the C_4 pathway in *Triodia irritans* (Spinifex), a grass adapted to arid environments. *Aust. J. Plant Physiol.* **1**:171-175.

McWilliam, J. R. and A. W. Naylor. 1967. Temperature and plant adaptation. I. Interaction of temperature and light in the synthesis of chlorophyll in corn. *Plant Physiol.* **42**:1711-1715.

McWilliam, J. R., K. Shanker, and R. B. Knox. 1970. Effects of temperature and photoperiod on growth and reproductive development in *Hyparrhenia hirta. Aust. J. Agric. Res.* **21**:557-569.

Medina, E. 1970. Effect of nitrogen supply and light intensity during growth on the photosynthetic capacity and carboxydismutase activity of leaves of *Atriplex patula* spp. *hastata. Carnegie Inst. Wash. Yearbook* **69**:551-559.

Meeker, G. B., A. C. Purvis, C. A. Neyra, and R. H. Hageman. 1974. Uptake and accumulation of nitrate as a major factor in regulation of nitrate reductase activity in corn (*Zea mays* L.) leaves: Effect of high ambient CO_2 and malate. In R. L. Bieleski, A. Ferguson, and M. M. Cresswell (eds.), *Mechanism of Regulation of Plant Growth. Roy. Soc. N. Z. Bull.* **12**:49-58.

Meiri, A., G. J. Hoffman, M. C. Shannon, and J. A. Poss. 1982. Salt tolerance of two muskmelon cultivars under two radiation levels. *J. Am. Soc. Hort. Sci.* **107**:1168-1172.

Melsted, S. W. 1953. Some observed calcium deficiencies in corn under field conditions. *Soil Sci. Soc. Am. Proc.* **17**:52-54.

Melsted, S. W., H. L. Motto, and T. R. Peck. 1969. Critical plant nutrition composition values useful in interpreting plant analysis data. *Agron. J.* **61**:17-20.

Mendelssohn, I. A., K. L. McKee, and W. H. Patrick, Jr. 1981. Oxygen deficiency in *Spartina alterniflora* roots: Metabolic adaptation to anoxia. *Science* **214**:439-441.

Menge, J. A., D. Steirle, D. J. Bagyeraj, E. L. V. Johnson, and R. T. Leonard. 1978. Phosphorus concentration in plant responsible for inhibition of mycorrhizal infection. *New Phytol.* **80**:575-578.

Mengel, D. B. and S. A. Barber. 1974. Development and distribution of the corn root system under field conditions. *Agron. J.* **66**:341-344.

Mengel, K. and E. A. Kirkby. 1980. Potassium in crop production. *Adv. Agron.* **33**:59-110.

Mengel, K. and E. A. Kirkby. 1982. *Principles of Plant Nutrition*, 3rd ed. International Potash Institute, Bern, Switzerland.

Mengel, K. and R. Pfluger. 1972. The release of potassium and sodium from young excised roots of *Zea mays* under various efflux conditions. *Plant Physiol.* **49**:16-19.

Menke, J. W. and M. J. Trlica. 1983. Effects of single and sequential defoliations on the carbohydrate reserves of four range species. *J. Range Manage.* **36**:70-74.

Meredith, H. L. and W. H. Patrick, Jr. 1961. Effects of soil compaction on subsoil root penetration and physical properties of three soils in Louisiana. *Agron. J.* **53**:163-167.

Merrill, L. B. and V. A. Young. 1962. Germination and root establishment in seedlings of curly mesquitegrass. MP-615, Tex. Agric. Exp. Stn., College Station.

Merwe, N. J. van der. 1982. Carbon isotopes, photosynthesis, and archaeology. *Am. Sci.* **70**:596-606.

Metcalfe, C. R. 1960. *Anatomy of the Monocotyledons. I. Gramineae.* Oxford University Press, London.

Metcalfe, R. A., A. Fernandez, and R. F. Williams. 1975. The genesis of form in bulrush millet (*Pennisetum americanum* (L.) K. Schum.). *Aust. J. Bot.* **23**:761-773.

Metraux, J. P. and H. Kende. 1983. The role of ethylene in the growth response of deep water rice. *Plant Physiol.* **72**:441-446.

Michelena, V. A. and J. S. Boyer. 1982. Complete turgor maintenance at low water potentials in the elongating region of maize leaves. *Plant Physiol.* **69**:1145-1149.

Middleton, C. H. 1982. Dry matter and nitrogen changes in five tropical grasses as influenced by cutting height and frequency. *Trop. Grassl.* **16**:112–117.

Midmore, D. J. 1980. Effects of photoperiod on flowering and fertility of sugarcane (*Saccharum* spp.). *Field Crops Res.* **3**:65–81.

Miedema, P. and J. Sinnaeva. 1980. Photosynthesis and respiration of maize seedlings at suboptimal temperatures. *J. Exp. Bot.* **31**:813–819.

Miller, G. W., A. Denney, J. Pushnik, and M. H. Yu. 1982. The formation of delta-aminolevulinate, a precursor of chlorophyll, in barley and the role of iron. *J. Plant Nutr.* **5**:289–300.

Miller, M. F. and F. L. Duley. 1925. The effect of a varying moisture supply upon the development and composition of the maize plant at different periods of growth. Mo. Agr. Exp. Stn. Res. Bull. 76.

Miller, M. H., C. P. Mamaril, and G. J. Blair. 1970. Ammonium effects on phosphorus absorption through pH changes and phosphorus precipitation at the soil-root interface. *Agron. J.* **62**:524–527.

Miller, R. D. and D. D. Johnson. 1964. Effect of soil moisture tension on carbon dioxide evolution, nitrification, and nitrogen mineralization. *Soil Sci. Soc. Am. Proc.* **28**:644–647.

Millerd, A., D. J. Goodchild, and B. Spencer. 1969. Studies on a maize mutant sensitive to low temperature. II. Chloroplast structure, development, and physiology. *Plant Physiol.* **44**:567–583.

Millerd, A. and J. R. McWilliam. 1968. Studies on a maize mutant sensitive to low temperature. I. Influence of temperature and light on the production of chloroplast pigments. *Plant Physiol.* **43**:1967–1972.

Mills, H. A. and W. S. McElhannon. 1982. Nitrogen uptake by sweet corn. *HortSci.* **17**:743–744.

Mirhadi, M. J. and Y. Kobayashi. 1979. Studies on the productivity of grain sorghum. II. Effects of wilting treatments at different stages of growth on the development, nitrogen uptake and yield of irrigated grain sorghum. *Jpn. J. Crop Sci.* **48**:531–542.

Mirhadi, M. J., S. Yoshida, and Y. Kobayashi. 1980. Protein fractions in sorghum grain under various levels of nitrogen application. *Jpn. J. Crop Sci.* **49**:502–503.

Mislevy, P., G. O. Mott, and F. G. Martin. 1982. Effect of grazing frequency on forage quality and stolon characteristics of tropical perennial grasses. *Soil Crop Sci. Soc. Fla. Proc.* **41**:77–83.

Mitchell, C. C., Jr., and W. G. Blue. 1981a. The sulfur fertility status of Florida soils. I. Sulfur distribution in Spodosols, Entisols, and Ultisols. *Soil Crop Sci. Soc. Fla. Proc.* **40**:71–76.

Mitchell, C. C., Jr., and W. G. Blue. 1981b. The sulfur fertility status of Florida soils. II. An evaluation of subsoil sulfur on plant nutrition. *Soil Crop Sci. Soc. Fla. Proc.* **40**:77–82.

Mitchell, C. C., Jr., and R. N. Gallaher. 1980. Sulfur fertilization of corn seedlings. *Soil Crop Sci. Soc. Fla. Proc.* **39**:40–44.

Mite, F. A. 1980. Redistribution of N from leaves and stalks to the ears as influenced by N-rate and corn genotype. M.S. thesis, North Carolina State University, Raleigh, NC.

Mock, J. J. and M. J. McNeill. 1979. Cold tolerance of maize inbred lines adapted to various latitudes in North America. *Crop Sci.* **19**:239–242.

Mock, J. J. and R. B. Pearce. 1975. An ideotype for maize. *Euphytica* **24**:613–623.

Mock, J. J. and W. H. Skrdla. 1978. Evaluation of maize plant introductions for cold tolerance. *Euphytica* **27**:27–32.

Mohr, P. J. and E. B. Dickinson. 1979. Mineral nutrition in maize. In *Maize*, pp. 26–32. Ciba-Geigy Agrochemicals Technical Monograph. Ciba-Geigy, New York.

Moll, R. H., E. J. Kamprath, and W. A. Jackson. 1982. Analysis and interpretation of factors which contribute to efficiency of nitrogen utilization. *Agron. J.* **74**:562–564.

Moloney, M. M., M. C. Elliott, and R. E. Cleland. 1981. Acid growth effects in maize roots: Evidence for a link between auxin-economy and proton extrusion in the control of root growth. *Planta* **152**:285–291.

Moncur. M. W. 1981. *Floral Initiation in Field Crops.* CSIRO, Melbourne.

Mondrus-Engle, M. 1981. Tetraploid perennial teosinte seed dormancy and germination. *J. Range Manage.* **34**:59-61.

Mongelard, J. C. and L. Mimura. 1972. Growth studies of the sugarcane plant. II. Some effects of root temperature and gibberellic acid and their interactions on growth. *Crop Sci.* **12**:52-58.

Monma, E. and S. Tsunoda. 1979. Photosynthesis heterosis in maize. *Jpn. J. Breed.* **29**:159-162.

Monsi, M., Z. Uchijuma, and T. Oikawa. 1973. Structure of foliage canopies and photosynthesis. *Ann. Rev. Ecol. Sys.* **4**:301-327.

Monsi, M. and T. Saeki. 1953. Uber den Lichtfaktor in den Pflanzengesellschaften und sein Bedeutung fur die Stoffproduktion. *Jpn. J. Bot.* **14**:22-52.

Monson, R. K., G. E. Edwards, and M. S. B. Ku. 1984. C_3-C_4 intermediate photosynthesis in plants. *BioScience* **34**:563-574.

Monson, R. K., R. O. Littlejohn, Jr., and G. J. Williams III. 1982. The quantum yield for CO_2 uptake in C_3 and C_4 grasses. *Photosynth. Res.* **3**:153-159.

Monson, R. K., R. O. Littlejohn, Jr., and G. J. Williams III. 1983. Photosynthetic adaptation to temperature in four species from the Colorado shortgrass steppe: A physiological model for coexistence. *Oecologia* **58**:43-51.

Monteith, J. L. 1965. Light distribution and photosynthesis in field crops. *Ann. Bot.* **29**:17-37.

Monteith, J. L. 1973. *Principles of Environmental Physics.* Edward Arnold, London.

Monteith, J. L. 1977. Climate and the efficiency of crop production in Britain. *Phil. Trans. R. Soc. Lond. B* **281**:277-329.

Monteith, J. L. 1978. Reassessment of maximum growth rates for C_3 and C_4 crops. *Expl. Agric.* **14**:1-5.

Monteith, J. L. 1979. Soil temperature and crop growth in the tropics. In R. Lal and D. J. Greenland (eds.), *Soil Physical Properties and Crop Production in the Tropics*, pp. 249-262. New York.

Monteith, N. H. and C. L. Banath. 1965. The effect of soil strength on sugarcane growth. *Trop. Agric.* **42**:293-296.

Monteith, N. H. and G. D. Sherman. 1962. The comparative effects of calcium carbonate and calcium silicate on the yield of sudan grass in a Humic Ferruginous Latosol and a Hydrol Humic Latosol. *Hawaii Agr. Exp. Stn. Tech. Bull.* **53**:1-40.

Moody, P. W. and J. Standley. 1980. Supernatant solution phosphorus concentrations required by some tropical pasture species. *Commun. Soil Sci. Plant Anal.* **11**:851-860.

Moore, D. P. 1972. Mechanism of micronutrient uptake by plants. In J. J. Mortvedt, P. M. Giordano, and W. L. Lindsay (eds.), *Micronutrients in Agriculture*, pp. 171-198. Soil Society of America, Madison, WI.

Moore, P. D. 1982. Evolution of photosynthetic pathways in flowering plants. *Nature* **295**:647-648.

Moore, P. H. and L. L. Buren. 1978. Gibberellin studies with sugarcane. I. Cultivar differences in growth responses to gibberellic acid. *Crop Sci.* **17**:443-446.

Moore, P. H. and D. J. Heinz. 1971. Increased post-inductive photoperiods for delayed flowering in *Saccharum* sp. hybrids. *Crop Sci.* **11**:118-121.

Moore, P. H., R. B. Osgood, J. B. Carr, and H. Ginoza. 1982. Sugarcane studies with gibberellin. V. Plot harvests vs. stalk harvests to assess the effect of applied GA_3 on sucrose yield. *J. Plant Growth Regul.* **1**:205-210.

Morgan, J. A. and R. H. Brown. 1983. Photosynthesis and growth of bermudagrass swards. I. Carbon dioxide exchange characteristics of swards mowed at weekly and monthly intervals. *Crop Sci.* **23**:347-352.

Morgan, M. A., R. J. Volk, and W. A. Jackson. 1973. Simultaneous influx and efflux of nitrate during uptake by perennial ryegrass. *Plant Physiol.* **51**:267-272.

Morilla, C. A., J. S. Boyer, and R. H. Hageman. 1973. Nitrate reduction activity and polyribosomal content of corn (*Zea mays* L.) having low leaf water potentials. *Plant Physiol.* **51**:817-824.

Morison, J. I. L. and R. M. Gifford. 1983. Stomatal sensitivity to carbon dioxide and humidity. A comparison of two C_3 and two C_4 species. *Plant Physiol.* **71**:789-796.

Morre, D. J., D. D. Jones, and H. H. Mollenhauer. 1967. Golgi apparatus mediated polysaccharide secretion by outer root cap cells of *Zea mays*. I. Kinetics and secretory pathway. *Planta* **74**:286-301.

Morris, C. J., J. F. Thompson, and C. M. Johnson. 1969. Metabolism of glutamic acid and *N*-acetyl glutamic acid in leaf discs and cell-free extracts of higher plants. *Plant Physiol.* **44**:1023-1026.

Morris, H. D. and J. Giddens. 1963. Response of several crops to ammonium and nitrate forms of nitrogen as influenced by soil fumigation and liming. *Agron. J.* **55**:372-374.

Morrow, L. A. and J. F. Power. 1979. Effect of soil temperature on development of perennial forage grasses. *Agron. J.* **71**:7-10.

Mortvedt, J. J. 1976. Soil chemical constraints in tailoring plants to fit problem soils. 2. Alkaline soils. In M. J. Wright (ed.), *Plant Adaptation to Mineral Stress in Problem Soils*, pp. 141-158. Cornell University, Ithaca, NY.

Mortvedt, J. J. 1982. Grain sorghum response to iron sources applied alone or with fertilizers. *J. Plant Nutr.* **5**:859-868.

Mortvedt, J. J. and P. M. Giordano. 1971. Response of grain sorghum to iron sources applied alone or with fertilizers. *Agron. J.* **63**:758-761.

Mosher, P. N. and M. H. Miller. 1972. Influence of soil temperature on the geotropic response of corn roots (*Zea mays* L.). *Agron. J.* **64**:459-462.

Moss, D. N. 1962. Photosynthesis and barrenness. *Crop. Sci.* **2**:366-367.

Moss, D. N., R. B. Musgrave, and E. R. Lemon. 1961. Photosynthesis under field conditions. III. Some effects of light, carbon dioxide, temperature, and soil moisture on photosynthesis, respiration, and transpiration of corn. *Crop Sci.* **1**:83-87.

Moss, G. I. and L. A. Downey, 1971. The influence of drought stress on female megagametophyte development in corn (*Zea mays* L.) and subsequent grain yield. *Crop Sci.* **11**:368-372.

Moss, R., P. J. Randall, and C. W. Wrigley. 1982. A simple test to detect sulfur deficiency in wheat. *Aust. J. Agric. Res.* **33**:443-452.

Mosse, B. 1973. Advances in the study of vesicular-arbuscular mycorrhizas. *Ann. Rev. Phytopath.* **11**:171-196.

Muchow, R. C., D. B. Coates, G. L. Wilson, and M. A. Foale. 1982. Growth and productivity of irrigated *Sorghum bicolor* (L. Moench) in Northern Australia. I. Plant density and arrangement effects on light interception and distribution, and grain yield, in the hybrid Texas 610SR in low and medium latitudes. *Aust. J. Agric. Res.* **33**:773-784.

Mukerji, N. and M. Alan. 1959. Roots as indicator to varietal behavior under different conditions. *Indian J. Sugarcane Res. Dev.* **3**:131-134.

Mulkey, T. J. and M. L. Evans. 1981. Geotropism in corn roots: Evidence for its mediation by differential acid efflux. *Science* **212**:70-71.

Mulkey, T. J., K. M. Kuzmanoff, and M. L. Evans. 1981. Correlations between proton-efflux patterns and growth patterns during geotropism and phototropism in maize and sunflower. *Planta* **512**:239-241.

Mulkey, T. J., K. M. Kuzmanoff, and M. L. Evans. 1982. Promotion of growth and hydrogen ion efflux by auxin in roots of maize pretreated with ethylene biosynthesis inhibitors. *Plant Physiol.* **70**:186-188.

Munchow, R. C. and G. L. Wilson. 1976. Photosynthetic and storage limitations to yield in *Sorghum bicolor* (L. Moench). *Aust. J. Agric. Res.* **27**:489-500.

Munns, R., C. J. Brady, and E. W. R. Barlow. 1979. Solute accumulation in the apex and leaves of wheat during water stress. *Aust. J. Plant Physiol.* **6**:379-389.

Munson, R. D. 1970. N-K balance—an evaluation. *Potash Rev.*, August 1970.

Murali, B. I. and G. M. Paulsen. 1981. Improvement of nitrogen use efficiency and its relationship to other traits in maize. *Maydica* **26**:63-73.

Murata, Y. 1981. Dependence of potential productivity and efficiency for solar energy utilization on leaf photosynthetic capacity in crop species. *Jpn. J. Crop Sci.* **50**:223-232.

Murdock, C. L., J. A. Jackobs, and J. W. Gerdemann. 1967. Utilization of phosphorus sources of different availability by mycorrhizal and non-mycorrhizal maize. *Plant Soil* **27**:329-334.

Murdock, J. T. and L. E. Engelbert. 1958. The importance of subsoil phosphorus to corn. *Soil Sci. Soc. Am. Proc.* **22**:53-57.

Murphy, L. S., R. Ellis, Jr., and D. C. Adriano. 1981. Phosphorus-micronutrient interaction effects on crop production. *J. Plant Nutr.* **3**:593-613.

Murphy, L. S. and L. M. Walsh. 1972. Correction of micronutrient deficiencies with fertilizers. In J. J. Mortvedt, P. M. Giordano, and W. L. Lindsay (eds.), *Micronutrients in Agriculture*, pp. 347-387. Soil Science Society of America, Madison, WI.

Musick, J. T. and D. A. Dusek. 1971. Grain sorghum response to number, timing, and size of irrigations in the southern high plains. *Trans. ASAE* **14**:401-405.

Mustardy, L. A., T. T. Vu, and A. Faludi-Daniel. 1982. Stomatal response and photosynthetic capacity of maize leaves at low temperature. A study on varietal differences in chilling sensitivity. *Physiol. Plant.* **55**:31-34.

Myers, R. J. K. 1980. The root system of a grain sorghum crop. *Field Crops Res.* **3**:53-64.

Nagahashi, G., W. W. Thomson, and R. T. Leonard. 1974. The casparian strip as a barrier to the movement of lanthanum in corn roots. *Science* **183**:670-671.

Naismith, R. W., M. W. Johnson, and W. I. Thomas. 1974. Genetic control of relative calcium, phosphorus, and manganese accumulation on chromosome 9 in maize. *Crop Sci.* **14**:845-849.

Nambudiri, E. M. V., W. D. Tidwell, B. N. Smith, and N. P. Hebbert. 1978. A C_4 plant from the Pliocene. *Nature* **275**:816-817.

Navarro-Chavira, G. and B. D. McKersie. 1983. Growth, development and digestibility of guinea grass (*Panicum maximum* Jacq.) in two controlled environments differing in irradiance. *Trop. Agric.* **60**:184-192.

Neales, T. F. and L. D. Incoll. 1968. The control of leaf photosynthesis rate by the level of assimilate concentration in the leaf: A review of the hypothesis. *Bot. Rev.* **34**:107-125.

Nelson, W. L. and R. F. Dale. 1978. Effect of trend or technology variables and record period on prediction of corn yields with weather variables. *J. Appl. Meteorol.* **17**:926-933.

Nemec, B. 1900. Ueber die Wahrnehmung des Schwerkraftreizes bie den Pflanzen. *Ber. Dtsch. Bot. Ges.* **18**:241-245.

Neubert, P., W. Wrazidlo, N. P. Vielmeyer, I. Hundt, F. Gullmick, and W. Bergman. 1969. Tabellen zur Pflanzenanelze—erste orientierrende Ubersicht. Inst. Pflanzen. Jena, Berlin.

Neuenschwander, L. F., H. A. Wright, and S. C. Bunting. 1978. The effect of fire on a tobosagrass-mesquite community in the Rolling Plains of Texas. *Southwest. Nat.* **23**:315-338.

Newman, E. I. 1969a. Resistance to water flow in soil and plant. I. Soil resistance in relation to amounts of root: Theoretical estimates. *J. Appl. Ecol.* **6**:1-12.

Newman, E. I. 1969b. Resistance to water flow in soil and plant. II. A review of experimental evidence on the rhizosphere resistance. *J. Appl. Ecol.* **6**:261-272.

Newman, E. I. 1976. Water movement through root systems. *Phil. Trans. R. Soc. Lond. B* **273**:463-478.

Newton, J. D. 1928. The selective absorption of inorganic elements by various crop plants. *Soil Sci.* **26**:85-91.

Ney, D. and P. E. Pilet. 1980. Importance of the caryopsis in root growth and georeaction. *Physiol. Plant.* **50**:166-168.

Neyra, C. A. and R. H. Hageman. 1975. Nitrate uptake and induction of nitrate reductase in excised corn roots. *Plant Physiol.* **56**:692-695.

Ng, T. T., J. R. Wilson, and M. M. Ludlow. 1975. Influence of water stress on water relations and growth on a tropical (C_4) grass, *Panicum maximum* var. *trichoglume. Aust. J. Plant Physiol.* **2**:581-595.

Nickerson, N. H. 1954. Morphological analysis of the maize ear. Am. J. Bot. **41**:87-91.

Nielsen, N. E. and S. A. Barber. 1978. Differences among genotypes of corn in the kinetics of P uptake. *Agron. J.* **70**:695-698.

Nielsen, N. E. and S. A. Barber. 1978. Differences among genotypes of corn in the kinetics of P uptake. *Agron. J.* **70**:695-698.

Nir, I., A. Poljakoff-Mayber, and S. Klein. 1970. The effect of water stress on the polysome population and ability to incorporate amino acids in maize root tips. *Isr. J. Bot.* **19**:451-462.

Nissen, P. 1974. Uptake mechanisms: Inorganic and organic. *Ann. Rev. Plant Physiol.* **25**:53-79.

Nobel, P. S. 1974. Temperature dependence of the permeability of chloroplasts from chilling-sensitive and chilling-resistant plants. *Planta* **115**:369-372.

Norden, A. J. 1964. Response of corn (*Zea mays* L.) to population, bed height, and genotype on poorly drained sandy soil. I. Root development. *Agron. J.* **56**:269-273.

Norman, M. J. T. 1963. The pattern of dry matter and nutrient content changes in native pastures at Katherine, N. T. *Aust. J. Exp. Agric. Anim. Husb.* **3**:119-124.

Norman, M. J. T. and R. Wetselaar. 1960. Performance of annual fodder crops at Katherine, N.T. CSIRO Aust., Div. Land Res. Reg. Surv. Tech. Paper No. 9.

Norris, F. De La M. 1913. Production of air passages in the root of *Zea mays* by variation of cultural media. *Proc. Bristol Nat. Soc. IV* **3**:134-136.

Nour, A. M. and D. E. Weibel. 1978. Evaluation of root characteristics in grain sorghum. *Agron. J.* **70**:217-218.

Nowakowski, T. Z. 1962. Effects of nitrogen fertilizers on total nitrogen, soluble nitrogen, and soluble carbohydrate contents of grass. *J. Agric. Sci.* **59**:387-392.

Nowakowski, T. Z. 1971. Effects of potassium on the contents of soluble carbohydrates and nitrogenous compounds in grass. In *Potassium in Biochemistry and Physiology*, pp. 45-49. Proceedings of 8th International Potash Institute Colloquium. Berne. Switzerland.

Nunez, R. and E. Kamprath. 1969. Relationships between N response, plant population, and row width on growth and yield of corn. *Agron. J.* **61**:279-282.

Nur, I., Y. Okon, and Y. Henis. 1980. Nitrogen fixation in grasses associated with various *Azospirillum* spp. *Can. J. Microbiol.* **26**:714-718.

Nursery, W. R. E. 1971. Starch deposits in *Themeda triandra* Forsk. *Proc. Grassld. Soc. S. Afr.* **6**:157-160.

Nye, P. H. 1977. The rate-limiting step in plant nutrient absorption from soil. *Soil Sci.* **123**:292-297.

Nye, P. H. 1981. Changes of pH across the rhizosphere induced by roots. *Plant Soil* **61**:7-26.

Nye, P. H. and P. B. Tinker. 1977. *Solute Movement in the Soil Root System*. Blackwells, Oxford.

Oaks, A. 1979. Nitrate reductase in roots and its regulation. In E. J. Hewitt and C. V. Cutting (eds.), *Nitrogen Assimilation in Plants*, pp. 217-226. Academic Press, New York.

Oaks, A., D. J. Mitchell, R. A. Barnard, and F. C. Johnson. 1970. The regulation of proline biosynthesis in maize roots. *Can. J. Bot.* **48**:2249-2258.

Obreza, T. A., F. M. Rhoads, and R. L. Stanley, Jr. 1982. Effect of phosphorus levels on phosphorus deficiency symptoms and growth of three corn hybrids. *Soil Crop Sci. Soc. Fla Proc.* **41**:185-188.

Ode, D. J., L. L. Tieszen, and J. C. Lerman. 1980. The seasonal contribution of C_3 and C_4 plant species to primary production in a mixed prairie. *Ecology* **61**:1304–1311.

Ogunkanmi, A. B., A. R. Wellburn, and T. A. Mansfield. 1974. Detection and preliminary identification of endogenous antitranspirants in water-stressed *Sorghum* plants. *Planta* **117**:293–302.

Ohlrogge, A. J. 1962. Some soil-root-plant relationships. *Soil Sci.* **106**:30–38.

Ohsugi, R. and T. Murata. 1980. Leaf anatomy, post-illumination CO_2 burst and NAD-malic enzyme activity of *Panicum dichoromiflorum*. *Plant Cell Physiol.* **21**:1329–1333.

Oke, O. L. 1970. Studies on the sulphur states of Nigerian soils and uptake by grasses. *J. Indian Soc. Soil Sci.* **18**:163–169.

Okon, Y. 1982. *Azospirillum*: Physiological properties, mode of association with roots and its application for the benefit of cereal and forage grass crops. *Israel J. Bot.* **31**:214–220.

Oliver, S. and S. A. Barber. 1966a. An evaluation of the mechanisms governing the supply of Ca, Mg, K, and Na to soybean roots (*Glycine max.*). *Soil Sci. Soc. Am. Proc.* **30**:82–86.

Oliver, S. and S. A. Barber. 1966b. Mechanisms for the movement of Mn, Fe, B, Cu, Zn, Al, and Sr from one soil to the surface of soybean roots (Glycine max). *Soil Sci. Soc. Am. Proc.* **30**:468–470.

Olmstead, C. E. 1941. Growth and development of range grasses. I. Early development of *Bouteloua curtipendula* in relation to water supply. *Bot. Gaz.* **102**:499–519.

Olmsted, C. E. 1944. Growth and development of range grasses. IV. Photoperiodic responses in twelve geographic strains of side-oats grama. *Bot. Gaz.* **106**:46–74.

Olsen, R. A., J. C. Brown, J. H. Bennett, and D. Blume. 1982. Reduction of Fe^{3+} as it relates to Fe chlorosis. *J. Plant Nutr.* **5**:433–445.

Olsen, S. R. 1972. Micronutrient interactions. In J. J. Mortvedt, P. M. Giordano, and W. L. Lindsay (eds.), *Micronutrients in Agriculture*, pp. 243–264. Soil Sci. Am. Inc., Madison, WI.

Olsen, S. R. and W. D. Kemper. 1968. Movement of nutrients to plant roots. *Adv. Agron.* **20**:91–151.

Olsen, S. R. and F. E. Khasawneh. 1980. Use and limitations of physical-chemical criteria for assessing the status of phosphorus in soils. In F. E. Khasawneh, E. C. Sample, and E. J. Kamprath (eds.), *The Role of Phosphorus in Agriculture*, pp. 361–410. American Society of Agronomy Crop Science Society of America, Soil Science Society of America, Madison, WI.

Olsen, S. R., F. S. Watanabe, and R. E. Danielson. 1961. Phosphorus absorption by corn roots as affected by moisture and phosphorus concentration. *Soil Sci. Soc. Am. Proc.* **25**:289–294.

Ong, C. K. and A. Everard. 1979. Short day induction of flowering in pearl millet (*Pennisetum typhoides*) and its effect on plant morphology. *Expl. Agric.* **15**:401–410.

Ortega, E. 1971. Correlation and calibration studies of chemical analysis in soils and plant tissues for nitrogen and available phosphorus. *J. Indian Soc. Soil Sci.* **19**:147–153.

Osborne, D. J. 1981. Dormancy as a survival strategem. *Ann. Appl. Biol.* **98**:525–531.

Osmond, C. B., K. Winter, and S. B. Powles. 1980. Adaptive significance of carbon dioxide cycling during photosynthesis in water-stressed plants. In N. C. Turner and P. J. Kramer (eds.), *Adaptation of Plants to Water and High Temperature Stress*, pp. 139–154. Wiley-Interscience, New York.

Oswalt, D. L., A. R. Bertrand, and M. R. Teel. 1959. Influence of nitrogen fertilization and clipping on grass roots. *Soil Sci. Soc. Am. Proc.* **23**:228–230.

Ottaviano, E. and A. Camussi. 1981. Phenotypic and genetic relationships between yield components in maize. *Euphytica* **30**:601–609.

Otto, H. J. and H. L. Everett. 1956. Influence of nitrogen and potassium fertilization on the incidence of stalk rot of corn. *Agron. J.* **48**:301–305.

Owen, D. F. and R. G. Wiegert. 1981. Mutualism between grasses and grazers: an evolutionary hypothesis. *OIKOS* **36**:376–378.

Owusu-Bennoah, E. and B. Mosse. 1979. Plant growth responses to vesicular-arbuscular mycorrhizas. XI. Field inoculation responses in barley, lucerne, and onion. *New Phytol.* **83**:671-679.

Pal, U. R., R. R. Johnson, and R. H. Hageman. 1976. Nitrate reductase activity in heat (drought) tolerant and intolerant maize genotypes. *Crop Sci.* **16**:775-779.

Pal, U. R. and H. S. Malik. 1981. Contribution of *Azospirillum brasilense* to the nitrogen needs of grain sorghum (*Sorghum bicolor* (L.) Moench) in humid sub-tropics. *Plant Soil* **63**:501-504.

Paleg, L. G., T. J. Douglas, A. van Daal, and D. B. Keech. 1981. Proline, betaine and other organic solutes protect enzymes against heat inactivation. *Aust. J. Plant Physiol.* **8**:107-114.

Palmer, A. F. E. 1973. Photoperiod and temperature effects on a number of plant characters in several races of maize grown in the field. In R. O. Slatyer (ed.), *Plant Response to Climatic Factors*, pp. 113-119. Proceedings of the Uppsala Symposium, 1970. UNESCO, Paris.

Palmer, A. E. F., G. H. Heichel, and R. B. Musgrave. 1973. Patterns of translocation, respiratory loss, and redistribution of ^{14}C in maize labeled after flowering. *Crop Sci.* **13**:371-376.

Pandley, D. P. and S. Kannan. 1979. Absorption and transport of iron in plants as influenced by the major nutrient elements. *J. Plant Nutr.* **1**:55-63.

Parker, D. T. and W. C. Burrows. 1959. Root and stalk rot in corn as affected by fertilizer and tillage treatment. *Agron. J.* **51**:414-417.

Parker, M. L. 1979. Morphology and ultrastructure of the gravity-sensitive leaf sheath base of the grass *Echinochloa colonum* L. *Planta* **145**:471-477.

Passioura, J. B. 1983. Roots and drought resistance. *Agric. Water Manage.* **7**:265-280.

Pasternak, D. and G. L. Wilson. 1969. Effects of heat waves on grain sorghum at the stage of head emergence. *Aust. J. Exp. Agr. Animal Husb.* **9**:636-638.

Pasternak, D. and G. L. Wilson. 1972. After-effects of night temperature on stomatal behavior and photosynthesis of *Sorghum*. *New Phytol.* **71**:683-689.

Pate, F. M. and G. H. Snyder. 1979. Effect of high water table in organic soil on yield and quality of forage grasses—lysimeter study. *Soil and Crop Sci. Soc. Fla., Proc.* **38**:72-75.

Patil, J. D. and N. D. Patil. 1981. Effect of calcium carbonate and organic matter on the growth and concentration of iron and manganese in sorghum (*Sorghum bicolor*). *Plant Soil* **60**:295-300.

Patrick, W. H., Jr., L. W. Sloane, and S. A. Phillips. 1959. Response of cotton and corn to deep placement of fertilizer and deep tillage. *Soil Sci. Soc. Am. Proc.* **23**:307-310.

Patriquin, D. G., F. M. M. Magalhaes, C. A. Scott, and J. Dobereiner. 1978. Infection of field grown maize in Rio de Janeiro by *Azospirillum*. In W. H. Orme-Johnson and W. E. Newton (eds.), *Steenbock–Kettering International Symposium on Nitrogen Fixation*, p. 29. Univ. Wis., Madison. WI.

Patterson, D. T. 1980a. Shading effects on growth and partitioning of plant biomass in Cogongrass (*Imperata cylindrica*) from shaded and exposed habitats. *Weed Sci.* **28**:735-740.

Patterson, D. T. 1980b. Light and temperature adaptation. In J. D. Hesketh and J. W. Jones (eds.), *Prediction Photosynthesis for Ecosystem Models*. Vol. I, pp. 205-235. CRC Press, Boca Raton, FL.

Patterson, D. T. and E. P. Flint. 1980. Potential effects of global atmospheric CO_2 enrichment on the growth and competitiveness of C_3 and C_4 weed and crop plants. *Weed Sci.* **28**:71-75.

Paull, R. E. and R. L. Jones. 1976. Studies on the secretion of maize root cap slime. IV. Evidence for the involvement of dictyosomes. *Plant Physiol.* **57**:249-256.

Pearson, C. J., D. G. Bishop, and M. Vesk. 1977. Thermal adaptation of *Pennisetum*: Leaf structure and composition. *Aust. J. Plant Physiol.* **4**:541-544.

Pearson, C. J. and G. A. Derrick. 1977. Thermal adaptation of *Pennisetum*: Leaf photosynthesis and photosynthate translocation. *Aust. J. Plant Physiol.* **4**:763-769.

Pearson, C. J. and S. G. Shah. 1981. Effects of temperature on seed production, seed quality, and growth of *Paspalum dilatatum*. *J. Appl. Ecol.* **18**:897-905.

Peaslee, D. E. and D. N. Moss. 1966. Photosynthesis in K- and Mg-deficient maize (*Zea mays* L.) leaves. *Soil Sci. Soc. Am. Proc.* **30**:220–223.

Peaslee, D. E., J. L. Ragland, and W. G. Duncan. 1971. Grain filling period of corn as influenced by phosphorus, potassium, and the time of planting. *Agron. J.* **63**:561–563.

Peck, T. R., W. M. Walker, and L. V. Boone. 1969. Relationship between corn (*Zea mays* L.) yield and leaf levels of ten elements. *Agron. J.* **61**:299–301.

Pendleton, J. W., D. B. Egli, and D. B. Peters. 1967. Response of *Zea mays* L. to a "light rich" field environment. *Agron. J.* **59**:395–397.

Pendleton, J. W., G. E. Smith, S. R. Winter, and T. J. Johnston. 1968. Field investigations of the relationship of leaf angle in corn (*Zea mays* L.) to grain yield and apparent photosynthesis. *Agron. J.* **60**:422–424.

Penny, L. H. 1981. Vertical-pull resistance of maize inbreds and their test crosses. *Crop Sci.* **21**:237–240.

Pepper, G. E. and G. M. Prine. 1972. Low light intensity effects on grain sorghum at different stages of growth. *Crop Sci.* **12**:590–593.

Perez-Escolar, R. and M. A. Lugo-Lopez. 1978. Effect of depth of lime incorporation on the yield of two corn hybrids grown on a typical Ultisol of Puerto Rico. *J. Agric. Univ. P. R.* **62**:203–213.

Perlin, D. S. and R. M. Spanswick. 1981. Characterization of ATPase activity associated with corn leaf plasma membranes. *Plant Physiol.* **68**:521–526.

Perrenoud, S. 1977. *Potassium and Plant Health.* International Potash Institute Research Topics No. 3, Berne, Switzerland.

Pesev, N. V. 1970. Genetic factors affecting maize tolerance to low temperatures at emergence and germination. *Theor. Appl. Genet.* **40**:351–356.

Peters, D. B., J. W. Pendleton, R. H. Hageman, and C. M. Brown. 1971. Effect of night air temperature on grain yield of corn, wheat, and soybeans. *Agron. J.* **63**:809.

Peterson, C. A., M. E. Emanuel, and G. B. Humphreys. 1981. Pathway of movement of apoplastic fluorescent dye tracers through the endodermis at the site of secondary root formation in corn (*Zea mays*) and broad bean (*Vicia faba*). *Can. J. Bot.* **59**:618–625.

Peterson, D. F. 1942. Duration of receptiveness in corn silks. *J. Am. Soc. Agron.* **34**:369–371.

Peterson, G., C. K. Suksayretrup, and D. E. Weibel. 1979. Inheritance and interrelationships of bloomless and sparce-bloom mutants in sorghum. *Sorgh. Newslett.* **22**:30.

Phillips, J. W., D. E. Baker, and C. O. Clagett. 1971. Kinetics of P absorption by excised roots and leaves of corn hybrids. *Agron. J.* **63**:517–520.

Phillips, P. J. and J. R. McWilliam. 1971. Thermal response of the primary carboxylating enzymes from C_3- and C_4-plants adapted to contrasting temperature environments. In M. D. Hatch, C. B. Osmond, and R. O. Slatyer (eds.), *Photosynthesis and Photorespiration*, pp. 97–104. Wiley-Interscience, New York.

Phillips, R. E. and D. Kirkham. 1962a. Mechanical impedance and corn seedling root growth. *Soil Sci. Soc. Am. Proc.* **26**:319–322.

Phillips, R. E. and D. Kirkham. 1962b. Soil compaction in the field and corn growth. *Agron. J.* **54**:29–34.

Pilet, P. E. and A. Chanson. 1981. Effect of abscisic acid on maize root growth. A critical examination. *Plant Sci. Lett.* **21**:99–106.

Pilet, P. E. and M. C. Elliott. 1981. Some aspects of the control of root growth and georeaction: The involvement of indoleacetic acid and abscisic acid. *Plant Physiol.* **67**:1047–1050.

Pilet, P. E. and D. Ney. 1981. Differential growth of georeacting maize roots. *Planta* **151**:146–150.

Pilet, P. E. and L. Rivier. 1980. Light and dark georeaction of maize roots: Effect and endogenous level of abscisic acid. *Plant Sci. Lett.* **18**:201–206.

Pilger, R. 1954. Das system der Gramineae. *Bot. Jahrb.* **76**:281–284.

Pinnell, E. L. 1949. Genetic and environmental factors affecting corn seed germination at low temperatures. *Agron. J.* **41**:562-568.

Pinto, C. M. 1980. Control of photosynthesis by photosynthate demand: Possible mechanism. *Photosynthetica* **14**:611-637.

Pitman, M. G. 1965a. Sodium and potassium uptake by seedlings of *Hordeum vulgare. Aust. J. Biol. Sci.* **18**:10-24.

Pitman, M. G. 1965b. Transpiration and the selective uptake of potassium by barley seedlings (*Hordeum vulgare* cv. Bolivia). *Aust. J. Biol. Sci.* **18**:987-998.

Pitman, M. G. 1971. Uptake and transport of ions in barley seedlings. I. Estimation of chloride fluxes in cells of excised roots. *Aust. J. Biol. Sci.* **24**:407-421.

Pitman, M. G. 1977. Ion transport into the xylem. *Ann. Rev. Plant Physiol.* **28**:71-88.

Pitman, W. D. and E. C. Holt. 1982. Environmental relationships with forage quality of warm-season perennial grasses. *Crop Sci.* **22**:1012-1016.

Place, G. A. and S. A. Barber. 1964. The effect of soil moisture and rubidium concentration on diffusion and uptake of rubidium-86. *Soil Sci. Soc. Am. Proc.* **28**:239-243.

Plant. Z., A. Blum, and I. Arnon. 1969. Effect of soil moisture regime and row spacing on grain sorghum production. *Agron. J.* **61**:344-347.

Plinthus, M. J. and J. Rosenblum. 1861. Germination and seedling emergence of sorghum at low temperatures. *Crop Sci.* **1**:293-296.

Pommer, G., W. Sanchez, and H. Hackel. 1981. Effects of variable row widths and plant densities on the yield of maize cultivars with horizontal and upright leaf orientation. *Z. Acker Pflanzenb.* **150**:113-128.

Poneleit, C. G. and D. B. Egli. 1979. Kernal growth rate and duration in maize as affected by plant density and genotype. *Crop Sci.* **19**:385-388.

Ponnaiya, B. W. X. 1960. Silica deposition in *Sorghum* roots and its possible roles. *Madras Agric. J.* **47**:31.

Poovaiah, B. W. 1979. Role of calcium in ripening and senescence. *Commun. Soil Sci. Plant Anal.* **10**:83-88.

Portas, C. A. M. and H. M. Taylor. 1976. Growth and survival of young plant roots in dry soil. *Soil Sci.* **121**:170-175.

Porter, L. K., F. G. Viets, Jr., and G. L. Hutchinson. 1972. Air containing nitrogen-15 ammonia: Foliar absorption by corn seedlings. *Science* **175**:759-761.

Prat, H. 1932. L'epiderme des Graminees: Etude anatomique et systematique. *Ann. Sci. Nat. Bot.* **14**:117-324.

Prat, H. 1936. La systematique des Gramineas. *Ann. Sci. Nat. Bot.* **18**:158-165.

Prest, T. J., R. P. Cantrell, and J. D. Axtell. 1983. Heritability of lodging resistance at its association with other agronomic traits in a diverse sorghum population. *Crop Sci.* **23**:217-221.

Prine, G. M. 1971. A critical period for ear development in maize. *Crop Sci.* **11**:782-786.

Prine, G. M. 1973. Critical period for ear development among different ear-types of maize. *Soil Crop Sci. Soc. Fla. Proc.* **33**:27-30.

Prine, G. M. and G. W. Burton. 1956. The effect of nitrogen rate and clipping frequency on the yield, protein content and certain morphological characteristics of Coastal bermudagrass [*Cynodon dactylon* (L.) Pers.]. *Agron. J.* **48**:296-301.

Purohit, A. N. and E. B. Tregunna. 1974. Carbon dioxide compensation and its association with the photoperiodic response of plants. *Can. J. Bot.* **52**:1146-1148.

Quarrie, S. A. 1980. Genotypic differences in leaf water potential, abscisic acid and proline concentrations in spring wheat during drought stress. *Ann. Bot.* **46**:383-394.

Quinby, J. R. 1966. Fourth maturity gene locus in sorghum. *Crop Sci.* **6**:516-518.

Quinby, J. R. 1967. The maturity genes of sorghum. *Adv. Agron.* **19**:267-305.

Quinby, J. R. 1972. Influence of maturity genes on plant growth in sorghum. *Crop Sci.* **12**:490–492.

Quinby, J. R., J. D. Hesketh, and R. L. Voigt. 1973. Influence of temperature and photoperiod on floral initiation and leaf number in sorghum. *Crop. Sci.* **13**:243–246.

Quinby, J. R. and R. E. Karper. 1945. Inheritance of three genes that influence time of floral initiation and maturity date in milo. *J. Am. Soc. Agron.* **37**:916–936.

Quinlan, T. J., K. A. Shaw, and W. H. R. Edgley. 1976. Kikuyu grass. *Queensl. Agric. J.,* **Sept.-Oct.**

Rabuffetti, A. and E. J. Kamprath. 1977. Yield, N, and S content of corn as affected by N and S fertilization on coastal plain soils. *Agron. J.* **69**:785–788.

Racusen, R. H. 1979. Plant bioelectricity. *What's New in Plant Physiol.* **10**:21–24.

Racusen, R. H. and A. W. Galston. 1977. Electrical evidence for rhythmic changes in the cotransport of sucrose and hydrogen ions in *Samanea* pulvini. *Planta* **135**:57–62.

Radin, J. W. and R. C. Ackerson. 1982. Does abscisic acid control stomatal closure during water stress? *What's New in Plant Physiol.* **13**:9–12.

Raghavendra, A. S. 1980. Characteristics of plant species intermediate between C_3 and C_4 pathways of photosynthesis: Their focus of mechanism and evolution of C_4 syndrome. *Photosynthetica* **14**:271–283.

Raheja, P. C. 1959. Performance of sugarcane varieties in relation to ecological habitat. *Indian J. Sugarcane Res. Devel.* **3**:127–130.

Raison, J. K. 1974. A biochemical explanation of low-temperature stress in tropical and sub-tropical plants. In R. L. Bieleski, A. R. Ferguson, and M. M. Cresswell (eds.), *Mechanisms of Regulation of Plant Growth. Royal Soc. N. Z. Bull.* **12**:487–496.

Raison, J. K., J. A. Berry, P. A. Armond, and C. S. Pike. 1980. Membrane properties in relation to the adaptation of plants to temperature stress. In N. C. Turner and P. J. Kramer (eds.), *Adaptation of Plants to Water and High Temperature Stress*, pp. 261–273. Wiley, New York.

Raison, J. K., E. A. Chapman, L. C. Wright, and S. W. L. Jacobs. 1979. Membrane lipid transitions: Their correlation with the climatic distribution of plants. In J. M. Lyons, D. Graham, and J. K. Raison (eds.), *Low Temperature Stress in Crop Plants: The Role of the Membrane*, pp. 177–186. Academic Press, New York.

Rajagopal, V. and A. S. Andersen. 1978. Does abscisic acid influence proline accumulation in stressed leaves? *Planta* **142**:85–88.

Randall, P. J., K. Spencer, and J. R. Freney. 1981. Sulfur and nitrogen fertilizer effects on wheat. I. Concentrations of sulfur and nitrogen and the nitrogen to sulfur ratio in grain, in relation to the yield response. *Aust. J. Agric. Res.* **32**:203–212.

Rao, K. C. and S. Asokan. 1978. Studies on free proline association to drought resistance in sugarcane. *Sugar J.*, January:23–24.

Rao, A. N., R. C. Trivedi, and P. S. Dubey. 1978. Primary production and photosynthetic pigment concentration of ten maize cultivars. *Photosynthetica* **12**:62–64.

Rao, R. C. N., K. S. Krishnasastry, and M. Udayakumar. 1981a. Role of potassium in proline metabolism. I. Conversion of precursors into proline under stress conditions in K-sufficient and K-deficient plants. *Plant Sci. Lett.* **23**:327–334.

Rao, R. C. N., K. S. Krishnasastry, and M. Udayakumar. 1981b. Role of potassium in proline metabolism. II. Activity of arginase in K-deficient and K-sufficient plants. *Plant Sci. Lett.* **23**:335–340.

Rapp, K. E. 1947. Carbohydrate metabolism of johnson grass. *J. Am. Soc. Agron.* **39**:869–873.

Raskin, I. and H. Kende. 1983. How does deep water rice solve its aeration problem. *Plant Physiol.* **72**:447–454.

Rasmussen, H. P. 1968. Entry and distribution of aluminum in *Zea mays*: Electron microprobe X-ray analysis. *Planta* **81**:28–37.

Ratanadilok, N. K. 1978. Salt tolerance in grain sorghum. Ph.D. dissertation, University of Ariz., Tucson.

Rathnam, C. K. M. 1978. Mechanism of C_4-acid decarboxylation in bundle sheath cells of C_4 dicarboxylic acid cycle plants. *What's New in Plant Physiol.* **9**:1-4.

Ratnayake, M., R. T. Leonard, and J. A. Menge. 1978. Root exudation in relation to supply of phosphorus and its possible relevance to mycorrhizal formation. *New Phytol.* **81**:543-552.

Ray, T. B. and C. C. Black. 1979. The C_4 pathway and its regulation. In M. Gibbs and E. Latzko (eds.), *Encyclopedia of Plant Physiology.* New Series. Vol. 6. *Photosynthesis. II. Photosynthetic Carbon Metabolism and Related Processes*, pp. 77-101. Springer-Verlag, Berlin.

Reardon, P. O. and L. B. Merrill. 1978. Response of sideoats grama grown in different soils to addition of thiamine and bovine saliva. In D. N. Hyder (ed.), Proceedings of the 1st International Rangeland Congress, Denver, CO pp. 396-397.

Redinbaugh, M. G. and W. H. Campbell. 1981. Purification and characterization of NAD(P)H: nitrate reductase and NADH:nitrate reductase from corn roots. *Plant Physiol.* **68**:115-120.

Reeves, S. A., Jr., and G. G. McBee. 1972. Nutritional influences on cold hardiness of St. Augustinegrass (*Stenotaphrum secundatum*). *Agron. J.* **64**:447-540.

Reeves, S. A., Jr., G. G. McBee, and M. E. Bloodworth. 1970. Effect of N, P, and K tissue levels and late fall fertilization on the cold hardiness of Tifgreen bermudagrass (*Cynodon dactylon* x *C. transvaalensis*). *Agron. J.* **62**:659-662.

Reichmann, G. A., D. L. Grunes, and F. G. Viets, Jr. 1966. Effect of soil moisture on ammonification and nitrification in two northern plains soils. *Soil Sci. Soc. Am. Proc.* **30**:363-366.

Reicosky, D. C. and J. T. Ritchie. 1976. Relative importance of soil resistance and plant resistance in root water absorption. *Soil Sci. Soc. Am. Proc.* **40**:293-297.

Remison, S. U. 1978. Effect of defoliation during the early vegetative phase and at silking on growth of maize (*Zea mays* L.). *Ann. Bot.* **42**:1439-1445.

Rench, W. E. and R. H. Shaw. 1971. Black layer development in corn. *Agron. J.* **63**:303-305.

Rendig, V. V. and F. Amparano. 1980. Nutrient deficiency diagnosis of corn (*Zea mays* L.) plants from yield and composition responses to additions of N and S. *Commun. Soil Sci. Plant Anal.* **11**:1181-1193.

Rennie, R. J. 1981. Diazotrophic biocoenosis—the workshop concensus paper. In P. B. Vose and A. P. Ruschel (eds.), *Associative N_2-fixation*, Vol. II, pp. 253-258. CRC Press, Boca Raton, FL.

Repka, J. and Z. Jurekova. 1981. Heterogeneity of the maize leaf blade in photosynthetic characteristics, respiration, mineral nutrient contents, and growth substances. *Biol. Plant (Prague)* **23**:145-155.

Rhue, R. D. and C. D. Grogan. 1977a. Screening corn for Al tolerance using different Ca and Mg concentrations. *Agron. J.* **69**:755-760.

Rhue, R. D. and C. P. Grogan. 1977b. Screening corn for aluminum tolerance. In M. J. Wright (ed.), *Plant Adaptation to Mineral Stress in Problem Soils*, pp. 297-310. Cornell University Press, Ithaca, NY.

Ricand, R. 1977. Effect of subsoiling on soil compaction and yield of sugarcane. *Int. Soc. Sugar Cane Technol. Proc.* **16**:1039-1047.

Ritchey, K. D. 1979. Potassium fertility in Oxisols and Ultisols of the humid tropics. Cornell International Agriculture Bulletin 37. Cornell University, Ithaca, NY.

Ritchey, K. D., J. E. Silva, and U. F. Costa. 1982. Calcium deficiency in clayey B horizons of savanna Oxisols. *Soil Sci.* **133**:378-382.

Ritchey, K. D., D. M. G. Souza, E. Lobato, and O. Correa. 1980. Calcium leaching to increase rooting depth in a Brazilian savannah Oxisol. *Agron. J.* **72**:40-44.

Ritter, W. F. and C. E. Beer. 1969. Yield reduction by controlled flooding of corn. *Trans. ASAE* **12**:46–48.

Rivier, L. and P. E. Pilet. 1981. Abscisic acid levels in the root tips of seven *Zea mays* varieties. *Phytochem.* **20**:17–19.

Robards, A. W. and M. Jackson. 1976. Root structure and function—an integrated approach. In N. Sunderland (ed.), *Perspectives in Experimental Biology*, Vol. 2, pp. 413–422. Pergamon Press, New York.

Robards, A. W., S. M. Jackson, D. T. Clarkson, and J. Sanderson. 1973. The structure of barley roots in relation to the transport of ions into the stele. *Protoplasma* **77**:291–312.

Robards, A. W. and M. E. Robb. 1974. The entry of ions and molecules into roots: An investigation using electron-opaque tracers. *Planta* **120**:1–12.

Robbins, J. S. and C. E. Domingo. 1953. Some effects of severe soil moisture deficits at specific growth stages in corn. *Agron. J.* **45**:618–621.

Robert, M. L., H. F. Taylor, and R. L. Wain. 1975. Ethylene production by cress root segments and its inhibition by 3,5-diiodo-4-hydroxybenzoic acid. *Planta* **126**:273–284.

Robertson, J. A., B. T. Kang, F. Ramirez-Paz, C. H. E. Werkhoven, and A. J. Ohlrogge. 1966. Principles of nutrient uptake from fertilizer bands. VII. P^{32} uptake by brace roots of maize and its distribution within the leaves. *Agron. J.* **58**:293–296.

Robertson, W. K., J. G. A. Fiskell, C. E. Hutton, L. G. Thompson, R. W. Lipscomb, and W. H. Lundy. 1957. Results from subsoiling and deep fertilization of corn for 2 years. *Soil Sci. Soc. Am. Proc.* **21**:340–346.

Robertson, W. K., L. C. Hammond, J. T. Johnson, and K. J. Boote. 1980. Effects of plant-water stress on root distribution of corn, soybeans, and peanuts in sandy soil. *Agron. J.* **72**:548–550.

Rogers, H. H., G. E. Bingham, J. D. Cure, J. M. Smith, and K. A. Surano. 1983. Responses of selected plant species to elevated carbon dioxide in the field. *J. Environ. Qual.* **12**:569–574.

Rogers, R. A., J. H. Dunn, and M. F. Brown. 1976. Ultrastructural characterization of the storage organs of zoysia and bermudagrass. *Crop Sci.* **16**:639–642.

Rogers, R. A., J. H. Dunn, and C. J. Nelson. 1977. Photosynthesis and cold hardening in zoysia and bermudagrass. *Crop Sci.* **17**:727–732.

Romheld, B. and H. Marschner. 1981. Effect of Fe stress on utilization of Fe chelates by efficient and inefficient plant species. *J. Plant Nutr.* **3**:551–560.

Romheld, V., H. Marschner, and D. Kramer. 1982. Responses to Fe deficiency in roots of "Fe-efficient" plant species. *J. Plant Nutr.* **5**:489–498.

Rood, S. B., T. J. Blake, and R. P. Pharis. 1983a. Gibberellins and heterosis in maize. II. Response to gibberellic acid and metabolism of [^{3}H] gibberellin A_{20}. *Plant Physiol.* **71**:645–651.

Rood, S. B. and D. J. Major. 1980. Response of early corn inbreds to photoperiod. *Crop Sci.* **20**:679–682.

Rood, S. B. and D. J. Major. 1981. Inheritance of tillering and flowering time in early maturing maize. *Euphytica* **30**:327–334.

Rood, S. B., R. P. Pharis, M. Koshioka, and D. J. Major. 1983b. Gibberellins and heterosis in maize. I. Endogenous gibberellin-like substances. *Plant Physiol.* **71**:638–644.

Rosario, E. L. 1972. Physiological aspects of photosynthesis differences in relation to yield exhibited by some sugarcane varieties. Ph.D. Dissertation, Cornell University.

Rosario, E. L., N. Chantha, and M. B. Lopez. 1977. Influence of fertility level on yield determining physiomorphological characteristics of some sugarcane varieties. *Int. Soc. Sugar Cane Technol. Proc.* **16**:1865–1884.

Rosario, E. L. and K. Sooksathan. 1977. Influence of fertility levels on nitrate reductase activity and its significance to sugar yield. *Int. Soc. Sugar Cane Technol. Proc.* **16**:1825–1841.

Rosenberg, N. J. 1981. The increasing CO_2 concentration in the atmosphere and its implications on agricultural productivity. I. Effects on photosynthesis, transpiration, and water use efficiency. *Climatic Change* **3**:265–279.

Rosenow, D. T., J. E. Quisenberry, C. W. Wendt, and L. E. Clark. 1983. Drought tolerant sorghum and cotton germplasm. *Agric. Water Manage.* **7**:207–222.

Rosler, P. 1928. Histologische Studien am Vegetation-punkt von *Triticum vulgare*. *Planta* **5**:28–69.

Ross, L., P. Nababsing, and Y. Wong You Cheong. 1974. Residual effect of calcium silicate applied to sugarcane soils. *Int. Soc. Sugar Cane Technol. Proc.* **15**:539–542.

Ross, W. M. 1972. Effect of bloomless (blbl) on yield in Combine-Kafir-60. *Sorgh. Newlett.* **15**:121.

Ross, W. M., K. D. Kofoid, J. W. Maranville, and R. L. Voigt. 1981. Selecting for grain protein and yield in sorghum random-mating populations. *Crop Sci.* **21**:774–777.

Roughton, P. G. and I. J. Warrington. 1976. Effect of nitrogen source on oxalate accumulation in *Setaria sphacelata* (cv. Kazungula). *J. Sci. Food Agric.* **27**:281–286.

Rouse, R. D. 1968. Soil test theory and calibration for cotton, corn, soybeans, and coastal bermudagrass. Auburn University (Ala.). Agr. Exp. Stn. Bull. 373, 67 pp.

Rowley, J. A. 1976. Development of freezing tolerances in leaves of C_4 grasses. *Aust. J. Plant Physiol.* **3**:597–603.

Roy, A. C., M. Y. Ali, R. L. Fox, and J. A. Silva. 1971. Influence of calcium silicate on phosphate solubility and availability in Hawaiian Latosols. *Int. Symp. Soil Fert. Eval. Proc.* **1**:757–765.

Roy, R. N. and B. C. Wright. 1973. Sorghum growth and nutrient uptake in relation to soil fertility. I. Dry matter accumulation patterns, yield, and N content of grain. *Agron. J.* **65**:709–711.

Roy, R. N. and B. C. Wright. 1974. Sorghum growth and nutrient uptake in relation to soil fertility. II. N, P, and K uptake pattern by various plant parts. *Agron. J.* **66**:5–10.

Rumpho, M. E. and R. A. Kennedy. 1983. Anaerobiosis in *Echinochloa crus-galli* (barnyard grass) seedlings. *Plant Physiol.* **72**:44–49.

Rundel, P. W. 1980. The ecological distribution of C_4 and C_3 grasses in the Hawaiian Islands. *Oecologia (Berl.)* **45**:354–359.

Ruschel, A. P. and R. Ruschel. 1977. Varietal differences affecting nitrogenase activity in the rhizosphere of sugarcane. *Int. Soc. Sugar Cane Technol. Proc.* **16**:1941–1947.

Ruschel, R. and A. P. Ruschel. 1981. Inheritance of N_2-fixing ability in sugarcane. In P. B. Vose and A. P. Ruschel (eds.), *Associative N_2-fixation*, Vol. II, pp. 133–140. CRC Press, Boca Raton, FL.

Russell, R. S. 1977. *Plant Root Systems: Their Function and Interaction with the Soil.* McGraw-Hill, London.

Russell, W. A. 1972. Effect of leaf angle on hybrid performance of maize (*Zea mays* L.). *Crop Sci.* **12**:90–92.

Russell, W. A. and A. R. Hallauer. 1980. Corn. In W. R. Fehr and H. H. Hadley (eds.), *Hybridization of Crop Plants*, pp. 299–312. American Society of Agronomy and Crop Science Society of America, Madison, WI.

Rykbost, L., L. Boersma, H. J. Mack, and W. E. Schmisseur. 1975. Yield response to soil warming: Agronomic crops. *Agron. J.* **67**:733–738.

Sabey, B. R. 1969. Influence of soil moisture tension on nitrate accumulation in soils. *Soil Sci. Soc. Am. Proc.* **33**:263–266.

Sachs, R. M., R. W. Kingsbury, and J. DeBie. 1971. Retardation by Carboxin of low-temperature induced discoloration in *Zoysia* and bermudagrass. *Crop Sci.* **11**:585–586.

Saeed, M. and R. L. Fox. 1977. Relations between suspension pH and zinc solubility in acid and calcareous soils. *Soil Sci.* **124**:199–204.

Safaya, N. M. and A. P. Gupta. 1979. Differential susceptibility of corn cultivars to zinc deficiency. *Agron. J.* **71**:132–136.

Saggar, S., J. R. Bettany, and J. W. B. Stewart. 1981. Sulfur transformations in relation to carbon and nitrogen in incubated soils. *Soil Biol. Biochem.* **13**:499–511.

Saif, S. R. and A. G. Khan. 1977. The effect of vesicular arbuscular mycorrhizal associations on growth of cereals. III. Effects on barley growth. *Plant Soil* **47**:17–26.

Saigusa, M., S. Shoji, and T. Takahashi. 1980. Plant root growth in acid andosols from northeastern Japan. 2. Exchange acidity Y as a realistic measure of aluminum toxicity potential. *Soil Sci.* **130**:242–250.

Sampaio, E. V. S. B. and E. R. Beatty. 1976. Morphology and growth of bahiagrass at three rates of nitrogen. *Agron. J.* **68**:379–381.

Samuels, G. 1969a. Silicon and sugar. *Sugar y Azucar* **64**:25–29.

Samuels, G. 1969b. *Foliar Diagnosis for Sugarcane.* Adams Press, Chicago.

Samuels, G., B. G. Capo, and I. S. Bangdiwala. 1953. The nitrogen content of sugarcane as influenced by moisture and age. *J. Agric. Res. Univ. P. R.* **37**:1–12.

Sanchez, P. A. and R. F. Isbell. 1979. A comparison of the soils of tropical Latin America and tropical Australia. In P. A. Sanchez and L. E. Tergas (eds.), *Pasture Production in Acid Soils of the Tropics*, pp. 25–53. Centro Internacional de Agricultura Tropical, Cali, Colombia.

Sandu, G. R., Z. Aslam, M. Salim, A. Sattar, R. H. Qureshi, N. Ahmad, and R. G. Wyn Jones. 1981. The effect of salinity on the yield and composition of *Diplachne fusca* (Kallar grass). *Plant Cell Environ.* **4**:177–181.

Sangster, A. G. 1977. Characteristics of silica deposition in *Digitaria sanguinalis* L. Scop. (Crabgrass). *Ann. Bot.* **41**:341–350.

Sangster, A. G. 1978. Silicon in the roots of higher plants. *Am. J. Bot.* **65**:929–935.

Sangster, A. G. and D. Wynn Parry. 1976a. Endodermal silicon deposits and their linear distribution in developing roots of *Sorghum bicolor* (L.) Moench. *Ann. Bot.* **40**:361–371.

Sangster, A. G. and D. Wynn Parry. 1976b. Endodermal silification in mature, nodal roots of *Sorghum bicolor* (L.) Moench. *Ann. Bot.* **40**:373–379.

Sangster, A. G. and D. Wynn Parry. 1976c. The ultrastructure and election-probe microassay of silicon deposits in the endodermis of the seminal roots of *Sorghum bicolor* (L.) Moench. *Ann. Bot.* **40**:447–459.

Sargent, E. and A. Arber. 1915. The comparative morphology of the embryo and seedling in the Gramineae. *Ann. Bot.* **24**:161–222.

Sass, J. E. 1960. The development of ear primordia of *Zea* in relation to position on the plant. *Proc. Iowa Acad. Sci.* **67**:82–85.

Sass, J. E. and F. A. Loeffel. 1959. Development of axillary buds in maize in relation to barrenness. *Agron. J.* **51**:484–486.

Savage, M. J. 1980. The effect of fire on the grassland microclimate. *Herb. Abst.* **50**:589–603.

Savile, D. B. O. 1979. Fungi as aids in higher plant classification. *Bot. Rev.* **45**:377–503.

Sayre, J. D. 1955. Mineral nutrition of corn. In G. F. Sprague (ed.), *Corn and Corn Improvement*, pp. 296–314. Academic Press, New York.

Schertz, K. F. and L. G. Dalton. 1980. Sorghum. In W. R. Fehr and H. H. Hadley (eds.), *Hybridization of Crop Plants*, pp. 577–588. American Society of Agronomy and Crop Science Society of America, Madison, WI.

Schirman, R. and K. P. Buchholtz. 1966. Influence of atrazine on control and rhizome carbohydrate reserves of quackgrass. *Weeds* **14**:233–236.

Schmehl, W. R. and R. P. Humbert. 1964. Nutrient deficiencies in sugar crops. In H. B. Sprague (ed.), *Hunger Signs in Crops*, pp. 415–450. David McKay, New York.

Schmidt, W. H. and W. L. Colville. 1967. Yield and yield components of *Zea mays* L. as influenced by artificially induced shade. *Crop Sci.* **7**:137–140.

Schmitt, M. R. and G. E. Edwards. 1981. Photosynthetic capacity and nitrogen use efficiency of maize, wheat, and rice: A comparison between C_3 and C_4 photosynthesis. *J. Exp. Bot.* **32**:459-466.

Schrader, L. E. 1978. Uptake, accumulation, assimilation, and transport of nitrogen in higher plants. In D. R. Nielsen and J. G. MacDonald (eds.), *Nitrogen in the Environment*, Vol. 2, *Soil-Plant-Nitrogen Relationships*, pp. 101-141. Academic Press, New York.

Schrader, L. E., D. Domska, P. E. Jung, Jr., and L. A. Peterson. 1972. Uptake and assimilation of ammonium-N and nitrate-N and their influence on the growth of corn (*Zea mays* L.). *Agron. J.* **64**:690-695.

Schwab, G. O., G. S. Taylor, J. L. Fouss, and E. Stibbe. 1966. Crop response from tile and surface drainage. *Soil Sci. Soc. Am. Proc.* **30**:634-637.

Schwartz, D. and T. Endo. 1966. Alcohol dehydrogenase polymorphism in maize—simple and compound loci. *Genetics* **53**:709-715.

Shackel, K. A., K. W. Foster, and A. E. Hall. 1982. Genotype differences in leaf osmotic potential among grain sorghum cultivars grown under irrigation and drought. *Crop Sci.* **22**:1121-1125.

Shafer, J., Jr., and R. G. Wiggans. 1941. Correlation of total dry matter with grain yield in maize. *J. Am. Soc. Agron.* **33**:927-932.

Shaner, D. L. and J. S. Boyer. 1976. Nitrate reductase activity in maize leaves. I. Regulation by nitrate flux. *Plant Physiol.* **58**:499-504.

Shaner, D. L., M. Mertz, and C. J. Arntzen. 1975. Inhibition of ion accumulation in maize roots by abscisic acid. *Planta* **122**:79-90.

Shank, D. B. 1943. Top-root ratios of inbred and hybrid maize. *J. Am. Soc. Agron.* **35**:976-987.

Shank, D. B. 1945. Effects of phosphorus, nitrogen, and soil moisture on top-root ratios of inbred and hybrid maize. *J. Agric. Res.* **70**:365-377.

Shapiro, R. E., G. S. Taylor, and G. W. Volk. 1956. Soil oxygen content and ion uptake by corn. *Soil Sci. Soc. Am. Proc.* **20**:193-197.

Sharif, R. and J. E. Dale. 1980. Growth-regulating substances and the growth of tiller buds in barley; effects of cytokinins. *J. Exp. Bot.* **31**:921-930.

Sharma, K. C., B. A. Krantz, A. L. Brown, and J. Quick. 1968. Interaction of Zn and P in top and root of corn and tomato. *Agron. J.* **60**:453-456.

Sharma, M. L. 1976. Interaction of water potential and temperature effects on germination of three semi-arid plant species. *Agron. J.* **68**:390-394.

Sharma, R. N., S. N. Singh, and R. S. Gupta. 1979. Evaluation of promising maize germplasms for response to nitrogen. *Indian J. Agr. Sci.* **49**:440-449.

Sharp, R. E. and W. J. Davies. 1979. Solute regulation and growth by roots and shoots of water-stressed maize plants. *Planta* **147**:43-49.

Sharpley, A. N. and L. W. Reed. 1982. Effect of environmental stress on the growth and amounts and forms of phosphorus in plants. *Agron. J.* **74**:19-22.

Shaw, R. H. and E. Loomis. 1950. Bases for the prediction of corn yields. *Plant Physiol.* **25**:225-244.

Shaw, R. H. and H. C. S. Thom. 1951. On the phenology of field corn, silking to maturity. *Agron. J.* **43**:541-546.

Shenk, M. K. and S. A. Barber. 1980. Potassium and phosphorus uptake by corn genotypes grown in the field as influenced by root characteristics. *Plant Soil* **54**:65-76.

Sherman, G. D. 1969. Crop growth response to application of calcium silicate to tropical soils in Hawaiian islands. *AGRI Dig.* **18**:11-19.

Shibles, R. 1976. Terminology pertaining to photosynthesis. *Crop Sci.* **16**:437-439.

Shinde, V. J. and P. Joshi. 1980. Dry matter production and distribution in grain sorghum. *Indian J. Genet. Pl. Breed.* **40**:490-495.

Shukla, U. C. and A. K. Mukhi. 1979. Sodium, potassium, and zinc relationship in corn. *Agron. J.* **71**:235-237.

Shukla, U. C. and A. K. Mukhi. 1980. Amerliorative role of Zn, K, and gypsum on maize growth under alkali soil conditions. *Agron. J.* **72**:85-88.

Shukla, U. C. and H. Raj. 1974. Influence of genetic variability on zinc response in wheat (*Triticum* spp.). *Soil Sci. Soc. Am. Proc.* **38**:477-479.

Shukla, U. C. and H. Raj. 1976. Zinc response in corn as influenced by genetic variability. *Agron. J.* **68**:20-22.

Siegel, S. M., B. Z. Siegel, J. Massey, P. Lahne, and J. Chen. 1980. Growth of corn in saline waters. *Physiol. Plant.* **50**:71-73.

Silberbush, M., B. Gornat, and D. Goldberg. 1979. Effect of irrigation from a point source (trickling) on oxygen flux and on root extension in the soil. *Plant Soil* **52**:507-514.

Silcock, R. G. and F. T. Smith. 1982. Seed coating and localized application of phosphate for improving seedling growth of grasses on acid, sandy red earths. *Aust. J. Agric. Res.* **33**:785-802.

Simmonds, N. W. 1976. Sugarcanes. In N. W. Simmonds (ed.), *Evolution of Crop Plants*, pp. 104-108. Longman, London.

Simpson, E., R. J. Cooke, and D. D. Davies. 1981. Measurement of protein degradation in leaves of *Zea mays* using [³H] acetic anhydride and tritiated water. *Plant Physiol.* **67**:214-219.

Singh, B. R. and K. Steenberg. 1974. Plant response to micronutrients. I. Uptake, distribution and translocation of zinc in maize and barley plants. *Plant Soil* **40**:637-646.

Singh, M., W. L. Ogren, and J. M. Widholm. 1974. Photosynthetic characteristics of several C_3 and C_4 plant species grown under different light intensities. *Crop Sci.* **14**:563-566.

Singh, P. M., J. R. Gilley, and W. E. Splinter. 1976. Temperature thresholds for corn growth in a controlled environment. *Trans. ASAE* **19**:1152-1155.

Singh, R. D. and B. N. Chatterjee. 1965. Tillering of perennial grasses in the tropics in India. Proceedings of the IX International Grassland Congress, Sao Paulo, Brazil, pp. 1075-1079.

Singh, R. P. and K. P. P. Nair. 1975. Defoliation studies in hybrid maize. II. Dry-matter accumulation, LAI, silking and yield components. *J. Agric. Sci.* **85**:247-254.

Singh, S. S. and W. L. Colville. 1962. Effect on yield and certain agronomic characters of irrigated grain sorghum. *Agron. J.* **54**:484-486.

Skene, K. G. 1975. Cytokinin production in roots. In J. G. Torrey (ed.), *The Development and Function of Roots*, pp. 365-396. Academic Press, New York.

Slack, C. R., P. G. Roughan, and H. C. M. Bassett. 1974. Selective inhibition of microphyll chloroplast development in some C_4-pathway species by low night temperatures. *Planta* **111**:57-73.

Sloger, C. and L. D. Owens. 1978. Field inoculation of cereal crops with N_2-fixing *Spirillum lipoferum*. *Plant Physiol.* **61**:5-7.

Sluijs, D. H. van der and D. N. Hyder. 1974. Growth and longevity of blue grama seedlings restricted to seminal roots. *J. Range Manage.* **27**:117-119.

Smid, A. E. and D. E. Peaslee. 1976. Growth and CO_2 assimilation by corn as related to potassium nutrition and simulated canopy shading. *Agron. J.* **68**:904-908.

Smith, A. M. and T. apRees. 1979. Pathways of carbohydrate fermentation in the roots of marsh plants. *Planta* **146**:327-334.

Smith, B. N. and S. Epstein. 1971. Two categories of $^{13}C/^{12}C$ ratios for higher plants. *Plant Physiol.* **47**:380-384.

Smith, D. 1972c. Carbohydrate reserves in grasses. In V. B. Youngner and C. M. McKell (eds.), *The Biology and Utilization of Grasses*, pp. 318-333. Academic Press, New York.

Smith, D. T. and N. A. Clark. 1968. Effect of soil nutrients and pH on nitrate nitrogen and growth of pearl millet [*Pennisetum typhoides* (Burm.) Staph and Hubbard] and sudangrass [*Sorghum sudanense* (Piper) Staph]. *Agron. J.* **60**:38-40.

Smith, F. A. and J. A. Raven. 1979. Intracellular pH and its regulation. *Ann. Rev. Plant Physiol.* **30**:289–311.

Smith, F. W. 1972a. Potassium nutrition, ionic relations, and oxalic acid accumulation in three cultivars of *Setaria sphacelata. Aust. J. Agric. Res.* **23**:969–980.

Smith, F. W. 1972b. Foliar symptoms of nutrient disorders in *Panicum maximum* var. *trichoglume* cv. Petrie. Division of Tropical Pastures Tech. Paper No. 9, C.S.I.R.O., Melbourne.

Smith, F. W. 1973. Foliar symptoms of nutrient disorders in *Chloris gayana.* Division of Tropical Pastures Tech. Paper No. 13., C.S.I.R.O., Melbourne.

Smith, F. W. 1974. Foliar symptoms of nutrient disorders in *Cenchrus ciliaris.* Division of Tropical Pastures Tech. Paper No. 16., C.S.I.R.O., Melbourne.

Smith, F. W. 1975. Tissue testing for assessing the phosphorus status of green panic, buffelgrass and setaria. *Aust. J. Exp. Agric. Anim. Husb.* **15**:383–390.

Smith, F. W. 1978. The effect of potassium and nitrogen on ion relations and organic acid accumulation in *Panicum maximum* var. *trichoglume. Plant Soil* **49**:367–379.

Smith, F. W. 1979. Tolerance of seven tropical pasture grasses to excess manganese. *Commun. Soil Sci. Plant Anal.* **10**:853–867.

Smith, F. W. 1981. Ionic relations in tropical pasture grasses. *J. Plant Nutr.* **3**:813–826.

Smith, F. W. and M. J. S. Verschoyle. 1973. Foliar symptoms of nutrient disorders in *Paspalum dilatatum.* Division of Tropical Pastures Tech. Paper No. 14., C.S.I.R.O., Melbourne.

Smith, R. C. and I. Majeed. 1981. Longitudinal gradients of ion transport in corn roots. *Am. J. Bot.* **68**:1257–1262.

Smith, R. L., J. H. Bouton, S. C. Shank, K. H. Quesenberry, M. E. Tyler, J. R. Milam, M. H. Gaskins, and R. C. Littell. 1976. Nitrogen fixation in grasses inoculated with *Spirillum lipoferum. Science* **193**:1003–1005.

Smith, S. N. 1934. Response of inbred lines and crosses in maize to variations of nitrogen and phosphorus supplied as nutrients. *J. Am. Soc. Agron.* **26**:785–804.

Smith, W. 1977. Aspects of drainage related to cultivation and harvesting of sugarcane. In *Proceedings of the Third National Drainage Symposium, 1976*, pp. 139–143. American Society of Agricultural Engineers.

Smittle, D. A., E. D. Threadgill, and W. E. Seigler. 1981. Sweet corn growth, yield, and nutrient uptake responses to tillage systems. *J. Am. Soc. Hort. Sci.* **106**:49–53.

Soil Survey Staff. 1975. Soil taxonomy, a basic system of soil classification for making and interpreting soil surveys. *Agricultural Handbook No. 436, USDA Soil Conservation Service*, U.S. Government Printing Office, Washington, D.C.

Sojka, R. E. and L. H. Stolzy. 1980. Soil-oxygen effects on stomatal response. *Soil Sci.* **130**:350–358.

Soltanpur, P. N., F. Adams, and A. C. Bennett. 1974. Soil phosphorus availability as measured by displaced soil solutions, calcium chloride extracts, dilute-acid extracts and labile phosphorus. *Soil Sci. Soc. Am. Proc.* **38**:225–228.

Somer, I. I. and J. W. Shive. 1942. The iron-manganese relation in plant metabolism. *Plant Physiol.* **17**:582–602.

Soni, S. L., P. B. Kaufman, and W. C. Bigelow. 1970. Electron microprobe analysis of the distribution of silicon in leaf epidermal cells of the oat plant. *Phytomorph.* **20**:350–363.

Soni, S. L., P. B. Kaufman, and W. C. Bigelow. 1972. Regulation of silicon deposition in *Avena* internodal epidermis by gibberellic acid and sucrose. *J. Exp. Bot.* **23**:787–791.

Soofi, G. A. and H. D. Fuehring. 1964. Nutrition of corn on a calcareous soil: I. Interrelationships of N, P, K, Mg, and S on the growth and composition. *Soil Sci. Soc. Am. Proc.* **6**:335–341.

Sorrells, M. E. and O. Myers, Jr. 1982. Duration of developmental stages of 10 milo maturity genotypes. *Crop Sci.* **22**:310–314.

Sosebee, R. E. and C. H. Herbel. 1969. Effects of high temperatures on emergence and initial growth of range plants. *Agron. J.* **61**:621–624.

Sosebee, R. E. and H. H. Wiebe. 1971. Effect of water stress and clipping on photosynthate translocation in two grasses. *Agron. J.* **63**:14–17.

South African Sugar Association. 1967. Plant physiology-root laboratory. Exp. Stn. Ann. Rep. 1966–67. pp. 32–38.

Spanswick, R. M. 1974. Evidence for an electrogenic pump in *Nitella translucens* II. *Biochim. Biophys. Acta* **332**:387–398.

Spiers, J. M. and E. C. Holt. 1971. Nitrogen utilization in blue panicgrass. *Agron. J.* **63**:309–312.

Spoehr, H. A. and H. W. Milner. 1939. Starch dissolution and amylolytic activity of leaves. *Proc. Am. Phil. Soc.* **81**:37–38.

Srinivasan, K. and M. B. G. R. Batcha. 1963. Performance of clones of *Saccharum* species and allied genera under conditions of water-logging. *Proc. Int. Soc. Sugar Cane Technol.* **11**:571–578.

Stamp, P. 1980. Activity of photosynthetic enzymes in leaves of maize (*Zea mays* L.) seedlings in relation to genotype and to temperature changes. *Z. Pflanzenphysiol.* **100**:229–239.

Stamp, P. 1981a. Activities of photosynthetic enzymes in leaves of maize seedlings (*Zea mays* L.) at changing temperature and light intensities. *Angew. Botanik* **55**:419–427.

Stamp, P. 1981b. Activities of PEP and RuBP carboxylase and pigment content in leaves of maize seedlings in relation to genotype and growing temperature. *Z. Pflanzenzuechtg.* **86**:20–32.

Stamp, P., G. Geisler, and R. Thiraporn. 1983. Adaptation to sub- and supraoptimal temperatures of inbred maize lines differing in origin with regard to seedling development and photosynthetic traits. *Physiol. Plant.* **58**:62–68.

Stanford, G. and A. S. Ayres. 1964. The internal nitrogen requirement of sugarcane. *Soil Sci.* **98**:338–344.

Stanford, G. and E. Epstein. 1974. Nitrogen mineralization-water relations in soils. *Soil Sci. Soc. Am. Proc.* **38**:103–107.

Stapf, O. 1917–1934. Gramineae. In D. Prain (ed.), *Flora of Tropical Africa*, Vol. 9, pp. 1–1100. (Published periodically in parts.) L. Reeve and Co., Kent.

Stebbins, G. L. 1956. Cytogenetics and the evolution of the grass family. *Amer. J. Bot.* **43**:89–905.

Stebbins, G. L. 1972. The evolution of the grass family. In V. B. Youngner and C. M. McKell (eds.), *The Biology and Utilization of Grasses*. Academic Press, New York.

Steck, P. and P. E. Pilet. 1979. Effect of the light on adenine nucleotide content of georeacting maize roots. *Plant Cell Physiol.* **20**:413–421.

Steele, K. W., S. J. McCormick, N. Percival, and N. S. Brown. 1981. Nitrogen, phosphorus, potassium, magnesium, and sulphur requirements for maize grain production. *N. Z. J. Exp. Agric.* **9**:243–249.

Stegman, E. C. 1982. Corn grain yield as influenced by timing of evapotranspiration deficits. *Irrig. Sci.* **3**:75–87.

Steinke, T. D. 1975. Effect of height of cut on translocation of ^{14}C-labelled assimilates in *Eragrostis curvula* (Schrad.) Nees. *Proc. Grassl. Soc. S. Afr.* **10**:41–47.

Steinke, T. D. and P. de V. Booysen. 1968. The regrowth and utilization of carbohydrate reserves of *Eragrostis curvula* after different frequencies of defoliation. *Proc. Grassl. Soc. S. Afr.* **3**:105–110.

Stelzer, R. and A. Lauchli. 1977. Salz- und Uberflutungstoleranz von *Puccinellia peisonis*. II. Strukturelle Differenzierung dur Wurzel in Beziehung zur Funktion. *Z. Pflanzen.* **84**:95–108.

Stencek, G. L. and H. W. Koontz. 1970. Phloem mobility of magnesium. *Plant Physiol.* **46**:50–52.

Stender, H. K. 1924. Some sugar cane growth measurements. *Hawaii. Plant Rec.* **28**:472–495.

Stenlid, G. 1982. Cytokinins as inhibitors of root growth. *Physiol. Plant.* **56**:500–506.

Stevenson, G. C. 1936. Investigations into the root development of the sugar cane in Barbados. (II) Further observations on root development in several varieties under one environment. Bull. 11, Brit. West Indies Cent. Sugar Cane Breed. Stn., Barbados.

Stevenson, G. C. and A. E. S. McIntosh. 1935. Investigations into the root development of the sugar cane in Barbados. (I) Root development in several varieties under one environment. Bull. 5, Brit. West Indies Cent. Sugar Cane Breed. Stn., Barbados.

Stevenson, J. C. and M. M. Goodman. 1972. Ecology of exotic races of maize. I. Leaf number and tillering of 16 races under four temperatures and two photoperiods. *Crop Sci.* **12**:864–868.

Stewart, B. A. and L. K. Porter. 1969. Nitrogen-sulfur relationships in wheat (*Triticum aestivum* L.), corn (*Zea mays*), and beans (*Phaseolus vulgaris*). *Agron. J.* **61**:267–271.

Stewart, J. A. and K. C. Berger. 1965. Estimation of available soil zinc using magnesium chloride as an extractant. *Soil Sci.* **100**:244–250.

Stibbe, E. and R. Terpstra. 1982. Effect of penetration resistance on emergence and early growth of silage corn in a laboratory experiment with sandy soil. *Soil Tillage Res.* **2**:143–153.

Stickler, F. C., A. W. Pauli, and A. J. Casady. 1962. Comparative responses of kaoliang and other grain sorghum types to temperature. *Crop Sci.* **2**:136–139.

Stiff, M. L. and J. B. Powell. 1974. Stem anatomy of turfgrass. *Crop Sci.* **14**:181–186.

Stocker, D. 1960. Physiological and morphological changes in plants due to water deficiency. In *Plant-Water Relationships in Arid and Semi-arid Conditions—A Review of Research*, pp. 63–104. UNESCO, Paris.

Stone, J. F. and B. B. Tucker. 1969. Nitrogen content of grain as influenced by water supplied to the plant. *Agron. J.* **61**:76–78.

Storey, R. and R. G. Wyn Jones. 1978. Salt stress and comparative physiology in the Gramineae. III. Effect of salinity upon ion relations and glycinebetaine and proline levels in *Spartina* X *townsendii*. *Aust. J. Plant Physiol.* **5**:831–838.

Stout, D. G., T. Kannangara, and G. M. Simpson. 1978. Drought resistance of *Sorghum bicolor.* 2. Waterstress effects on growth. *Can. J. Plant Sci.* **58**:225–233.

Stowe, L. G. and J. A. Teeri. 1978. The geographic distribution of C_4 species of the Dicotyledonae in relation to climate. *Am. Natur.* **112**:609–613.

Stribley, D. P., P. B. Tinker, and J. H. Rayner. 1980. Relation of internal phosphorus concentrations and plant weight in plants infected by vesicular-arbuscular mycorrhizas. *New Phytol.* **86**:261–266.

Stryker, R. B., J. W. Gilliam, and W. A. Jackson. 1974. Nonuniform phosphorus distribution in the root zone of corn: Growth and phosphorus uptake. *Soil Sci. Soc. Am. Proc.* **38**:334–340.

Stukenholtz, L. D., R. J. Olsen, G. Gogan, and R. A. Olson. 1966. On the mechanism of phosphorus-zinc interaction in corn nutrition. *Soil Sci. Soc. Am. Proc.* **30**:759–763.

Subramonia Iyer, P. R. V. and M. C. Saxena. 1977. Kinetic studies on the inducement in uptake and transport of P by maize (*Zea mays* L.) as a result of pre-uptake NO_3-N supply. *Plant Soil* **46**:55–67.

Suehisa, R. H., D. R. Younge, and G. D. Sherman. 1963. Effects of silicates on phosphorus availability to sudan grass grown on Hawaiian soils. *Hawaii Agr. Exp. Stn. Tech. Bull.* **51**:1–40.

Suh, H. W., A. J. Casady, and R. L. Vanderlip. 1974. Influence of sorghum seed weight on the performance of the resultant crop. *Crop Sci.* **14**:835–836.

Sullivan, C. Y. 1972. Mechanisms of heat and drought resistance in grain sorghum and methods of measurement. In N. G. P. Rao and L. R. House (eds.), *Sorghum in Seventies*, pp. 247–264. Oxford IBH Publishing, New Delhi.

Sullivan, C. Y., N. V. Norcio, and J. D. Eastin. 1977. Plant responses to high temperatures. In A. Muhammed, R. Aksel, and R. C. von Borstel (eds.), *Genetic Diversity in Plants*, pp. 301–317. Plenum Press, New York.

Sumner, D. C., W. E. Martin, and H. S. Etchegaray. 1965. Dry matter and protein yields and nitrate content of Piper sudangrass [*Sorghum sudanense* (Piper) Stapf.] in response to nitrogen fertilization. *Agron. J.* **57**:351-354.

Sumner, M. E. 1970. Aluminum toxicity—a growth limiting factor in some Natal sands. *Proc. S. Afr. Sugar Technol. Assoc.* June:176-182.

Sun, V. G. and N. P. Chow. 1949. The effect of climatic factors on the yield of sugarcane in Tainan, Taiwan. Part 2. Multiple factors investigation. *Rep. Taiwan Sug. Exp. Stn.* **4**:1-40 (Chinese with English summary).

Sung, F. J. M. and D. R. Krieg. 1979. Relative sensitivity of photosynthetic assimilation and translocation of ^{14}carbon to water stress. *Plant Physiol.* **64**:852-856.

Suzuki, A., P. Gadal, and A. Oaks. 1981a. Intracellular distribution of enzymes associated with nitrogen assimilation in roots. *Planta* **151**:457-461.

Suzuki, T., Y. Shimazaki, T. Fujii, and M. Furuya. 1980. Photoreversible and photoirreversible absorbance changes in the red and far-red spectral regions in *Zea* primary roots. *Plant Cell Physiol.* **21**:1309-1317.

Suzuki, T., M. Tanaka, and T. Fujii. 1981b. Function of light in the light-induced geotropic response in *Zea* roots. *Plant Physiol.* **67**:225-228.

Swanson, A. F. 1941. Relation to leaf area to grain yield in sorghum. *J. Am. Soc. Agron.* **33**:908-914.

Swanson, A. F. and R. Hunter. 1937. Effect of germination and seed size on sorghum stands. *J. Am. Soc. Agron.* **29**:997-1004.

Sweeney, F. C. and J. M. Hopkinson. 1975. Vegetative growth of nineteen tropical and sub-tropical pasture grasses and legumes in relation to temperature. *Trop. Grassl.* **9**:209-217.

Syed, M. M. and S. A. El-Swaify. 1972. Effect of saline water irrigation on N.Co. 310 and H50-7209 cultivars of sugar-cane. I. Growth parameters. *Trop. Agric.* **49**:337-346.

Takeoka, Y., P. B. Kaufman, and O. Matsumara. 1979. Comparative microscopy of idioblasts in lemma epidermis of some 3 carbon pathway and 4 carbon pathway grasses Poaceae using the Suzuki universal micro-print method. *Phytomorphology* **29**:330-337.

Takkar, P. N., M. S. Mann, R. L. Bansal, N. S. Randhawa, and H. Singh. 1976. Yield and uptake response of corn to zinc, as influenced by phosphorus fertilization. *Agron. J.* **68**:942-946.

Tamimi, Y. N. and D. T. Matsuyama. 1973. The effect of calcium silicate and calcium carbonate on the growth of sorghum. *AGRI Dig.* **25**:37-44.

Tanner, P. D. 1977. Toxic effect of manganese on maize as affected by calcium and molybdenum application in sand culture. *Rhod. J. Agric. Res.* **15**:25-32.

Tarchevskii, I. A. and Y. S. Karpilov. 1963. On the nature of products of short term photosynthesis. *Sov. Plant Physiol.* **10**:183-184.

Tateoka, T. 1957. Miscellaneous papers on the phylogeny of Poaceae (10.) Proposition of a new phylogenetic system of Poaceae. *J. Jpn. Bot.* **32**:275-287.

Taylor, A. O. and A. S. Craig. 1971. Plants under climatic stress. II. Low temperature, high light effects on chloroplast ultrastructure. *Plant Physiol.* **47**:719-725.

Taylor, A. O., C. R. Slack, and H. G. McPherson. 1974. Plants under climatic stress. VI. Chilling and light effects on photosynthetic enzymes of sorghum and maize. *Plant Physiol.* **54**:696-701.

Taylor, A. O., G. Halligan, and J. A. Rowley. 1975. Faris banding in panicoid grasses. *Aust. J. Plant Physiol.* **2**:247-251.

Taylor, H. M. and R. R. Bruce. 1968. Effects of soil strength on root growth and crop yield in the Southern United States. *Trans. 9th Int. Cong. Soil Sci.* **1**:803-811.

Taylor, H. M. and B. Klepper. 1971. Water uptake by cotton roots during an irrigation cycle. *Aust. J. Biol. Sci.* **24**:853-859.

Taylor, H. M. and B. Kepper. 1973. Rooting density and water extraction patterns for corn (*Zea mays* L.). *Agron. J.* **65**:965–968.

Taylor, H. M. and B. Klepper. 1975. Water uptake by cotton root systems: An examination of assumptions in the single root model. *Soil Sci.* **120**:57–67.

Taylor, H. M., L. F. Locke, and J. E. Box, Jr. 1964a. Pans in Southern Great Plains soils. III. Their effects on yield of cotton and grain sorghum. *Agron. J.* **56**:542–545.

Taylor, H. M., A. C. Mathers, and F. B. Lotspeich. 1964b. Pans in Southern Great Plains soils. I. Why root-restricting pans occur. *Agron. J.* **56**:328–332.

Taylor, H. M., J. J. Parker, Jr., and G. M. Roberson. 1966. Soil strength and seedling emergence relations. II. A generalized relation for Gramineae. *Agron. J.* **56**:393–395.

Teare, I. D., R. Manam, and E. T. Kanemasu. 1974. Diurnal and seasonal trends in nitrate reductase activity in field grown sorghum plants. *Agron. J.* **66**:733–736.

Teeri, J. A. 1980. Adaptation of kinetic properties of enzymes to temperature variability. In N. C. Turner and P. J. Kramer (eds.), *Adaptation of Plants to Water and High Temperature Stress*, pp. 251–260. Wiley, New York.

Teeri, J. A. and L. G. Stowe, 1976. Climatic patterns and the distribution of C_4 grasses in North America. *Oecologia* **23**:1–12.

Teeri, J. A., L. G. Stowe, and D. A. Livingston. 1980. The distribution of C_4 species of the Cyperaceae in North America in relation to climate. *Oecologia* **47**:307–310.

Teeri, J. A., D. T. Patterson, R. S. Alberte, and R. M. Castleberry. 1977. Changes in the photosynthetic apparatus of maize in response to simulated natural temperature fluctuations. *Plant Physiol.* **60**:370–373.

Tenhunen, J. D., J. D. Hesketh, and D. M. Gates. 1980. Leaf photosynthesis models. In J. D. Hesketh and J. W. Jones (eds.), *Predicting Photosynthesis for Ecosystem Models*, Vol. I, pp. 123–181. CRC Press, Boca Raton, FL.

Terman, G. L. and S. E. Allen, 1974a. Yield-nutrient concentration relationships in young maize, as affected by applied nitrogen. *J. Sci. Food Agric.* **25**:1135–1142.

Terman, G. L. and S. E. Allen. 1974b. Accretion and dilution of nutrients in young corn, as affected by yield response to nitrogen, phosphorus, and potassium. *Soil Sci. Soc. Am. Proc.* **38**:455–460.

Terman, G. L., P. M. Giordano, and N. W. Christenson. 1975. Corn hybrid yield affects on phosphorus, manganese, and zinc absorption. *Agron. J.* **67**:182–184.

Terry, N. and G. Low. 1982. Leaf chlorophyll content and its relation to the intracellular localization of iron. *J. Plant Nutr.* **5**:301–310.

Thakur, P. S. and V. K. Rai. 1981. Growth characteristics and proline content in relation to water status in two *Zea mays* L. cultivars during rehydration. *Biol. Plant.* **23**:98–103.

Thakur, P. S. and V. K. Rai. 1982. Dynamics of amino acid accumulation of two differentially drought resistant *Zea mays* cultivars in response to osmotic stress. *Environ. Exp. Bot.* **22**:221–226.

Theodorides, T. N. and C. J. Pearson. 1982. Effect of temperature on nitrate uptake, translocation and metabolism in *Pennisetum americanum*. *Aust. J. Plant Physiol.* **9**:309–320.

Thiagarajah, M. R., L. A. Hunt, and J. D. Mahon. 1981. Effects of position and age on leaf photosynthesis in corn (*Zea mays*). *Can. J. Bot.* **59**:28–33.

Thomas, G. W. and D. E. Peaslee. 1973. Soil testing for phosphorus. In L. M. Walsh and J. D. Beaton (eds.), *Soil Testing and Plant Analysis*, pp. 115–132. Soil Sci. Soc. Am., Madison, WI.

Thomas, J. C., K. W. Brown, and W. R. Jordan. 1976. Stomatal response to leaf water potential as affected by preconditioning water stress in the field. *Agron. J.* **68**:706–709.

Thomas, J. R. and G. W. Langdale. 1980. Ionic balance in coastal bermudagrass influenced by nitrogen fertilizer and soil salinity. *Agron. J.* **72**:449–452.

Thomasson, J. R. 1978. Observation on the characteristic of the lemma and palea of a late Cenozoic grass *Panicum elegans* Elias. *Am. J. Bot.* **65**:34–39.

Thomasson, J. R. 1980. *Archaeolersia nebraskensis* gen. et sp. nov. (Gramineae-Oryzeae), a new fossil grass from the late Tertiary of Nebraska. *Am. J. Bot.* **67**:876–882.

Thompson, D. L. 1982. Grain yield of two synthetics of corn after seven cycles of selection for lodging resistance. *Crop Sci.* **22**:1207–1210.

Thorne, G. N. 1962. Survival of tillers and distribution of dry matter between ear and shoot of barley varieties. *Ann. Bot. (N. S.)* **26**:37–54.

Thornley, J. H. M. 1976. *Mathematical Models in Plant Physiology.* Academic Press, New York.

Tien, T. M., M. H. Gaskins, and D. H. Hubbell. 1979. Plant growth substances produced by *Azospirillum brasilense* and their effect on the growth of pearl millet (*Pennisetum americanum* L.). *Appl. Environ. Microbiol.* **37**:1016–1024.

Tinker, P. B. 1975. Effects of vesicular arbuscular mychorrizas on higher plants. In *Symbiosis: Symposia of the Society for Experimental Biology.* The University Press, Cambridge. **29**:325–349.

Tinker, P. B. H. and F. E. Sanders. 1975. Rhizosphere microorganisms and plant nutrition. *Soil Sci.* **119**:363–367.

Tieszen, L. L., M. M. Senyimba, S. K. Imbamba, and J. H. Troughton. 1979. The distribution of C_3 and C_4 grasses and carbon isotope discrimination along an altitudinal and moisture gradient in Kenya. *Oecologia* **37**:337–350.

Todd, G. W. and D. L. Webster. 1965. Effects of repeated drought periods on photosynthesis and survival of cereal seedlings. *Agron. J.* **57**:399–404.

Toit, J. L. Du, B. E. Beater, and R. R. Maud. 1963. Available soil phosphate and yield response in sugar cane. *Int. Soc. Sugar Cane Technol. Proc.* **11**:101–111.

Tollenaar, M. 1977. Sink-source relationships during reproductive development of maize. A review. *Maydica* **22**:49–75.

Tollenaar, M. and T. B. Daynard. 1978. Effect of defoliation on kernel development in maize. *Can. J. Plant Sci.* **58**:207–212.

Tollenaar, M. and T. B. Daynard. 1982. Effect of source-sink ratio on dry matter accumulation and leaf senescence of maize. *Can. J. Plant Sci.* **62**:855–860.

Tollenaar, M., T. B. Daynard, and R. B. Hunter. 1979. Effect of temperature on rate of leaf appearance and flowering date of maize. *Crop Sci.* **19**:363–366.

Tollenaar, M. and R. B. Hunter. 1983. A photoperiod and temperature sensitive period for leaf number of maize. *Crop Sci.* **23**:457–460.

Tompsett, P. B. 1976. Factors affecting the flowering of *Andropogon gayanus* Kunth. Responses to photoperiod, temperature, and growth regulators. *Ann. Bot.* **40**:695–705.

Toole, V. K. 1973. Effects of light, temperature and their interactions on the germination of seeds. *Seed Sci. Technol.* **1**:339–396.

Toole, V. K. and H. A. Borthwick. 1968a. The photoreaction controlling seed germination in *Eragrostis curvula. Plant Cell Physiol.* **9**:125–136.

Toole, V. K. and H. A. Borthwick. 1968b. Light responses of *Eragrostis curvula* seed. *Proc. Int. Seed Test. Assoc.* **33**:515–550.

Torrey, J. G. and D. T. Clarkson. 1975. *The Development and Function of Roots.* Academic Press, London.

Toth, S. J. 1939. The stimulating effects of silicates on plant yields in relation to anion displacements. *Soil Sci.* **47**:123–141.

Tothill, J. C. and R. B. Knox. 1968. Reproduction in *Heteropogon contortus*. I. Photoperiodic effects on flowering and sex expression. *Aust. J. Agric. Res.* **19**:869–878.

Trenbath, B. R. and J. F. Angus. 1975. Leaf inclination and crop production. *Field Crop Abstr.* **28**:231–244.

Tripathy, P. C., J. A. Eastin, and L. E. Schrader. 1972. A comparison of ^{14}C-labeled photosynthate export from two leaf positions in a corn (*Zea mays* L.) canopy. *Crop Sci.* **12**:495–497.

Troughton, J. H. 1971. Aspects of the evolution of the photosynthetic carboxylation reaction in plants. In M. D. Hatch, C. B. Osmond, and R. O. Slatyer (eds.), *Photosynthesis and Photorespiration*, pp. 124–127. Wiley-Interscience, New York.

Troughton, J. H. 1979. δ^{13}C as an indicator of carboxylation reactions. In M. Gibbs and E. Latzko (eds.), *Encyclopedia of Plant Physiol. New Series.* Vol. 6. *Photosynthesis.* II. *Photosynthetic Carbon Metabolism and Related Processes*, pp. 140–149. Springer-Verlag, Berlin.

Trouse, A. C., Jr. 1964. Effects of compression of some subtropical soils on the soil properties and upon root development. Ph.D. dissertation, University of Hawaii, Honolulu.

Trouse, A. C., Jr. 1965. Effects of soil compression on the development of sugarcane roots. *Int. Soc. Sugar Cane Technol. Proc.* **12**:137–152.

Trouse, A. C., Jr. 1971. Soil conditions as they affect plant establishment, root development, and yield. (A) Present knowledge and need for research. In *Compaction of Agricultural Soils*, pp. 225–240. American Society of Agricultural Engineers, St. Joseph, MI.

Trouse, A. C., Jr. 1978. Root tolerance to soil impediments. In G. A. Jung (ed.), *Crop Tolerance to Suboptimal Land Conditions*, pp. 193–232. American Society of Agronomy, Crop Science Society of America, Soil Science Society of America, Madison, WI.

Trouse, A. C., Jr. and L. D. Baver. 1962. The effect of soil compaction on root development. Trans. Joint Meeting Com. IV and V., Intern. Soc. Soil Sci., New Zealand. pp. 258–263.

Trouse, A. C., Jr., and R. P. Humbert. 1961. Some effects of soil compaction on the development of sugarcane roots. *Soil Sci.* **91**:208–217.

Tuil, H. D. W. van. 1965. Organic salts in plants in relation to nutrition and growth. Agric. Res. Rep. 657. PUDOC, Wageningen, Netherlands.

Tully, R. E., A. D. Hanson, and C. E. Nelson. 1979. Proline accumulation in water-stressed barley leaves in relation to translocation and the nitrogen budget. *Plant Physiol.* **63**:518–523.

Turitzin, S. N. and B. G. Drake. 1981. The effect of a seasonal change in canopy structure on the photosynthetic efficiency of a salt marsh. *Oecologia (Berl.)* **48**:79–84.

Turner, N. C. 1975. Concurrent comparison of stomatal behavior, water status, and evaporation of maize in soil at high and low water potential. *Plant Physiol.* **55**:932–936.

Turner, N. C. and J. E. Begg. 1973. Stomatal behavior and water status of maize, sorghum, and tobacco under field conditions. I. At high soil water potential. *Plant Physiol.* **51**:31–36.

Turner, N. C., J. E. Begg, H. M. Rawson, S. D. English, and A. B. Hearn. 1978a. Agronomic and physiological responses of soybean and sorghum crops to water deficits. III. Components of leaf water potential, leaf conductance, ^{14}CO$_2$ photosynthesis, and adaptation to water deficits. *Aust. J. Plant Physiol.* **5**:179–194.

Turner, N. C., J. E. Begg, and M. L. Tonnet. 1978b. Osmotic adjustment of sorghum and sunflower crops in response to water deficits and its influence on the water potential at which stomata close. *Aust. J. Plant Physiol.* **5**:597–608.

Turner, N. C. and M. M. Jones. 1980. Turgor maintenance by osmotic adjustment: A review and evaluation. In N. C. Turner and P. J. Kramer (eds.), *Adaptation of Plants to Water and High Temperature Stress*, pp. 87–103. Wiley, New York.

Tyner, E. H. 1946. The relation of corn yields to leaf nitrogen, phosphorus, and potassium content. *Soil Sci. Soc. Am. Proc.* **11**:317–323.

Uchijima, Z., T. Udagawa, T. Horie, and K. Kobayashi. 1968. The penetration of direct solar radiation into corn canopy and the intensity of direct radiation on the foliage surface. *J. Agron. Met., Tokyo* **3**:141–151.

Ulrich, A. 1941. Metabolism of non-volatile organic acids in excised barley roots as related to cation-anion balance during salt accumulation. *Am. J. Bot.* **28**:526-537.

Umali-Garcia, M., D. H. Hubbell, and M. H. Gaskins. 1978. Process of infection of *Panicum maximum* by *Spirillum lipoferum*. *Ecol. Bull. (Stockholm)* **26**:373-379.

Umali-Garcia, M., D. H. Hubbell, M. H. Gaskins, and F. B. Dazzo. 1980. Association of *Azospirillum* with grass roots. *Appl. Environ. Microbiol.* **39**:219-226.

United States Salinity Laboratory Staff. 1954. *Diagnosis and Improvement of Saline and Alkali Soils.* Handbook No. 60., U.S. Dept. Agric.

Uren, N. C. 1981. Chemical reduction of an insoluble higher oxide of manganese by plant roots. *J. Plant Nutr.* **4**:65-71.

Uren, N. C. 1982. Chemical reducation at the root surface. *J. Plant Nutr.* **5**:515-520.

Usberti, J. A., Jr., and S. K. Jain. 1978a. Variation in *Panicum maximum*: A comparison of sexual and asexual populations. *Bot. Gaz.* **139**:112-116.

Usberti, J. A., Jr., and S. K. Jain. 1978b. The role of sexuality in the response of guinea grass populations to heat stress. *J. Hered.* **69**:188-190.

Vallejos, C. E. 1979. Genetic diversity of plants for response to low temperatures and its potential use in crop plants. In J. M. Lyons, D. Graham, and J. K. Raison (eds.), *Low Temperature Stress in Crop Plants. The Role of the Membrane*, pp. 473-489. Academic Press, New York.

Vanderlip, R. L., J. D. Ball, P. J. Banks, F. N. Reece, and S. J. Clark. 1977. Flaming grain sorghum to delay flowering. *Crop Sci.* **17**:902-904.

Veen, B. W. 1981. Relation between root respiration and root activity. *Plant Soil* **63**:73-76.

Venkatraman, T. S. 1930. Problems of the sugar cane breeder. *Int. Soc. Sugar Cane Technol. Proc.* **3**:429-445.

Venkatraman, T. S. and R. Thomas. 1922. Sugarcane and root systems. Studies in development and anatomy. *Agric. J. India* **17**:381-388.

Verduin, J. and W. E. Loomis. 1944. Absorption of CO_2 by maize. *Plant Physiol.* **19**:278-293.

Vicente-Chandler, J., F. Abruna, R. Caro-Costas, J. Figarella, S. Silva, and R. W. Pearson. 1974. Intensive grassland management in the humid tropics of Puerto Rico. Bull. 233. Agric. Exp. Sta., Univ. P. R., Rio Piedras, Puerto Rico.

Vicente-Chandler, J., R. W. Pearson, F. Abruna, and S. Silva. 1962. Potassium fertilization of intensively managed grasses under humid tropical conditions. *Agron. J.* **54**:450-453.

Vicente-Chandler, J., S. Silva, and J. Figarella. 1959. The effect of nitrogen fertilization and frequency of cutting on the yield and composition of three tropical grasses. *Agron. J.* **51**:202-206.

Vidal, D. R. and T. C. Broyer. 1962. Effect of high levels of Mg on the Al uptake and growth of maize in nutrient solutions. *An. Edafol. Agrobiol.* **21**:13-30.

Vietor, D. M. 1982. Variation of ^{14}C-labeled photosynthate recovery from roots and rooting media of warm season grasses. *Crop Sci.* **22**:362-366.

Vietor, D. M., R. P. Areyanagan, and R. B. Musgrave. 1977. Photosynthetic selection of *Zea mays* L. I. Plant age and leaf position effects and a relation between leaf and canopy rates. *Crop Sci.* **17**:567-573.

Vietor, D. M. and R. B. Musgrave. 1979. Photosynthetic selection of *Zea mays* L. II. The relationship between CO_2 exchange and dry matter accumulation of canopies of two hybrids. *Crop Sci.* **19**:70-75.

Viets, F. G., Jr., L. C. Boawn, C. L. Crawford, and C. E. Nelson. 1953. Zinc deficiency in corn in central Washington. *Agron. J.* **45**:559-565.

Viets, F. G., Jr. and W. L. Lindsay. 1973. Testing soils for zinc, copper, manganese, and iron. In *Soil Testing and Plant Analysis*, pp. 153-172. Soil Science Society of America, Madison, WI.

Viets, F. G., Jr., C. E. Nelson, and C. L. Crawford. 1954. The relationships among corn yields, leaf composition, and fertilizers applied. *Soil Sci. Soc. Am. Proc.* **18**:297-301.

Vigh, L., I. Horvath, T. Farkas, L. A. Mustardy, and A. Faludi-Daniel. 1981. Stomatal behavior and cuticular properties of maize leaves of different chilling resistance during cold treatment. *Physiol. Plant.* **51**:287–290.

Vine, P. N., R. Lal, and D. Payne. 1981. The influence of sands and gravels on root growth of maize seedlings. *Soil Sci.* **131**:124–129.

Visser, J. H. 1965. Root exudates of *Eragrostis curvula* as an ecological factor. In *Proceedings of the 9th International Grassland Congress, Sao Paulo, Brazil*, pp. 453–455.

Vizarova, G. 1978. Effect of water stress on the endogenous cytokinins in the apical root meristems of the maize. *Biologia (Bratislava)* **33**:591–596.

Vogel, J., A. Fuls, and R. Ellis. 1978. The geographic distribution of Kranz grasses in South Africa. *S. Afr. J. Sci.* **74**:209–215.

Vogel, J. C. and N. J. van der Merwe. 1977. Isotopic evidence for early maize cultivation in New York State. *Am. Antiquity* **42**:238–242.

Voigt, P. W. 1973. Induced seed dormancy in weeping lovegrass, *Eragrostis curvula. Crop. Sci.* **13**:76–79.

Voigt, P. W., C. L. Dewald, J. E. Matocha, and C. D. Foy. 1982. Adaptation of iron-efficient and iron-inefficient lovegrass strains to calcareous soils. *Crop Sci.* **22**:672–676.

Volk, G. M. 1947. Significance of moisture translocation from soil zones of low-moisture tension to zones of high-moisture tension by plant roots. *Agron. J.* **39**:93–107.

Vong, N. Q. and Y. Murata. 1977. Studies on the physiological characteristics of C_3 and C_4 crop species. I. The effects of air temperature on the apparent photosynthesis, dark respiration, and nutrient absorption of some crops. *Jpn. J. Crop Sci.* **46**:45–52.

Voorhees, W. B. 1977a. Soil compaction: Our newest natural resource. *Crops Soils Mag.* **29**(5):13–15.

Voorhees, W. B. 1977b. Soil compaction: How it influences moisture, temperature, yield, root growth. *Crops Soils Mag.* **29**(6):7–10.

Vorm, P. D. J. van der. 1980. Uptake of Si by five plant species, as influenced by variations in Si-supply. *Plant Soil* **56**:153–156.

Vose, P. B. and A. P. Ruschel. 1981. *Associative N_2-fixation*, Vol. I and II. CRC Press, Boca Raton, FL.

Waldren, R. P. and I. D. Teare. 1974. Free proline accumulation in drought-stressed plants under laboratory conditions. *Plant Soil* **40**:689–692.

Waldren, R. P., I. D. Teare, and S. W. Ehler. 1974. Changes in free proline concentration in sorghum and soybean plants under field conditions. *Crop Sci.* **14**:447–450.

Waldron, J. C. 1976. Nitrogen compounds transported in the xylem of sugar cane. *Aust. J. Plant Physiol.* **3**:415–419.

Waldron, J. C., K. T. Glasziou, and T. A. Bull. 1967. The physiology of sugar-cane. IX. Factors affecting photosynthesis and sugar storage. *Aust. J. Biol. Sci.* **20**:1043–1052.

Walker, R. B. and A. K. Sarkar. 1979. Iron uptake and utilization in maize. *Mineral Nutrition of Plants.* Vol. I, pp. 39–45. Proceedings of the First International Symposium, Plant Nutrition. Publishing House of the Central Coop. Union, Sofia, Bulgaria.

Walker, W. M. and T. R. Peck. 1972. A comparison of the relationship between corn yield and nutrient concentration in whole plants and different plant parts at early tassel at two locations. *Commun. Soil Sci. Plant Anal.* **3**:513–523.

Walker, W. M. and T. R. Peck. 1974. Relationship between corn yield and plant nutrient content. *Agron. J.* **66**:253–256.

Wallace, L. L. 1981. Growth, morphology and gas exchange of mycorrhizal and nonmycorrhizal *Panicum coloratum* L., a C_4 grass species, under different clipping and fertilization regimes. *Oecologia* **49**:272–278.

Walters, C. L. and R. Walker. 1977. Consequence of accumulation of nitrate in plants. In E. J. Hewitt and C. V. Cutting (eds.), Proceedings of the Sixth Long Ashton Symposium, *Nitrogen Assimilation of Plants*, pp. 637-648. Academic Press, London.

Ward, C. Y. and R. E. Blaser. 1961. Carbohydrate food reserves and leaf area in regrowth of orchardgrass. *Crop Sci.* **1**:366-370.

Wardlaw, I. F. 1967. The effect of water stress in relation to photosynthesis and growth. *Aust. J. Biol. Sci.* **20**:25-39.

Warncke, D. D. and S. A. Barber. 1973. Ammonium and nitrate uptake by corn (*Zea mays* L.) as influenced by nitrogen concentration and NH_4^+/NO_3^- ratio. *Agron. J.* **65**:950-953.

Warnock, R. E. 1970. Micronutrient uptake and mobility within corn plants (*Zea mays* L.) in relation to phosphorus-induced zinc deficiency. *Soil Sci. Soc. Am. Proc.* **34**:765-769.

Warsi, A. S. and B. C. Wright. 1973. Sorghum growth and composition in relation to ratio and methods of nitrogen application. I. Pattern of dry matter accumulation. *Indian J. Agron.* **18**:273-276.

Watanabe, F. S., W. L. Lindsay, and S. R. Olsen. 1965. Nutrient balance involving phosphorus, iron, and zinc. *Soil Sci. Soc. Am. Proc.* **29**:562-565.

Watson, L. 1972. Smuts on grasses: Some general implications of the incidence of Ustilaginales on the genera of Gramineae. *Quart. Rev. Bot.* **47**:46-62.

Watts, S., J. L. Rodriguez, S. E. Evans, and W. J. Davies. 1981. Root and shoot growth of plants treated with abscisic acid. *Ann. Bot.* **47**:595-602.

Watts, W. R. 1971. Role of temperature in the regulation of leaf extension in *Zea mays*. *Nature* **229**:46-47.

Watts, W. R. 1972. Leaf extension in *Zea mays*. II. Leaf extension in response to independent variation of the temperature of the apical meristem, of the air around the leaves, and of the rootzone. *J. Exp. Bot.* **23**:713-721.

Watts, W. R. 1973. Soil temperature and leaf expansion in *Zea mays*. *Expl. Agric.* **9**:1-8.

Watts, W. R. 1974. Leaf extension in *Zea mays*. III. Field measurements of leaf extension in response to temperature and leaf water potential. *J. Exp. Bot.* **25**:1085-1096.

Wear, J. L. 1956. Effect of soil pH and calcium on uptake of zinc by plants. *Soil Sci.* **81**:311-315.

Weatherly, A. B. and J. H. Dane. 1979. Effect of tillage on soil-water movement during corn growth. *Soil Sci. Soc. Am. J.* **43**:1222-1225.

Weatherwax, P. 1923. *The Story of the Maize Plant*. University of Chicago Press, Chicago.

Weaver, R. W., S. F. Wright, M. W. Varanka, O. E. Smith, and E. C. Holt. 1980. Dinitrogen fixation (C_2H_2) by established forage grasses in Texas. *Agron. J.* **72**:965-968.

Webster, O. J. 1977. Sorghum studies in Arizona. *Sorgh. Newlett.* **20**:81.

Webster, P. W. D. and B. W. Eavis. 1971. Effects of flooding on sugarcane growth. I. Stage of growth and duration of flooding. *Int. Soc. Sugar Cane Technol. Proc.* **14**:708-714.

Weier, K. L. 1980. Nitrogen fixation associated with grasses. *Trop. Grassl.* **14**:194-201.

Weier, K. L., I. C. Macrae, and J. Whittle. 1981. Seasonal variation in the nitrogenase activity of a *Panicum maximum* var. *trichoglume* pasture and identification of associated bacteria. *Plant Soil* **63**:189-197.

Weigl, J. and U. Lüttge. 1962. Mikroautoradiographische Untersuchungen über die Aufnahme von $^{35}CO_4^{2-}$ durch Wurzeln von *Zea mays* L. Die Funktion per primäen Endoderis. *Planta* **59**:15-28.

Weihing, R. M. 1935. The comparative root development of regional types of corn. *J. Am. Soc. Agron.* **27**:526-537.

Weiland, R. T. and C. A. Stutte. 1979. Pyro-chemiluminescent differentiation of oxidized and reduced N forms evolved from plant foliage. *Crop Sci.* **19**:545-547.

Weimberg, R., H. R. Lerner, and A. Poljakoff-Mayber. 1982. A relationship between potassium and proline accumulation in salt-stressed *Sorghum bicolor*. *Physiol. Plant.* **55**:5-10.

Weinmann, H. 1940. Storage of root reserves in Rhodes grass. *Plant Physiol.* **15**:467-484.

Weinmann, H. 1961. Total available carbohydrate in grasses and legumes. *Herb. Abstr.* **31**:255-261.

Weiser, G. C., R. L. Smith, and R. J. Varnell. 1979. Spikelet abscission in guineagrass as influenced by auxin and gibberellin. *Crop Sci.* **19**:231-235.

Welburn, A. R., A. B. Ogunkanmi, R. Fenton, and T. A. Mansfield. 1974. *All-trans* farnesol: A naturally occurring antitranspirant? *Planta* **120**:255-263.

Welsh, L. F., D. L. Mulvaney, L. V. Boone, G. E. McKibben, and J. W. Pendleton. 1966. Relative efficiency of broadcast versus banded phosphorus for corn. *Agron. J.* **58**:283-287.

Wenkert, W. 1981. The behavior of osmotic potential in leaves of maize. *Environ. Exptl. Bot.* **21**:231-239.

Went, R. W. 1944. Plant growth under controlled conditions. II. Thermoperiodicity in growth and fruiting of the tomato. *Am. J. Bot.* **31**:135-140.

Werker, E. and M. Kislev. 1978. Mucilage on the root surface and root hairs of *Sorghum*: Heterogeneity in structure, manner of production and site of accumulation. *Ann. Bot.* **42**:809-816.

West, S. H. 1970. Biochemical mechanism of photosynthesis and growth depression in *Digitaria decumbens* when exposed to low temperatures. In *Proceedings of the 11th International Grassland Congress, Surfers Paradise, Australia*, pp. 514-517.

West, S. H. 1973. Carbohydrate metabolism and photosynthesis of tropical grasses subjected to low temperatures. In *Plant Response to Climatic Factors*, pp. 165-168. Proceedings of the Uppsala Symposium, 1970. UNESCO.

Westgate, M. E. and J. S. Boyer. 1984. Transpiration- and growth-induced water potentials in maize. *Plant Physiol.* **74**:882-889.

Wetselaar, R. and G. D. Farquhar. 1980. Nitrogen losses from tops of plants. *Adv. Agron.* **33**:263-302.

Whalley, D. B., C. M. McKell, and L. R. Green. 1966. Seedling vigor and the early non-photosynthetic stage of seedling growth in grasses. *Crop Sci.* **6**:147-150.

Whelan, T., W. M. Sackett, and C. R. Benedict. 1970. Carbon isotope discrimination in a plant possessing the C_4 dicarboxylic acid pathway. *Biochem. Biophys. Res. Comm.* **41**:1205-1210.

Whiteman, P. C. 1968. The effects of temperature on the vegetative growth of six tropical legume pastures. *Aust. J. Exp. Agric. Anim. Husb.* **8**:528-532.

Whiteman, P. C. 1980. *Tropical Pasture Science*. Oxford University Press, Oxford, U.K.

Whyte, R. O., T. R. G. Moir, and J. P. Cooper. 1959. *Grasses in Agriculture*. Food and Agriculture Organization of the United Nations, Rome.

Wier, B. L., K. N. Paulson, and O. A. Lorenz. 1972. The effect of ammoniacal nitrogen on lettuce (*Lactuca sativa*) and radish (*Raphanus sativa*) plants. *Soil Sci. Soc. Am. Proc.* **36**:462-465.

Wilkins, H., A. Larque'-Saavedra, and R. L. Wain. 1974. Control of *Zea* root elongation by light and the action of 3, 5-diiodo-4-hydroxybenzoic acid. *Nature* **248**:449-450.

Wilkins, H. and R. L. Wain. 1974. The root cap and control of root elongation in *Zea mays* L. seedlings exposed to white light. *Planta* **121**:1-8.

Wilkins, H., S. M. Wilkins, and R. L. Wain. 1977. Studies on plant growth-regulating substances. XLVIII. The effects of 3,5-diiodo-hydroxybenzoic acid on seedling growth in compact and loose soils. *Ann. Appl. Biol.* **87**:415-422.

Wilkins, M. B. 1977. Gravity and light-sensing guidance systems in primary roots and shoots. In D. H. Jennings (ed.), *Integration of Activity in the Higher Plant*, pp. 295-335. University Press, Cambridge.

Wilkins, M. B. 1979. Growth-control mechanisms in gravitropism. In W. Haupt and M. E. Feinleib (eds.), *Encyclopedia of Plant Physiology, New Series*, Vol. 7, *Physiology of Movements*, pp. 601-626. Springer, Berlin.

Wilkins, S., H. Wilkins, and R. L. Wain. 1976. Chemical treatment of soil alleviates effects of soil compaction on pea seedling growth. *Nature* **259**:392-394.

Wilkins, S., H. Wilkins, and R. L. Wain. 1978. Studies on plant growth-regulating substances. LI. The effects of 3,5-cibromo- and 3,5-dichloro-4-hydroxygenzoic acids on seedling growth in compact and loose soils. *Ann. Appl. Biol.* **89**:265-269.

Williams, E. P., R. B. Clark, Y. Yusuf, W. M. Rose, and J. W. Maranville. 1982. Variability of sorghum genotypes to tolerate iron deficiency. *J. Plant Nutr.* **5**:553-567.

Williams, W. A., R. S. Loomis, W. G. Duncan, A. Dovrat, and F. Nunez A. 1968. Canopy architecture at various population densities and the growth and grain yield of corn. *Crop Sci.* **8**:303-308.

Williamson, R. E. and J. van Schilfgaarde. 1965. Studies of crop response to drainage. II. Lysimeters. *Trans. of the ASAE.* **8**:98-102.

Williamson, R. E. and G. J. Kriz. 1970. Response of agricultural crops to flooding, depth-of water table and soil gaseous composition. *Trans. ASAE* **13**:216-220.

Williamson, R. E., C. R. Willey, and T. N. Gray. 1969. Effect of water table depth and flooding on yield of millet. *Agron. J.* **61**:310-313.

Wilson, A. M. and D. D. Briske. 1979. Seminal and adventitious root growth of blue grama seedlings on the Central Plains. *J. Range Manage.* **32**:209-213.

Wilson, D. R., C. J. Fernandez, and K. J. McCree. 1978. CO_2 exchange of subterranean clover in variable light environments. *Crop Sci.* **18**:19-22.

Wilson, J. H. and J. C. S. Allison. 1978. Effect of plant population on ear differentiation and growth in maize. *Ann. Appl. Biol.* **90**:127-132.

Wilson, J. R. 1975. Comparative response to nitrogen deficiency of a tropical and temperate grass in the interrelation between photosynthesis, growth and the accumulation of non-structural carbohydrate. *Neth. J. Agric. Sci.* **23**:104-112.

Wilson, J. R. and R. H. Brown. 1983. Nitrogen response of *Panicum* species differing in CO_2 fixation pathways. I. Growth analysis and carbohydrate accumulation. *Crop Sci.* **23**:1148-1153.

Wilson, J. R., R. H. Brown, and W. R. Windham. 1983. Influence of leaf anatomy on the dry matter digestibility of C_3, C_4 and C_3/C_4 intermediate types of *Panicum* species. *Crop Sci.* **23**:141-146.

Wilson, J. R. and W. J. Davies. 1979. Farnesol-like antitranspirant activity and stomatal behavior in maize and *Sorghum* lines of differing drought tolerance. *Plant, Cell Environ.* **2**:49-57.

Wilson, J. R. and C. W. Ford. 1971. Temperature influences on the growth, digestibility, and carbohydrate composition of two tropical grasses, *Panicum maximum* var. *trichoglume* and *Setaria sphacelata*, and two cultivars of the temperate grass *Lolium perenne*. *Aust. J. Agric. Res.* **22**:563-571.

Wilson, J. R. and K. P. Haydock. 1971. The comparative response of tropical and temperate grasses to varying levels of nitrogen and phosphorus nitrition. *Aust. J. Agric. Res.* **22**:573-587.

Wilson, J. R., M. M. Ludlow, M. J. Fisher, and E.-D. Schulze. 1980. Adaptation to water stress of the leaf water relations of four tropical forage species. *Aust. J. Plant Physiol.* **7**:207-220.

Wilson, J. R., A. O. Taylor, and G. R. Dolby. 1976. Temperature and humidity effects on cell wall content and dry matter digestibility of some tropical and temperate grasses. *N. Z. J. Agric. Res.* **19**:41-46.

Wilson, R. H. and H. J. Evans. 1968. The effect of potassium and other univalent cations on the conformation of enzymes. In V. J. Kilmer, S. E. Younts, and N. C. Brady (eds.), *The Role of Potassium in Agriculture*, pp. 189-202. American Society of Agronomy, Madison, WI.

Winter, S. R. and A. J. Ohlrogge. 1973. Leaf angle, leaf area, and corn (*Zea mays* L.) yield. *Agron. J.* **65**:395-397.

Withee, L. V. and E. W. Carlson. 1959. Foliar and soil applications of iron compounds to control iron chlorosis of grain sorghum. *Agron. J.* **51**:474-476.

Wolfe, J. 1978. Chilling injury in plants—the role of membrane lipid fluidity. *Plant, Cell Environ.* **1**:241-247.

Wong, C. C. and J. R. Wilson. 1980. Effects of shading on the growth and nitrogen content of green panic and siratro in pure and mixed swards defoliated at two frequencies. *Aust. J. Agric. Res.* **31**:269-285.

Wong, S. C., I. R. Cowan, and G. D. Farquhar. 1979. Stomatal conductance correlates with photosynthetic capacity. *Nature* **282**:424-426.

Wong, W. W., W. M. Sackett, and C. R. Benedict. 1975. Isotope fractionation in photosynthetic bacteria during carbon dioxide assimilation. *Plant Physiol.* **55**:475-479.

Wong You Cheong, Y. and P. Halais. 1970. Needs of sugarcane for silicon when growing in highly weathered latosols. *Exp. Agric.* **6**:99-106.

Wong You Cheong, Y., A. Heitz, and J. Deville. 1971a. Foliar symptoms of silicon deficiency in the sugarcane plant. *Int. Soc. Sugar Cane Technol. Proc.* **14**:766-776.

Wong You Cheong, Y., A. Heitz, and J. Deville. 1971b. The effect of silicon on enzyme activity in vitro and sucrose production in sugarcane leaves. *Int. Soc. Sugar Cane Technol. Proc.* **14**:777-785.

Wood, G. H. and R. A. Wood. 1967. The estimation of cane root development and distribution using radiophosphorus. In *Proceedings of the South African Sugar Technology Association, April, 1967*, pp. 1-8.

Woodhouse, W. W., Jr. 1968. Long-term fertility requirements of coastal bermudagrass. I. Potassium. *Agron. J.* **60**:508-512.

Woodhouse, W. W., Jr. 1969a. Long-term fertility requirements of coastal bermuda. II. Nitrogen, phosphorus and lime. *Agron. J.* **61**:251-256.

Woodhouse, W. W., Jr. 1969b. Long term fertility requirements of coastal bermudagrass. *Agron. J.* **61**:705-708.

Woodruff, J. R. and E. J. Kamprath. 1965. Phosphorus absorption maximum as measured by the Langmuir isotherm and its relationship to phosphorus availability. *Soil Sci. Am. Proc.* **29**:148-150.

Woodruff, J. R. and C. L. Parks. 1980. Topsoil and subsoil potassium calibration with leaf potassium for fertility rating. *Agron. J.* **72**:392-396.

Woolley, D. J. and P. F. Wareing. 1972. The interaction between growth promotors in apical dominance. 2. Environmental effects on endogenous cytokinin and gibberellin levels in *Solanum andigena*. *New Phytol.* **71**:1015-1025.

Worker, G. F., Jr., and J. Ruckman. 1968. Variations in protein levels in grain sorghum grown in the southwest desert. *Agron. J.* **60**:485-488.

Wright, G. C. and R. C. G. Smith. 1983. Differences between two grain sorghum genotypes in adaptation to drought stress. II. Root water uptake and water use. *Aust. J. Agric. Res.* **34**:627-636.

Wright, G. C., R. C. G. Smith, and J. R. McWilliam. 1983a. Differences between two grain sorghum genotypes in adaptation to drought stress. I. Crop growth and yield responses. *Aust. J. Agric. Res.* **34**:615-626.

Wright, G. C., R. C. G. Smith, and J. M. Morgan. 1983b. Differences between two grain sorghum genotypes in adaptation to drought stress. III. Physiological responses. *Aust. J. Agric. Res.* **34**:637-651.

Wright, H. A. and A. W. Bailey. 1982. *Fire Ecology, United States and Southern Canada*. Wiley-Interscience, New York.

Wright, L. N. 1973. Seed dormancy, germination environment, and seed structure of Lehmann lovegrass, *Eragrostis lehmanniana* Nees. *Crop Sci.* **13**:432-435.

Wright, L. N. 1976. Recurrent selection for shifting gene frequency of seed weight in *Panicum antidotale* Retz. *Crop Sci.* **16**:647-649.

Wright, L. N. 1977. Germination and growth response of seed weight genotypes of *Panicum antidotale* Retz. *Crop Sci.* **17**:176-178.

Wright, M. J. and K. L. Davison. 1964. Nitrate accumulation in crops and nitrate poisoning in animals. *Adv. Agron.* **16**:197-214.

Wyn Jones, R. G. and O. R. Lunt. 1967. The function of calcium in plants. *Bot. Rev.* **33**:407-426.

Wynn Parry, D. and M. Kelso. 1975. The distribution of silicon deposits in the roots of *Molinia caerulea* (L.) Moench. and *Sorghum bicolor* (L.) Moench. *Ann. Bot.* **39**:995-1001.

Wynn Parry, D. and M. Kelso. 1977. The ultrastructure and analytical microscopy of silicon deposits in the roots of *Saccharum officinarum* (L.). *Ann. Bot.* **41**:855-862.

Wynn Parry, D. and F. Smithson. 1966. Opaline silica in the inflorescences of some British grasses and cereals. *Ann. Bot.* **30**:525-538.

Wynn Parry, D. and S. L. Soni. 1972. Electron-probe microanalysis of silicon in the roots of *Oryza sativa* L. *Ann. Bot.* **36**:781-783.

Yadov, R. L. 1981. Leaf area index and functional leaf area duration of sugarcane as affected by N-rates. *Indian J. Agron.* **26**:130-136.

Yanuka, M., Y. Leshem, and A. Dovrat. 1982. Forage corn response to several trickle irrigation and fertilization regimes. *Agron. J.* **74**:736-740.

Yates, R. A. 1972. Effects of environmental conditions and the coadministration of growth retardants on the response of sugarcane to foliar treatment with gibberellin. *Agron. J.* **64**:31-35.

Yeo, A. R., D. Kramer, A. Lauchli, and J. Gullasch. 1977. Ion distribution in salt stressed mature *Zea mays* roots in relation to ultrastructure and relation of sodium. *J. Exp. Bot.* **28**:17-29.

Yeon, H. H., N. E. Stone, and L. Watson. 1982. Taxonomic variation in the subunit amino acid compositions of RuBP carboxylase from grasses. *Phytochemistry* **21**:71-80.

York, E. T., Jr., R. Bradfield, and M. Peech. 1954. Influence of lime and potassium on yield and cation composition of plants. *Soil Sci.* **77**:53-63.

Yost, R. S. and R. L. Fox. 1979. Contribution of mycorrhizae to P nutrition of crops growing on an Oxisol. *Agron. J.* **71**:903-908.

Yost, R. S., E. J. Kamprath, E. Lobato, and G. Naderman. 1979. Phosphorus response of corn on an Oxisol as influenced by rates and placement. *Soil Sci. Soc. Am. J.* **43**:338-343.

Yost, R. S., E. J. Kamprath, G. C. Naderman, and E. Lobato. 1981. Residual effects of phosphorus applications on a high phosphorus adsorbing Oxisol of central Brazil. *Soil Sci. Soc. Am. J.* **45**:540-543.

Youngdahl, L. J., L. V. Svec, W. C. Liebhardt, and M. R. Tiel. 1977. Changes in the zinc-65 distribution in corn root tissue with a phosphorus variable. *Crop Sci.* **17**:66-68.

Youngner, V. B. and O. R. Lunt. 1967. Salinity effects on roots and tops of bermuda grass. *J. Brit. Grassl. Soc.* **22**:257-259.

Younts, S. E. and R. B. Musgrave. 1958. Chemical composition, nutrient absorption, and stalk rot incidence of corn as affected by chloride in potassium fertilizer. *Agron. J.* **50**:426-429.

Yu, P. T., L. H. Stolzy, and J. Letey. 1969. Survival of plants under prolonged flooded conditions. *Agron. J.* **61**:844-847.

Zartman, R. E. and R. T. Woyewodzic. 1979. Root distribution patterns of two hybrid grain sorghums under field conditions. *Agron. J.* **71**:325-328.

Zelitch, I. 1973. Plant productivity and the control of photorespiration. *Proc. Nat. Acad. Sci., U.S.A.* **70**:579-584.

Zelitch, I. 1982. The close relationship between net photosynthesis and crop yield. *BioScience* **32**:796-802.

Ziegler, H. 1975. Nature of transported substances. In M. H. Zimmerman and J. A. Milburn (eds.), *Transport in Plants I. Phloem transport*, pp. 59-100. Springer-Verlag, Berlin.

Zimmerman, R. P. and L. T. Kardos. 1961. Effect of bulk density on root growth. *Soil Sci.* **91**:280-288.

Zimmerman, U. 1978. Physics of turgor and osmoregulation. *Ann. Rev. Plant Physiol.* **29**:121-148.

INDEX

Abscisic acid, 123–124, 200, 210–211, 213, 227
Abutilon theophrasti, 166
Acid soil tolerance, 13–21, 289–311
Aerenchyma, 226–227
After-ripening, 57
Alang-alang, 20
Alcohol, 9
Alkaligrass, nuttall, 317
Alkaline soil tolerance, 14
Aluminum, 127, 138
Aluminum toxicity, 118, 291–298, 308
Ammonium, 126, 136, 234, 240, 243–247, 249–250, 283, 303
Amylopectin, 4, 112
Amylose, 4, 112
Andropogon gayanus, see Gambagrass
Andropogon gerardii var. *gerardii, see* Bluestem, big
Andropogon sorghum, 6
Andropogon virginicus, see Bluestem, broomsedge
Angolagrass, 13
Apomixis, 99
Auxins, 66, 80, 123–124, 252

Bacterial diseases, 11
Bagasse, 9
Bahiagrass, 17, 99, 103, 133, 251, 307
Bajra, 7
Baksha, 15

Bamboo, 2
Barley, 2, 147, 177, 208, 221, 282, 305, 316–318
Barnyardgrass, 19, 221
Bentgrass, 316
Betaine, 207–209
Bermudagrass:
  characteristics, 14–15, 17
  cold tolerance, 151, 153, 157, 286–287
  drought tolerance, 208
  nitrogen, 235, 240, 244, 248, 251
  phosphorus, 264, 268
  potassium, 275, 286–288
  salinity, 316–318
  silicon, 307
  sulfur, 310
  vegetative growth, 66, 71
Biological recognition, 47
Bird damage, 7–21
Black layer, 108–110
Bladygrass, 20
Bluestem:
  angleton, 90, 99
  broomsedge, 99
  big, 12, 181
  little, 183
Boron, 127, 308, 313
Bouteloua curtipendula, see Grama, sideoats
Bouteloua eriopoda, see Grama, black
Bouteloua gracilis, see Grama, blue
Brachiaria decumbens, see Surinamgrass